Unconventional Green Synthesis of Inorganic Nanomaterials

Inorganic Materials Series

Series editors:
Duncan W. Bruce, *University of York, UK*
Dermot O'Hare, *University of Oxford, UK*
Richard I. Walton, *University of Warwick, UK*

Titles in the Series:

1: Pre-combustion Carbon Dioxide Capture Materials
2: Post-combustion Carbon Dioxide Capture Materials
3: Solar Energy Capture Materials
4: Bioinspired Inorganic Materials: Structure and Function
5: Energy Storage and Conversion Materials
6: Two-dimensional Inorganic Nanomaterials for Conductive Polymer Nanocomposites
7: The Chemistry of Inorganic Biomaterials
8: Computer Simulation of Porous Materials: Current Approaches and Future Opportunities
9: Inorganic Thermoelectric Materials: From Fundamental Concepts to Materials Design
10: Biomedical Applications of Inorganic Materials
11: Photocatalysis Using 2D Nanomaterials
12: Mechanical Behaviour of Metal–Organic Framework Materials
13: Flexible Metal–Organic Frameworks: Structural Design, Synthesis and Properties
14: Unconventional Green Synthesis of Inorganic Nanomaterials

How to obtain future titles on publication:
A standing order plan is available for this series. A standing order will bring delivery of each new volume immediately on publication.

For further information please contact:
Book Sales Department, Royal Society of Chemistry, Thomas Graham House, Science Park, Milton Road, Cambridge, CB4 0WF, UK
Telephone: +44 (0)1223 420066, Fax: +44 (0)1223 420247
Email: booksales@rsc.org
Visit our website at books.rsc.org

Unconventional Green Synthesis of Inorganic Nanomaterials

Edited by

Silvia Gross
Università degli Studi di Padova, Italy
Email: silvia.gross@unipd.it

ROYAL SOCIETY OF CHEMISTRY

Inorganic Materials Series No. 14

Print ISBN: 978-1-78801-775-6
PDF ISBN: 978-1-83916-575-7
EPUB ISBN: 978-1-83916-576-4
Print ISSN: 2472-3819
Electronic ISSN: 2472-3827

A catalogue record for this book is available from the British Library

The Royal Society of Chemistry is a charity, registered in England and Wales, Number 207890, and a company incorporated in England by Royal Charter (Registered No. RC000524), registered office: Burlington House, Piccadilly, London W1J 0BA, UK, Telephone: +44 (0) 20 7437 8656.

For further information see our website at www.rsc.org

Printed in the United Kingdom by CPI Group (UK) Ltd, Croydon, CR0 4YY, UK

Dedication

Gewidmet Giulia und Klaus.

Inorganic Materials Series No. 14
Unconventional Green Synthesis of Inorganic Nanomaterials
Edited by Silvia Gross

Published by the Royal Society of Chemistry, www.rsc.org

Series Preface

Materials chemistry has grown hugely in the past few decades and continues to evolve rapidly. This is especially the case since the subject now finds itself central to many of the major challenges that face global society, such as those concerning energy and the environment; but applications of materials are also found in many everyday situations, with ongoing developments in technology around electronics and the miniaturisation of devices. Although the study of materials is a highly interdisciplinary topic, the role of innovative, fundamental chemistry continues to lie at the heart of the discovery of new materials, the understanding of their atomic-scale structure and the relationship of structure to a useful property.

With the *Inorganic Materials Series*, our aim is to provide a series of themed volumes, each reflecting the diversity of the topic as it continues to develop over the coming years, with many chosen to relate closely to emerging practical applications. The use of the title *Inorganic Materials* is by no means limiting, but instead emphasises the study of the chemistry of all elements in the Periodic Table and how they impact the development of new materials. The chapters in each book will have a pedagogical flavour with a target audience from first-year postgraduate student upwards, but with wide subject coverage ranging from continuous inorganic solids, to molecular and soft matter systems. We will use expert, active researchers as editors and contributors to provide an up-to-date perspective of their field. At the same time our aim is to provide an international perspective, so to reflect the diversity and interdisciplinarity of the now very broad area of the subject.

Duncan W. Bruce, York, UK
Dermot O'Hare, Oxford, UK
Richard I. Walton, Warwick, UK

Inorganic Materials Series No. 14
Unconventional Green Synthesis of Inorganic Nanomaterials
Edited by Silvia Gross

Published by the Royal Society of Chemistry, www.rsc.org

Preface

Inorganic nanomaterials are an extremely broad, diverse and versatile class of materials, often endowed with outstanding functional properties (*e.g.*, optical, magnetic, semiconducting, electric, thermoelectric, electronic, *etc.*). They are typically characterised by enhanced chemical, thermal and mechanical stability with respect to their organic counterparts, which make them appealing candidates for a very wide range of technological applications. The current energy transition, renewable energy sources, advanced catalysis and electronics are all technologies that rely strongly on the availability of inorganic materials, typically in their nanostructured form.

Green synthesis, being more environmentally sustainable and lowering the environmental burden of materials production, could represent a required competitive advantage. As the recent scientific literature demonstrates, there are increasing endeavours to develop sustainable, green and easy syntheses of inorganic nanomaterials using approaches relying on unconventional conditions (*e.g.*, confined spaces, non-standard temperature and pressure, *etc.*). Indeed, although several chemical processes have been established for the synthesis of the ever-increasing and expanding portfolio of nanomaterials, there has been a considerable amount of recent interest in developing more sustainable routes to produce inorganic nanomaterials, and also to explore unconventional synthesis pathways.

The motivation behind the planning of this book was many-fold. From one perspective, the book aims to cover an emerging area of inorganic synthesis, that is, the green and/or unconventional synthesis of metal nanoparticles, metal oxides, chalcogenides and halides, as well as composite and hybrid materials (see Chapter 1, for example). All of these are interesting for the different properties they display and the very broad range of applications they exhibit, ranging from catalysis, to electronics, optics and nanomedicine. The second motivation is to fill a cultural and technological gap, showing

Inorganic Materials Series No. 14
Unconventional Green Synthesis of Inorganic Nanomaterials
Edited by Silvia Gross

Published by the Royal Society of Chemistry, www.rsc.org

how it is possible to combine environmental and economical sustainability. The third and final aim is to provide the reader with a state-of-the-art and updated overview of less-known and emerging synthesis routes for inorganic nanomaterials, and to stimulate the inorganic chemistry community in relation to topics such as sustainability in synthesis, as well as the exploration of novel and challenging preparation methods. As such, we believe this book offers something quite different to the many other volumes regarding synthetic methods.

Selected green and/or unconventional wet-chemistry routes for the synthesis of functional inorganic nanostructures will be thoroughly described and discussed, by outlining their green and/or unconventional features. For each explored method, a brief background and state-of-the-art of the synthetic route are provided, followed by a detailed description of the procedure and a discussion of the roles of the synthetic parameters involved. Where possible, insight into how these parameters influence the final products will be given. Each chapter also includes a detailed description of the state-of-the-art concerning the compounds that can be obtained, as well as the potential scalability in an industrial environment.

The book is organised into six chapters. Chapters 1–4 are each devoted to a specific synthetic approach, each being characterised by some unconventional feature: nucleation and growth under confined space afforded by micro- and miniemulsion droplets (Chapter 1), use of biogenic agents (Chapter 2), use of supercritical fluids (Chapter 3), and use of microwaves (Chapter 4). Chapter 5 deals with advanced approaches for the *in situ* synthesis of hybrid nanocomposites based on graphene derivatives and inorganic nanoparticles for advanced applications. Chapter 6 collects together, in a short and concise form, further synthesis methods that are not covered in the previous chapters, *i.e.*, hydrothermal, polyol-assisted, continuous-flow and sonochemical methods, as well as radiochemistry and laser ablation in liquid media. Also in Chapter 6, a theoretical background is provided for each method, along with a detailed survey of the applications for the synthesis of different classes of inorganic nanomaterials.

Silvia Gross

Contents

Inorganic Materials Series No. 14
Unconventional Green Synthesis of Inorganic Nanomaterials
Edited by Silvia Gross

Published by the Royal Society of Chemistry, www.rsc.org

CHAPTER 1

Inorganic Chemistry Within Nanoreactors

RAFAEL MUÑOZ-ESPÍ[a*], INÉS ADAM-CERVERA[a] AND KATHARINA LANDFESTER[b]

[a]Institute of Materials Science (ICMUV), Universitat de València, c/Catedràtic José Beltrán 2, 46980 Paterna, Spain; [b]Max Planck Institute for Polymer Research, Ackermannweg 10, 55128 Mainz, Germany
*E-mail: rafael.munoz@uv.es

1.1 Introduction

The confinement provided by colloidal systems, in particular by droplets in emulsions, has been extensively exploited in polymer science for the synthesis of organic polymers through so-called emulsion polymerisation techniques.[1–5] In an analogous way, emulsions can also be employed for confining inorganic reactions, typically of precipitation, oxidation/reduction or sol–gel processes.[6] Several publications in recent years, including some from us, have periodically reviewed the use of droplets and nanodroplets for the synthesis of inorganic materials.[6–9] Our aim in this chapter, rather than to provide an exhaustive research review, is to cover in an explanatory fashion the possibilities offered by microemulsions and, especially, by miniemulsions for obtaining nanostructured inorganic materials, and providing thereby examples of different synthetic strategies. Based on a didactic perspective, we will also review some related general concepts of colloidal chemistry, but specifically focused on the field of inorganic reactions within nanodroplets.

Inorganic Materials Series No. 14
Unconventional Green Synthesis of Inorganic Nanomaterials
Edited by Silvia Gross

Published by the Royal Society of Chemistry, www.rsc.org

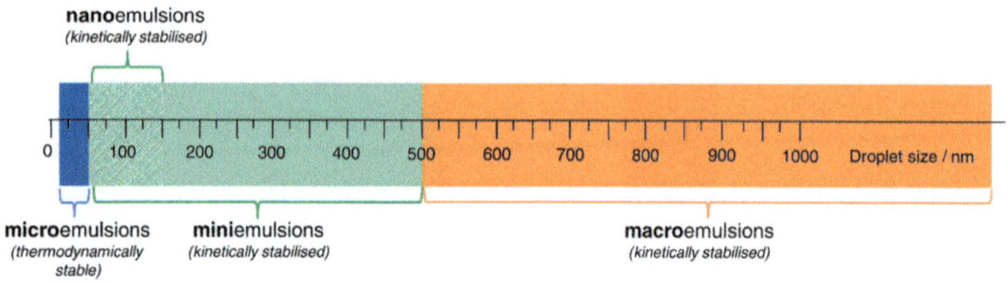

Figure 1.1 Schematic representation of the different types of emulsions.

Before starting the description of the synthetic strategies, it is convenient to make a few initial remarks about the confusing terminology in the field of emulsions. Traditionally, emulsions have been classified in *macro*emulsions (or simply emulsions), *mini*emulsions and *micro*emulsions.[1] This nomenclature has been followed by polymer scientists and part of the colloid community.[5,10] However, another partially overlapping term is very often found in the literature, especially—although not only—in the fields of food science and pharmacy: *nano*emulsions.[11,12] We will revisit here the most relevant aspects of the differences between these emulsion terms, which are discussed in more detail elsewhere.[5,9]

Macroemulsions are thermodynamically unstable emulsions with droplet sizes above 500 nm (usually >1 μm). *Microemulsions*, in contrast, are characterised by being thermodynamically stable and have droplet sizes below 50 nm. *Miniemulsions* are in between these two types and are kinetically stabilised emulsions but thermodynamically unstable (like macroemulsions), with droplet sizes in the range between 50 and 500 nm.[2,3] The term *nanoemulsion* (systematically and intentionally hyphenated by some authors[13–15] as nano-emulsion) should in principle be used only for miniemulsions with sizes below 100 nm (or 200 nm, at most). Although the term 'nanoemulsion' is often used interchangeably with 'miniemulsion', nanoemulsions are a subdivision of miniemulsions, corresponding to nanosized (below 100 nm) and thermodynamically unstable emulsions. This classification of emulsions is schematically represented in Figure 1.1. The inclusion of microemulsions within the label 'nanoemulsions' is discouraged, because microemulsions are supposed to be thermodynamically stable, by definition. Although some contradictory examples are found from time to time in the literature, we encourage the reader to follow this recommendation, as it is the consensus preference of most of the research groups working in the field.

1.2 General Aspects of Inorganic Reactions Within Droplets

When a droplet is formed, there is an intrinsic pressure difference between the inner (P_{inn}) and the outer (P_{ext}) parts of the curve interface, ΔP_L, given by the Young–Laplace equation,

$$\Delta P_L = P_{inn} - P_{ext} = \frac{2\gamma}{r} \tag{1.1}$$

where γ is the surface tension and r is the radius of the droplet. As an example, according to this expression, for water droplets with a radius of 1 mm, the pressure difference is 145.6 Pa = 1.09 mm Hg (considering $\gamma = 72.8$ mN m^{-1} at 20 °C for water). By reducing the radius by a factor of 10 (*i.e.*, 0.1 mm = 100 μm), the pressure is multiplied by the same factor, so that $\Delta P = 10.9$ mm Hg. If we go to droplets with a size of 100 nm ($r = 50$ nm), the pressure difference would be as high as 2.2×10^5 mm Hg (*ca.* 29.1 bar). This simple calculation makes very obvious the great effect that the formation of a curved interface has on the pressure of the confined space within a nanodroplet.

In addition to the advantages related to the formation of smaller particles in smaller spaces (higher specific surface area)[16] and the possible finite-size effects in crystallisation and selection of polymorphs,[17] the high pressure within nanodroplets can also allow the formation of crystal phases of inorganic materials that otherwise would require higher temperatures.[18] A shift in the phase diagram through the pressure resulting from the space confinement could even allow the stabilisation of certain polymorphs.[9] Also, other properties of the solvent (typically water in inorganic synthesis), which are relevant for solubilisation and crystallisation phenomena (*e.g.*, viscosity, relative permittivity), dramatically change when confined in constrained environments.[19–21] Emulsions containing nanodroplets (*i.e.*, miniemulsions and microemulsions) have, therefore, significant potential in inorganic synthesis. It is very important to remark that, in most cases, in miniemulsion and microemulsions, crystallisation of the materials takes place under very mild conditions, often at room temperature or close to it.[22]

As mentioned in Section 1.1, the use of emulsions in synthesis has been reviewed in different works. Some representative publications are contained in Table 1.1, with a brief indication of the focus in each case.

The title of this chapter contains the term 'nanoreactor', which has been very commonly linked to nanodroplets in miniemulsions.[3,23,24] The term can also be used for other types of colloidal confinement, such as micelles, colloidal assemblies (see the review by Swisher *et al.*[25]) or nano-/mesopores, but we will focus our attention here on nanodroplets. If we move away for a moment from the inorganic synthesis to emulsion polymers, and we compare conventional emulsion polymerisation with miniemulsion polymerisation, the former is characterised by diffusion from monomer droplets, acting as reservoirs, to the growing particles being formed. By contrast, in miniemulsion polymerisation, there is no diffusion between droplets, and the polymerisation reaction takes place (at least ideally) within the droplets, which act as independent 'nanoreactors'. To maintain the stability of the system, two types of agent are normally used: effective surfactants, which stabilise the droplets (sterically, electrostatically or electrosterically) and minimise the coalescence between them; and osmotic-pressure agents, which help to avoid Ostwald ripening.[2–7] For inorganic synthesis, the idea

Table 1.1 Reviews focusing (at least in part) on the use of emulsions in inorganic synthesis.

Authors and reference	Year	Type of emulsion	Focus
Pileni[53]	1993	Microemulsions	Water-in-oil systems
Sanchez-Dominguez *et al.*[86]	2012	Microemulsions	Inorganic nanoparticles (specific for oil-in-water systems)
Muñoz-Espí *et al.*[6]	2012	Miniemulsions	Inorganic nanoparticles
Diodati *et al.*[8]	2015	Micro- and miniemulsions	General on wet-chemistry and colloidal routes
Muñoz-Espí *et al.*[7]	2013	Miniemulsions and microemulsions	Crystalline materials
Wolf and Feldmann[59]	2016	Microemulsions	Inorganic nanoparticles
Boutonnet and Sánchez-Domínguez[68]	2017	Microemulsions	Synthesis of catalytic nanoparticles
Álvarez-Bermúdez and Muñoz-Espí[5]	2018	Nanoemulsions	Nanoparticle synthesis (not only inorganic)
Muñoz-Espí and Landfester[9]	2020	Miniemulsions	Metal oxides
López *et al.*[71]	2020	Microemulsions	Metal and metal oxide nanoparticles
Elzayat *et al.*[85]	2021	Nanoemulsions	Biomedical nanocarriers (mostly polymeric, but also inorganic)
Swisher *et al.*[25]	2022	Nanoemulsions	Nanoreactors for particle synthesis (not only emulsions)
van Embden *et al.*[87]	2023	Miniemulsions and microemulsions	Very specific on colloidal synthesis of ZnO (not only emulsions)

is essentially the same: in an ideal case, the formation of the desired compound should take place within a droplet containing the precursor, while avoiding diffusion of the contained matter towards other 'nanoreactors' (see cartoon in Figure 1.2). In a real system, however, this situation is very difficult to reach, and diffusion and interaction between droplets and growing particles occurs in most of cases, and is more likely than in polymerisation reactions for organic polymers.

1.2.1 Formation and Stability of Emulsions

From a thermodynamic point of view, in a first approximation, the change in the Gibbs energy associated with the formation of an emulsion, ΔG_{form}, can be considered as the sum of an interfacial free energy term (ΔG_{interf}) and a negative configuration entropic term ($-T\Delta S_{\text{config}}$):[26]

$$\Delta G_{\text{form}} = \Delta G_{\text{interf}} - T\Delta S_{\text{config}} \tag{1.2}$$

If temperature, pressure and interfacial chemical potential remain constant, ΔG_{interf} can be expressed as the change in the contact area

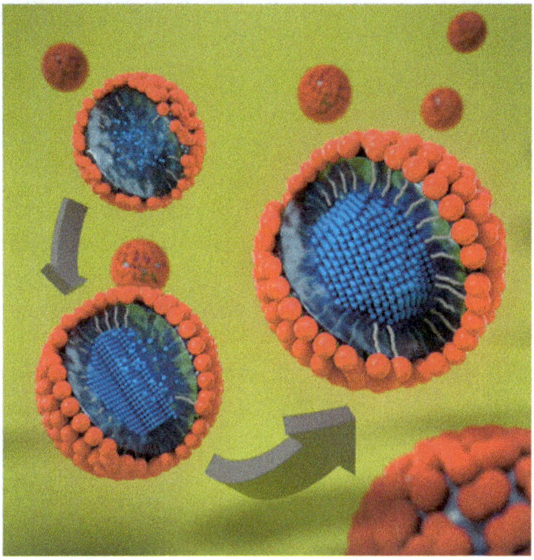

Figure 1.2 Graphical representation of emulsion droplets acting as nanoreactors for the formation of inorganic matter. Reproduced from ref. 9, https://doi.org/10.1002/chem.202001246, under the terms of the CC BY 4.0 license https://creativecommons.org/licenses/by/4.0/.

between the oil and the aqueous phase (ΔA) multiplied by the interfacial tension between these two phases, γ:

$$\Delta G_{\text{interf}} = \gamma \Delta A \quad (1.3)$$

Since both γ and ΔA are positive (the formation of an emulsion involves an increase of the overall surface area), ΔG_{interf} is always positive and it becomes increasingly positive with decreasing droplet size (the interfacial area increases with decreasing size). By contrast, the configuration free energy, $-T\Delta S_{\text{config}}$, becomes increasingly negative with decreasing droplet size (the number of different ways in which the droplets can be organised increases). For spontaneous formation of an emulsion, ΔG_{form} has to be <0, so that ΔG_{interf} will oppose the formation of the emulsion, whereas the configuration term will favour it. Since the interfacial term dominates the configuration term, the formation of an emulsion should be thermodynamically less favourable when the droplet size decreases. However, this is only true if we assume that the interfacial tension remains constant with the curvature radius, which is not the case when the size decreases. Indeed, the interfacial tension does not have a trivial dependence on the droplet size. Therefore, the dependence of the overall formation free energy with the size is more complex than initially assumed. For detailed explanations about the contributions of the interfacial and the configuration entropic terms, the work by McClements[26] is recommended.

For macroemulsions, the interfacial term clearly dominates (the increase in the interfacial area is much more relevant than the increase in the possible configurations), so that $\gamma\Delta A \gg T\Delta S_{\text{config}}$, and $\Delta G_{\text{form}} > 0$. For miniemulsions, the interfacial area increases, but so does ΔS_{config}; eventually, the former continues to dominate the latter, and $\gamma\Delta A > T\Delta S_{\text{config}}$, so that $\Delta G_{\text{form}} > 0$ also. Accordingly, macroemulsions and miniemulsions are both thermodynamically unstable and work has to be done to form them. From a kinetic point of view, both need to be stabilised, but miniemulsions achieve much longer stability than macroemulsions.

The situation for microemulsions is more complex. When the droplets become sufficiently small, the interfacial tension of a droplet can no longer be assumed to be the same as for planar interfaces. As mentioned previously, the interfacial tension depends on the curvature of the surfactant monolayer and so at a certain small droplet size (typically <10 nm), the value of the interfacial tension as a function of the droplet radius reaches a minimum. Around that minimum the interfacial free energy term, which usually dominates, is substantially reduced. Under these conditions, $\gamma\Delta A < T\Delta S_{\text{config}}$ and ΔG_{form} becomes negative, so that the formation of microemulsions is spontaneous. In other words, the resulting system is thermodynamically stable.

As for any colloidal system, many processes can affect the stability of emulsions. Figure 1.3 depicts these processes, together with the driving forces in each case. First of all, gravity leads to settling (if the density of the disperse phase is higher than that of the continuous phase) or

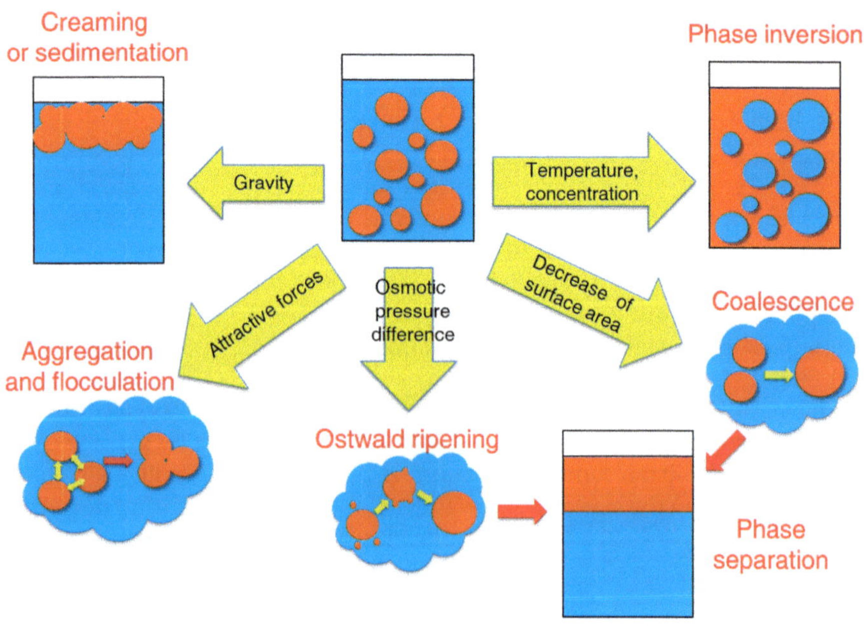

Figure 1.3 Factors affecting the stability of emulsions.

creaming (if the density of the disperse phase is lower) and, eventually, to phase separation. For spherical particles of radius r and volume V_p, the gravitational force contributing to settling down in a suspension, F_{grav}, can be expressed as

$$F_{grav} = mg = V_p(\rho_p - \rho_m)g = \frac{4}{3}\pi r^3(\rho_p - \rho_m)g \quad (1.4)$$

where m is the mass of the particles, g is the gravitational constant, and ρ_p and ρ_m are the densities of the particles and the dispersing medium, respectively. The reaction force opposing gravity is the viscous drag force, F_{drag}, given by Stokes' law:

$$F_{drag} = mg = 6\pi r v \eta \quad (1.5)$$

where r is the radius of the particle, v is the settling rate and η is the viscosity of the medium. By equating eqn (1.4) and (1.5) and clearing v, the following expression for the settling rate is obtained:

$$v = mg = \frac{2r^2(\rho_p - \rho_m)g}{9\eta} \quad (1.6)$$

According to this eqn (1.6), general ways to reduce sedimentation/creaming are either by decreasing the droplet size or increasing the viscosity of the continuous phase.

In addition to gravity, under certain conditions, temperature (affecting solubility) and concentration of the disperse phase with respect to the continuous one, may be the origin of a phase inversion. Furthermore, attractive forces between droplets, typically from van der Waals interactions or introduced by charges in the system resulting from components of the formulation (charged precursors or additives), can lead to aggregation (flocculation, coagulation). Finally, two other effects are also crucial in the destabilisation of emulsions: coalescence and Ostwald ripening. Since they are especially relevant, we will discuss in more detail the driving forces behind them and what strategies can be used to minimise them.

1.2.1.1 *Coalescence and Use of Surfactants*

Coalescence is a direct consequence of the minimisation of surface area and, therefore, the minimisation of the related surface energy. Let us make a simple exemplifying calculation by considering two spherical droplets of equal radius (r_1), both with mass m_1, volume V_1 and area A_1 (with ρ being the density of the substance in the droplet):

$$m_1 = \frac{4}{3}\pi r_1^3 \rho \quad (1.7)$$

$$V_1 = \frac{4}{3}\pi r_1^3 \quad (1.8)$$

$$A_1 = 4\pi r_1^2 \tag{1.9}$$

When the two smaller droplets of radius r_1 are combined to form a bigger one of radius r_2, the resulting mass and volume, m_2 and V_2, respectively, will clearly be just the addition of the original two values:

$$m_2 = 2m_1 = 2\left(\frac{4}{3}\pi r_1^3 \rho\right) \tag{1.10}$$

$$V_2 = 2V_1 = 2\left(\frac{4}{3}\pi r_1^3\right) = \frac{4}{3}\pi r_2^3 \rightarrow r_2 = 1.59 r_1 \tag{1.11}$$

The radius of the resulting droplet is a factor 1.59 bigger than the initial one. Accordingly, the area will be

$$A_2 = 4\pi r_2^2 = 4\pi(1.59 r_1)^2 \tag{1.12}$$

which is smaller than double the area given by eqn (1.9), $2(4\pi r_1^2)$. The new droplet has about 20.9% less area than the sum of the two initial droplets.

With the exception of surfactant-free systems (which are rather unusual) and cases of spontaneous emulsification or the 'ouzo effect'[27,28] (also rare for inorganic compounds), any emulsion strategy implies the use of surfactants, which stabilise the emulsions against coalescence, but also against aggregation and flocculation. Non-ionic surfactants provide a steric stabilisation, while ionic surfactants provide electrostatic or electrosteric (electrostatic and steric simultaneously) stabilisation. The mechanism of emulsification is schematically depicted in Figure 1.4. In a first step, a pre-emulsion is obtained by mechanical or magnetic stirring (droplets with sizes of several micrometres and polydispersed are initially formed). Afterwards, disruption (*i.e.*, comminution) of the droplets is achieved by the

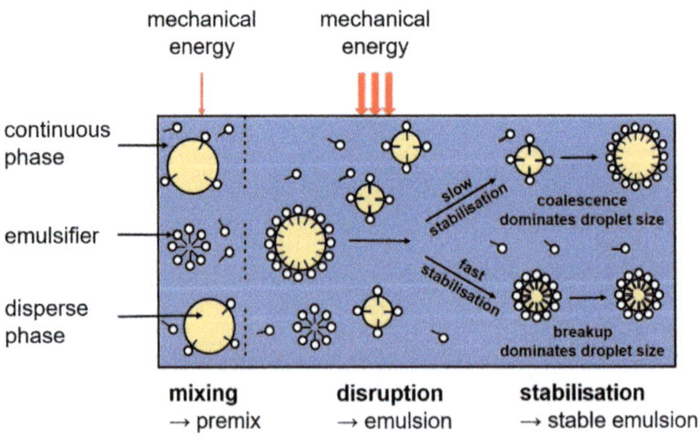

Figure 1.4 Schematic diagram of droplet breakup and coalescence in emulsification. Reproduced from https://publikationen.bibliothek.kit.edu/1000120408, under the terms of the CC BY-SA 4.0 license https://creativecommons.org/licenses/by-sa/4.0/deed.de.

application of mechanical energy. The formation of miniemulsions requires high shear forces (ultrasonication or high-pressure homogenisation are the most common methods) to reach stable nanosized droplets. Achieving a stable emulsion will require efficient droplet breakup and minimisation of coalescence.

Representative examples of surfactants used in micro- and miniemulsions are shown in Figure 1.5. Poly(ethylene-*co*-butylene)-*block*-poly(ethylene oxide) (P(E/B-*b*-PEO)), often referred to as 'KLE' (an acronym comprised of 'KL', for Kraton Liquid, the company that produced the ethylene/butylene block, and 'E' for ethylene oxide), has been widely used for stabilising, very effectively, water-in-oil systems. Unfortunately, in spite of its great stabilisation ability, this copolymer is not commercially available, requires a complex and tedious synthesis,[29] and so is not a very convenient option for most research groups. In this respect, surfactants such as polyglycerol

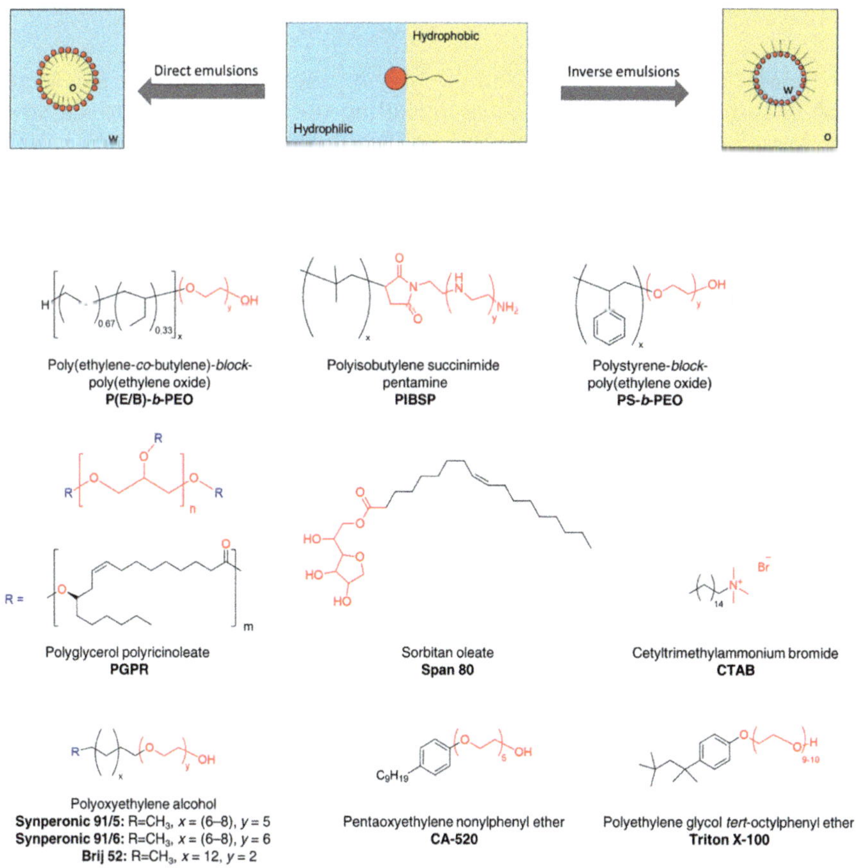

Figure 1.5 Schematic representation of how a surfactant stabilises direct and inverse systems (top) and chemical structures of common surfactants used in direct and inverse emulsions (bottom).

polyricinoleate (PGPR) or sorbitan oleate (commercially known as Span 80) are much more readily available for stabilising inverse systems, even if they are sometimes less effective than P(E/B-*b*-PEO). The stabilisation of oil-in-water (direct) systems is generally easier. Although ionic (anionic or cationic) surfactants are normally very efficient, they may interact easily with other charges in the systems, leading to coagulation. Therefore, especially when metal complexes are involved as precursors, non-ionic surfactants may be preferred. In this context, polyoxyethylene alcohols linked to apolar chains are a common choice.

In general, the suitability of surfactants for stabilising direct (oil-in-water) or inverse (water-in-oil) systems relies on the so-called hydrophilic–lipophilic balance (HLB). The initial HLB index was introduced by Griffin for non-ionic emulsifiers,[30] and it was defined as an arbitrary number between 0 and 20 calculated according to the formula

$$\mathrm{HLB} = 20 \times \frac{M_{\mathrm{hydro}}}{M_{\mathrm{total}}} \tag{1.13}$$

where M_{hydro} is the molar mass of the hydrophilic segment of the surfactant and M_{total} is the total molar mass, including the hydrophilic and the lipophilic segments. An index between 0 and 8 indicates that the surfactant is oil soluble and suitable for stabilising water-in-oil systems, and surfactants with indices above 11 are water soluble and normally good as oil-in-water emulsifiers. Indices between 8 and 11 indicate poor solubility in both oil and water, and may result in the formation of lamellar structures or vesicles.

In 1957, Davies proposed an alternative method to calculate the HLB index, which could also be applied to ionic surfactants,[31]

$$\mathrm{HLB} = 7 + \sum mH_{\mathrm{hydro}} + \sum nH_{\mathrm{lipo}} \tag{1.14}$$

where m and n are the number of hydrophilic and lipophilic groups, respectively, and H_{hydro} and H_{lipo} are arbitrary values assigned to the different chemical groups, of hydrophilic and lipophilic (*i.e.*, hydrophobic) nature, respectively. Table 1.2 lists the H values proposed by Davies for some hydrophilic and lipophilic groups, making some special considerations for ether groups (ethylene oxide is hydrophilic, while propylene oxide is considered lipophilic). It should be noted that eqn (1.14) may sometimes be found in the literature with a negative sign before the lipophilic term; in this case, the values in Table 1.2 for lipophilic groups should be positive instead of negative, so that the final result for the HLB index is the same.

Although HLB values provide indications related to application in direct and inverse systems, in practice the identification of suitable surfactants is slightly more complex, especially for microemulsions. Surfactants such as cetyltrimethylammonium bromide (CTAB), which is ionic and with a HLB index above 20, can be used in water-in-oil microemulsions.[32–34] In miniemulsions, in contrast to microemulsions, high or low HLB values are quite a good estimation of the suitability for oil-in-water or water-in-oil systems, respectively.

Table 1.2 Values proposed by Davies for some hydrophilic and hydrophobic groups for the calculation of HLB indices[31]

Chemical groups	Group value (*H*)
$-OSO_3^- Na^+$	38.7
$-COO^- K^+$	21.1
$-COO^- Na^+$	19.1
N (tertiary amine)	9.4
Ester (sorbitan ring)	6.8
Ester (free)	2.4
–COOH	2.1
–OH (free)	1.9
–OH (sorbitan ring)	0.5
–O–	1.3
=CH–	−0.475
$-CH_2-$	−0.475
$-CH_3$	−0.475
$-(CH_2-CH_2-O)-$	+0.33
$-(CH_2-CH_2-CH_2-O)-$	−0.15

1.2.1.2 *Ostwald Ripening and Use of Osmotic-pressure Agents*

As a consequence of eqn (1.1), the Laplace pressure for droplets of different size will also be different. Smaller droplets (with higher radius of curvature) will have a higher pressure difference between the inner and the outer part, being energetically less favoured than bigger droplets, for which the pressure difference is lower. Thermodynamically, the system will tend to compensate these differences, and smaller droplets will merge with bigger ones, leading to, in the limiting situation, phase separation and destabilisation of the emulsion. For supressing—or at least minimising—Ostwald ripening, so-called osmotic-pressure agents are used in emulsions. These agents are compounds that are highly soluble in the disperse phase and nearly immiscible in the continuous phase. To understand how osmotic-pressure agents work, one has to take into account the driving force of the Ostwald ripening, which is related to the Laplace pressure difference and the different osmotic pressure of droplets of different size: smaller particles (higher radius of curvature) are more soluble than larger ones (with lower radius of curvature), so that smaller droplets dissolve with time and diffuse to the bulk, to eventually be integrated in larger droplets. This phenomenon, in general when a curved interface is formed, is sometimes known as the Kelvin effect, and is described by the Kelvin equation,[35]

$$S(r) = S(\infty)\exp\left(\frac{2\gamma V_m^L}{rRT}\right) \tag{1.15}$$

where $S(r)$ is the solubility of a particle or droplet of radius r, $S(\infty)$ is the solubility of the bulk ($r \rightarrow \infty$), γ is the surface tension of the liquid, V_m^L is the molar volume of the liquid, R is the gas constant and T is the absolute temperature. For efficiency, the added osmotic-pressure agent should have a lower solubility in the continuous phase than the predominant compound of the disperse phase, so that the overall solubility will be slightly lower than without it. When the predominant compound diffuses from smaller droplets to bigger ones, the solubility of the small ones will be further reduced. Consequently, the increased solubility from the Kelvin effect will be balanced by the decrease in solubility when adding the additional compound.[36] In this way, the addition of a third component to emulsion systems is able to minimise Ostwald ripening. In the case of inorganic reactions within droplets, inorganic precursors themselves (*e.g.*, inorganic salts in the case of water-in-oil emulsions) may directly have the effect of osmotic-pressure agents.

1.3 Strategies for the Synthesis of Inorganic Nanoparticles from Emulsions

As an attempt at systematisation of the possible routes for synthesising inorganic nanoparticles from emulsions, we classify the strategies into three groups, which are schematically represented in Figure 1.6: (i) the

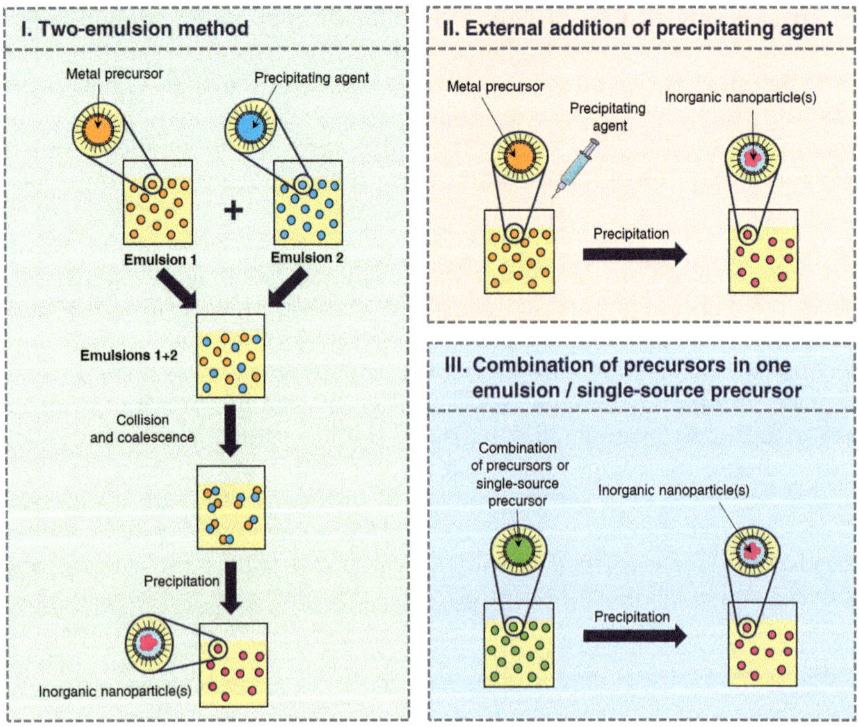

Figure 1.6 Different strategies of inorganic synthesis in micro- and miniemulsions.

'two-emulsion method'; (ii) external addition of a precipitating agent; and (iii) combination of precursors in the disperse phase, or use of single-source precursors. For completeness, some examples of strategies falling outside this classification will also be discussed at the end of this section.

1.3.1 The 'Two-emulsion Method'

This method involves mixing two separate miniemulsions, one containing a first reactant (typically a metal precursor) and another containing a second reactant (normally a precipitating agent). Upon application of high shear forces (produced, for example, by ultrasound), the droplets of both components fuse, coalesce and their contents react, leading to the formation of the desired compound, or of an intermediate that eventually converts to it (*e.g.*, hydroxide species that transform to an oxide). This strategy has been used, for example, to prepare ZnO nanoparticles by mixing two emulsions containing a zinc salt and a base, such as NaOH or ammonia.[37,38]

1.3.2 External Addition of a Precipitating Agent

In this case, the second reactant (precipitating agent) is added externally to an emulsion in which the first reactant (metal precursor) is contained. The added reactant dissolves in the continuous phase and diffuses to the droplets containing the first component, driving the reaction. Since the interface between the droplet and the continuous phase is the first place where the two components meet, hollow structures can also be formed, as discussed in Section 1.5, depending on the reaction kinetics and how fast the diffusion takes place.[22,39–41] The concentration of the precursor within the droplets can be crucial for obtaining solid or hollow particles, as reported for $Ce_{1-x}Cu_xO_2$ systems, in which hollow structures are formed at lower precursor concentrations and solid particles at higher concentrations.[40] The chemistry of the precipitating agent (more or less soluble in the disperse phase, depending on polarity), as well as the rate of addition, can also determine the formation, or not, of solid particles.[41]

1.3.3 Combination of Precursors in the Disperse Phase or Use of Single-source Precursors

If the reaction leading to the desired compound is sufficiently slow or requires a stimulus, such as temperature, the precursor and the second reactant can be combined in the disperse phase of a single microemulsion/miniemulsion, which is then allowed to react. The formation of some systems by a sol–gel route can take place through this method. For instance, TiO_2 and zirconium-doped anatase ($Zr_xTi_{1-x}O_2$, 0–7.1 mol% Zr) nanoparticles were prepared under acidic catalysis starting from an inverse miniemulsion containing the precursor bis(2-hydroxyethyl)titanate ($C_4H_{12}O_5Ti$) in aqueous solution in the disperse phase.[42,43] In some systems, the desired compound can be produced from just one precursor, or so-called *single-source precursor*. A single-source precursor is typically a metal complex that contains all necessary elements (for instance, the metal and oxygen, in the case of a metal oxide) to obtain an inorganic compound through a thermal- or light-activated decomposition process. The preparation of molybdic acid (hydrated forms of molybdenum trioxide, $MoO_3 \cdot nH_2O$) from peroxo complexes is one example of this kind of approach:

$$H_2Mo_2O_3(O_2)_4 \rightleftharpoons 2MoO_3 \cdot \tfrac{1}{2}H_2O + 2O_2 \tag{1.16}$$

$$2H_2MoO_2(O_2)_2 \rightleftharpoons 2MoO_3 \cdot H_2O + O_2 \tag{1.17}$$

$$2MoO_2(OH)(OOH) \rightleftharpoons 2MoO_3 \cdot H_2O + O_2 \tag{1.18}$$

In the previous reactions, the peroxo species on the left-hand side decompose to produce molybdenum trioxide hydrates. Such processes have been investigated in miniemulsion systems to produce molybdic acid and, analogously, tungstic acid.[44]

Another example of the use of a complex single-source precursor in a miniemulsion was reported for the preparation of Au/TiO_2 materials starting with a gold-containing titanium peroxo complex with the structure $AuCl_4(NH_4)_7[Ti_4(O_2)_4(cit)(Hcit)_2]_2 \cdot 12H_2O$, which was photochemically decomposed in the confined space of the droplets.[45]

1.3.4 Further Synthetic Strategies

In addition to the three previously described groups of strategies, a different approach for the formation of inorganic nanoparticles is the use of polymer–inorganic hybrid nanoparticles as precursors that can transform to the targeted material by oxidation or reduction of the metal (*e.g.*, through thermolysis or plasma steps). For example, different organotin compounds were encapsulated in polystyrene nanoparticles by a miniemulsion polymerisation procedure; after thermolysis of the films formed with the resulting suspension of organotin/polystyrene nanoparticles, tetragonal SnO_2 (cassiterite) nanoparticles were obtained.[46] In a slightly more sophisticated way, plasma treatment can be applied to hybrid nanoparticles containing metal complexes to transform them into highly ordered arrangements of metal nanoparticles.[47] In this process, hexagonal close-packed nanoparticles deposited on a surface are initially treated with oxygen plasma to remove the organic parts of the precursor particles, and subsequently thermally annealed to obtain the metal nanoparticles. This approach, depicted graphically in Figure 1.7, has been used for the preparation of platinum[47–50] and Fe–Pt alloy[49,51] nanoparticles. An analogous process was also reported for the preparation of ZnO nanoparticles from hybrid precursor particles containing zinc complexes.[52]

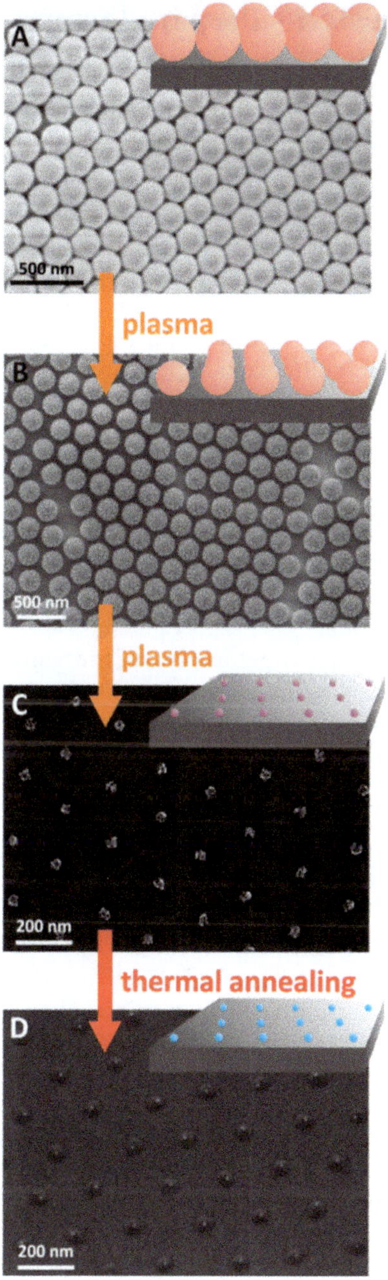

Figure 1.7 Process for the generation of arrays of metal nanoparticles with hexagonal symmetry. Reproduced from ref. 6 with permission from Elsevier, Copyright 2012.

1.4 Variety of Systems Prepared from Microemulsions and Miniemulsions

Most of the work in the field of inorganic synthesis in emulsions has been conducted in microemulsions. In this context, it is important to mention the pioneering work of Pileni in the application of nanodroplets and colloidal soft templates (micelles and colloidal associations) to prepare inorganic nanoparticles.[53–55] Many other groups have also made significant advancements in this area, with especially remarkable contributions from Feldmann[34,56–60] and Sánchez-Domínguez[61–68] and their respective teams. Both groups have been very active in recent years and have prepared a wide range of materials by using microemulsions.

Since many inorganic precursors are water soluble, it is common to use inverse emulsions, that is, water-in-oil microemulsions in which the disperse phase is aqueous (or based on a polar solvent or mixture of polar solvents) and the continuous phase is apolar. In this case, the inorganic precursor is solubilised in the aqueous phase, with inorganic salts being very common reagents. However, oil-soluble precursors, such as metal acetylacetonates, can also be used in direct (*i.e.*, oil-in-water) emulsions for the synthesis of inorganic compounds.[61–63,69,70]

As reported in Table 1.3, a large number of inorganic systems have been prepared in microemulsions, including metal oxides, mixed oxides, metal hydroxides, chalcogenides and more complex systems.[59,65,71] For the case of miniemulsions, the diversity of inorganic substances prepared is slightly more modest, but it has been increasing in the last few years, in parallel with the development of the technique for inorganic synthesis. Most of the substances prepared have been metal oxides (see Muñoz-Espí and Landfester[9]), but some chalcogenides (*e.g.*, CuS[72] and ZnS[73]) have also been reported. The preparation of ternary systems is also possible, including, among others, $Ca(OH)_2{:}Ln$ and $Mg(OH)_2{:}Ln$ ($Ln = Eu^{III}, Sm^{III}, Tb^{III}$),[74] lanthanide-doped CaF_2,[75] MFe_2O_4 (M: Co^{II}, Cu^{II}, Ni^{II}, Zn^{II}),[18] $YCrO_3$[76] or $Ce_{1-x}Cu_xO_2$.[40]

The majority of work conducted in miniemulsions has applied water-soluble inorganic precursors (typically metal salts). Consequently, inverse (*i.e.*, water-in-oil) miniemulsions have almost always been used. Common apolar solvents in the continuous phase are cyclohexane (b.p. 80.7 °C, which allows for easy evaporation), toluene (b.p. 100.6 °C) or oils, such as *n*-decane or Isopar M (b.p. >150 °C, which allows for higher synthesis temperatures, but makes it very difficult to remove the solvent).

Table 1.3 contains a collection of the inorganic compounds prepared so far at the nanoscale by using emulsion systems (both microemulsion and miniemulsion), together with a selection of representative references. We include here nanoparticles formed by precipitation within droplets. Particular cases involving interfacial precipitation to form hollow structures are considered in Section 1.5 (see Table 1.4).

Most studies on the use of micro- and miniemulsions for confining inorganic synthesis have been conducted under ambient pressure. In the

Table 1.3 Inorganic nanoparticles prepared by using nanodroplets as templates. The works are listed in alphabetical order of the inorganic system.

System	Precursor	Continuous phase	Type of emulsion	Surfactant	Reference
Ag_2S	$AgNO_3$ and Na_2S	Water and ethanol	Microemulsion	Linoleic acid	88
Ag_2Se	$AgNO_3$ and Na_2SeSO_3	Water and ethanol	Microemulsion	Linoleic acid	88
Au/TiO_2	Single-source precursor[a]	Pentanol/Heptane	Miniemulsion	SDS[b] and Triton X100	45
$BaCrO_4$	$Ba(NO_3)_2$ and K_2CrO_4	Water and Ethanol	Microemulsion	Linoleic acid	88
BaO	$BaCl_2$	Hexane	Miniemulsion	CTATos[c]	89
$CaCO_3$	$CaCl_2$ and $(NH_4)_2CO_3$	Cyclohexane	Microemulsion	CTAB[d]	32
$CaCO_3$	$NaHCO_3$ and $Ca(CH_3COO)_2$	*n*-Dodecane	Microemulsion	CTAB[d]	34
$Ca(OH)_2:Ln$ and $Mg(OH)_2:Ln$[e]	$Ca(NO_3)_2{\cdot}4H_2O$, $Mg(NO_3)_2{\cdot}6H_2O$, $Eu(NO_3)_3{\cdot}5H_2O$, $Tb(NO_3)_3{\cdot}5H_2O$, $C_6H_9O_6Sm{\cdot}xH_2O$	Cyclohexane	Miniemulsion	Igepal CO630, Brij® 52, Span 80	74
CdS	$Cd(NO_3)_2$ and Na_2S	Water and ethanol	Microemulsion	Linoleic acid	88
CdSe	$Cd(NO_3)_2$ and Na_2SeSO_3	Water and ethanol	Microemulsion	Linoleic acid	88
CeO_2	$Ce(NO_3)_3{\cdot}6H_2O$	Cyclohexane	Miniemulsion	P(E/B-*b*-EO)[f] and PIBSP[g]	90
CeO_2	$Ce[CH_3(CH_2)_3CH(C_2H_5)CO_2]_3$	Water	Microemulsion	Brij® 96V or Synperonic® 10/6	61 and 62
$CePO_4$	$Ce(NO_3)_3{\cdot}6H_2O$	Water and ethanol	Microemulsion	Linoleic acid	88
$Ce_{0.5}Zr_{0.5}O_2$	$Ce[CH_3(CH_2)_3CH(C_2H_5)CO_2]_3$ and $Zr(C_8H_{15}O_2)_4$	Water	Microemulsion	Synperonic® 10/6	62
$CoFe_2O_4$	$Co_2[Fe(CN)_6]$	*n*-Dodecane	Microemulsion	CTAB[d]	34
$CoFe_2O_4$–TiO_2	$C_{24}H_{45}FeO_6$, $C_{16}H_{30}CoO_4$, $C_{32}H_{50}O_8Ti$	Water	Microemulsion	Synperonic® 91/5	70

(continued)

Table 1.3 (*Continued*)

System	Precursor	Continuous phase	Type of emulsion	Surfactant	Reference
Co_3O_4	$Co_3[Co(CN)_6]_2$	*n*-Dodecane	Microemulsion	CTAB[d]	34
$Cu_xCe_{1-x}O_{2-\delta}$	$Ce[CH_3(CH_2)_3CH(C_2H_5)CO_2]_3$ and $Cu[CH_3(CH_2)_3CH(C_2H_5)CO_2]_2$	Isooctane	Microemulsion	Synperonic® 91/5	67
$CuFe_2O_4$	$Cu_2[Fe(CN)_6]$	*n*-Dodecane	Microemulsion	CTAB[d]	34
CuO	Cu(CN)	*n*-Dodecane	Microemulsion	CTAB[d]	34
CuS	$CuCl_2$ and $CS(NH_2)_2$	*n*-Dodecane	Microemulsion	CTAB[d]	34
CuS	$Cu(NO_3)_2$ and Na_2S	Cyclohexane	Miniemulsion	Brij® 52 and Igepal CO630	72
Fe_2O_3	$FeCl_2\cdot 4H_2O$	Cyclohexane or Isopar M	Miniemulsion	P(S/EO)[h]	91
Fe_2O_3	$FeCl_3$	*n*-Decane	Miniemulsion	Glissopal EM-23	92
$Fe_3Mn_3O_8$, Fe_3O_4, MFe_2O_4[i]	$MCl_2\cdot nH_2O$	Cyclohexane	Miniemulsion	PGPR[j]	18 and 93
Fe_3O_4	$C_{24}H_{45}FeO_6$	Hexane	Microemulsion	Synperonic® 10/6	64
Fe_3O_4	$FeCl_3$ and $FeCl_2$	*n*-Octane	Microemulsion	CTAB[d]	64
Fe_3O_4 or Fe_2O_3	$Fe[CH_3(CH_2)_3CH(C_2H_5)CO_2]_3$	Water	Microemulsion	Synperonic® 10/6	62
$HoPO_4$	$Ho(NO_3)_3$ and NaH_2PO_4	Water and ethanol	Microemulsion	Linoleic acid	88
$K_3[Co(NO_2)_6]$	$CoCl_2$, CH_3COOH and KNO_2	*n*-Dodecane	Microemulsion	CTAB[d]	34
MgAl, NiAl, and ZnAl hydrotalcites	$M(NO_3)_2\cdot 6H_2O$ (M = Mg, Zn or Ni) and $Al(NO_3)_3\cdot 9H_2O$	*n*-Butanol and isooctane	Microemulsion	CTAB[d]	94
$(NH_4)Y(C_2O_4)_2$	$Y(NO_3)_3$ and $(NH_4)_2C_2O_4$	*n*-Dodecane	Microemulsion	CTAB[d]	34
$(NH_4)_3PO_4(MoO_3)_{12}\cdot 4H_2O$	NaH_2PO_4, $NH_4H_2PO_4$ or organophosphate precursors[l]	Cyclohexane	Miniemulsion	PGPR[j]	16
$NiFe_2O_4$	$Ni_2[Fe(CN)_6]$	*n*-Dodecane	Microemulsion	CTAB[d]	34
NiO	$Ni(OCOCH_3)_2\cdot 4H_2O$	Isooctane	Miniemulsion	Aerosol-OT	95
NiO	$Ni(CN)_2$	*n*-Dodecane	Microemulsion	CTAB[d]	34
$PbCrO_4$	$Pb(CH_3COO)_2\cdot 3H_2O$ and K_2CrO_4	Water and ethanol	Microemulsion	Linoleic acid	88

PbS	$Pb(CH_3COO)_2 \cdot 3H_2O$ and Na_2S	Water and ethanol	Microemulsion	Linoleic acid	88
PbSe	$Pb(CH_3COO)_2 \cdot 3H_2O$ and Na_2SeSO_3	Water and ethanol	Microemulsion	Linoleic acid	88
TiO_2	TIP[k]	Isopar M	Miniemulsion	P(E/B-*b*-EO)[e]	42
TiO_2	$Ti(C_8H_{15}O_2)_4$	Water	Microemulsion	Synperonic® 10/6	62
$Y(OH)_3$	$Y(NO_3)_3$	*n*-Dodecane	Microemulsion	CTAB[d]	34
$Zn(Mn_{0.5}Zn_{0.5}Fe_2O_4)$	$FeSO_4 \cdot 7H_2O$, $MnSO_4 \cdot H_2O$ and $ZnSO_4 \cdot 7H_2O$	Isooctane	Microemulsion	Synperonic® 13/6.5	66
ZnO	$ZnSO_4$	*n*-Decane	Miniemulsion	Glissopal EM-23	96
ZnO and M : ZnO[m]	$Zn(NO_3)_2 \cdot 6H_2O$	Cyclohexane	Miniemulsion	Triton X100	37 and 97
ZnO	$ZnSO_4$	*n*-Decane	Miniemulsion	PGPR[j] and Span 80	38
ZnO	$Zn(AcO)_2 \cdot 2H_2O$	*n*-Decane	Miniemulsion	Span 20	98
ZnO	$ZnCl_2$	*n*-Dodecane	Microemulsion	CTAB[d]	34
ZnS	$Zn(NO_3)_2 \cdot 6H_2O$	Cyclohexane	Miniemulsion	Brij® 52 and Igepal CO630	73
ZnS	$Zn(NO_3)_2 \cdot 6H_2O$ and Na_2S	Water and ethanol	Microemulsion	Linoleic acid	88
ZnO_2	$Zn[CH_3(CH_2)_3CH(C_2H_5)CO_2]_2$	Hexane	Microemulsion	Synperonic® 91/6	99
ZrO_2	$Zr(C_8H_{15}O_2)_4$	Water	Microemulsion	Synperonic® 10/6	62
$Zr(HPO_4)_2 \cdot nH_2O$	$ZrOCl_2 \cdot 8H_2O$	Cyclohexane	Microemulsion	Igepal CO520	100
$Zr_xTi_{1-x}O_2$, $ZrTiO_4$	$Zr(O^iPr)_4 \cdot {}^nPrOH$ and TIP[j]	Isopar M	Microemulsion	P(E/B-*b*-EO)[e]	43

[a]Single-source precursor: $AuCl_4(NH_4)_7[Ti_4(O_2)_4(cit)(Hcit)_2]_2 \cdot 2H_2O$.
[b]SDS: sodium dodecyl sulfate.
[c]CTATos: hexadeciltrimetilamonium-*p*-toluensulfonate.
[d]CTAB: cetyltrimethylammonium bromide.
[e]Ln = Eu^{III}, Sm^{III}, Tb^{III}.
[f]P(E/B-*b*-EO): poly(ethylene/butylene-*block*-ethylene oxide).
[g]PIBSP: polyisobutylene succinimide pentaamine.
[h]P(S/EO): poly(styrene/ethylene oxide).
[i]M: Co^{II}, Cu^{II}, Ni^{II}, Zn^{II}.
[j]PGPR: polyglycerol polyricinoleate.
[k]TIP: titanium isopropoxide.
[l]Organophosphate precursors: phytic acid sodium salt hydrate, D-glucose-6-phosphate sodium salt, *O*-phospho-DL-serine.
[m]M = Ag^{I}, Co^{II}, Cu^{II}, Eu^{III}, Mg^{II}, Mn^{II}.

Table 1.4 Inorganic nanocapsules prepared at the droplet interface in micro- and miniemulsions. The works are listed in alphabetical order of the inorganic system.

System	Precursor	Continuous phase	Emulsion type	Surfactant	Reference
Ag_2S	$[Ag(Ph_3P)_4]NO_3$	Toluene	Microemulsion	CTAB[a]	57
$AlO(OH)$	$Al[OCH(CH_3)C_2H_5]_3$	*n*-Dodecane	Microemulsion	CTAB[a]	33
$CaCO_3$	$CaCl_2/Na_2CO_3$	*n*-Dodecane	Microemulsion	CTAB[a]	79
$Ce_{1-x}Cu_xO_2$	$Cu(NO_3)_2 \cdot 3H_2O$ and $Ce(NO_3)_3 \cdot 6H_2O$	Toluene	Miniemulsion	PGPR[b]	40
CeO_2	$Ce(NO_3)_3 \cdot 6H_2O$ and $FeCl_2$	Toluene	Miniemulsion	PIBSP[c], PGPR[b] and P(S-*b*-AA)[d]	22
CuO	$Cu(NO_3)_2 \cdot 3H_2O$	Toluene	Miniemulsion	PIBSP[c] and PGPR[b]	39
CuS	$[C_6H_{11}(CH_2)_3CO_2]_2Cu$	Toluene	Microemulsion	CTAB[a]	58
$Cu_{1.8}S$	$CuCl(Ph_3P)_3$	Toluene	Microemulsion	CTAB[a]	58
Cu_2S	$CuCl(Ph_3P)_3$	Toluene	Microemulsion	CTAB[a]	58
Fe_2O_3	$FeCl_3 \cdot 6H_2O$	Toluene	Miniemulsion	PIBSP[c], PGPR[b] and P(S-*b*-AA)[d]	22
HfO_2, ZrO_2, $HfZr_{1-x}O_2$	$ZrOCl_2 \cdot 8H_2O$ and $HfOCl_2 \cdot 8H_2O$	Toluene	Miniemulsion	PIBSP[c]	41
$La(OH)_3$	$La(C_5H_5)_3$	*n*-Dodecane	Microemulsion	CTAB[a]	56
SiO_2	Tetraethoxysilane (TEOS)	Cyclohexane, *n*-hexane, *n*-octane, *n*-decane, *n*-dodecane	Microemulsion	CA-520 and/or Triton X100	101
SiO_2	Tetraethoxysilane (TEOS)	Cyclohexane	Microemulsion	CTAB[a]	102
SiO_2	Tetraethoxysilane (TEOS)	Water	Miniemulsion	CTMA-Cl[e]	81
SnO_2	$Sn(Ot\text{-}Bu)_4$	*n*-Dodecane	Microemulsion	CTAB[a]	77
TiO_2	$TiCl_4$	*n*-Dodecane	Microemulsion	CTAB[a]	60
$YCrO_3$	$Cr(NO_3)_3 \cdot 9H_2O$ and YCl_3	Toluene	Miniemulsion	PIBSP[c]	76
ZnO	$Zn(C_5(CH_3)_5)_2$	Toluene	Microemulsion	CTAB[a]	78
ZrO_2	$Zr(O^tC_4H_9)_4$	*n*-Dodecane	Microemulsion	CTAB[a]	79

[a]CTAB: cetyltrimethylammonium bromide.
[b]PGPR: polyglycerol polyricinoleate.
[c]PIBSP: polyisobutylene succinimide pentaamine.
[d]P(S-*b*-AA): poly(styrene-*block*-acrylic acid).
[e]CTMA-Cl: cetyltrimethylammonium chloride.

last few years, some studies have started to explore the combination of miniemulsions with hydrothermal conditions (*i.e.*, temperatures above room temperature and pressures above the ambient one). In other words, miniemulsions are placed under hydrothermal conditions, so that there is a double pressure effect: the intrinsic pressure resulting from the formation of the curved interface of the droplets (Laplace pressure) and the external one applied during the hydrothermal reaction. Different spinel ferrites, including Fe_3MnO_8, $CoFe_2O_4$, $CuFe_2O_4$, $NiFe_2O_4$ and $ZnFe_2O_4$ have been prepared under hydrothermal conditions by adding NaOH to an inverse emulsion of the metal precursors (metal salts) and ultrasonicating again to allow coalescence and precipitation.[18] This study demonstrated that for some of the zinc ferrite systems, the materials obtained by combining miniemulsion and hydrothermal conditions resulted in much more crystalline ferrites than the analogous ones obtained by miniemulsions at ambient pressure or under bulk conditions, either at ambient pressure or under hydrothermal conditions.

1.5 Reactions at the Liquid–Liquid Droplet Interface: Formation of Inorganic Nanocapsules

Nanocapsules of inorganic materials can be produced when the formation reaction of the target substance takes place at the liquid–liquid droplet interface. A variety of metal oxides, metal hydroxides and metal chalcogenides have been produced in microemulsion in the form of hollow particles.[33,56–58,60,77–79]

In miniemulsions, one of the most common inorganic capsular systems is silica, prepared by a sol–gel process starting at the droplet interface. In a typical oil-in-water formulation, a silicon alkoxide, such as tetraethyl orthosilicate (TEOS), is dissolved in an oil (*e.g.*, octane) and dispersed in a continuous aqueous phase.[80] The alkoxide groups of the silane precursor are hydrolysed to silanol groups upon contact with water under acidic or basic catalysis. A silica network is produced when the silanol groups condense with each other, leading to the capsule formation. The surface chemistry of the capsule can be tuned by adding functionalised silanes (such as 3-aminopropyltriethoxysilane or 3-mercaptopropyltrimethoxysilane) during the process.[81] The strategy is similar to the copolymerisation procedures used to prepare functionalised polymer particles with hydrophilic monomers in miniemulsion polymerisation.[82] The silanes with polar functional groups tend to be placed at the surface of the resulting particles.

An analogous strategy for the preparation of silica can also be carried out in inverse systems.[83] In this case, the silicon alkoxide is added to the apolar continuous phase, and hydrolyses upon contact with the water droplets of the disperse phase. This situation, together with the formation reactions of silica by hydrolysis and condensation of silanes, is presented in Figure 1.8. By adding silanes with different polarities during the synthesis, the surface hydrophobicity of the resulting capsules can be adjusted.[84]

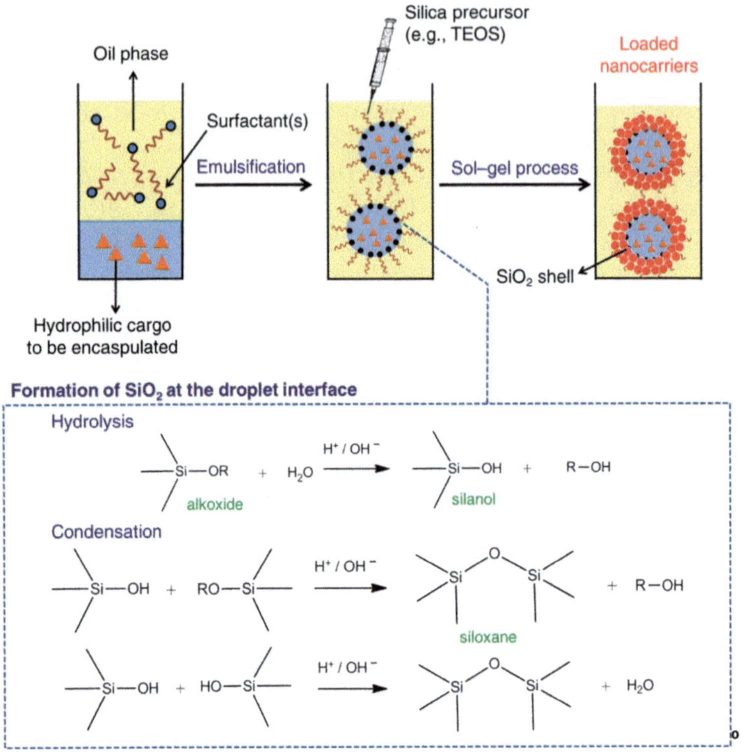

Figure 1.8 Scheme of the formation of silica nanocapsules in inverse miniemulsions. The reactions present the hydrolysis and condensation of a general alkoxide with rest R. Reproduced from ref. 85, https://doi.org/10.1016/j.colsurfb.2021.111764, under the terms of the CC BY 4.0 license https://creativecommons.org/licenses/by/4.0/.

In a similar fashion to silica, sol–gel processes for group 4 metals (Zr, Hf) have also been carried out in inverse miniemulsions to yield hydrous zirconia ($ZrO(OH)_2 \cdot nH_2O$) and hafnia ($HfO(OH)_2 \cdot nH_2O$).[41] Metal oxychloride precursors ($ZrOCl_2 \cdot 8H_2O$ and $HfOCl_2 \cdot 8H_2O$) are dissolved in water and emulsified in toluene. A schematic representation of the mechanism is depicted in Figure 1.9. Triethylamine (Et_3N), a base that is soluble in organic solvents, but also in water to a certain extent, is added to the continuous phase, generating hydroxide ions at the interface and within the droplets:

$$Et_3N + H_2O \rightleftharpoons Et_3NH^+OH \tag{1.19}$$

The oxychloride precursor reacts with the hydroxide ions to form zirconia or hafnia:

$$2Et_3NH^+OH^- + ZrOCl_2 \rightleftharpoons ZrO(OH)_2 + 2Et_3NH^+Cl^- \tag{1.20}$$

$$2Et_3NH^+OH^- + HfOCl_2 \rightleftharpoons HfO(OH)_2 + 2Et_3NH^+Cl^- \tag{1.21}$$

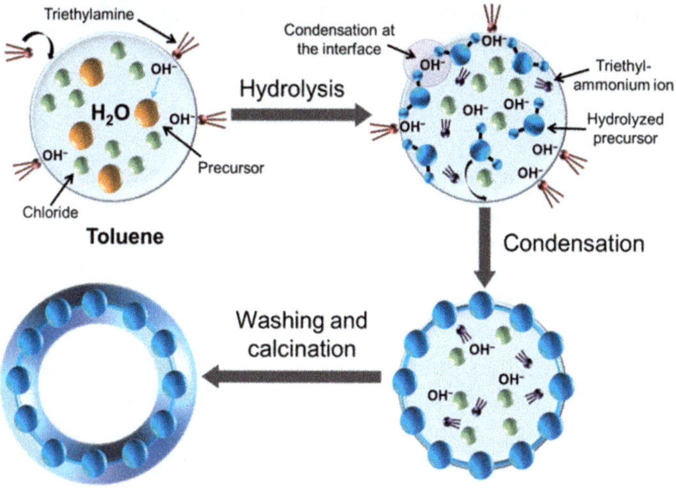

Figure 1.9 Scheme of the formation of zirconia and hafnia capsules by interfacial precipitation in the presence of triethylamine. Reproduced from ref. 41 with permission from the Royal Society of Chemistry.

For typical sol–gel systems, such as silica, zirconia, hafnia or titania, the materials produced are amorphous and only crystallise upon calcination. By contrast, crystallisation pathways can occur under mild miniemulsion conditions for various transition metal oxide and lanthanide oxide systems, including γ-Fe_2O_3, CeO_2 and CuO.[22,39]

Some of the most representative systems reported in the literature for the production of capsules in micro- and miniemulsions are listed in Table 1.4. This compilation complements and completes the inorganic systems given in Table 1.3.

1.6 Concluding Remarks

In this chapter we have seen that nanodroplets in miniemulsions and microemulsions can create enclosed areas where chemical reactions can occur. For this reason, nanodroplets can be considered as 'nanoreactors', especially for miniemulsion systems, in which the diffusion between droplets is supposed to be ideally suppressed. The formation of the curved interface at the droplet surface provides an intrinsic increase of the inner pressure, which may be advantageous for obtaining certain crystal phases under milder temperature conditions. Unusual crystallisation pathways may be pursued under such conditions, even allowing the potential stabilisation of certain metastable phases. The solubility of the precursor and some properties (*e.g.*, vapour pressure, viscosity, relative permittivity) of the solvent in the disperse phase, most commonly—but not only—water in the case of inorganic systems, may also be significantly affected by the very small size of the droplets and the related confinement. The confinement can be utilised to tune the size of the particles obtained. In the simplest case, the crystalline size and the number of crystalline domains that can aggregate to form polycrystalline particles will be influenced by the restricted space available for crystallisation.

We have classified the methods for producing inorganic materials in emulsions in three groups: (1) the 'two-emulsion method', based on the coalescence of droplets containing the two reactants leading to the desired compound; (2) addition of the second reactant to an emulsion containing the first reactant in the disperse phase; and (3) emulsions in which both reactants are simultaneously present in the disperse phase or, alternatively, emulsions containing a single-source precursor that yields, upon thermally or photochemically induced decomposition, the target materials. Furthermore, besides these strategies, we have also seen that highly structured arrangements of metal oxide and metal nanoparticles can be produced by oxidation or reduction of films of polymer/inorganic precursor particles, after thermolysis or plasma treatment to remove the organic matter, to achieve the desired compound.

In addition to the confinement within droplets to form nanoparticles, the space constriction can also take place at the droplet interface, leading to the formation of hollow structures or capsules. The variety of inorganic compounds reported in the literature is very broad, including silica, metal oxides, hydroxides, chalcogenides, phosphates or even ternary or more complex systems.

References

1. *Emulsion Polymerization and Emulsion Polymers*, ed. P. A. Lovell, and M. S. El-Aasser, Wiley, Chichester, 1997.
2. M. Antonietti and K. Landfester, *Prog. Polym. Sci.*, 2002, **27**, 689.
3. K. Landfester, *Angew. Chem., Int. Ed.*, 2009, **48**, 4488.
4. K. Landfester and C. K. Weiss, *Adv. Polym. Sci.*, 2010, **229**, 1–49.
5. O. Álvarez-Bermúdez and R. Muñoz-Espí, in *Nanoemulsions*, ed. S. M. Jafari, and D. J. McClements, Academic Press, 2018, pp. 477–515.
6. R. Muñoz-Espí, C. K. Weiss and K. Landfester, *Curr. Opin. Colloid Interface Sci.*, 2012, **17**, 212.
7. R. Muñoz-Espí, Y. Mastai, S. Gross and K. Landfester, *CrystEngComm*, 2013, **15**, 2175.
8. S. Diodati, P. Dolcet, M. Casarin and S. Gross, *Chem. Rev.*, 2015, **115**, 11449.
9. R. Muñoz-Espí and K. Landfester, *Chem. – A Eur. J.*, 2020, **26**, 9304.
10. D. Crespy and K. Landfester, *Beilstein J. Org. Chem.*, 2010, **6**, 1132.
11. A. Gupta, H. B. Eral, T. A. Hatton and P. S. Doyle, *Soft Matter*, 2016, **12**, 2826.
12. *Nanoemulsions*, ed. S. M. Jafari, and D. J. McClements, Academic Press, 2018.

13. C. Solans, P. Izquierdo, J. Nolla, N. Azemar and M. J. Garcia-Celma, *Curr. Opin. Colloid Interface Sci.*, 2005, **10**, 102.
14. C. Solans and I. Sole, *Curr. Opin. Colloid Interface Sci.*, 2012, **17**, 246.
15. J.-L. Blin, M.-J. Stébé and B. Lebeau, *Curr. Opin. Colloid Interface Sci.*, 2016, **25**, 75.
16. A. Antonello, C. Benedetti, F. F. Pérez-Pla, M. Kokkinopoulou, K. Kirchhoff, V. Fischer, K. Landfester, S. Gross and R. Muñoz-Espí, *ACS Appl. Mater. Interfaces*, 2018, **10**, 23174.
17. J. Lam and J. F. Lutsko, *Nanoscale*, 2018, **10**, 4921.
18. A. Antonello, G. Jakob, P. Dolcet, R. Momper, M. Kokkinopoulou, K. Landfester, R. Muñoz-Espí and S. Gross, *Chem. Mater.*, 2017, **29**, 985.
19. M. F. Chaplin, in *Adsorption and Phase Behaviour in Nanochannels and Nanotubes*, ed. L. J. Dunne, and G. Manos, Springer Netherlands, Dordrecht, 2010, pp. 241–255, DOI: 10.1007/978-90-481-2481-7_11.
20. D. Wang, Y. Tian and L. Jiang, *Small*, 2021, **17**, 2100788.
21. E. Breynaert, M. Houlleberghs, S. Radhakrishnan, G. Grübel, F. Taulelle and J. A. Martens, *Chem. Soc. Rev.*, 2020, **49**, 2557.
22. H. S. Varol, O. Alvarez-Bermudez, P. Dolcet, B. Kuerbanjiang, S. Gross, K. Landfester and R. Muñoz-Espí, *Langmuir*, 2016, **32**, 13116.
23. K. Landfester, *Annu. Rev. Mater. Res.*, 2006, **36**, 231.
24. D. Crespy, M. Stark, C. Hoffmann-Richter, U. Ziener and K. Landfester, *Macromolecules*, 2007, **40**, 3122.
25. J. H. Swisher, L. Jibril, S. H. Petrosko and C. A. Mirkin, *Nat. Rev. Mater.*, 2022, **7**, 428.
26. D. J. McClements, *Soft Matter*, 2012, **8**, 1719.
27. C. Solans, D. Morales and M. Homs, *Curr. Opin. Colloid Interface Sci.*, 2016, **22**, 88.
28. F. Ganachaud and J. L. Katz, *ChemPhysChem*, 2005, **6**, 209.
29. A. Thomas, H. Schlaad, B. Smarsly and M. Antonietti, *Langmuir*, 2003, **19**, 4455.
30. W. C. Griffin, *J. Soc. Cosmet. Chem.*, 1954, **5**, 249.
31. J. T. Davies, in *Proceedings of the International Congress of Surface Activity*, 1957, pp. 426–438.
32. D. Liu and M. Z. Yates, *Langmuir*, 2006, **22**, 5566.
33. D. H. M. Buchold and C. Feldmann, *Nano Lett.*, 2007, **7**, 3489.
34. D. H. M. Buchold and C. Feldmann, *Adv. Funct. Mater.*, 2008, **18**, 1002.
35. T. Tadros, in *Encyclopedia of Colloid and Interface Science*, ed. T. Tadros, Springer Berlin Heidelberg, Berlin, Heidelberg, 2013, p. 820, DOI: 10.1007/978-3-642-20665-8_124.
36. S. S. Davis, H. P. Round and T. S. Purewal, *J. Colloid Interface Sci.*, 1981, **80**, 508.
37. P. Dolcet, M. Casarin, C. Maccato, L. Bovo, G. Ischia, S. Gialanella, F. Mancin, E. Tondello and S. Gross, *J. Mater. Chem.*, 2012, **22**, 1620.
38. M. Winkelmann, E.-M. Grimm, T. Comunian, B. Freudig, Y. Zhou, W. Gerlinger, B. Sachweh and H. P. Schuchmann, *Chem. Eng. Sci.*, 2013, **92**, 126.

39. I. Singh, K. Landfester, A. Chandra and R. Muñoz-Espí, *Nanoscale*, 2015, **7**, 19250.
40. I. Singh, K. Landfester, R. Muñoz-Espí and A. Chandra, *Nanotechnology*, 2017, **28**, 075601.
41. M. Hajir, P. Dolcet, V. Fischer, J. Holzinger, K. Landfester and R. Muñoz-Espí, *J. Mater. Chem.*, 2012, **22**, 5622.
42. R. Rossmanith, C. K. Weiss, J. Geserick, N. Hüsing, U. Hormann, U. Kaiser and K. Landfester, *Chem. Mater.*, 2008, **20**, 5768.
43. R. Schiller, C. K. Weiss and K. Landfester, *Nanotechnology*, 2010, **21**, 405603.
44. R. Muñoz-Espí, F. Graf, L. Preiss and K. Landfester, in *Proceedings BIWIC 2010*, ed. T. Stelzer, and J. Ulrich, Cuvillier Verlag, Göttingen, 2010, pp. 71–77.
45. N. A. Heutz, P. Dolcet, A. Birkner, M. Casarin, K. Merz, S. Gialanella and S. Gross, *Nanoscale*, 2013, **5**, 10534.
46. R. Muñoz-Espí, P. Dolcet, T. Rossow, M. Wagner, K. Landfester and D. Crespy, *ACS Appl. Mater. Interfaces*, 2011, **3**, 4292.
47. A. Manzke, C. Pfahler, O. Dubbers, A. Plettl, P. Ziemann, D. Crespy, E. Schreiber, U. Ziener and K. Landfester, *Adv. Mater.*, 2007, **19**, 1337.
48. A. Manzke, N. Vogel, C. K. Weiss, U. Ziener, A. Plettl, K. Landfester and P. Ziemann, *Nanoscale*, 2011, **3**, 2523.
49. E. Schreiber, U. Ziener, A. Manzke, A. Plettl, P. Ziemann and K. Landfester, *Chem. Mater.*, 2009, **21**, 1750.
50. N. Vogel, U. Ziener, A. Manzke, A. Plettl, P. Ziemann, J. Biskupek, C. K. Weiss and K. Landfester, *Beilstein J. Nanotechnol.*, 2011, **2**, 459.
51. A. Manzke, A. Plettl, U. Wiedwald, L. Han, P. Ziemann, E. Schreiber, U. Ziener, N. Vogel, C. K. Weiss, K. Landfester, K. Fauth, J. Biskupek and U. Kaiser, *Chem. Mater.*, 2012, **24**, 1048.
52. E. Kobitskaya, D. Ekinci, A. Manzke, A. Plettl, U. Wiedwald, P. Ziemann, J. Biskupek, U. Kaiser, U. Ziener and K. Landfester, *Macromolecules*, 2010, **43**, 3294.
53. M. P. Pileni, *Adv. Colloid Interface Sci.*, 1993, **46**, 139.
54. M. P. Pileni, *Langmuir*, 1997, **13**, 3266.
55. M.-P. Pileni, *Nat. Mater.*, 2003, **2**, 145.
56. P. Leidinger, R. Popescu, D. Gerthsen and C. Feldmann, *Small*, 2010, **6**, 1886.
57. P. Leidinger, R. Popescu, D. Gerthsen and C. Feldmann, *Chem. Mater.*, 2013, **25**, 4173.
58. P. Leidinger, R. Popescu, D. Gerthsen, H. Lünsdorf and C. Feldmann, *Nanoscale*, 2011, **3**, 2544.
59. S. Wolf and C. Feldmann, *Angew. Chem., Int. Ed.*, 2016, **55**, 15728.
60. C. Zurmühl, R. Popescu, D. Gerthsen and C. Feldmann, *Solid State Sci.*, 2011, **13**, 1505.
61. M. Sánchez-Domínguez, M. Boutonnet and C. Solans, *J. Nanopart. Res.*, 2009, **11**, 1823.

62. M. Sánchez-Domínguez, L. F. Liotta, G. Di Carlo, G. Pantaleo, A. M. Venezia, C. Solans and M. Boutonnet, *Catal. Today*, 2010, **158**, 35.
63. M. Sánchez-Domínguez, H. Koleilat, M. Boutonnet and C. Solans, *J. Dispersion Sci. Technol.*, 2011, **32**, 1765.
64. C. Okoli, M. Sánchez-Domínguez, M. Boutonnet, S. Jaras, C. Civera, C. Solans and G. R. Kuttuva, *Langmuir*, 2012, **28**, 8479.
65. M. Sánchez-Domínguez, K. Pemartin and M. Boutonnet, *Curr. Opin. Colloid Interface Sci.*, 2012, **17**, 297.
66. C. Aubery, C. Solans, S. Prevost, M. Gradzielski and M. Sánchez-Domínguez, *Langmuir*, 2013, **29**, 1779.
67. K. Pemartin-Biernath, A. V. Vela-González, M. B. Moreno-Trejo, C. Leyva-Porras, I. E. Castañeda-Reyna, I. Juárez-Ramírez, C. Solans and M. Sánchez-Domínguez, *Materials*, 2016, **9**, 480.
68. M. Boutonnet and M. Sánchez-Domínguez, *Catal. Today*, 2017, **285**, 89.
69. M. Sánchez-Domínguez, K. Pemartin, C. Solans and M. Boutonnet, in *Colloid and Interface Chemistry for Nanotechnology*, ed. P. Kralchevsky, R. Miller, and F. Ravera, Routledge, Abingdon-on-Thames, 2014, pp. 177–197.
70. A. A. Rodríguez-Rodríguez, S. Martínez-Montemayor, C. C. Leyva-Porras, F. E. Longoria-Rodríguez, E. Martínez-Guerra and M. Sánchez-Domínguez, *J. Nanomater.*, 2017, **2017**, 2367856.
71. I. López, L. Garza Tovar, E. T. Adesuji and M. Sánchez-Domínguez, in *Colloidal Metal Oxide Nanoparticles*, ed. S. Thomas, A. Tresa Sunny, and P. Velayudhan, Elsevier, Amsterdam, 2020, pp. 125–181.
72. G. Morgese, P. Dolcet, A. Feis, C. Gellini, S. Gialanella, A. Speghini, D. Badocco, P. Pastore, M. Casarin and S. Gross, *Eur. J. Inorg. Chem.*, 2017, **2017**, 2745.
73. P. Dolcet, C. Maurizio, M. Casarin, L. Pandolfo, S. Gialanella, D. Badocco, P. Pastore, A. Speghini and S. Gross, *Eur. J. Inorg. Chem.*, 2015, **2015**, 706.
74. E. Butturini, P. Dolcet, M. Casarin, A. Speghini, M. Pedroni, F. Benetti, A. Motta, D. Badocco, P. Pastore, S. Diodati, L. Pandolfo and S. Gross, *J. Mater. Chem. B*, 2014, **2**, 6639.
75. P. Dolcet, A. Mambrini, M. Pedroni, A. Speghini, S. Gialanella, M. Casarin and S. Gross, *RSC Adv.*, 2015, **5**, 16302.
76. I. Singh, A. K. Nigam, K. Landfester, R. Muñoz-Espí and A. Chandra, *Appl. Phys. Lett.*, 2013, **103**, 182902.
77. F. Gyger, M. Hübner, C. Feldmann, N. Barsan and U. Weimar, *Chem. Mater.*, 2010, **22**, 4821.
78. P. Leidinger, N. Dingenouts, R. Popescu, D. Gerthsen and C. Feldmann, *J. Mater. Chem.*, 2012, **22**, 14551.
79. C. Zurmühl, R. Popescu, D. Gerthsen and C. Feldmann, *Z. Anorg. Allg. Chem.*, 2014, **640**, 2669.
80. B. Peng, M. Chen, S. Zhou, L. Wu and X. Ma, *J. Colloid Interface Sci.*, 2008, **321**, 67.

81. J. Fickert, P. Rupper, R. Graf, K. Landfester and D. Crespy, *J. Mater. Chem.*, 2012, **22**, 2286.
82. R. Muñoz-Espí, Y. Qi, I. Lieberwirth, C. M. Gomez and G. Wegner, *Chemistry*, 2006, **12**, 118.
83. M. A. Hood, U. Paiphansiri, D. Schaeffel, K. Koynov, M. Kappl, K. Landfester and R. Muñoz-Espí, *Chem. Mater.*, 2015, **27**, 4311.
84. M. A. Hood, N. Encinas, D. Vollmer, R. Graf, K. Landfester and R. Muñoz-Espí, *Colloids Surf., A*, 2017, **532**, 172.
85. A. Elzayat, I. Adam-Cervera, O. Álvarez-Bermúdez and R. Muñoz-Espí, *Colloids Surf., B*, 2021, **203**, 111764.
86. M. Sanchez-Dominguez, K. Pemartin and M. Boutonnet, *Curr. Opin. Colloid Interface Sci.*, 2012, **17**, 297.
87. J. van Embden, S. Gross, K. R. Kittilstved and E. Della Gaspera, *Chem. Rev.*, 2023, **123**, 271.
88. J. P. Ge, W. Chen, L. P. Liu and Y. D. Li, *Chem. – Eur. J.*, 2006, **12**, 6552.
89. R. Gomes, S. Roming, A. Przybilla, M. A. R. Meier and C. Feldmann, *J. Mater. Chem. C*, 2014, **2**, 1513.
90. N. Nabih, R. Schiller, I. Lieberwirth, E. Kockrick, R. Frind, S. Kaskel, C. K. Weiss and K. Landfester, *Nanotechnology*, 2011, **22**, 135606.
91. M. Willert, R. Rothe, K. Landfester and M. Antonietti, *Chem. Mater.*, 2001, **13**, 4681.
92. M. Winkelmann and H. P. Schuchmann, *Particuology*, 2011, **9**, 502.
93. P. Dolcet, K. Kirchberg, A. Antonello, C. Suchomski, R. Marschall, S. Diodati, R. Muñoz-Espí, K. Landfester and S. Gross, *Inorg. Chem. Front.*, 2019, **6**, 1527.
94. F. Bellezza, A. Cipiciani, U. Costantino, M. Nocchetti and T. Posati, *Eur. J. Inorg. Chem.*, 2009, **2009**, 2603.
95. F. Fazlali, A. r. Mahjoub and R. Abazari, *Solid State Sci.*, 2015, **48**, 263.
96. M. Winkelmann, T. Schuler, P. Uzunogullari, C. A. Winkler, W. Gerlinger, B. Sachweh and H. P. Schuchmann, *Chem. Eng. Sci.*, 2012, **81**, 209.
97. P. Dolcet, F. Latini, M. Casarin, A. Speghini, E. Tondello, C. Foss, S. Diodati, L. Verin, A. Motta and S. Gross, *Eur. J. Inorg. Chem.*, 2013, **2013**, 2291.
98. M. Fricke, A. Voigt, P. Veit and K. Sundmacher, *Ind. Eng. Chem. Res.*, 2015, **54**, 10293.
99. K. Pemartin, C. Solans, G. Vidal-Lopez and M. Sánchez-Domínguez, *Chem. Lett.*, 2012, **41**, 1032.
100. F. Bellezza, A. Cipiciani, U. Costantino, F. Marmottini and M. A. Quotadamo, *Colloid Polym. Sci.*, 2006, **285**, 19.
101. C.-H. Lin, J.-H. Chang, Y.-Q. Yeh, S.-H. Wu, Y.-H. Liu and C.-Y. Mou, *Nanoscale*, 2015, **7**, 9614.
102. K. M. Fuentes, D. Onna, T. Rioual, M. A. L. Huvelle, F. Britto, M. Simian, M. Sánchez-Domínguez, G. Soler-Illia and S. A. Bilmes, *Microporous Mesoporous Mater.*, 2021, **327**, 111391.

CHAPTER 2

Biogenic Synthesis of Inorganic Materials

JOSEPH M. SLOCIK[a], MARC R. KNECHT[b] AND RAJESH R. NAIK[a*]

[a]711th Human Performance Wing, Air Force Research Lab, WPAFB, OH 45433-7750, USA; [b]Department of Chemistry, University of Miami, Coral Gables, FL, USA
*E-mail: rajeshnaik.phd@gmail.com

2.1 Introduction

Sustainable production of materials remains a pressing goal in light of the consumption of limited available resources. Functional materials have a broad scope of applications that range from biomedical diagnostics and therapies to energy harvesting/storage and electronic device components where current production methods require significant energy input. Such synthesis methods can also be inefficient during their application, necessitating additional resources and energy, exacerbating the problems with their continued use. Consequently, new and sustainable methods for both material production and application are required that overcome these limitations. Ideally, these should limit the amount of resources required for production, require minimal to no energy input for production *and* application, have long application lifetimes, and be readily recyclable into new systems for enhanced materials use. Advances in any of these four key areas could be pivotal for long-term materials production and application, thus presenting a set of grand challenges for the 21st century.

Inorganic Materials Series No. 14
Unconventional Green Synthesis of Inorganic Nanomaterials
Edited by Silvia Gross

Published by the Royal Society of Chemistry, www.rsc.org

As an alternative to conventional approaches, natural systems have exploited the precision of biorecognition and biotemplating for the production of inorganic materials for numerous functions at low temperatures and in aqueous solvents.[1,2] These bio-based approaches typically result in the formation of a complex biomolecular interaction due to the binding of a biological material (amino acids, peptides, proteins, DNA, *etc.*) to an inorganic component (SiO_2, $CaCO_3$, *etc.*). Numerous examples are well known, ranging from more commonplace inorganic structures, such as bones, teeth, silica cell walls (aquatic diatoms and sponges), and nacre (molluscs), to more exotic nanostructured compositions, such as Au (alfalfa sprouts), CdS/ZnS (bacteria), and Fe_3O_4/Fe_3S_4 (magnetotactic bacteria).[3] These inorganic structures, constructed by biological processes, are mainly for structural support, bioremediation in the presence of environmental toxins, and for protection from predation. However, a limited set of materials, which include about 60 known biomineral forms, are used in nature to produce biogenic structures, thus limiting their direct translation from biological systems to marketable applications.[4]

While nature can produce fascinating inorganic structures, direct routes to commercially relevant and non-naturally occurring materials remain elusive. To circumvent this limitation, researchers have used biology as an inspiration for the biomimetic production and application of complex, functional materials for compositions not routinely observed in nature. Such approaches have greatly expanded the toolbox of bioinspired materials, which are still produced under sustainable conditions, for applications ranging from catalytic transformations to biosensing.[5,6] These materials exploit the principles identified in nature for controlled molecular recognition and translate them to abiological materials. A unique example of this approach would be the generation of highly complex assemblies of inorganic nanoparticles using DNA as the templating agent.[14] These highly organised arrays of materials can be assembled into a variety of known and new crystalline orientations due to DNA duplex formation.[15] Such structures could prove to be critically important as metamaterials for optical cloaking, biomedical diagnostics, photonics, *etc.*; however, the fundamental basis for these structures is achieved due to the functional bio-interface formed between the abiological nanoparticle and the biological DNA ligands attached to the inorganic surface.

In this chapter, we focus our discussion on bioinspired methods for the synthesis and application of inorganic nanostructures, paying particular attention to nanoparticle structures. This area of research has been extensively studied for the past 30 years and numerous review articles[7,8] have been written focused on materials for specific applications, compositions, bioligand identity, *etc.*; thus we have focused on more recent advances over the last decade to highlight the current state of bioinspired materials for both synthesis and applications. We also direct the reader to earlier volumes in this series concerned with bioinspired materials and biomaterials.[9,10] Initially we focus on simple small biomolecules, such as amino acids and peptides, and then consider large structures, such as proteins and viruses, as the templates for nanoparticle production. A succinct discussion of controlled fabrication by microorganisms and mammalian cells is also included as this represents a burgeoning area of great research to advance sustainable nanoparticle synthesis. Advances in synthetic biology and artificial intelligence/machine learning can also result in new and much more complex material designs due to better control.[16] We conclude each section with a discussion of the great potential these biomimetic approaches have for future applications, which represent key grand challenges from both a fundamental and broad-scale use perspective. We envisage that this review will highlight the state-of-the-art, providing the reader with the opportunity to design the next steps in achieving long-term sustainable routes towards functional inorganic materials.

2.2 Synthesis of Nanoparticles Using Biomolecules

Many small biological molecules, such as amino acids, sugars, and metabolites, have been exploited for the production of inorganic nanoparticles *via* biomimetic syntheses. For instance, amino acids provide a diverse palette of functional groups for coordinating to the inorganic surface (*e.g.*, thiols from cysteine and methionine, amines from lysine and arginine, imidazole from histidine, carboxylic acids from glutamic acid and aspartic acid, *etc.*). This allows for tuning of the interaction between the inorganic and biological components, thus altering the biomolecular overlayer structure. Additionally, these biomolecules tend to be highly water soluble due to their zwitterionic head group, promoting the use of sustainable conditions for material production. Since these biomolecules possess multiple functional groups, chemical modification once they are bound to the particle surface is also possible, providing a handle for secondary functionalisation post-material production. Beyond their molecular simplicity, these biomolecules can be commercially available at low costs, making them attractive ligands for biomimetic production of inorganic nanomaterials.

As the complexity of the biomolecules increases from simple amino acids to short chain peptides, enhanced molecular recognition capabilities are possible. Peptides with affinity for a variety of materials have been identified from both natural sources (*e.g.*, diatom-produced silica) and from biocombinatorial selection methods, such as phage and cell surface display. The use of biocombinatorial methods is particularly attractive as it allows for identification of peptides with affinity for wide-ranging compositions of inorganic materials, including those not biomineralised in nature.[17,18] By having these short chain peptides, increased affinity for the target inorganic structure is possible, including selectivity between two different material compositions.[19] Such selectivity could prove to be important for both material production and application, especially in the area of nanoparticle assembly, where organisation of multicomponent nanoparticles could be instrumental in the final ensemble properties.

Beyond short peptides, proteins and viruses have also been exploited for the production of complex material structures.[7,9] A well-studied example of this is the use of the iron storage protein ferritin, which has been exploited for the fabrication of numerous inorganic nanoparticles with a variety of compositions.[11,20,21] Its interior cavity is ideal as a template for the fabrication of nanoparticles of controlled sizes.[22] Additional proteins have also been extensively studied that can be used as surface passivants, paving the way toward multifunctional materials between the activity of the core inorganic nanoparticle and the surface-bound protein/enzyme.[12] These large and complex structures can be difficult to achieve, but provide unique opportunities for specific applications.

In this section, we consider the fabrication and production of inorganic nanoparticles based upon the increasing complexity of biomolecules, from single amino acids to short chain peptides and finally on to large protein structures. The complexity of the biomolecule, and the resultant bio-interface generated, provides intriguing opportunities to tune the material properties for specific, and potentially multiple, functionalities and applications from a single system.

2.2.1 Amino Acids

Amino acids have been used for the production of inorganic nanoparticles of various compositions. Some of the earliest work performed by Sastry's group,[22–28] as well as by other research teams,[29,30] focused on the fabrication of amino acid-capped inorganic nanoparticles. In this research, the amino acid served as the surface passivating ligand to stabilise the colloidal suspension, but it could also be exploited as the reducing reagent under specific conditions. For instance, cysteine was used for the fabrication of Au, Ag, Pt, and Cu nanoparticles.[28,30] Using this approach, the size of the metallic nanoparticles ranged from 3.5 to 17 nm in diameter based upon the metallic composition of the material. In general, the metal ions were reduced with $NaBH_4$, where cysteine was added to bind to the metal surface through the terminal thiol group, leading to colloidal stability. Similar approaches were employed in these early works for the preparation of amino acid-capped metal nanoparticles; however, various different residues were used, including lysine,[24,26] valine,[27] aspartic acid,[26] tyrosine,[25,30] and histidine.[30] It is anticipated that nearly all amino acids are probably capable of binding to the nanoparticle surface through either the side chain moiety and/or the carboxyl or amine groups, which is known to have affinity for metallic materials, as confirmed through computational modelling of amino acid adsorption onto Au and Ag.[31,32]

Interestingly, two amino acids, tryptophan and tyrosine, were shown to not only passivate the metallic nanoparticle surface, but also reduce metal ions, leading to nanoparticle nucleation and growth.[23,29] For instance,

Selvakannan *et al.* demonstrated that tryptophan leads to Au^{3+} reduction for Au^0 nanoparticle production at a reaction temperature of 50 °C.[23] In this case, tryptophan drives metal ion reduction *via* oxidation of the indole side chain and passivates the growing nanoparticle surface to prepare materials of ~31 nm in size. Wangoo *et al.* observed similar effects for nanoparticle production using glutamic acid.[29] In this case, Au^{3+} was heated under reflux in the presence of the amino acid, leading to the generation of ~40 nm Au^0 nanoparticles. The authors speculated that the amine group of the bioligand was oxidised to drive Au^{3+} reduction.

Selvakannan *et al.* exploited this reductive capability to generate Au@Ag core@shell nanoparticles capped with the amino acid lysine.[25] In this study, the team demonstrated that tyrosine could reduce Ag^+ to Ag^0, leading to nanoparticle formation and stabilisation by the biomolecule. The reaction was processed under basic conditions at reflux to ensure efficient metal ion reduction, where the authors reasoned that the phenol side chain of the residue was oxidised for Ag^+ reduction.[25] This Ag^+ reduction was exploited to prepare the bimetallic materials (see Figure 2.1). To this end, tyrosine-capped Au nanoparticles were prepared *via* $NaBH_4$ reduction of Au^{3+}. This resulted in the bioligands being present on the Au nanoparticle surface to generate a bio-interface. To these materials, Ag^+ ions were added to generate the composite structures, followed by KOH to raise the pH of the mixture. The solution was subsequently refluxed for tyrosine-driven Ag^+ reduction, leading to Ag^0 deposition on the Au nanoparticle for core@shell formation.[25]

2.2.1.1 *Amino Acid-derived Au Nanomaterials*

One of the most well-studied amino acid-derived nanomaterials is Au. This noble metal composition has been extensively examined for both biological and abiological synthetic methods,[19,33,35] thus providing a wealth of understanding for how amino acids affect the synthesis. For instance, Maruyama *et al.* examined the ability of the 20 canonical amino acids in the synthesis of Au nanoparticles.[35] For their reaction, they exploited the bioligand as both the reductant and surface passivant by heating a mixture of $HAuCl_4$ with the amino acid at 80 °C. As discussed above, such a temperature is required to

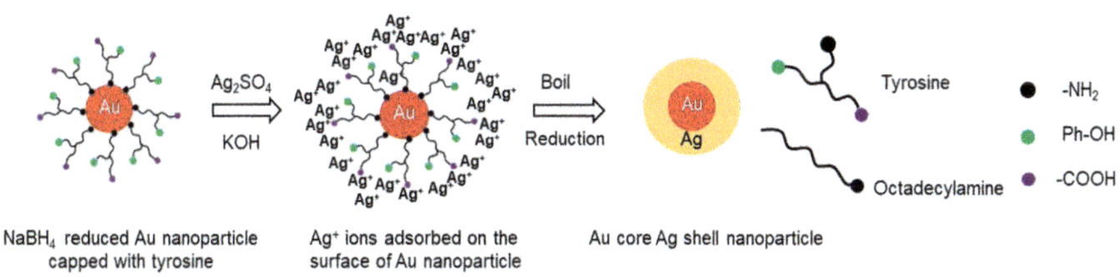

Figure 2.1 Scheme for the use of tyrosine to generate Au@Ag core@shell nanoparticles. Reproduced from ref. 25 with permission from American Chemical Society, Copyright 2004.

allow for Au^{3+} reduction by the various ligands. Using this synthetic approach, asparagine, alanine, glycine, histidine, leucine, lysine, serine, and tyrosine generated colloidal nanoparticle suspensions of varying colours (ranging from red to purple) and colour intensities.[35] Several of the residues demonstrated no colour changes (arginine, cysteine, and threonine) or resulted in bulk material precipitation (methionine, phenylalanine, and tryptophan). Interestingly, alanine and lysine did generate a suspension of particles; however, a degree of material precipitation was evident, suggesting only partial stabilisation of the materials. This suggests that the residues possess inherent differences in their affinity for the growing noble metal surface, resulting in various degrees of colloidal stability of the nanoparticles. Such effects were confirmed *via* computational modelling analysis by Heinz *et al.* using molecular dynamics simulations.[31]

Maruyama *et al.* further examined the effects of reaction conditions of histidine-driven reduction and particle stabilisation for Au nanomaterials.[35] These results demonstrated that the synthesis was sensitive to the amount of Au^{3+} ions in the reaction and the reaction pH. To this end, when the histidine concentration was held constant and the concentration of Au^{3+} increased, the aggregate size of the materials also increased. While transmission electron microscopy (TEM) evidence suggested that negligible changes in the individual particle diameter were present over the Au^{3+} concentration range studied, dynamic light scattering (DLS) data demonstrated large increases in the particle aggregation state at high metal ion concentrations. The authors reasoned that this was due to the oxidation of the bioligand during the amino acid-driven reduction process, where more histidine is oxidised at high Au^{3+} concentrations. In turn, this led to a lower surface density of histidine (*i.e.*, molecules per nm^2), resulting in particle aggregation. For pH studies, the authors demonstrated that the degree of histidine-driven reduction was optimised under acidic conditions and inhibited at basic pH values.[35] Interestingly, at low pH (1.2) material precipitation was evident; however, TEM analysis demonstrated the formation of Au nanoplates and nanorods in the precipitate that was adsorbed onto the glass vial wall. This suggests that the reaction pH influences Au^{3+} reduction, colloidal stability, and the nanoparticle shape, when using histidine as both reductant and surface passivant.

The use of tyrosine was further exploited by Shankar *et al.* for the generation of enzyme-modified Au nanoparticles.[36] Similar to the approach described above, these researchers heated Au^{3+} under reflux with tyrosine in water under basic conditions, resulting in the production of Au nanoparticles with an average hydrodynamic diameter of 14.8 nm from DLS measurements. After dialysis, the tyrosine-capped Au nanoparticle suspension had a pH of 8.4, resulting in a negatively charged material (zeta potential = −29 mV). Upon addition of the enzyme solution (phytase I in pH 2.5 Gly–HCl buffer) to the colloidal suspension of nanoparticles, the pH of the reaction mixture dropped to 4.6, thus changing the nanoparticle surface charge to +19.3 mV. As a result, electrostatic deposition of the negatively charged enzyme to the tyrosine-capped Au nanoparticle surface occurred, generating the final composite material. From this, the enzyme/nanoparticle superstructure displayed an average hydrodynamic diameter of 39 nm, an increase of more than double compared to the tyrosine-capped Au nanoparticle core, as anticipated. Upon optimisation of the enzyme deposition process, the catalytic ability of the immobilised enzyme was compared to the free biocatalyst.[36] In general, the immobilised phytase enzyme demonstrated notably enhanced thermal stability for the hydrolysis of phytic acid displaying 73% reactivity retention at 65 °C, compared to 45% for the free enzyme. Interestingly, the K_m values for the surface-adsorbed and free enzyme under optimal reaction conditions were 2.56×10^{-3} M and 0.76×10^{-3} M, respectively, suggesting that the immobilised enzyme had a weaker interaction with the substrate compared to the free enzyme. These two key points suggest that the enzyme on the nanoparticle surface may be conformationally more rigid and recalcitrant to structural changes, leading to enhanced stability, diminished substrate affinity, and preserved active site.

In a highly elegant approach, Lee *et al.* were able to generate chiral nanostructures using cysteine.[37] For this, they used cubic Au nanoparticles as seeds to generate the final structure using a seeded growth mechanism. To prepare the materials, cetyltrimethylammonium bromide (CTAB)-coated nanocubes were incubated in a solution of Au^+ ions in the presence of cysteine. The reaction was processed for 2 h at 30 °C allowing for slow reduction of Au^+ to Au^0, which resulted in the production of a 432 helicoid I structure. A helicoid structure is a high-index curved surface defined by its point group symmetry and atomic arrangement along a specific direction. In general, the underlying cubic morphology of the nanoparticle seed was evident; however, Au atom deposition at the particle vertices occurred, resulting in the formation of the chiral nanostructure. Remarkably, the chirality of the final particles could be directly tuned by the chirality of the cysteine used in the reaction (L-cysteine *versus* D-cysteine).[37] To this end, the handedness of the particles was the opposite of that of the amino acid used to template the structures. Such changes in chirality are evident in the scanning electron microscopy (SEM) images and the circular dichroism (CD) spectra of the two materials. In general for the CD, the same peaks were present in both systems, but opposite in intensity, confirming the change in handedness based upon the chirality of the amino acid.

While it is clear that the thiol of the cysteine residue binds the Au surface, it is likely that the amine moiety also plays a role in the binding event.[37] High-resolution TEM (HRTEM) imaging of the materials during particle growth indicated that the amino acids were adsorbing to the high-index facets of the growing chiral structures and not necessarily to the low-index facets of the cubic seeds. In addition, blocking of the α head group functionalities of the amino acids prevented their binding to Au, resulting in changes to the final material morphology.[37] This was confirmed by blocking of the amine group, which inhibited chiral particle formation; however, blocking of the carboxylic acid only reduced the *g*-factor (Kuhn's dissymmetry factor) of the final materials.[13] As such, chirality is still engendered into the system when carboxylic acid binding is prevented, but handedness is lost without amine binding. This suggests that both thiol and amine binding of the cysteine residue to the Au surface is required to generate chiral nanostructures, most likely to the (3 2 1) facets of the particle.[37] As a result of switching the chirality of the amino acid, changes in its adsorption to the growing Au nanoparticle is likely, giving rise to the variation in particle chirality.

In a separate approach by Bergquist and Hegmann, spherical cysteine-capped Au nanoparticles were used to alter the chirality of lyotropic chromonic liquid crystals (LCLCs).[38] In this case, the chiral amino acids are adsorbed to the Au nanoparticle surface *via* the thiol residue, allowing for interactions between the nanoparticles and the LCLC molecule disodium cromoglycate (DSCG) *via* the exposed α head group moieties. In general, these interactions led to enhanced chirality in the liquid crystal assembly, which was amplified using the cysteine-capped Au nanoparticles.

Amino acid-capped Au nanoparticles can also be used for the design of more complex, multicomponent composite structures. For instance, Liu *et al.* used a lysine to both passivate Au nanoparticles and deposit the materials onto a ZnO microrod surface.[39] For this, the ZnO component was mixed with $HAuCl_4$ and lysine in water, followed by reduction with $NaBH_4$. From this reaction, the formation of Au nanoparticles occurred, which were capped with lysine. The materials were subsequently deposited onto the oxide surface, generating a Schottky barrier between the two inorganic components (see Figure 2.2). This allows for rapid electron transfer from the ZnO component to the lysine-capped Au nanoparticle. Such effects can be highly important for photocatalytic reactions to reduce charge recombination that leads to diminished reactivity. Additionally, the deposition of Pt nanoparticles onto the ZnO microrods can also be processed using the same approach,[39] suggesting that the lysine-based method could be applicable to a variety of material compositions.

Once prepared, the Au/ZnO materials were subsequently examined as a gas-phase sensor for ethanol. For this, the lysine residues were removed *via* thermal treatment, thus exposing the Au surface to bind ethanol vapour.[39] To detect the ethanol, an electrochemical system was employed that measured changes in the electrical resistance of the sensor due to ethanol binding. In general, the Au/ZnO composite system demonstrated substantially enhanced sensing capabilities compared to the bare ZnO structure under all conditions. Furthermore, the sensing platform was highly reproducible up to eight cycles. Taken together, this enhancement effect is probably due to enhanced adsorption to the Au component and the formation of an electron depletion layer at the Au/ZnO interface. Nevertheless, enhanced sensing capabilities were observed from the composite structure formed using lysine.

In separate work, Zhang *et al.* used lysine again to generate a composite material using semiconducting Fe_2O_3 materials as the core component.[40] Identical approaches were employed as described for the ZnO system, where Au^{3+} was reduced using $NaBH_4$ in the presence of lysine and the ~200 nm Fe_2O_3 particles. Using this approach, Au-decorated Fe_2O_3 structures were

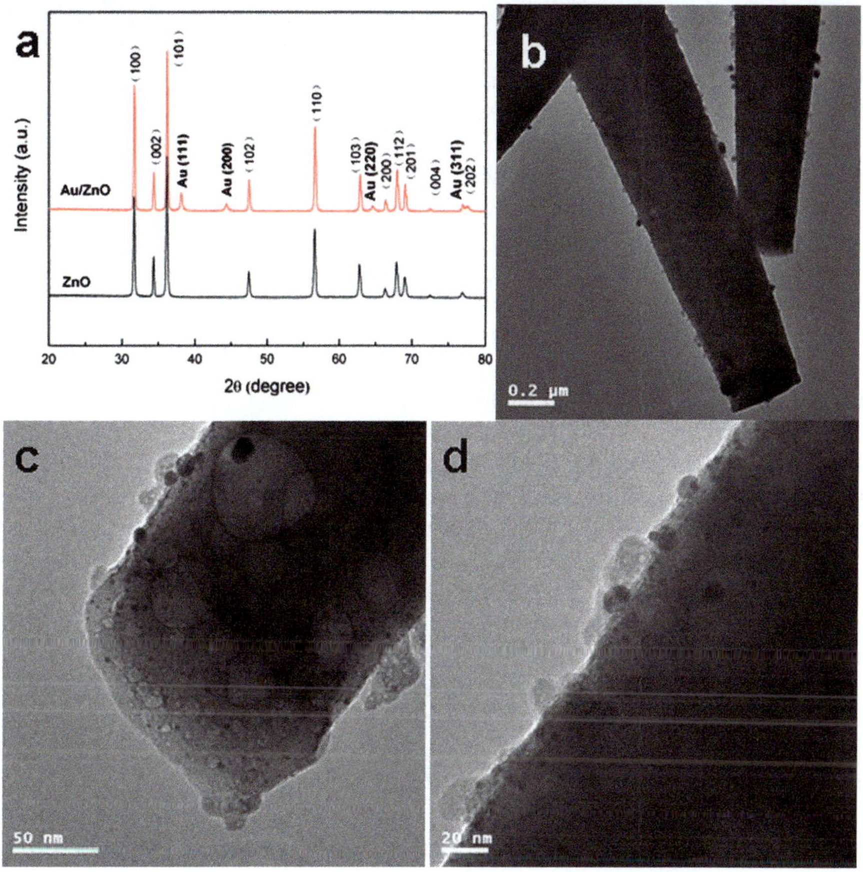

Figure 2.2 ZnO microrods surface decorated using lysine-capped Au nanoparticles. (a) Powder X-ray diffraction (XRD) analysis of the materials. (b–d) TEM images of the structures at different magnifications. Reproduced from ref. 39 with permission from the Royal Society of Chemistry.

prepared where the amino acid templated the formation of the Au layer. When the reaction was processed in the absence of the biomolecule, Au nanoparticles did deposit on the Fe_2O_3 materials; however, they were poorly distributed on the oxide and many of the Fe_2O_3 structures remained uncoated. This demonstrates the importance of the lysine residue to ensure maximal oxide coverage. These structures were subsequently employed for ethanol gas sensing, which demonstrated enhanced electrochemical ethanol detection compared to the bare Fe_2O_3.[38] Similar to the ZnO system, the approach using lysine and Fe_2O_3 was adaptable to other noble metal compositions, including monometallic Pt and Ag, as well as bimetallic Pt/Au. This further reinforces the suggestion that this approach could be universally applied for a variety of inorganic compositions. Similarly, graphene and aluminosilicate (montmorillonite clay) decorated with small engineered peptides were used to template gold nanoparticles.[41,42] In both materials, the two-dimensional (2D) surfaces of graphene and aluminosilicate were uniformly covered with a dense layer of gold nanoparticles.

Amino acid-capped Au nanoparticles were also used to selectively deposit at the nodes in carbon nanotube networks. For this, Velamakanni *et al.* used cysteine-capped Au nanoparticles where the exposed amine could be used to covalently link the materials specifically at the nodes.[43] The authors exploited a standard synthetic preparation method for the Au nanoparticles, where $HAuCl_4$ was reduced with $NaBH_4$ in the presence of cysteine. In this case, the terminal thiol group rapidly coordinates to the Au surface, arresting particle growth and displaying the amine functionality to solution at the bio-interface. After synthesis, the materials were adjusted to pH 10, where the amine group was deprotonated to form NH_2 for eventual coupling. To generate the nanotube network, carbon nanotubes with acid chloride functionalities present at the tube ends were added to the lysine-capped Au nanoparticles.[43] The acid chlorides readily react with the amines on the nanoparticles, thus binding the inorganic materials at the nanotube tips only. Multiple nanotubes can react with the Au nanoparticles, thus forming the networked morphology and locking the particles at the nodes. These complex structures were formed due to the bio-interface generated on the Au nanoparticle, thus positioning the reactive amine groups to react with the nanotubes to generate the nanotube network.

While amino acids are readily known to bind to the Au nanoparticle surface, this newly formed bio-interface can also be exploited to drive nanoparticle assembly. For instance, Sethi *et al.* exploited the binding of arginine to citrate-capped Au nanoparticles for the formation of linear chains of materials.[44] This was achieved *via* place exchange of the arginine with the original citrate ligand adsorbed onto the Au nanoparticle surface, wherein the arginine : Au nanoparticle molar ratio was varied from 0 to 100 000 (referred to as 100k). Since the particle concentration remained constant, changes in the molar ratio were achieved through addition of increasing amounts of the biological ligand. Interestingly, as arginine was added to the Au nanoparticles, changes in the optical properties were observed over time, as monitored by UV-visible (UV–vis) spectroscopy. Before biomolecule addition, the Au nanoparticles displayed a single surface plasmon band at 520 nm; however, when arginine was added to the mixture, the 520 nm peak intensity decreased, concomitant with the formation of a new peak at 665 nm.[44] This new surface plasmon band continually grew in intensity over the 6 h study, with the rate of peak growth at 665 nm increasing with increasing amounts of arginine in the system. Additional work demonstrated that the rate of 665 nm peak growth was affected by the reaction temperature and solvent composition.[45]

Based upon the changes in the plasmonic properties of the materials, it was likely that some type of material aggregation was occurring upon arginine addition. Interestingly, since a single plasmon band was generated for the assembled materials, this suggested that a controlled assembly process was occurring in suspension to form a one-dimensional type structure. Should uncontrolled aggregation in all directions be occurring, a substantially broadened plasmon that was shifted further into the infrared region would have been observed. DLS confirmed that the formation of larger structures was occurring in the presence of the amino acid;[44] however, it was not possible to ascertain the morphology of the assemblies. For this, Sethi *et al.* exploited TEM imaging of the materials 6 h post-arginine addition (see Figure 2.3). As anticipated, prior to arginine addition, independent Au nanoparticles were observed. As the arginine : Au nanoparticle molar ratio increased from 20 000 (20k) to 60k, the formation of linearly branched chains of Au nanoparticles was evident. Above this ratio (80k and 100k) regions of chained nanoparticles were observed in the presence of bulk material aggregates.[44] Such linear structures are consistent with the UV–vis results, which displayed a well-revolved second plasmon band associated with one-dimensional Au nanostructures.

From these arginine-based assembly studies, it is clear that incorporation of the amino acid onto the particle surface drives material organisation. Sethi *et al.* hypothesised that incomplete citrate replacement with arginine occurred, whereby segregation of the bioligand from the citrate was possible.[44,45] This could be achieved through zwitterionic interactions between multiple surface-bound arginines, as confirmed through

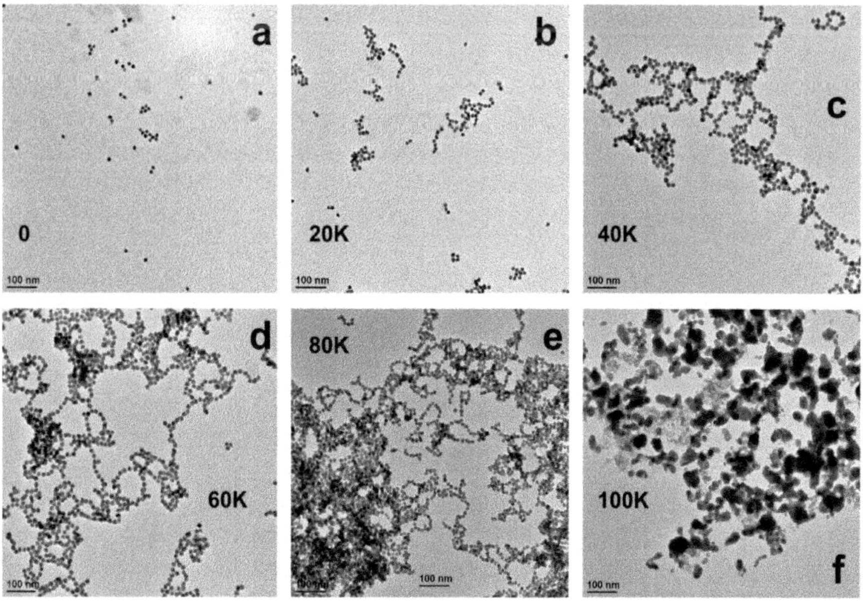

Figure 2.3 TEM imaging analysis of the arginine-controlled assembly of Au nanoparticles at arginine : Au nanoparticle molar ratios of (a) 0, (b) 20k, (c) 40k, (d) 60k, (e) 80k, and (f) 100k. Reproduced from ref. 45 with permission from American Chemical Society, Copyright 2009.

computational modelling.[46] Should this ligand segregation event occur, this would generate a dipole moment across the particles, arising from regions of negative charge (citrate domain) and neutral zones (arginine domain). From this morphology, dipole-based material assembly could occur, giving rise to the linear structures.

When the same assembly process was examined, but with cysteine, changes in the material organisation were noted.[47] In this regard, using the thiol-based residue, nanoparticle assembly was observed; however, the assemblies were generally random in organisation and not linear. Additionally, very little change was observed spectroscopically in the materials after cysteine addition. This suggested that a second mechanism of material aggregation was present with cysteine. Sethi *et al.* hypothesised that the cysteine molecules were able to readily replace the weakly bound citrate ligands *via* a thiol bond to the Au surface.[47] This notably stronger interaction, compared to arginine, probably locked the cysteine residues in place, prohibiting ligand rearrangement on the particles. As such, the formation of a dipole was not possible; however, the zwitterions were presented to suspension. In this arrangement, two nanoparticles could aggregate *via* interactions between the zwitterions on the particle surface. Once sufficiently close, the amines in the bio-interface could bind to the Au surface on the opposite particle, leading to a strongly assembled set of materials. Such effects were supported by the work of Majzik *et al.*, in which similar assemblies were observed using cysteine.[48] This work also supported the formation of interactions across zwitterions on two different cysteine-capped Au nanoparticles, where they were able to observe linear assemblies from these materials. Interestingly, they demonstrated that the process was pH dependent, suggesting that differences in assembly morphologies may be based upon changes in the reaction pH.[48]

In further work with cysteine, Sethi *et al.* were able to drive linear assembly of Au nanorods.[49] In general, the Au nanorods were fabricated with a CTAB bilayer adsorbed to the inorganic surface. Caswell *et al.* had previously shown that the bilayer at the tips is notably more susceptible to ligand displacement, thus ligand exchange reactions preferentially occur at this location.[50] Upon introduction of cysteine to a suspension of the Au nanorods, changes in the UV–vis spectrum of the materials were observed, similar to the spherical Au nanoparticles. To this end, in the absence of cysteine, two plasmon bands are evident: one at ~515 nm and a second band at 770 nm arising from the transverse and longitudinal plasmons, respectively. Upon introduction of cysteine, the 770 nm band decreases in intensity with the formation of a new band at ~950 nm.[49] The reaction is highly sensitive to pH, where spectral changes are faster at low pH values. In fact, the reaction is generally inhibited under basic conditions. TEM imaging of the materials after introduction of cysteine demonstrated the formation of linear nanorod chains, which arose from the interaction of the cysteine on the nanorod tips driving material assembly. This was achieved through crosslinking of the Au nanorods through binding of the cysteine thiol on the first particle and the amine of the α head group on a second nanorod.

Beyond self-assembly, amino acid-capped Au nanoparticles have demonstrated applications for sensing of small molecules and toxic metal ions. For instance, Devi *et al.* employed ~16 nm Au particles with cysteine passivated on the surface for the detection of trinitrotoluene (TNT), a well-known explosive.[51] For this, the materials were prepared *via* the citrate-reduction approach, where the initial citrate adsorbed to the particle surface was exchanged for cysteine. Prior to addition of the TNT analyte, the Au nanoparticle displayed a plasmon band at 522 nm, whereupon incorporation of the analyte into the reaction mixture resulted in a decrease in the intensity of this band along with the formation of a shoulder band at 580–620 nm. TEM and DLS analyses of the materials in the presence of TNT indicated that the analyte drove nanoparticle aggregation, most likely through the formation of a Jackson–Meisenheimer complex.[51] From this aggregation, sensing of the TNT was possible *via* changes in the UV–vis spectrum, with a limit of detection down to 4 nM (1 ppb). Furthermore, the approach was selective to TNT over other nitroaromatic compounds (including dinitrotoluene and 2,4-dinitrotoluic acid). Such a low limit is intriguing using such a simple system as the citrate-capped Au nanoparticles, especially for the detection of an explosive agent.

Xianyu *et al.* have used the aggregation/sensing of Hg^{2+} by arginine and Au nanoparticles for the generation of a logic gate.[52] For this, the researchers used citrate-capped Au nanoparticles with a plasmon band at 520 nm. When the nanoparticles were mixed in suspension with either arginine or Hg^{2+} ions alone, the materials remained unaggregated, thus minimal changes in the plasmon absorbance intensity were noted; however, when both arginine and Hg^{2+} were mixed with the nanoparticles, particle aggregation occurred, resulting in a red to blue visible colour change of the suspension. This resulted in the generation of an 'AND' logic gate.[52] More quantitatively, the ratio of the absorbance intensity at 650 nm (observed due to aggregation) *versus* the plasmon absorbance at 520 nm was compared, where a ratio of >1 was required for a positive output. Similar approaches were employed for the generation of 'OR' logic gates employing Cr^{3+} and Hg^{2+} as the assembly elements with the citrate-capped Au nanoparticles and arginine. In this case, aggregation was observed in the presence of both metal ions, and was dependent upon their individual concentration.[52] Finally, 'INHIBIT' logic gates can also be achieved using the Hg^{2+} system. In this case, the incorporation of cysteine residues to the 'AND' system is used, which bind to the Hg^{2+} ions in suspension. This inhibits the metal ions from driving arginine-based nanoparticle aggregation, and thus no colour change is present in the sample. This approach is dependent upon the concentration of all components in the reaction mixture to achieve the intricate logic gate design. Interestingly, Xianyu *et al.* further enhanced the amino acid/nanoparticle-based logic gates through incorporation into a microfluidic-based device, thus allowing for automation of the approach using substantially smaller reaction volumes.[52]

While amino acid-capped Au nanoparticles have shown remarkable properties and applications, their use in biomedical technologies could be inhibited should these materials be inherently cytotoxic.[34] To address this issue, Carnovale *et al.* explored the toxicity of tyrosine- and tryptophan-capped Au nanoparticles and compared these results to similar structures capped with commonplace non-biological ligands (CTAB and citrate).[53] To ensure comparable results, Au nanoparticles of similar sizes (between 20 and 27 nm in diameter) were employed, avoiding any complications of size effects in the analysis. Interestingly, the amino acid-stabilised structures demonstrated enhanced cellular uptake, thus increasing the amount of nanoparticles per cell, compared to the abiological structures. For instance, the tryptophan-capped Au nanoparticles could reach maximum levels of ~3.5×10^5 nanoparticles per cell, while the tyrosine-based structures had a maximum of ~3.25×10^5 nanoparticles per cell. Such values are remarkably higher than the citrate- and CTAB-capped materials, which demonstrated maximum cellular uptakes of ~5.0×10^4 and ~6.0×10^4 nanoparticles per cell, respectively. When studied for their specific toxicity, the citrate-, tyrosine-, and tryptophan-capped materials demonstrated low to no toxicity, even at high concentrations, depending upon the cell culture conditions (*i.e.*, serum-free, serum pre-incubated, and serum-supplemented media). As anticipated, the CTAB-capped materials demonstrated significant toxicity down to concentrations of 0.5 μM, probably arising from the toxic CTAB surfactant.

2.2.1.2 *Amino Acid-derived Ag Nanomaterial*

While extensive work has been performed on the production and application of amino acid-capped Au nanoparticles, considerably less information is available for their Ag-based counterparts. This probably arises from the great stability that is accessible from Au nanoparticles; however, a variety of unique applications can be achieved from Ag structures. For instance, Ag nanoparticles are routinely used as substrates for surface enhanced Raman spectroscopy (SERS) or as antimicrobial agents. Such applications could prove to be important for biological applications; thus amino acid-capped Ag nanoparticles have been explored for these approaches.

Work by Hu *et al.* used two basic amino acids (arginine and lysine) as Ag^+ reductants to generate Au^0 nanoparticles that were stabilised by soluble starch.[54] While the amino acid was not directly used to stabilise the final materials, it was instrumental in their production. To achieve this reduction, the system was microwaved to a temperature of 150 °C. Using lysine as the reductant, Ag^0 nanoparticles were generated with an average size of ~26.3 nm; however, when arginine replaced lysine in the reaction, the Ag nanoparticle size was highly similar (~26.7 nm). Interestingly, when other amino acids were used in the reaction that did not contain amine-based side chains (aspartic acid, phenylalanine, and alanine), either highly aggregated Ag^0 structures or a lack of Ag^+ reduction was observed. This suggested that the amine-containing side chain was instrumental in the reducing capability of the residue. In addition, it may also play a role, along with the soluble starch, in stabilising the final nanostructures.

Once prepared, the Ag nanoparticles were deposited onto a glass slide *via* solvent evaporation.[54] This resulted in the production of a Ag nanoparticle film where the thickness of the film was controlled by the nanoparticle concentration in the starting suspension. Using these Ag nanoparticle films, their ability to serve as a SERS substrate was explored using the adsorption of 4-mercaptobenzoic acid on the metallic surface. Interestingly, the enhancement factor of the substrates increased as the thickness of the Ag film decreased. This suggests that optimisation of the substrate surface is required to enhanced SERS-based detection using these materials. Nevertheless, this approach demonstrates a simple and green method for the production of important surface-based materials.

Beyond simple surface evaporation to form a Ag film, Nanda *et al.* demonstrated the ability to integrate amino acid-capped Ag nanoparticles into peptide-based hydrogels.[55] In this regard, the gelator peptide Boc-Phe-Phe-Ala-COOH was employed using a thermally assisted process under basic conditions. For this, the peptide was dissolved in a NaOH solution and heated to dissolve the gelator. Upon cooling to room temperature, the formation of an opaque gel was achieved. SEM imaging of the gel demonstrated that it was highly fibrillar in structure with fibre widths of 80–150 nm; however, their lengths were up to several micrometres in size.

To integrate Ag nanoparticles into the hydrogel, Nanda *et al.* used cysteine-capped materials.[55] To generate the materials, Ag^+ ions were coordinated to the thiol side chain. Upon reduction with $NaBH_4$, cysteine-capped Ag nanoparticles were prepared with sizes of <15 nm in diameter. The nanoparticles were then added to the gelating system at elevated temperatures. Under these conditions, the hydrogel was in the sol state, thus allowing for formation of a homogenous suspension. Upon cooling and gelation, the Ag nanoparticles were non-covalently adsorbed onto the fibre surface. Additional Ag nanoparticles capped with two different dipeptides (Cys-Leu and Cys-Phe) were also adsorbed to the hydrogels. Interestingly, measurement of the hydrogel containing the Cys-Phe-capped Ag nanoparticles demonstrated diminished stress values as the amount of Ag nanoparticle incorporated into the system increased.[55] Similarly, the stress values changed when the ligand on the Ag nanoparticle surface changed, such that the stress values decreased in the order of Cys > Cys-Leu > Cys-Phe. This suggests that modulation of the hydrogel strength is highly sensitive to the amount of nanoparticles incorporated and the type of interactions between the nanoparticles and the gel fibres.

In a separate work, Shankar and Rhim introduced tryptophan and tyrosine-capped Ag nanoparticles into agar for use as an antimicrobial film.[56] For this, the aromatic residues were used as both the capping agent and Ag^+ reductant. For this, Ag^+ ions were refluxed in the presence of the amino acid until the solution turned yellow, which was indicative of nanoparticle formation. The reaction was processed in 2.5 mM KOH to ensure a highly basic solution. This pH was required to deprotonate the phenyl and indole ring of tyrosine and tryptophan, respectively, which are required for the reduction process. TEM analysis of the amino acid-capped structures demonstrated the formation of spherical nanoparticles with sizes of 10–15 nm for the tyrosine particles and 15–20 nm for the tryptophan-capped materials. Once the amino acid-capped Ag nanoparticles were incorporated into agar films, the Ag-containing films demonstrated antimicrobial activity against both Gram-positive and Gram-negative bacteria. To this end, cell viability of both *Escherichia coli* and *Listeria monocytogenes* was greatly diminished after 15 h of exposure to the Ag-containing thin films compared to controls without Ag.

Based upon their brilliant optical properties, lysine-capped Ag nanoparticles have been exploited as plasmonic-based sensors. To achieve this capability, Bae *et al.* fabricated the materials first *via* the reduction of $AgNO_3$ in the presence of sodium citrate, $NaBH_4$, and polyvinylpyrrolidone at reflux.[57] Once the particles were prepared, *N*-bis(carboxymethyl)-L-lysine was place-exchanged onto the surface of the Ag nanoparticle to prepare the amino acid-capped material. Using this approach, zerovalent materials were prepared with a diameter of 3–5 nm and a vibrant plasmon band at ~400 nm, which presented the lysine residues to solution. Interestingly, incubation of the lysine-capped Ag nanoparticles with histidine resulted in a solution colour change from yellow to red, concomitant with growth of a new peak in the UV–vis spectrum of the structures at 520 nm. Such effects are consistent with nanoparticle aggregation, which was confirmed *via* DLS analysis of the materials before and after histidine addition. Using this system, histidine could be detected down to a concentration of 5.0 μM. Testing of the sensor with the remaining 19 canonical amino acids demonstrated no changes in the aggregation state of the material nor in the plasmonic properties, confirming that the system was highly selective for histidine.

This ability to discriminate histidine amino acids was advanced for the detection of His-tagged proteins. For this, Bae *et al.* challenged the lysine-capped Ag nanoparticle sensor with two proteins: ubiquitin, which did not contain a His-tag, and Cdc20, which did possess the His-tag.[57] As anticipated, incubation of ubiquitin with the sensor resulted in no solution colour change; however, introduction of the His-tag-containing Cdc20 lead to nanoparticle aggregation and the growth of a new peak at 520 nm in the UV–vis spectrum. Polyacrylamide gel electrophoresis (PAGE) confirmed that the Cdc20 protein was bound to the Ag nanoparticles, but the ubiquitin species was not, consistent with the lack of the His-tag. Gel electrophoresis is commonly used in biochemistry to separate, identify, and determine purity of proteins by size (*i.e.*, molecular weight), and in this case was used to confirm the presence of protein on the nanoparticle surface. Such results demonstrate powerful sensing capabilities, which arise from the amino acid ligand on Ag acting as the binding site *via* electrostatic interactions with His and the aggregation state of the Ag nanoparticle serving as the reporting element of plasmonic properties.

2.2.1.3 *Amino Acid-derived Cu-based Nanomaterials*

In addition to Au and Ag, some advancement in the production of nanoparticles containing copper has been achieved using amino acids. Cu is known to be easily oxidised to two different forms: CuO and Cu_2O. Remarkably, amino acids have been exploited for the production of all three materials, including Cu metal, demonstrating the versatility of these biomolecules for the production of an important inorganic composition. Yu *et al.* demonstrated the production of zerovalent Cu nanomaterials using amino acids.[58] In this approach, the researchers reacted Cu^{2+} ions with the amino acid and hydrazine under basic conditions at 80 °C. In this case, the hydrazine was the reductant, driving Cu nanomaterial production, which was capped by the amino acid in solution. Interestingly, when glycine was employed as the passivant, instead of preparing the typical spherical nanoparticle structure, long nanowires were generated with dimensions of 301 nm in width and 21.2 μm in length (see Figure 2.4). Further characterisation demonstrated that the nanowires presented a rough metal surface; however, the materials were single-crystalline Cu metal with no evidence of oxidation in the form of CuO or Cu_2O observed in the sample. When the same reaction was processed with the amino acid switched for histidine, polycrystalline spherical Cu microparticles were generated with an average diameter of 2.47 μm.

Using their synthetic approach, Yu *et al.* examined the ability of additional amino acids to generate Cu materials, to identify the effect of the side chain on the final structure of the materials.[58] Based upon the hydrophobic side chain of glycine, the authors used other hydrophobic amino acids, which resulted in the production of nanowires (alanine and proline), hedgehog-like structures (valine), a mixture of nanoparticles and nanorods (leucine), and nanoparticles (phenylalanine). From this, it is evident that as the length of the hydrophobic side chain increased, the propensity to form nanowires decreased. When hydroxylated amino acids were employed in the reaction (serine, threonine, and tyrosine), spherical nanoparticles were again generated; however, cysteine, which replaces the hydroxyl functionality for a thiol, was unable to produce Cu materials. For negatively charged aspartic and glutamic acid, a mixture of short nanorods and nanoparticles was obtained. Interestingly, use of asparagine, which converts the side chain carboxylic acid of aspartic acid to an amide, resulted in the restored of formation of Cu nanowires. For the final amino acids studied with amine-containing side chains, a variety of structures were observed that included nanowires (lysine and tryptophan) and nanoparticles (histidine). When arginine was employed, like cysteine, no materials were produced, as noted by a lack of a colour change. Presumably, this was due to the poor ability of arginine and cysteine to serve as capping agents. Note that while similar structures of Cu materials were prepared with different amino acids, variation in their specific dimensions was observed, possibly presenting an approach to control the particle size and morphology through selection of the amino acid in the reaction.

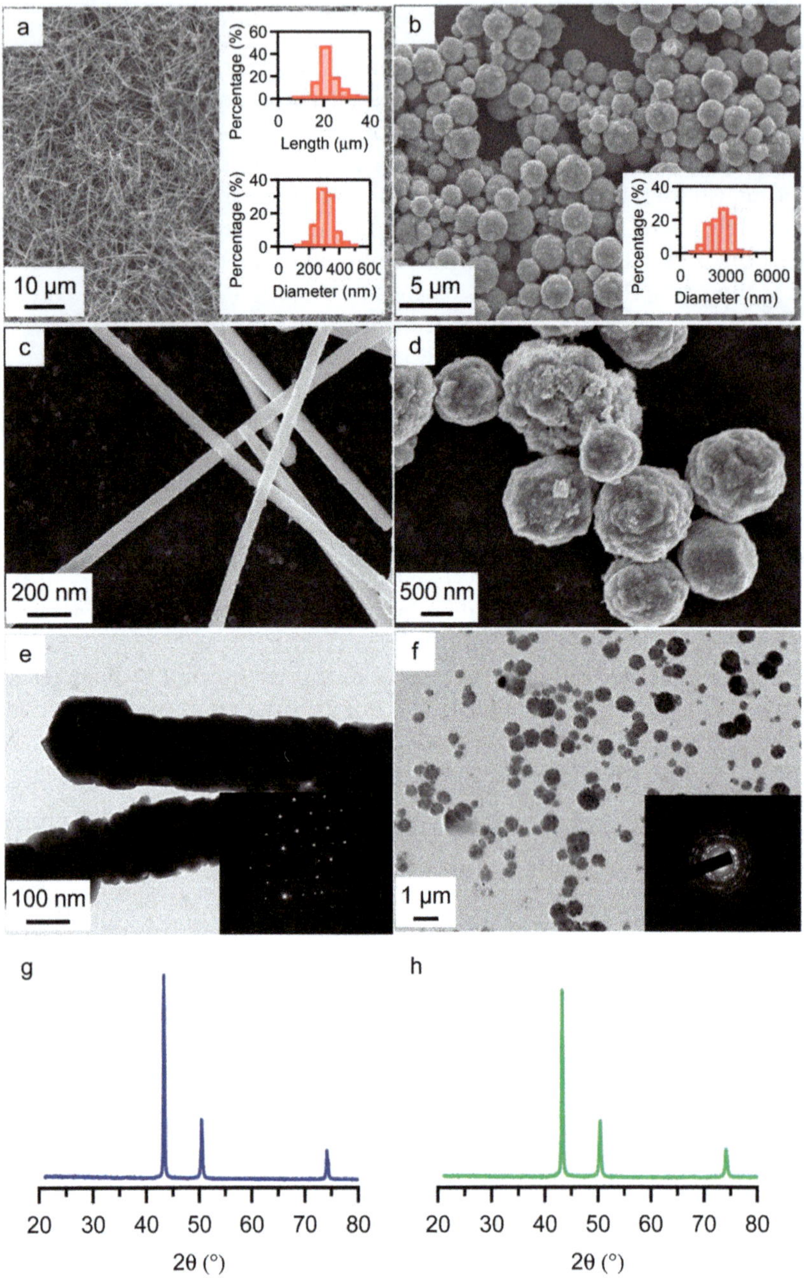

Figure 2.4 TEM analysis of the production of Cu nanowires (a, c and e) and microparticles (b, d and f) using glycine and histidine, respectively. (g) XRD profile of Cu nanowires and (h) XRD profile of Cu microparticles. Reproduced from ref. 58 with permission from the Royal Society of Chemistry.

The change in morphology of the Cu materials was proposed to arise from changes in the binding of the amino acid at the metallic surface.[58] In many cases, the amino acids bound to the inorganic material through the amine and carboxylic acid; however, if the biomolecule possessed two acid groups (*e.g.*, aspartic acid), the two acids could preferentially adsorb to the particle surface. These changes in adsorption motif were envisaged to alter material growth, leading to changes in the final material morphology.

In separate work, Munro *et al.* explored the use of the 20 canonical amino acids for the production of Cu_2O, where the copper component was present in the CuI oxidation state.[59] Such materials are intriguing for use as photocatalysts in that they are activated using visible light to drive a variety of reactions of environmental and energy importance. To prepare the Cu_2O materials, the authors heated a solution of $CuSO_4$, amino acid, sodium citrate, sodium carbonate, and glucose to 70 °C. At this temperature, the glucose is activated as the reductant to drive the reduction of Cu^{2+} to Cu^{+}, resulting in production of the Cu_2O materials. As a control, the reaction was conducted without the amino acid, which resulted in the production of a variety of different particle morphologies. In general, the majority of the amino acids produced a set of materials that mirrored the structures prepared in the absence of a biomolecule, indicating that they had no effect on the reaction process; however, arginine, cysteine, glutamic acid, lysine, tryptophan, and histidine displayed the ability to control the particle size and structure. To this end, histidine was able to produce small particles of Cu_2O, which were significantly smaller than the control materials, while the other residues listed generated only spherical nanostructures. The spherical structures were notable, as such a morphology was not observed in the analysis of the control in the absence of amino acid. This change in morphology was attributed to the interactions of the amino acids with either the Cu ions in solution or the growing material surface, suggesting that these species could be incorporated into materials directing peptides to control particle growth. In addition, the final materials demonstrated a notable degree of photocatalytic reactivity for the degradation of organic dyes. These results support the use of simple biomolecules to control the production of materials with potentially wide-ranging applications.

Beyond Cu metal and Cu_2O materials, Bhattacharjee and Ahmaruzzaman employed two amino acids, lysine and glycine, to prepare CuO nanorods and nanosheets, respectively.[60] In this specific case, CuO possesses copper in the Cu^{2+} oxidation state, spanning the most commonplace oxidation states of copper. For this approach, a microwave-based synthesis was employed. To this end, $CuSO_4$ was mixed with the amino acid under basic conditions. The reaction was then microwaved 30 times at 10 s intervals to reach a temperature of 100 °C, resulting in the precipitation of a black CuO solid from the reaction mixture. From the synthesis using lysine, CuO nanorods were prepared of ~30 nm in diameter. For the glycine-prepared structures, nanosheets were observed that were ~300–600 nm in length, ~200 nm in width, and 30–60 nm thick. The materials were fully characterised to confirm their shape and conformation and the authors ascribed the different structures to the changes in the specific adsorption of the two amino acids to the growing CuO material. Once fully characterised, the reactivity of the different amino acid-prepared CuO structures was examined for the reduction of 4-nitrophenol to 4-aminophenol and for photocatalytic dye degradation, where both systems displayed varying degrees of reactivity. For the reduction of 4-nitrophenol specifically, the CuO nanorods displayed higher reactivity over their nanosheet counterparts.

2.2.1.4 Amino Acid-derived Cd(S or Se) Nanomaterials

Beyond noble metal nanoparticles, amino acids have been employed to generate highly fluorescent CdS and CdSe quantum dots. Such materials are highly intriguing for their brilliant optical emission properties, which have found applications as, for example, solar cells and biosensors. Samadi-Maybodi and Akhoondi used a straightforward solution-based approach for the production of CdS quantum dots where alanine, histidine, and serine were employed to cap the growing particles in suspension.[61] In their synthesis, $CdCl_2$ was co-mixed with the amino acid, to which thioacetamide was added to serve as the sulfide source. From this approach, CdS nanoparticles rapidly grew in suspension, and were capped by the amino acids in the reaction mixture; all three biomolecules were able to stabilise a colloidal suspension of CdS quantum dots. Interestingly, the materials were highly fluorescent, with the emission profile (*e.g.*, peak position and full-width at half-maximum) being highly dependent upon the reaction pH. Once characterised, the amino acid-capped CdS materials were photocatalytically reactive for the degradation of the organic dye alizarin. This generally occurred through hydroxyl radical formation from the oxidation of water *via* photocatalytic processes at the CdS surface and formation of electron/hole pairs. In this case, the reaction was able to be recycled over many catalytic cycles.

To advance amino acid-capped materials to a new composition of quantum dots, Oluwafemi *et al.* reported a method for the production of CdSe quantum dots capped with cysteine.[62] This reaction was processed in water at relatively low temperatures (90 °C), which is unique compared to the traditional synthetic approach that requires the use of organic solvents at high reaction temperatures (>300 °C). For this synthesis, selenium was added to water in the presence of $NaBH_4$ to generate NaHSe. Subsequently, addition of $CdCl_2$ and cysteine resulted in the production of cysteine-capped CdSe quantum dots. TEM analysis confirmed the production of the spherical nanomaterials; however, the particles appeared to be partially aggregated. HRTEM confirmed that the final materials were highly crystalline, which was supported by powder X-ray diffraction (XRD) analysis. The cysteine-capped CdSe quantum dots were highly fluorescent, displaying a strong fluorescence band at ~400 nm, consistent with the known properties of comparable CdSe quantum dots capped with various ligands. This approach for CdSe production is unique and attractive as it presents a potentially green and sustainable route to these important materials, with key applications for future use in solar cells and biomedical technologies. This is enhanced, potentially, by the biofriendly ligand coat compared to most common organic ligands used to stabilise CdSe materials (*e.g.*, trioctylphosphine oxide (TOPO)).

2.2.1.5 Additional Amino Acid-capped Materials

In addition to the conventional materials discussed above, amino acids have been employed for the production of more exotic material compositions. For instance, Tiwari *et al.* used cysteine for the fabrication of $La(OH)_3$ nanoparticles.[63] Their impetus for this was to exploit the materials as an electrochemical sensor of a biomarker associated with oral cancer. A simple solution-based approached was used to prepare the materials. For this, a NaOH solution was added to a solution of lanthanum nitrate, leading to precipitation of the $La(OH)_3$. When the precipitate was formed, cysteine was added to the mixture under sonication, allowing for chemisorption of the amino acid to the material surface *via* the carboxylate group instead of the free thiol. This allows for disulfide formation between the adsorbed residues, as confirmed by infrared (IR) spectroscopy, thus exposing the amine group to solution for coupling with secondary species. From this approach, cysteine-capped $La(OH)_3$ nanoparticles were prepared with an average diameter of ~45 nm.

The cysteine-capped $La(OH)_3$ materials were subsequently modified with antibodies to bind Cyfra-21-1, an oral cancer biomarker.[63] For this, antibodies with affinity for the biomarker (anti-Cyfra-21-1) were chemically conjugated to the exposed amines on the nanoparticle surface. Once prepared, the conjugated materials were deposited onto an indium-tin-oxide electrode surface for use in differential pulse voltammetry-based biosensing. In this arrangement, the Cyfra-21-1 biomarker was flowed over the sensor surface, allowing for binding at the antibodies. Changes in the system current were observed associated with the binding event, allowing for sensing of the target species. The system proved to be highly sensitive, with a lower detection limit of 0.001 ng mL^{-1}.

Beyond $La(OH)_3$-based materials, Han *et al.* demonstrated the ability to prepare unique upconversion nanoparticles (UCNPs) using amino acids.[64] In general, $NaLuF_4$ doped with Yb and/or Er was prepared using aspartic acid, serine, or lysine, while $NaLuF_4$ doped with Gd, Yb, and/or Tm was generated using aspartic acid as a proof of concept. To prepare the materials, the rare earth metals, amino acids, NH_4F, and NaCl were co-mixed in an aqueous/ethanol solution at room temperature. The system was subsequently autoclaved at 200 °C for 24 h, resulting in material precipitation after cooling to 60 °C. In general, this approach delivered nanoparticles of ~30 nm diameter, which were fully characterised with a variety of approaches. Based upon their composition, the materials possessed fluorescence upconversion capabilities, which is highly attractive for both energy harvesting and nanomedicine applications. For instance, when the $NaLuF_4$ materials doped with Yb or Er capped with the amino acids were excited with a laser set at 980 nm, three emission peaks were noted at 520, 540, and 653 nm with wavelengths independent of the amino acid. All three of these peaks arise from emissions from the Er^{3+} ions. Remarkably, the UCNPs demonstrated significant resistance to photobleaching, even after 1 h of laser excitation.

Once prepared, the materials were exploited for bioimaging in two modalities: upconversion luminescence (UCL) or magnetic resonance (MR).[64] This was demonstrated using the aspartic acid-capped $NaLuF_4$ materials in BALB/c nude mice. To this end, the UCNPs were injected into the mice *via* the tail vein and allowed to accumulate in the livers. Note that the materials were confirmed to be non-toxic prior to their use as imaging agents. Imaging of the mice using UCL demonstrated emission from the liver and spleen, identifying the nanoparticle locations *in vivo*. In addition, MR-based imaging was also used, confirming the location of the UCNPs within the liver and spleen. Such dual modality-based imaging is highly attractive and could be adapted for specific use in a variety of biomedical-based imaging applications, thus proving the effectiveness of the synthetic approach in obtaining functional optical materials.

In a final area of work, Ghorbani-Choghamarani and Azadi used arginine as a ligand to direct the production of a complex, multicomponent material composed of Fe_3O_4, SiO_2, and Pd^0.[65] The structures were prepared to act as a nanocatalyst where the Fe_3O_4 component can be used as a means to magnetically separate the materials from the reaction medium. To prepare the materials, the Fe_3O_4 was fabricated *via* a standard co-precipitation method, and the Fe_3O_4 was then coated with SiO_2 using tetraethylorthosilicate. Once these components were prepared, arginine was added to the reaction, allowing for coordination of the amino acid *via* the acid functionality. This resulted in exposure of the guanidinium group to solution, which could coordinate to Pd^{2+} ions added to the mixture. Upon reduction with $NaBH_4$, Pd^0 nanoparticles were prepared at the surface of the magnetic structures, generating the final Fe_3O_4@SiO_2@arginine@Pd^0 materials. Once prepared, the nanostructures demonstrated reactivity for C–C coupling reactions where the materials could be readily removed from the reaction system *via* magnetic separations. Such effects were important for the recyclability of the structures over multiple reaction cycles.

2.2.2 Peptides

While the amino acids provide a diverse palette of functional groups for binding to inorganic nanoparticles, leading to their colloidal stability, combining these residues into a peptide chain can prove to be highly advantageous. This allows for the integration of multiple functional groups into a single molecule in specific arrangements, leading to the potential to achieve material-specific binding of the peptides to the nanoparticle surface. This is highly intriguing and could be exploited for a variety of applications, especially for the design of multicomponent assemblies; however, the basis of this potential specificity remains an ongoing challenge.

Peptides with affinity for target surfaces have been the subject of intense research for several decades. These biomolecules are typically identified *via* two different pathways: from individual organisms or through bioselection processes. For the first type of peptides, biological organisms have used evolution over millennia to develop these peptides to generate inorganic nanoparticles for three purposes: defence mechanisms, structural support, and detoxification. For instance, peptides with affinity for CdS quantum dots have been extracted from yeast and other cells, which are used for the bioproduction of these highly fluorescent nanoparticles.[66,67] This process is used to detoxify the cells from toxic Cd^{2+} ions that are present in a specific location. In the second approach, biocombinatorial libraries in the form of phage or cell surface-displayed peptides are used to isolate peptides with affinity for targeted inorganic or biological materials from libraries containing 10^9 random peptide sequences (discussed in more detail below). Such approaches are attractive as they allow for the on-demand isolation of peptides that bind to materials not typically observed in nature. In both approaches, the peptides can bind to the nanoparticle to generate a biotic/abiotic interface that leads to colloidal suspension of the materials.

A variety of high-quality reviews on the areas of peptide-capped inorganic nanoparticles have recently been published that highlight this area of biogenic materials.[19,68,69] We recommend that the reader consider these sources, which provide a discussion of the rich history of this field. As such, this section has been prepared to highlight the most cutting-edge studies, focusing on peptide-capped metallic materials.

2.2.2.1 Advances in Peptide Adsorption on Au Nanoparticles

One of the key aspects with peptide adsorption to nanoparticle surfaces is the molecular recognition event between the biological molecule and the inorganic material. This is typically programmed at the biological level, especially for peptides identified *via* biocombinatorial selection methods, such as phage display. In this regard, the peptide sequences, typically composed of 12 amino acid residues, use non-covalent interactions to drive surface adsorption, which in turn leads to colloidal suspension of the metallic particles. This was examined extensively recently by Hughes *et al.* using the Pd4 peptide (TSNAVHPTLRHL) adsorbed to Au.[70] While the Pd4 sequence was originally isolated with affinity for Pd materials,[71] previous studies confirmed that it was able to adsorb to Au, where the two histidine residues served as anchoring points to the metallic surface.[72] Hughes *et al.* exploited a peptide library that used synthetic mutations and alterations to the sequence that either mutated one or both of the histidine residues to an alanine, scrambled the sequence, or clustered the two histidine residues at the N-terminus (see Table 2.1).[70] Using quartz crystal microbalance (QCM) analysis, they

Table 2.1 Peptide mutation library of the Pd4 peptide, their free energy of binding (ΔG) and fractional surface coverage (θ).

Peptide	Sequence	ΔG/kJ mol^{-1}	θ/%
Pd4	TSNAVHPTLRHL	−31.4 + 0.1	79.8 + 0.5
Pd4-1	TSNAVAPTLRHL	−27.2 + 0.2	41.2 + 1.6
Pd4-2	TSNAVHPTLRAL	−30.7 + 0.1	75.0 + 0.9
Pd4-3	TSNAVAPTLRAL	−32.0 + 0.8	82.8 + 5.0
Pd4-4	**SPHTTNHRALLV**	−33.1 + 0.2	88.8 + 0.8
Pd4-5	HHTSNAVPTLRL	−35.5 + 0.2	95.3 + 0.4

demonstrated that these variations could lead to differences in the free energy of binding (ΔG). Interestingly, mutation of either histidine residue to an alanine led to diminished binding affinity; however, mutation of both residues led to a new sequence with slightly enhanced affinity compared to the parent Pd4. This suggests that, while the histidine resides are important to binding, they are not absolutely necessary to maintain affinity for the biomolecule to the Au surface. Remarkably, scrambling of the sequence or clustering the histidines at the N-terminus led to new peptides with further enhanced ΔG values, indicating that the sequence order of the residues was not critical in controlling the binding event.[70]

When these mutated/altered peptides were employed to generate peptide-capped Au nanoparticles, all of the sequences were able to prepare colloidally stable particles of sizes <8 nm, depending upon the Au : peptide mole ratio employed; however, only the parent Pd4 and the sequence that clustered the histidines at the N-terminus (Pd4-5) fabricated particles that were stabile for at least 24 h.[70] All of the other sequences with histidine to alanine mutations or scrambled sequence led to materials that eventually precipitated over time due to poor binding affinity of alanine. Interestingly, these stability results were not directly related to the overall binding strength of the peptide to the particle. The Pd4-5 sequence revealed the highest reported binding free energy ΔG value, which led to stable Au nanoparticles; however, the parent Pd4 sequence, which also generated stable materials, had a lower affinity for Au, as measured by QCM, than other sequences that resulted in materials that aggregated and precipitated in less than 24 h.[70] This suggests that other factors beyond absolute binding free energy are important in controlling nanoparticle stability.

To further examine the Pd4 system, Hughes *et al.* used replica exchange with solvent temperature molecular dynamics (REST-MD) simulations of the Pd4 peptide and mutants on the Au surface.[70] This was employed to identify changes in the overall bound peptide conformation, which could be used to extract the configurational entropy of the peptides in the surface-adsorbed stated. Remarkably, these studies suggested that the alanine mutations to the sequence resulted in peptides with highly similar surface-adsorbed structures, indicating that the peptide sequence was recalcitrant to conformational changes *via* mutation of anchoring residues. For the biomolecules that more substantially altered the sequence (scrambled or clustering of histidines at the N-terminus), more variation in the adsorbed structure was observed, as would be anticipated. In addition, from the simulation studies, the conformational entropy (S_{conf})[32] for each peptide in the surface-adsorbed state could be determined. Interestingly, the Pd4 and Pd4-5 peptides both had the lowest S_{conf} values, where these sequences were also the only two that prepared stable Au peptide-capped nanoparticles. This suggests that for the Pd4 system, nanoparticle stability is more likely to be linked to the conformational entropy of the biomolecules once adsorbed rather than the overall affinity of the peptide for the target surface. Subsequent work from Nguyen *et al.*[73] examined the mutation analysis for Au-binding affinity on two additional Au-binding peptides: A3 (AYSSGAPPMPPF) and AuBP1 (WAGAKRLVLRRE).[74,75] This work further supported the effects of the S_{conf} value in determining the colloidal stability of the peptide-capped Au nanoparticles. Across all of the sequences studied, lower degrees of conformational entropy tended to lead to higher nanoparticle stability.

Beyond the ability to stabilise Au nanoparticles, Munro and Knecht demonstrated that peptide sequences can be modified to also drive Au^{3+} ion reduction for Au nanoparticle production.[76] As such, the sequence would be responsible for all steps in nanoparticle synthesis, from metal ion reduction to colloidal stabilisation *via* surface passivation. To achieve such capabilities, tryptophan residues were incorporated into Au-binding peptides that did not possess the amino acids, as previous studies had confirmed the reductive capability of the residue.[77,78] For instance, the AgBP2 peptide (EQLGVRKELRGV) isolated by Hnilova *et al.*[79] was able to bind Au nanoparticles,[80] but was unreactive for the reduction of Au^{3+} to Au^0 as the sequence did not possess a tryptophan. After incorporation of the indole-based residue at the N-terminus, the sequence became quite reactive for Au^{3+} reduction, resulting in the production of Au nanoparticles of ~14 nm in diameter.[76] The peptides in the reaction mixture not only drove the reduction process, but also stabilised the growing particles to prevent bulk aggregation. Interestingly, Munro and Knecht demonstrated that the rate of the reaction process was highly dependent upon the position of the synthetically added tryptophan residue, the overall peptide sequence and composition, and the binding affinity of the biomolecule for the metal ions and metal surface.[76] Nevertheless, the generation of a multifunctional sequence to control multiple aspects of nanoparticle synthesis is attractive for sustainable material production.

In separate work, Slocik *et al.* explored the effects of residue chirality for peptides adsorbed to Au nanoparticles.[81] In this work, they used the FlgA3 peptide (DYKDDDDKPAYSSGAPPMPPF) to generate peptide-capped Au nanoparticles with optical chirality.[74] In a particularly creative approach, they employed a set of peptide enantiomers of FlgA3 composed of either all L-amino acids or all D-amino acids. For the L-FlgA3 peptide, both the free biomolecule in solution and the peptide adsorbed onto the Au nanoparticle demonstrated negative ellipticity at ~203 nm, as measured *via* CD spectroscopy (see Figure 2.5). Interestingly for the D-flga3 peptide, the free peptide

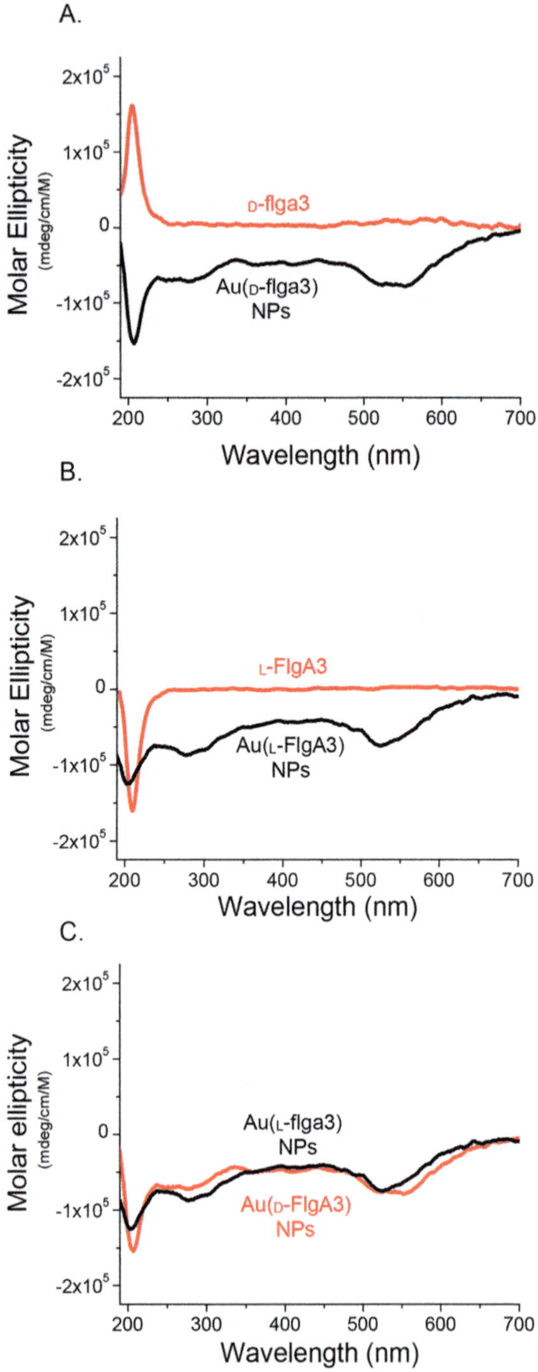

Figure 2.5 CD analysis of (a) L-FlgA3 and (b) D-flga3 peptides both free in solution and adsorbed to Au nanoparticles. (c) Comparison of the two peptide-capped materials, demonstrating nearly identical spectra for the peptides when bound to the inorganic material. Reproduced from ref. 81 with permission from American Chemical Society, Copyright 2019.

presented a typical positive ellipticity at this wavelength; however, when this sequence was adsorbed to the Au nanoparticle surface, inversion of the peak was observed, producing a negative ellipticity (see Figure 2.5).[81] This suggests that conversion of the residues in the D-flga3 to their L form occurred, resulting in the formation of the L-FlgA3 peptide on the Au nanoparticle surface. This probably occurred *via* an optical transition of the chemical chirality of the peptide from the D to L form.[81]

The effect of the chirality conversion was supported by enzymatic digestion of the surface-adsorbed peptides.[81] For this, two enzymes were employed, proteinase K and chymotrypsin, which digest the biomolecules at different locations in the sequence. For proteinase K, the enzyme was able to digest the L-FlgA3 sequence that was free in solution; however, no reactivity was observed for the free D-flga3 based upon the chirality of the residues. When the enzymatic digestion process was applied to both the L-FlgA3- and D-flga3-capped Au nanoparticles, degradation of both peptide sequences was observed.[81] This confirms the inversion of the residue chirality on the nanoparticle surface, as the enzyme is unable to digest D-based residues. Beyond enzymatic digestion, high-resolution characterisation using pair distribution function (PDF) analysis of high-energy X-ray diffraction data suggested a difference in the surface atomic structure of the materials based upon the chirality of the peptide used. This variation in surface structure was reflected in differences in the catalytic reactivity of the two nanoparticles for the reduction of 4-nitrophenol to 4-aminophenol.[81] Taken together, this suggests that while chirality inversion occurs on the Au nanoparticle surface for D-peptides, the surface atomic structure is different between the two nanoparticles templated by L-FlgA3 or D-flga3 peptides.

2.2.2.2 Peptide-based Metal Nanoparticle Assembly

Owing to their inherent affinity for nanoparticles and their ability to controllably assemble, peptides have been exploited for the controlled organisation of metal nanoparticles into exquisite three-dimensional morphologies.[19] For instance, Rosi and co-workers have elegantly demonstrated this ability using the A3 peptide, modified with aliphatic moieties, for the organisation of Au nanoparticles into complex three-dimensional structures.[82–84] In this approach, the integrated hydrophobic domain facilitates peptide assembly, where growing Au nanoparticles can be adsorbed to the exposed peptide domain in solution. Changes in the chemical structure of the aliphatic component can be exploited to vary the final assembly morphology, from helices to ribbons to hollow spherical structures, with precise control over the dimensions of the assembly.

In the most recent studies of this capability, Merg *et al.* demonstrated that hollow spherical assemblies of Au nanoparticles using a peptide template could be surface modified using thiols in solution.[85] Such capabilities could prove to be important to chemically modify the structures for a variety of applications, post-assembly. For this, the team used the C6-AA-A3 peptide, where C6 is a six-carbon aliphatic chain, A represents alanine, and A3 is the A3 peptide discussed above. To generate the hollow spherical assembly, the C6-AA-A3 peptide was co-mixed with Au^{3+} ion in an *N*-(2-hydroxyethyl)-piperazine-*N*′-(2-ethanesulfonic acid) (HEPES) buffered solution. In this approach, the HEPES drives Au^{3+} reduction to Au^0 for nanoparticle formation. The peptides bind to the particles in suspension, which leads to their stabilisation and assembly into the spherical hollow structure *via* formation of a tightly packed peptide bilayer structure. To maintain particle stability, free A3 peptide was added to the mixture 20 min after reaction initiation, which stabilised the solvent-exposed surface of the organised structure to prevent material aggregation.[85]

Based upon this peptide bilayer structure (interior ← **A3**-AA-C6/C6-AA-**A3** → exterior), A3 peptides are constrained at the interior of the hollow spherical assembly and exposed on the exterior surface. For the exterior, A3 peptides are passivated at the solvent-exposed interface by gold nanoparticles.[85] As such, two different peptide-binding domains would be present, potentially allowing for selective exchange of the solvent-exposed peptides. In this case, conjugation of different functionalities post-assembly could be possible, without degrading the spherical hollow structure. Merg *et al.* demonstrated this capability through exchanging the A3 peptides with thiolated polyethylenegylcol (PEG) molecules.[85] For the exchange, the strongly binding thiol could displace the more weakly bound solvent-exposed peptide, attaching the hydrophilic polymer to the assembled material surface. NMR characterisation of the assemblies confirmed the ligand exchange process, with negligible changes to the organised structure.[85] Such approaches could prove to be important for biomedical applications of these assemblies, as nanoparticle PEG modifications are routinely used to minimise non-specific adsorption of biomacromolecules.

Rosi and co-workers have further exploited this peptide-based assembly approach for the production of chiroptically active helical assemblies.[86,87] As shown previously, when using the C_{18}-$(A3)_2$ biomolecule, the formation of chiroptically active single helical assemblies of Au nanoparticles could be achieved.[86] In this arrangement, the 18-carbon aliphatic chain was responsible for the assembly process; however, the methionine at the 9-position in the peptide was required to be oxidised to the sulfoxide form. A recent study from Mokashi-Punekar *et al.* demonstrated that the oxidation state of the methionine (oxidised form is methionine sulfoxide) and its position within the sequence can be exploited to control Au nanoparticle size and aspect ratio within the helical assemblies, leading to control over the chiroptical properties.[87] In this regard, when reduced methionine was present in the sample at any position, only irregular double helices of spherical Au nanoparticle were produced; however, when the oxidised form of the peptide was employed with the methionine sulfoxide in either the 9 or 11 position, single helices of oblong Au nanoparticles were prepared. In addition, the dimensions of the oblong particles varied based on the methionine sulfoxide position (see Figure 2.6), resulting in measurable differences in the chiroptical signals of the different assemblies.[87] These differences in material morphology arise from the peptide/Au nanoparticle interactions, which could prove to be important for a variety of different property control capabilities based upon the material composition in the assembly.

Further studies by Mokashi-Punekar and Rosi exploited the C_{18}-$(A3)_2$ peptides with the oxidised methionine. In the conventional synthesis, spherical Au nanostructures are generally prepared; however, the addition of secondary shape-directing agents can be used to integrate particle anisotropy into the assembly.[88] To test this effect, when the authors added 0.1 M NH_4Br into the A3 synthesis system (in the absence of assembly motifs), the percentage

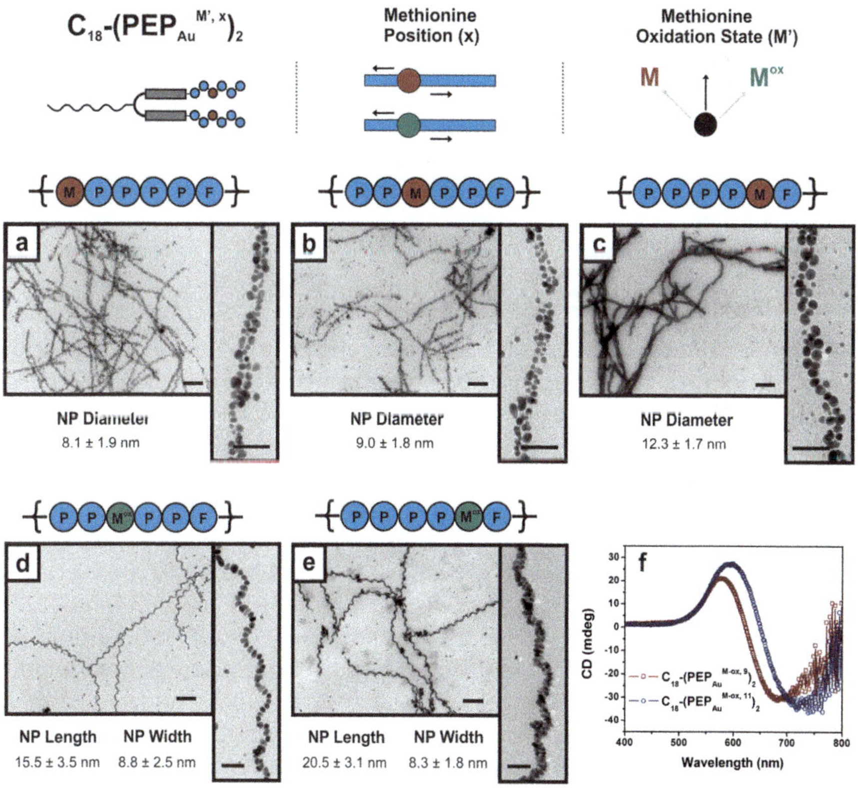

Figure 2.6 TEM analysis of the assemblies generated with the C_{18}-$(A3)_2$ peptide. Note that the original work referred to A3 as PEP_{Au}. (a–c) The assemblies generated with a reduced methionine positioned at the (a) 7, (b) 9, and (c) 11 positions in the peptide sequence. (d and e) The single helices prepared with an oxidised methionine at the (d) 9 and (e) 11 positions. (f) Comparison of the CD spectra of the materials prepared in (d) and (e). Reproduced from ref. 87 with permission from American Chemical Society, Copyright 2019.

of discreet anisotropic particles reached ~13% of the total materials. For this, large Au nanoparticles with sharp facets were observed that diverged from the typical spherical structures prepared in the absence of Br^-. In addition, CTAB-based surfactants, known to be instrumental for Au nanorod syntheses,[89] were also added to the peptide-based synthesis process, which varied the carbon chain length (C_{10}TAB, C_{12}TAB, C_{14}TAB, and C_{16}TAB, where C_x represents the aliphatic carbon chain length).[88] For the shorter chains (C_{10}TAB and C_{12}TAB), the final discreet particles were dominated by spherical structures with a few hexagonal Au materials; however, for the longer chain lengths (C_{14}TAB and C_{16}TAB), Au nanoprisms at higher yields were observed (~27% and ~30%, respectively). Interestingly, when the same experiment was conducted using the C_{18}-$(A3)_2$ biomolecule with oxidised methionine in the presence of C_{16}TAB, the number of anisotropic materials integrated into the helices exceeded that of the discreet particle syntheses, reaching ~60% anisotropic particles in the assemblies.[88] This may arise from the nanoparticle growth process at the helical interface, which may facilitate anisotropic growth. Nevertheless, this ability to control nanoparticle shape in these peptide-based assemblies provides a unique opportunity to control the resultant optical properties for applications in, for example, plasmonics and metamaterials.

2.2.2.3 *Peptide-based Metal Nanoparticle Catalysis*

Many metal nanoparticles are inherently catalytically reactive, where peptide-based structures have demonstrated significant reactivity. Such bio-based structures have been used to drive reactions such as C–C couplings, olefin hydrogenation, 4-nitrophenol reduction and electrocatalytic reactions. Most of these reactions are highly dependent upon the biotic/abiotic interface at the nanoparticle surface, where such material reactivity has been recently reviewed.[19,68]

In very recent studies, Zhu *et al.* used the T7 peptide (Ac-TLTTLTN-$CONH_2$) to fabricate electrocatalytically active Pt- and PtPd-based nanoplates.[90] The T7 peptide was originally isolated with affinity for the Pt(100) surface *via* phage display methods.[91] This method utilises phage-displayed peptide libraries (10^9 random sequences) for identification of peptides with high binding affinity to inorganic or biological materials. Libraries are constructed from phage viruses (infect bacteria) that are genetically engineered to display five copies of short peptides (7-mer or 12-mer). In a typical experiment, a phage-displayed peptide library is incubated with target material, subjected to multiple rounds of bio-panning in order to remove weak peptide binders, and sequenced to identify peptide sequences with high binding affinities. To generate the Pt nanoparticles, small colloidal Pt crystals are prepared in the presence of the peptide and $AlCl_3$. These crystals spontaneously assemble in two dimensions, generating triangular nanoparticles that can reach thicknesses of ~3.5 nm with edge lengths of ~75 nm (see Figure 2.7).[90] Once prepared, the peptide-derived Pt nanoplates were examined for their

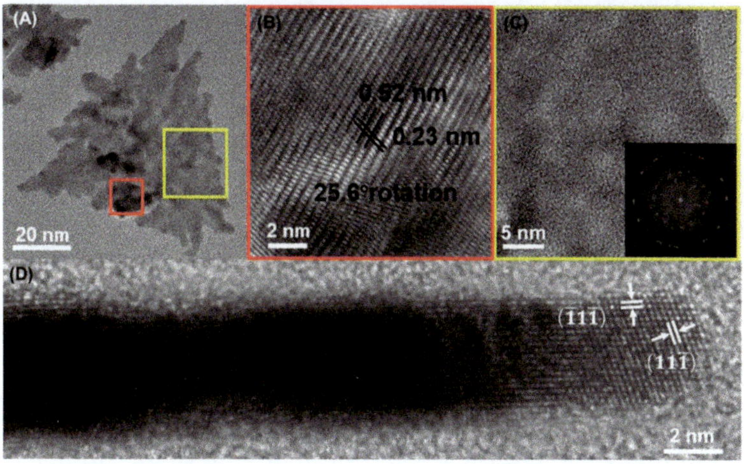

Figure 2.7 TEM analysis of the Pt nanoplates prepared using the T7 peptide: (a) TEM image of a representative nanoparticle, (b) image of Moiré fringes from two overlapping nanoplates, (c) HRTEM image of a single plate, and (d) HRTEM image of the edge of a single plate. Reproduced from ref. 90 with permission from American Chemical Society, Copyright 2019.

electrocatalytic reactivity, in which these structures displayed significantly enhanced specific activity and mass activity compared to state-of-the-art Pt/C catalysts. The monometallic Pt plates were also highly reactive over multiple electrocatalytic cycles; after 30 000 cycles in the accelerated durability test, the structures retained ~57% of the mass activity.[90]

To further enhance the reactivity, Zhu *et al.* used Pd doping of the Pt nanoplates.[90] For this, Pd atoms were deposited at open coordination sites on the Pt material surface, thus stabilising the structures. Using the Pd-doped Pt nanoparticles (~4.6 mass% Pd in the structures), enhanced specific activity and mass activity was observed; however, the Pd-doped materials demonstrated enhanced reactivity after 30 000 cycles compared to the native Pt structures.[90] To this end, the Pd-doped materials preserved 70.7% of the initial mass activity, compared to ~57% for the monometallic Pd materials. Such reactivity enhancement, especially after extensive use, is highly important for sustainable reactivity, in light of growing energy concerns.

Zong *et al.* used the FlgA3 peptide to generate AuPd bimetallic nanoparticles for electrocatalytic oxygen reduction and hydrogenation evolution reactions (ORR and HER).[92] To prepare the materials, the researchers initially generated the peptide-capped Au nanoparticles using standard approaches, and then reduced varying amounts of Pd^{2+} in the presence of the peptide-capped particles. This resulted in the deposition of Pd nanoparticles onto the Au surface, resulting in a surface-decorated structure, consistent with previous results.[93] Once prepared, the materials were examined for their electrocatalytic reactivity.[92] To this end, the Au@Pd1.0 structures were highly reactive for ORR, with enhanced reactivity over Pt/C and Pd/C controls, where the $E_{1/2}$ value for the materials was 0.85 V; however, the Pt/C and Pd/C structures demonstrated $E_{1/2}$ values of 0.82 V and 0.78 V, respectively. In addition, the peptide-based particles were remarkably stable compared to the control structures. This was evidenced by only a 3.5% loss in cathodic current for the Au@$Pd_{1.0}$ structures, compared to 20% for the Pt/C and 15% for the Pd/C materials after 8 h of continuous operation.

Finally here, Parab *et al.* exploited the A3 peptide for the production of catalytically reactive Rh nanoplates.[94] Using this peptide, Rh^{3+} ions were reduced with $NaBH_4$ in the presence of the peptide, which resulted in the formation of the plate-like Rh^0 materials. Such a structure was unanticipated as the A3 sequence is known to prepare spherical metal nanoparticles. The morphology of the structures was dependent upon the peptide : Rh mole ratio employed; at a mole ratio of 1 : 1, square-like nanoplates of the metallic structures are prepared with edge lengths of ~44 nm and thicknesses of ~11 nm, but significantly aggregated structures are generated at a ratio of 1 : 3 (thicknesses of ~8 nm) due to less peptide being present to stabilise Rh plates.[94] TEM imaging of the structures demonstrated the formation of nanoplates that appeared to be generated from the aggregation of individual structures, similar to the Pt plates prepared by Zhu *et al.* (discussed above).[90] Once fully characterised, the Rh materials were tested for their catalytic reactivity for olefin hydrogenation.[94] Interestingly, the materials prepared at a peptide : Rh mole ratio of 1 : 3 demonstrated the greatest reactivity with a turnover frequency value of ~3700 mol product per mol Rh per h for the hydrogenation of allyl alcohol to 1-propanol. These structures displayed catalytic reactivity for a variety of olefinic substrates; however, negligible reactivity for alkyne hydrogenation was observed. It is possible that the alkyne poisoned the Rh surface; however, additional studies are required to confirm such effects.[94]

2.2.2.4 Peptide-based Metal Nanoparticles for Biological Applications

Based upon the biotic/abiotic interface generated at the particle surface, peptide-capped nanomaterials present intriguing opportunities for biological and biomedical applications. This could include integration of receptors or other molecules for biosensing, targeting agents such as antibodies for drug delivery, *etc.* at the particle surface. An additional intriguing factor is the fact that the peptides are non-covalently adsorbed to the metallic materials, potentially allowing for release under select conditions, and further enhancing the opportunities available from peptide-based materials.

Many biological applications have been recently reviewed;[18,95,96] however, as a unique recent demonstration of potential biological applications of peptide-capped nanoparticles, Bansal *et al.* used a cell-penetrating peptide known as TAT (YGRKKRRQRRR)[97] to produce peptide-capped Au nanoparticles.[98] For their unique synthesis, the peptide and Au^{3+} ions were mixed in aqueous solution and heated at 80 °C. At this elevated temperature, the peptide itself was able to reduce the Au^{3+} *via* the tyrosine residues of the biomolecule, driving Au^0 nanoparticle nucleation and growth.[98] The remaining TAT peptides in solution were subsequently able to passivate the particle surface, resulting in peptide-capped Au nanoparticles with an average size of 10–30 nm in diameter, depending upon the peptide concentration used and the reaction pH. Using this approach, a significantly high loading of TAT on the Au nanoparticle surface was achieved, reaching values of ~427 peptides per particle.[98] Such substantially high peptide loadings could prove to be important to aid in intracellular nanoparticle delivery.

Once the materials were fully characterised, Bansal *et al.* explored the ability of the TAT-capped Au nanoparticle for intracellular delivery using HeLa cancer cells.[98] As anticipated, treatment of the cancer cells with the bio-based materials resulted in only 70% and 50% of the cells remaining alive after 6 and 24 h of exposure to the nanoparticles, respectively. As a control, the same experiment was processed using free TAT peptide and bare Au nanoparticles, which demonstrated negligible cell death, confirming the role of the high loading of the TAT peptide on the Au nanoparticles driving material internalisation into the cells *via* penetration of cell membrane. The intracellular delivery of the peptide-capped structures was further confirmed using confocal *Z*-stack microscopy.[98] From this analysis, no delivery of the bare Au nanoparticles was observed; however, a significant number of materials were present inside the cells when the TAT-capped Au particles were used.

2.2.3 Proteins

Proteins have also been used extensively for the production of a wide range of inorganic nanoparticles.[99] Specific protein-based templates include cage proteins, globular proteins, enzymes, viral capsid proteins, *etc.*, which all give rise to nanomaterials of different properties depending on the protein used to stabilise the system. A wide variety of different proteins have been used previously, as discussed in numerous review articles.[22,100–104] To provide a background for this area, we have focused on two specific proteins that have been well studied for inorganic nanomaterial production: ferritin and the tobacco mosaic virus (TMV) capsid protein. These two systems are unique, which allows for the production of discrete nanoparticles within the ferritin protein cage, or for the fabrication of nanoparticle assemblies or nanowires using the linear TMV structure. Such a discussion provides an overview of the breadth of materials produced using proteins.

2.2.3.1 Ferritin-based Nanoparticles

Ferritin is an ideal protein for the fabrication of inorganic nanoparticles based upon its cage-like structure. It is composed of 24 protein subunits that assemble to generate the spherical-shaped protein (see Figure 2.8). The

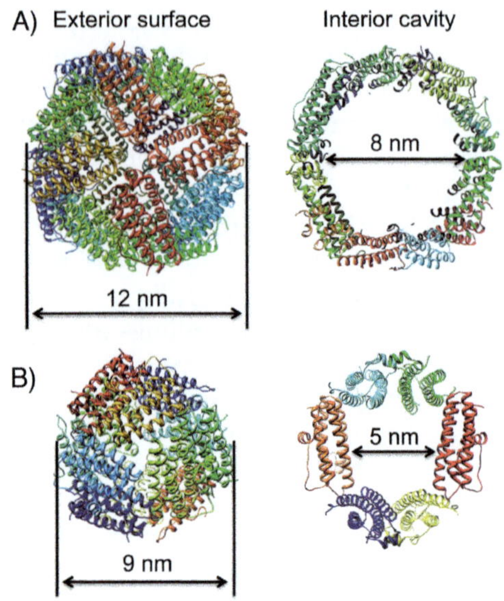

Figure 2.8 Protein structures of (a) human heavy-chain ferritin and (b) *Listeria innocua* DNA-binding protein from starved cells (Dps). Reproduced from ref. 102 with permission from Elsevier, Copyright 2010.

empty cavity is an ideal location to template inorganic nanoparticles, which can be grown directly inside the protein. Based upon the stoichiometry used in the reaction, the number of metal atoms can be highly controlled in the final nanoparticle, providing a unique handle on controlled material growth. Furthermore, these ferritin-templated structures are quite stable and can be prepared with a variety of ferritins from different species, including thermophilic organisms,[20,99,105] thus extending the range of conditions in which the protein-templated materials can be grown or used in.

To fabricate the ferritin-based inorganic nanoparticles, metal ions are co-mixed in solution with the apoferritin protein. This allows for the metal ions to be uploaded into the interior cavity of the biomolecular template, where these ions can be reduced to form zerovalent particles or reacted with oxygen or sulfide to make the corresponding metal oxides or metal sulfides, respectively. Using this approach, numerous compositions of materials have been prepared, including noble metals (Au, Ag, Cu, Pd, and Pt),[99,106–109] non-noble metals (Co and Ni), metal oxides (Fe_3O_4, Co_3O_4, Mn_3O_4, Fe_3O_4/γ-Fe_2O_3, Eu(O)OH, $Ti(O)(OH)_2$, Fe(O)OH),[110–114] and metal sulfides or selenides (CdS, AuS_2, Fe_2S_3, CdSe, and ZnSe).[21,115–118] Note that, in some cases, the existing ferrihydrite (Fe(O)OH) core of ferritin was converted into new mineral forms (iron(III) sulfide) or used as a photocatalyst to generate new nanoparticle compositions (Cu^0).[21,108] Based upon the size of the protein, the final particle sizes are typically <15 nm in diameter; however, the size of the particles can be directly controlled by the amount of metal loaded into the ferritin templates. Based upon the wide variety of compositions prepared, numerous properties can be engendered into the materials for applications in plasmonics, magnetics, catalysis, energy harvesting/storage, *etc.*

Jolley *et al.* used advanced X-ray-based analytical methods to characterise the ferritin-templated nanoparticles with atomic-scale resolution.[119,120] Such studies are highly important as they can probe the inorganic component, as well as the interface between the metal and protein shell, where fascinating chemistries are possible to control the resultant properties. For instance, Jolley *et al.* used pair distribution function (PDF) analysis of total X-ray scattering to characterise five different nanoparticles (ferrihydrite, γ-Fe_2O_3, Mn_3O_4, CoPt, and FePt) capped with ferritin from two different species (*Homo sapiens* and *Pyrococcus furiosus*).[120] In general, the results indicated that the material composition and loading played key roles in determining the crystallite size. In addition, the crystalline domain size for some of the materials was smaller when confined within the cavity of ferritin, compared to structures prepared *via* more conventional approaches. For instance, Mn_3O_4 templated by ferritin demonstrated crystalline domains of 3.5 nm, which is substantially smaller than the size observed for the same materials prepared without the ferritin template (17.5 nm).[120] This suggests that the protein cage itself is playing a role in the crystallisation process, potentially through interactions between the inorganic and biological components.

In separate work, Jolley *et al.* advanced this characterisation analysis to explore the synthesis of Mn ferrites, which are mixed spinel structures containing Fe and Mn.[119] They examined synthesis of these mixed ferrites by varying the proportion of $[Fe^{2+}]/[Mn^{2+}]$ from 0–100% and analysing the changes in material structure as a function of composition. To confirm the effects of the protein template, they also synthesised the same materials in the absence of ferritin to see whether changes in the material structure occurred based upon the biomolecule interacting with the growing inorganic structures. Once prepared, the inorganic structures were characterised using X-ray absorption spectroscopy (XAS), X-ray magnetic circular dichroism, and PDF of high-energy total X-ray scattering.[119] In combination, the results confirmed changes in the material morphology based upon the protein template at higher mole fractions of Mn. For instance, for the materials generated without ferritin, γ-Fe_2O_3 (maghemite) was prepared in the absence of Mn, which turned into Mn_3O_4 (hausmannite) as the faction of Mn in the sample increased to 100% Mn. Remarkably, when ferritin was employed to template the materials, maghemite was again prepared at 0% atomic Mn; however, as the system increased to 100% atomic Mn in the reaction, a new Mn_3O_4 material was generated.[119] To this end, the protein-encapsulated Mn_3O_4 structure was more similar to chalcophanite, which is a layered manganese oxide structure with multiple oxidation states (Mn^{2+}, Mn^{3+}, and Mn^{4+}) depending on atomic position within the crystal lattice. Such a structure was significantly different to the materials prepared in the absence of the protein, suggesting that ferritin is altering the inorganic material growth process. Such capabilities could have profound structural effects on the inorganic component, altering the resultant material properties.

Owing to the structure of the materials, an inorganic core with protein shell, this provides great opportunities for a variety of applications. For instance, Uchida *et al.* used a ferritin magnetite particle templated by ferritin to target cancer cells.[121] This was achieved *via* manipulation of the ferritin structure to present RGD-4C targeting peptide (CDCRGDCFC) on the protein surface. While the protein cage was readily able to template the magnetic Fe_3O_4 particles, the solvent-exposed protein coat displayed the newly added peptide, which was able to bind $\alpha_v\beta_3$ and $\alpha_v\beta_5$ integrins, which have been shown to be upregulated on cancer cell surfaces.[122] When the nanoparticles were incubated with C32 amelanotic melanoma cells, enhanced binding of the RGD-4C-containing ferritin magnetite nanoparticles was observed, compared to control particles prepared without the targeting peptide.[121] While some of the control particles did adsorb to the cells, the targeting peptide lead to ~50% more particles adsorbed, supporting the role of the integrated RGD-4C component. This result provides unique opportunities to modify the protein shell for applications in drug delivery, biosensing, *etc.*

Beyond biomedical applications, ferritin-templated metal nanoparticles have been used in elegant self-assembly processes to form complex organised three-dimensional structures. In initial work, Okuda *et al.* used Fe and In nanoparticles prepared within the ferritin cavity to form two-dimensional lattices.[123] The nanoparticles were prepared *via* standard approaches and then used to drive two-dimensional assembly at the air/water interface. For this process, the nanoparticles were injected into a solution of glucose and $CdSO_4$, buffered at a pH of 5.8. Due to the buoyancy of the ferritin protein in the glucose solution, the nanoparticles rose to the top of the solution. Once at the top of the solution, some of the proteins denatured, generating a biomacromolecular film to which the intact ferritin nanoparticles could adsorb, leading to the formation of the two-dimensional material array.[123] In addition, the Cd^{2+} ions in the solution were able to form salt bridges between the nanoparticles by interacting with adjacent ferritin templates, aiding in the self-assembly process. Using this approach, highly ordered hexagonal two-dimensional arrays of the Fe and In nanoparticles could be achieved, as imaged by TEM, with lattice constants of 13 nm. It is interesting to note that assemblies of >1 μm^2 could be achieved for the ferritin-templated Fe nanoparticles; however, larger assemblies of >4 μm^2 were observed for the In-based materials.[123] In addition, fewer defects (*e.g.*, ferritin with no nanoparticles) were observed for the In structures compared to the Fe nanoparticles. This may arise from a more efficient uptake of In by the ferritin over Fe, thus fewer apoferritin samples would be present in the In sample, leading to fewer defects.

In separate work by Matsui *et al.*, two-dimensional hexagonal arrays of ferritin-templated Fe and In nanoparticles were also prepared; however, the arrays were formed using a genetically modified ferritin protein.[124] In this case, the researchers incorporated a peptide identified with affinity for carbon nanohorns onto the ferritin outer surface,[125] thus engendering the protein with affinity for carbonaceous supports. Once expressed and purified, the modified proteins were used to generate the Fe and In nanoparticles using typical methods, after which the protein-based nanoparticles were allowed to adsorb to the carbonaceous surfaces, which were designed to be either hydrophobic or hydrophilic.[124] Interestingly, in both cases, the nanoparticles adsorbed to the surface; however, disordered structures were noted for the hydrophobic surfaces, while ordered hexagonal nanoparticle arrays were observed on the hydrophilic support. The adsorbed nanoparticles on the hydrophobic surfaces resembled control experiments of Fe nanoparticles prepared in the parent ferritin protein without the carbon-binding peptides, suggesting that non-specific adsorption was occurring on the hydrophobic surfaces. This was supported by studies demonstrating that adsorption on the hydrophilic surface was actually weaker on the hydrophobic support,[124] suggesting that the weaker interactions allowed for surface rearrangement of the bound nanoparticles to generate the ordered array of materials. Alternatively, iron nanoparticles synthesised within the DNA-binding protein from starved cells (Dps) protein cage of *Listeria innocua* (smaller than ferritin) were used as a catalyst to grow single-walled carbon nanotubes.[126]

While the above assemblies resulted in the formation of two-dimensional arrays, Künzle *et al.* used ferritin-templated materials to generate large three-dimensional assemblies.[127] To drive three-dimensional assembly, the researchers used two different genetically modified ferritin proteins (see Figure 2.9). In one case, the outer protein surface was modified to incorporate additional lysine and arginine residues to give rise to a positively charged protein, while a negatively charged ferritin was prepared *via* the incorporation of aspartic and glutamic acids. This allows for electrostatic-based assembly of the proteins, where control over the precise conditions was used to generate a tetragonal protein lattice with a *P*4 space group for the apoproteins.[127] Using this system, crystals of edge lengths of up to 200 μm could be grown rapidly (<1 day).

With confirmation of apoprotein electrostatic assembly, Künzle *et al.* used the oppositely charged proteins to incorporate inorganic nanoparticles.[127]

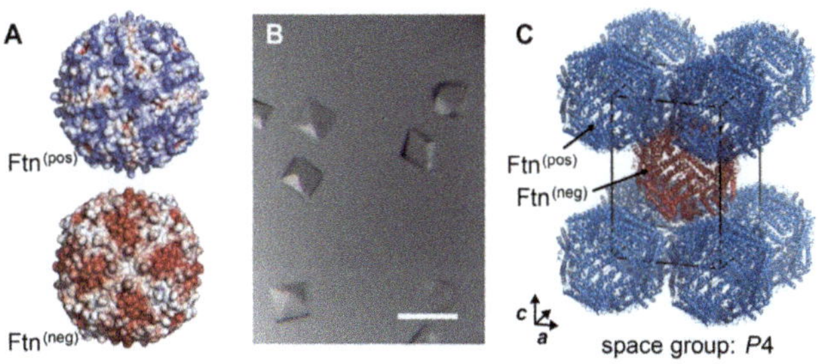

Figure 2.9 Electrostatic-based assembly of ferritin-templated nanoparticles. (a) The negatively (red) and positively (blue) charged ferritin proteins used in the assembly. (b) The crystals formed using these positively and negatively charged proteins. (c) The crystal structure of the assembled materials. Reproduced from ref. 127 with permission from American Chemical Society, Copyright 2016.

To this end, CeO_2 nanoparticles were grown within the positively charged ferritin, while Co_3O_4 was prepared inside the negatively charged protein. These two materials could then be used to drive electrostatic assembly, forming the mixed material crystals.[127] The formation of the binary material was confirmed *via* elemental analysis in SEM and through small-angle X-ray scattering analysis. To increase the complexity of the system, the assembly process could be driven using one of the ferritin-templated nanoparticles with the apoferritin protein of the opposite charge (*e.g.*, CeO_2 nanoparticles in the positive ferritin with the negatively charged apoferritin).[127] This allowed for the growth of crystals with only one type of nanoparticle incorporated into the assembly, thus controlling the interparticle distances between the inorganic components. Interestingly, additional work from Künzle *et al.* demonstrated that the negatively charged ferritin particles could also be used to generate three-dimensional assemblies individually, through the addition of high concentrations of Mg^{2+} in the reaction mixture.[128] This was achieved through the formation of salt bridges between the nanoparticles, thus changing the material arrangement compared to the electrostatic assemblies using the binary proteins. Taken together, these studies by Künzle *et al.* provide an elegant approach to achieve complex material assemblies of multiple compositions and particle arrangements. Such approaches could be highly attractive for a variety of applications, including optics, plasmonics, and catalysis.

2.2.3.2 Tobacco Mosaic Virus-based Nanoparticles

As a second protein-based system for nanoparticle production, the TMV represents an intriguing biological construct for nanostructure production.[129] The TMV is a rod-shaped virus with dimensions of ~18 nm in width and ~300 nm in length.[130] The viral capsid is composed of 2130 protein molecules that form a tube-like structure surrounding a single RNA strand of ~6400 base pairs in length.[130] Based upon its highly repetitive structure and relative ease in use in recombinant processes, the TMV serves as a ripe template for the production of inorganic nanoparticles. It is also an attractive species to control nanoparticle assembly for the formation of higher ordered structures (*e.g.*, nanowires), which can prove to be challenging to fabricate using biomolecules.

To generate inorganic materials on the TMV surface, a variety of approaches have been employed. For instance, Shenton *et al.* were able to deposit CdS and PbS onto the virus through bubbling of H_2S gas through a solution of TMV loaded with Cd^{2+} or Pb^{2+} ions, respectively.[131] For the CdS materials, imaging *via* TEM demonstrated that the structures were 50 nm wide with a 16 nm thick inorganic core shell surrounding an 18 nm core. Interestingly, these structures were comprised of 5 nm CdS nanoparticles that were aggregated along the TMV surface. For the PbS materials, a similar coating was observed; however, these structures were 40 nm wide and were composed of 30 nm aggregated crystallites.[131]

Beyond the metal sulfide materials, Shenton *et al.* were able to deposit SiO_2 *via* sol–gel chemistries on TMV, which resulted in the deposition of a 3 nm silica shell.[131] Interestingly, the silicified tubes were noticeably longer than the parent template, suggesting that tip-to-tip assembly of the tubes occurred in the sol–gel process. Finally, iron oxide materials were also deposited.[131] This was achieved by suspending the TMV templates in an Fe^{2+}/Fe^{3+} solution under basic conditions. In this approach, ferrihydrite was deposited on the virus; however, if the reaction was processed using only Fe^{2+} at pH 6.5 in the absence of O_2, an ultrathin 2 nm amorphous iron oxide shell could be deposited.[131] Similar to the silica materials, tip-to-tip aggregation of the individual TMV templates was observed to produce final materials with lengths greater than the parent virus.

Beyond these earlier materials, a variety of different structures have been templated *via* the TMV. For instance, Dujardin *et al.* templated nanomaterials of Pt, Au (see Figure 2.10), and Ag using TMV.[132] For all three materials, the

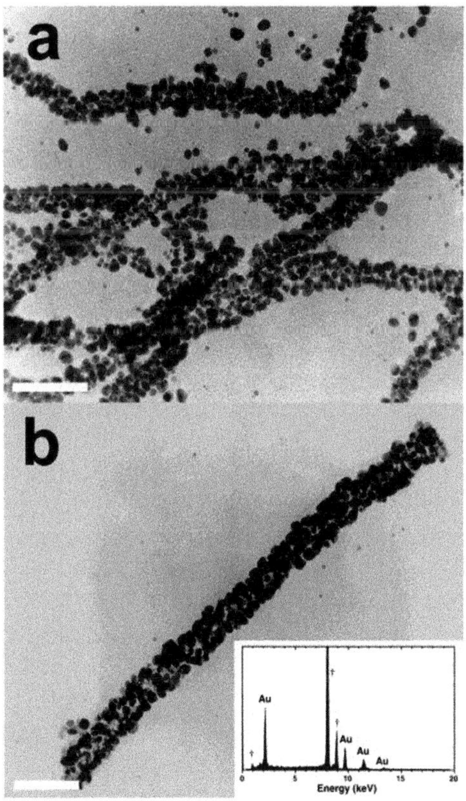

Figure 2.10 TEM images of Au nanoparticles deposited onto the TMV template (a) at low magnification (scale bar = 100 nm) and (b) at higher magnification (scale bar = 50 nm). Reproduced from ref. 132 with permission from American Chemical Society, Copyright 2003.

metal ions were allowed to complex with the virus, followed by either chemical reduction for Pt^{4+} and Au^{3+}, or photochemically for Ag^{+}. Interestingly, this approach resulted in the deposition of the Pt (2.5–5 nm in diameter) and Au (~8.6 nm in diameter) nanoparticles on the TMV outer surface, while the Ag nanoparticles (<5 nm in diameter) were preferentially incorporated at the TMV inner channel.[132] The Ag species were suggested to have preferential interactions with the acidic residues on the inner channel, resulting in their specific incorporation at that domain. Similar approaches have also been reported for the production of Au[133] and Pt[134] materials deposited onto the TMV surface. In these cases, chemical reduction-based approaches were used, albeit with different reducing agents, resulting in the deposition of high levels of metal nanoparticles for the production of metal nanowires. Finally, Pd nanomaterials have also been deposited onto the TMV surface.[135] In this case, sorption of the Pd ions to the virus occurred, where spontaneous reduction to form Pd metal was observed. Such reduction may be driven by the virus itself; however, a thin PdO layer was observed on the external surface of the material. This oxide could be easily removed *via* H_2 treatment.

Magnetic nanomaterials have also been prepared using the TMV. For this, Kobayashi *et al.* used a genetically modified TMV that increased the number of positively and negatively charged resides in the inner channel; the numbers of aspartic acid or lysine residues were increased to make the channel more negative or positive, respectively.[136] This resulted in increased nucleation sites, which allowed for complexation of Pt^{2+} and Co^{2+} at the inner channel. Upon reduction, the nucleation of magnetic Co/Pt nanowire was evident within the virus interior.

Knez *et al.* have demonstrated an altered approach for the incorporation of metal oxide materials on the TMV template. For this, the researchers used atomic layer deposition (ALD) methods to coat the virus with TiO_2 and Al_2O_3.[137] For this process, TMV was dried onto a Si wafer. Once under ultrahigh vacuum conditions, ALD methods were used to deposit the oxide-based materials on the viral capsid. Using this process, both the outer surface and internal channel can be coated with the oxide.[137] Interestingly, sonication-based methods can be used to remove the oxide on the exterior; however, the inner channel of inorganic material can remain. Such an approach is intriguing for material fabrication, especially for compositions not amenable for production in aqueous environments.

Beyond the fabrication of the extended rod structure, TMV assembly can result in the formation of nanodiscs by controlling solution pH and ionic strength.[138] From this, the TMV nanodiscs are 18 nm in diameter with a 4 nm interior pore, which can be used to generate metallic nanorings. For instance, Zahr and Blum demonstrated this capability for Au-based materials (see Figure 2.11).[139] To generate the materials, the TMV nanodiscs were incubated with negatively charged Au nanoparticles. This allows for electrostatic interactions between the nanoparticles and the positively charged residues on the protein disc (*via* arginines), to template the formation of the Au nanorings. The number of rings on the outer edge of the template and

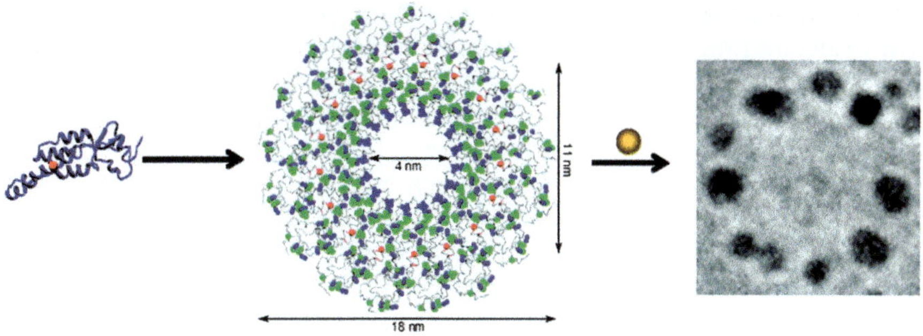

Figure 2.11 Scheme for the production of Au nanorings using a TMV nanodisc. Au NPs are ~4 nm in diameter. Reproduced from ref. 139 with permission from American Chemical Society, Copyright 2011.

the interparticle distance changed as a function of the system pH.[139] For instance, at pH values <7.0, ten particles per ring were observed, on average; however, at pH 7.5, only nine particles per ring were noted. In general, the interparticle spacing also decreased as the pH increased. Interestingly, the incorporation of a Au nanoparticle at the central pore of the ring was also pH sensitive, demonstrating increased incorporation under acidic conditions.[139] Such effects are probably due to changes in the charge state of the pore as a function of pH, which leads to diminished electrostatic interactions with the negatively charged Au nanoparticles.

Using the TMV nanodiscs, Ag nanorings have also been prepared.[140,141] For this approach, Ag^+ ions are photochemically reduced in the presence of the disc-based template. This results in reduction of the Ag^+ to form Ag^0 atoms, which grow to generate the Ag^0 nanoparticles on the surface of the biomolecular template. Using this approach, an average of 5–6 nanoparticles of Ag are typically bound to the template,[140] where the average particle size is ~7 nm in diameter.[140] The morphology and size distribution of the Ag materials is broad, compared to the Au nanodiscs discussed above, as these materials are fabricated directly on the template surface. In general, the number of Ag nanoparticles on the disc surface is controlled by the number of metal particle-binding sites on the template. To enhance this number, Pan *et al.* genetically incorporated an additional 17 cysteine residues, which are known to bind Ag.[140] From this modified TMV nanodisc, additional Ag^0 metal could be incorporated into the structures, resulting in the formation of continuous metallic nanorings.

Zhang *et al.* further expanded the use of the TMV nanodisc template through genetic manipulation of the protein structure.[142] For this, they incorporated cysteines at the interior pore and His-tags on the outer ring of the assembled template. This allowed for highly precise incorporation of size-selected Au nanoparticles based upon their known affinity for these residues. For the interior cavity, 3.5 nm Au nanoparticles could fit within the pore; however, larger particles (5 nm) were able to cap the pore surface.[142]

With incorporation of the His-tags at the outer edge, Au nanoparticles could be site selectively incorporated at these domains to prepare highly precise Au nanorings.[142] Based upon the size of the materials selected, a Au nanoparticle could be selectively positioned at the central pore domain to form a Au nanoring with a central Au nanoparticle.

Of particular note, Zhang *et al.* demonstrated that these genetically modified TMV nanodiscs could self-assemble to form hexagonally arrayed structures over micrometre scale distances.[142] This assembly was achieved through chelation of Cu^{2+} ions *via* the His-tags, which led to long-range assembly when crosslinking between two nanodiscs was achieved when bound to the Cu^{2+} ions. The complexity of the system changed based upon the incorporation of Au nanoparticles into the individual discs.[142] For instance, hexagonally arrayed Au nanoparticles could be assembled using nanodiscs with Au structures only at the central pore; however, closer packed hexagonal outlines of particles could be prepared using the templates with Au nanoparticles along the outer ring of the TMV discs. Further complexity was achieved with Au nanoparticles at both the inner pore and outer ring.[142] Such an approach is quite attractive to organise structures in two dimensions, demonstrating the remarkable capabilities of biotemplates for material production and assembly.

While the TMV template has demonstrated great opportunities for the fabrication of important inorganic materials, these structures display important properties for a variety of applications. For instance, many of the materials discussed above are composed of noble metal materials with vibrant plasmons. Coupling of these plasmons between the assembled nanoparticles is possible, providing attractive optical properties. To highlight other unique opportunities available *via* the TMV-templated materials, two specific applications are discussed below: catalytic reduction of toxic Cr^{6+} to Cr^{3+} and electrodes for sodium-ion batteries.

Two common oxidation states of Cr have very different toxicological properties; Cr^{6+} is highly toxic and carcinogenic, while Cr^{3+} is an essential nutrient.[143] To address removal of Cr^{6+} from the environment, Sadeghi *et al.* incorporated Pd nanoparticle-decorated TMV templates into a hydrogel-based membrane.[144] This allows for the catalytically active Pd nanoparticles to be exposed to the solution as it is passed through the hydrogel membrane, to drive Cr^{6+} reduction to Cr^{3+}. The poly(ethylene glycol) diacrylate (PEGDA) hydrogel was used on a polysulfone ultrafiltration membrane (PS35) with the TMV-templated Pd materials incorporated into the hydrogel. The reduction reaction was monitored through passing of a Cr^{6+} solution through the membrane, where the rate constant (k_{app}) varied as a function of the Pd concentration used to prepare the membranes. To this end, at the highest Pd concentration studied (10 mg mL^{-1}), a k_{app} of 0.18 min^{-1} was reported.[144] Note that, in these experiments, formic acid was used as the reductant. More importantly, these structures retained their reactivity for long durations (three days of continuous flow/reactivity), thus suggesting the possibility of their use for long-term environmental remediation applications.

In separate work by Liu *et al.*, the researchers used the TMV capsid for the production of multicomponent anodes for sodium-ion batteries.[145] This was achieved using a virus template modified to present cysteine residues at the 3′ terminus of the rod-like biostructure. This allows for the TMV templates to bind a stainless steel disc, resulting in their vertical alignment due to the binding motif. Once deposited, an ~50 nm thick Ni layer was incorporated onto the TMV template using electroless plating methods, followed by the deposition of a 20 nm Sn layer *via* physical vapour deposition. Finally, a thin layer of carbon (5 nm) was incorporated onto the surface of the multicomponent materials using radiofrequency magnetron sputtering.[145] Using these Sn nanorod forests, substantial activity for use in sodium-ion batteries was observed. For instance, these structures provided 722 mA h g^{-1} during initial cycles. Even after 150 cycles, the material retained 405 mA h g^{-1},[145] which is extremely high after a long number of cycles for sodium-ion batteries. Such capabilities are quite challenging to achieve, with the structures prepared providing unique opportunities based upon the biomolecular template to prepare the composite materials.

2.3 Synthesis of Nanoparticles by Plants and Microorganisms

Biology has mastered the creation and development of complex materials and structures in nature. To accomplish this feat, biological systems exert an unparalleled level of molecular and genetic control, employ chemical specificity, and utilise a large diversity of highly functional biological templates. Therefore, unlike the use of single biotemplates (amino acids, peptides, and proteins) discussed above, the ability to control and harness complex living biological systems (*e.g.*, plants and microorganisms) offers an alternative path to achieving more complex and advanced biofunctionalised materials with enhanced properties and precise nanostructures for sensing, catalysis, and biomedical applications. For example, the biomineralisation of gold nanoparticles by bacteria and the controlled biosynthesis of silica by the marine diatom have resulted in new green routes to the synthesis of functional nanomaterials.[146] In this section, we will describe the advantages of using complex biological systems (plants and microorganisms) for nanoparticle synthesis and highlight some of the key features that living biological systems offer. Also, through the use of different microorganisms, functionality has been successfully imparted to nanoparticle surfaces in the form of building blocks for assembly, molecular recognition, and/or biological activity.

While nature has perfected the synthesis of a variety of inorganic-based nanostructures, the formation of novel nanoparticles not found in nature is currently a challenge using biological systems, as emphasised above. This applies to exotic types of non-natural inorganic nanoparticles that involve multi-elemental compositions (*e.g.*, $BaTiO_3$) or rare earth elements. In this case, the need for dissimilar metal salt precursors, the inability of biological systems to manage complex stoichiometries and mixed-metal oxidation states, and the different metal toxicities have limited the use of living plants and microorganisms for the production of multi-elemental nanomaterials. Additional requirements, such as high temperatures and non-aqueous media, also have prevented biological systems from producing nanomaterials with increased complexities and multi-elemental compositions, given that the majority of biological systems cannot operate or survive in extreme conditions. More likely though is that the processing of multiple metal salts by living plants or microorganisms into a functional nanomaterial is simply beyond their genetically programmed biological functions. Although this concept is quickly becoming antiquated as the limits of biology are constantly being expanded through advancements in genetic engineering, synthetic biology, cell-free systems, non-canonical amino acid translation systems, and discovery of extremophilic organisms that can handle high temperatures and non-standard biological conditions (*i.e.*, high or low pH, anaerobic). Consequently, factories of microbial communities are being utilised to synthesise materials on demand using synthetic biology, to achieve precise control over properties.

2.3.1 Plant- and Plant Extract-mediated Nanoparticle Synthesis

Plants are vital to human survival and for creating a functional and sustainable ecosystem; but, at the same time, they represent an abundant, diverse, accessible, and renewable resource of biologically active components/materials for nanoparticle remediation, synthesis, and biotransformation. In total, there are approximately 400 000 different terrestrial and marine plant species in the world, which contribute 560 billion tons of global biomass.[147,148] As a result, the great abundance, accessibility, and diversity of plants and plant extracts make them one of the most widely utilised and studied materials for processing and/or synthesising a variety of inorganic nanoparticles. In this section, we have judicially selected an assortment of different examples to highlight the role of unique plants and bioactive components in nanoparticle synthesis, given the exponential abundance of published articles. For processing and synthesis of nanoparticles, plants offer a unique composition of highly functional molecules and biological components that, in most cases, are only found in the plant kingdom. Compositionally, plants contain numerous metal-chelating peptides and proteins, complex photosystems, chlorophyll, amino acids, and vitamins.[149] These components primarily function in light harvesting, energy production, and metal sequestration in response to exposure and/or stress of toxic metal ions. In addition, plants also contain numerous important secondary metabolites, which include polysaccharides, polyphenols, terpenoids, alkaloids, flavonoids, and organic acids (tannic acid and gallic acid).[150] Functionally, these metabolites impart unique flavours and fragrances characteristic to plants, which serve as attractants or deterrents for insects and animals. Plants are excellent at remediating heavy metals due to an extensive network of porous root systems and abundance of metal-chelating peptides and proteins. For example, living plants grown in the presence of Cd^{2+} were shown to accumulate and precipitate chemically inert fluorescent CdS nanoparticles through the use of cysteine-rich metal-binding phytochelatin peptides ($(\gamma EC)_nG$);[151] while, live alfalfa plants (*Medicago sativa*) exposed to high concentrations of $K[AuCl_4]$ were the first plant material shown to absorb Au^{3+} and produce high-quality Au^0 nanoparticles internally.[152] Consequently, plants have been used as sponges to absorb toxic heavy metals for environmental cleanup of toxic metal contamination.

2.3.1.1 Plant Extracts for Nanoparticle Synthesis

Plants and plant extracts have been predominantly used for the synthesis of monometallic nanoparticles compared to multicompositional nanoparticles, as mentioned above (see Figure 2.12).[149] As a result, the synthesis of pure metal nanoparticles is generally preferred and chosen as a model nanoparticle reaction in order to evaluate the performance, reduction potential capacity, and properties of unknown biological systems. This is ultimately due to the well-characterised nature of metal nanoparticles (Au, Ag, Pd, Pt, Cu, Rh) and their formation, the straightforwardness and simplicity of the reaction, the requirement of only a single metal salt, and the reliability of the measurable, quantifiable, and characteristic properties of metallic nanoparticles (*e.g.*, size, shape, crystallinity, optical, catalytic, and electronic properties). More importantly, the reduction of metal salt precursors is well established and readily achieved by the use of biomolecules (*e.g.*, sugars, polyphenols, amino acids, ascorbic acid), inorganic reducing agents (*e.g.*, borohydride, hydrazine, citrate, sodium hydroxide, hydroquinone), and in some cases under light irradiation (*e.g.*, photoreduction of Ag^+ to Ag^0), as described in Section 2.2. In total, this section broadly covers reducing agents that range in strength from mild to strong. However, this large set of reducing agents makes it difficult to identify and confirm the redox-active molecules/components responsible in plant-mediated

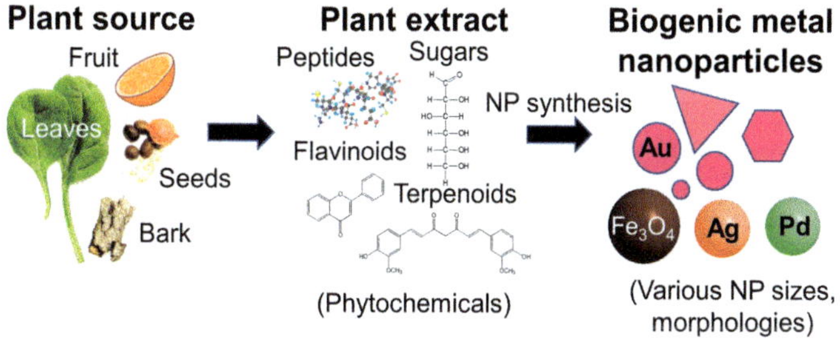

Figure 2.12 Plant-mediated nanoparticle synthesis.

reduction of metal salts, as well as difficult to determine any additional contributions from co-reducing agents present. In general, plants contain multiple reducing agents along with a variety of redox-active biomolecules/metabolites, which probably contribute equally to reduction of metal salts.

The plant-mediated synthesis of gold and silver nanoparticles has involved the use of many different species of plants and/or extracts. Viable sources of plant extracts have been obtained and isolated from orange and banana peel, onion, tea leaves (black and green), potato, woody plant bark, and plant seeds. In all, this has resulted in the favourable formation of gold and silver nanoparticles bearing a broad range of sizes, morphologies, and properties (see Table 2.2). Plant extracts are generally favoured over living plants for nanoparticle synthesis due to challenges associated with cultivation, maintaining consistent growing conditions, supplying an appropriate source of nutrients and supplements (light, water, phosphates, nitrates) for plant growth, and, more importantly, difficulties associated with extraction and purification of synthesised nanoparticles from living plants. To address this latter point in more detail, given its significance, nanoparticles synthesised by living plants are localised internally and can be embedded in plant cell walls, chloroplasts, and be associated with other internal structures. Another significant aspect that makes extraction and purification of nanoparticles

Table 2.2 Plant-mediated synthesis of nanoparticles.

Plant	Extract source	Nanoparticle type	Nanoparticle size/nm	Bioreducing agents	Reference
Coffea arabica	Bean	Au	NA	Alkaloids	155
Murraya koeniggi	Leaf	Au	NA	Quercetin	159
Green tea	Leaf	Au	Up to 84 nm	ND	154
Black tea	Leaf	Au	NA	Alkaloids	153
Zimbro tea	Leaf	Au	NA	ND	154
Cinnamon extract	Bark	Pd	15–20 nm	Terpenoids	163
Boswellia ovalifoliolata	Stem bark	Ag	30–40 nm	ND	164
Aegle marmelos	Leaf	Ag and Au	13, 23 nm	Acacia	156
Navel orange extract	Peel	Ag	7.4 nm	Citronellal	165
Apples	Fruit + skin	Ag	30 nm	ND	166
Cumin	Seeds	Au	5–400 nm	ND	168
Fenugreek	Seeds	Se	50–150 nm	Phenols	169
Cacao	Beans	Ag	35–42 nm	Oxalic acid	171
Nigella sativa	Seeds	ZnO	~24 nm	ND	172
Seaweed	Macroalgae	Au	6–10 nm	Sulfonated polysaccharides	161
Onion	Bulb	Ag and Au	19–37 nm Ag 30–113 nm Au	Sugars, quercetin	167
Potato	Buds	Fe_3O_4	~40 nm	Starch	157

challenging is the low mass percentage of nanoparticle products (microgram quantities) produced by plants compared to the excess amount of biomass present. This essentially corresponds to the analogy of finding a few needles in a haystack. By comparison, plant extracts are easily prepared and obtained from virtually any plant source by boiling dried leaves or dried fruits/vegetables in water for a determined amount of time. The advantages of extracts include cost, space, and time savings of not having to wait weeks or months for a plant to fully mature, simplicity of preparation, robustness, and ease of separating nanoparticles from extracts as opposed to separating them from inside a living plant, as mentioned above. Also, the functionality and activity of biological components present within extracts are comparable to living plants, but are obtained in a much more practical, simple, and concentrated form, which is conducive to nanoparticle synthesis. In total, plant extracts serve as a good analogue to living plants and provide ample benefits. Thus, in the following examples, we will mainly describe the use of plant extracts unless noted otherwise. For example, tea extracts rich in phytochemicals from black, green, or zimbro tea leaves have been used to rapidly synthesise gold nanoparticles at room temperature from the bioreduction of Au^{3+} salt precursors ($HAuCl_4$).[153,154] Experimentally, $HAuCl_4$ is added to different concentrations of tea solutions (1, 5, and 10 wt%) to produce water-stable Au nanoparticles of various sizes, shapes, and morphologies with distinct colours and plasmon resonances. In this case, the polyphenols and phytochemicals present in tea extracts are multifunctional and contribute to all aspects of Au nanoparticle synthesis (*i.e.*, the reduction of Au^{3+} to Au^0 and passivation of the gold nanoparticle surfaces). However, comparison of the different varieties of tea extracts revealed that green tea extracts produced multifaceted gold nanoparticles with the largest sizes (average size of 84 nm). As a result, the different types of phytochemicals present in tea leaf extracts have a significant impact on the final gold nanoparticle properties. Circumstantially, these differences could presumably be attributed to the presence or absence of alkaloids (*i.e.*, caffeine) in black tea- *vs.* green tea-mediated nanoparticle synthesis. Similarly, coffee extract from *Coffea arabica* beans also provides a naturally concentrated source of alkaloids, similar to black tea leaves. In this example, coffee extract was used to synthesise gold nanoparticles following a factorial guided approach.[155] This factorial approach was designed to examine the effects of time, pH, temperature, extract concentration, and agitation on gold nanoparticle properties and optimise synthesis conditions. Of these factors, temperature showed the largest influence on properties statistically. Similarly, the Taguchi method (factorial design method) was employed to optimise pH, reaction temperature, colour of light illumination, and metal ion precursor concentration for the synthesis of Au and Ag nanoparticles using *Aegle marmelos* leaf extract and addition of different sources of plant-based surfactants (*i.e.*, saponins).[156] Statistical analysis of the parameter set revealed that pH and temperature had the largest impact on Ag nanoparticle size, while temperature and the presence of a co-plant-derived surfactant had the greatest influence on Au properties. In this case, the addition of plant-derived surfactants (Acacia or Reetha) with *Aegle marmelos* extract helped to reduce final nanoparticle size (Ag: ~13 nm and Au: ~23 nm) and to control shape (*i.e.*, Au nanoplates or Au pyramids). In both statistical design approaches using *Coffea arabica* or *Aegle marmelos*, the design of experiments enabled the identification of variables over a specific range of synthesis conditions, to yield the fewest number of experiments with each portraying the largest amount of information.

In a similar example, potato extracts rich in starch were used to synthesise and stabilise magnetic nanoparticles of Fe_3O_4 with high quality.[157] Briefly, extracts were first prepared by boiling sun-dried fresh potatoes in water. Next, ferrous sulfate was added to the potato extract at pH 8 to form water-stable Fe_3O_4 nanoparticles. The resulting nanoparticles were superparamagnetic, face-centred cubic, ~40 nm in size by TEM, and exhibited catalytic activity, as shown by the degradation of methylene blue at room temperature.[157] Mechanistically, the starch component in the potato extract functioned in a dual capacity to both reduce Fe^{3+} precursors and stabilise the surface of growing Fe_3O_4 nanoparticles (see Figure 2.13). Compositionally, starch is made up of

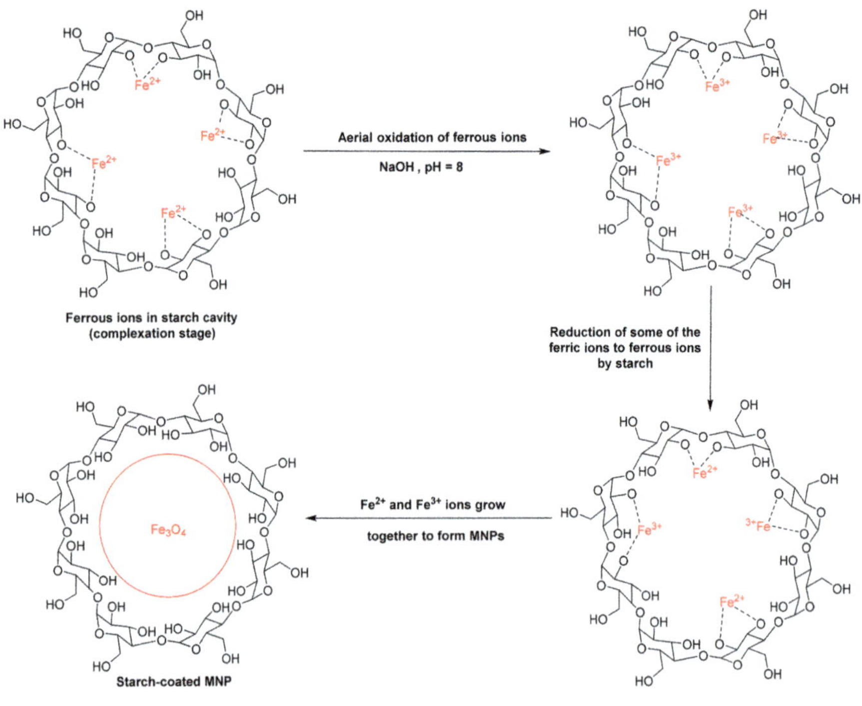

Figure 2.13 Synthesis of magnetic nanoparticles using starch. Potato extract rich in starch is complexed with Fe^{2+} ions, oxidised by air at pH 8 to Fe^{3+}, partially reduced back to Fe^{2+} by starch molecule, and then grown within the starch cavity to form starch-coated magnetic nanoparticles. Reproduced from ref. 158 with permission from American Chemical Society, Copyright 2019.

a complex mixture of polyhydroxylated macromolecules (~20% amylose and ~80% amylopectin) that form cavities for Fe^{2+} complexation with hydroxyl groups and templates for nanoparticle growth.[158] In all, the simplicity of these bio-mediated processes and instant appearance of a colour change due to formation of optically active gold nanoparticles or magnetic iron oxide nanoparticles has formed the basis for undergraduate laboratory experiments (tea extracts and potato extracts) and is currently used to demonstrate the importance and benefits of biological approaches *vs.* traditional material science methods.[158]

Additional examples include the use of *Murraya koenigii* Spreng. leaves, onion extracts, medicinal plants and herbs, and bark from woody vascular trees and plants for the synthesis of gold nanoparticles. Notably, the *Murraya koenigii* Spreng. plant is indigenous to India, Sri Lanka, and Bangladesh and, historically, has provided an important source of curry leaves for dried spices, antioxidants, essential oils, and medicines.[159] Besides being a main ingredient in Asian cuisine and a highly valued commodity, the aqueous extract of *Murraya koenigii* Spreng. leaves was successfully used to produce a variety of gold nanoparticles consisting of multiple sizes, shapes, and morphologies.[160] Moreover, these nanoparticle properties were shown to be highly tuneable by simply varying the reaction conditions (*i.e.*, concentration of curry extract and gold salt precursor, pH, and temperature). Of these parameters, the resultant gold nanoparticle properties were most affected by the concentration of bioactive components in the extract used during gold salt reduction and synthesis. For instance, higher concentrations of extract yielded more gold nanoparticles with better homogeneous sizes and shapes. Analysis of the extract revealed that curry contained high amounts of polyhydroxy flavonoids (81.9 mg g^{-1} of extract) and glycosides (39.98 mg g^{-1} of extract), which are well known to effectively reduce Au salts and form stable gold nanoparticles. Consequently, the pure forms of these compounds were individually tested and confirmed to exhibit gold-reducing activity, and, at the same time, play a significant role in the synthesis and stabilisation of gold particles. By comparison, the use of these pure compounds (quercetin and quercetin-3-glucoside) produced higher quality gold nanoparticles with better size and shape distributions compared to using crude curry extracts that contained complex mixtures of bioactive components. Seaweed represents another important example of an abundant plant that provides a diverse source of bioactive constituents for nanoparticle synthesis. Naturally, it contains sterols, vinyl and ethyl cholesterol, cyclohexane, sulfonated polysaccharides, rhamnose, fucose, and uronic acids. For clarification, seaweed is a marine macroalgae that is phototrophic and is generally found in every ocean on the planet. Furthermore, seaweed can be subdivided into red, green, and brown algae depending on the type of photosynthetic pigments it contains. For gold nanoparticle synthesis, extracts from brown algae (*Turbinaria conoides*) prepared by boiling 5 g of biomass was added to 1 mM $HAuCl_4$.[161] After 1 h, gold nanoparticles were quickly formed, as indicated by the presence of a dark pink colour; and after 24 h, the formation of gold nanoparticles was completed. Seaweed-mediated synthesis produced crystalline and small-sized gold nanoparticles of 6–10 nm.

2.3.1.2 Nanoparticle Synthesis Using Extracts from Wood Bark

Based on plant morphology, the anatomy of plant leaves is ideally suited for extraction of biologically active molecules, discovery of new therapeutics, and synthesising nanoparticles, as described above; whereas the protective nature of bark from woody vascular trees and plants is largely considered an industrial waste product by lumber and construction companies, with little or no economic value besides being utilised as firewood.[162] The reasons for this preference result from the fact that plant leaves are composed of soft biological tissues, possess easy to lyse plant cell walls, and can be processed through simple extraction techniques by virtually anyone (*i.e.*, by boiling). Wood bark on the other hand is a much harder and more durable material that can require multiple processing and elaborate extraction steps depending on the source (*e.g.*, mechanical grinding, Soxhlet extraction, acid hydrolysis, and/or boiling). However, wood bark represents an alternative, inexpensive, and abundant material that contains an assortment of unlimited bioactive components that are not typically found in the soft tissues of plant leaves. These include tannins, aldehydes and ketones, aromatics, and azo- and nitro-containing compounds.[162] Unfortunately, this processing aspect, combined with fewer sources of bark, has undoubtedly limited their use in nanoparticle synthesis and produced fewer reports of nanoparticle products compared to the much simpler extraction of plant leaves and overwhelming accounts of leaf extract-synthesised nanoparticles. In general, bark extract or bark powder has been harvested from multiple sources and used for nanoparticle synthesis. These sources have included coconut husks, cinnamon bark extract from *Cinnamon zeylanicum*,[163] and stem bark from *Boswellia ovalifoliolata* and *Shorea tumbuggaia*.[164] In particular, the stem bark from *Boswellia ovalifoliolata* and *Shorea tumbuggaia* plants was used for the synthesis of antimicrobial silver nanoparticles. Notably, these plants are highly treasured and endangered medicinal plants used in the daily rituals of many tribal communities in India to treat and cure diseases (*i.e.*, rheumatism, duodenal ulcers, earache). Nevertheless, stem bark extract from both species was capable of rapidly producing silver nanoparticles with sizes ranging from 30 to 40 nm within 10–15 minutes of addition of silver ions.[164] However, the contents of these crude extracts have not yet been fully characterised, but have been shown to contain bioactive agents capable of reducing silver ions. Citrus fruit peel represents another under-utilised plant material and/or waste product that is known to provide some health benefits (*e.g.*, lower cholesterol, improve digestion), contains high amounts of ascorbic acid (vitamin C), and offers a fragrant scent (citronellal). Similarly, citrus fruit peel also functions as a tough protective barrier for the edible parts of fruit, but, overall, is softer and easier to extract than wood bark. As a result, aqueous citrus peel extracts were obtained from navel oranges, ruby red grapefruit, Minneola tangelo, lemon, and lime peels and tested for their ability to form silver nanoparticles using microwave-assisted synthesis.[165] In the case of silver nanoparticle synthesis, microwave heating was used to rapidly heat reactions (citrus peel extracts plus Ag^+ ions) to high temperatures in a pressure-controlled environment. Beneficially, microwave heating produces nanoparticles with narrower size distributions and higher crystallinity compared to the use of traditional heating for nanoparticle synthesis, due to the more controlled and uniform heating. From the set of citrus peel extracts, the microwave-assisted synthesis of stable silver nanoparticles was only achieved successfully using orange peel extracts (see Figure 2.14). Using orange peel extract, Ag^+ ions were reduced by the fragrant compound of citronellal present in the extract to form ~7.4 nm silver nanoparticles after 15 min of microwave heating at 90 °C and 15 psi.[165] Also, extracts from apples (*Malus pumila*) were used for the synthesis of silver nanoparticles. Apple extracts were prepared by heating small pieces of apple, including the skin, in water for 1 h at 80 °C and incubating with Ag^+ ions. After 60 min, the apple extract effectively reduced Ag^+ to form silver nanoparticles with average sizes of 30.25 + 5.26 nm.[166] Similarly, onion extract was used for the synthesis of silver and gold nanoparticles. In this example, quercetin and sugars present in onion extract were probably involved in mediating the reduction of Ag^+ or Au^{3+} salts to form nanoparticles.[167] Nanoparticles were produced with sizes of 19–37 nm for silver and 30–113 nm for gold. Notably, onion extract-synthesised gold nanoparticles were used to prime new onion seeds for germination and plant seedling growth. As a result, onion seeds treated with gold nanoparticles showed significant enhancement in seed germination, an increase in seedling emergence percentage (63.2% for gold-primed seeds *vs.* 37.4% for unprimed control seeds), and an increased yield of 23.9% compared to unprimed seeds.

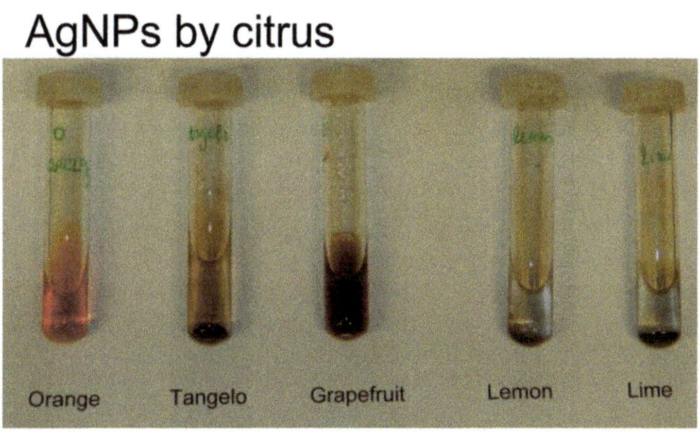

Figure 2.14 Synthesis of silver nanoparticles using citrus peel extract. Reproduced from ref. 165 with permission from American Chemical Society, Copyright 2013.

2.3.1.3 Plant Seed Extracts for Synthesis

Plant seeds are also capable of serving as an alternative source for obtaining extracts for nanoparticle synthesis. In terms of material hardness, plant seeds are encased in a hard durable coat that enables long-term dry storage and provides protection for plant embryos (*i.e.*, against digestive enzymes if consumed and environmental factors). Favourably, seeds contain a composition of bioactive components that is representative of the entire plant and/or individual parts, such as leaves and stems, albeit at a lower percentage biomass and packaged in a much smaller, compact form. Consequently, these qualities are well suited to nanoparticle synthesis as was demonstrated by the synthesis of gold nanoparticles using cumin seed powder extract.[168] Cumin seed extract was prepared by grinding seeds into a fine powder, dispersing in water, and boiling for 2 min. After extraction, a solution of 1 mM $HAuCl_4$ was added to cumin extract and incubated for 24 h. In the presence of extract, gold nanoparticles were synthesised with sizes ranging from 5 to 400 nm, platelet or spherical nanoparticle morphologies, and surface plasmon resonance bands at 540 nm, or shifted to 800–1200 nm depending on pH and temperature during the nanoparticle synthesis. Another example of seed extract-mediated synthesis involves the synthesis of selenium nanoparticles using fenugreek seed extract.[169] Fenugreek seeds are commonly used as herbs and spices and were found to contain high amounts of polysaccharide galactomannan, saponins, oils, alkaloids (choline and trigonelline), flavonoids, amino acids, and proteins. Notably, selenium nanoparticles (SeNPs) are highly desired for their unique therapeutic properties and would greatly benefit from plant-mediated synthesis approaches, as well as gold and silver nanoparticles. SeNPs are known to exhibit anticancer activity by induction of apoptosis in tumours, inhibit bacterial growth *via* antimicrobial activity, and impart bioprotective properties in the form of antioxidants.[170] For the synthesis of selenium nanoparticles, a 1 wt% fenugreek seed extract was mixed with selenous acid (H_2SeO_3) and ascorbic acid. Ascorbic acid was used as a co-reductant to initiate reduction of Se^{4+} with fenugreek seed extract to form Se^0 nanoparticles. The biosynthesis of Se nanoparticles was confirmed by the appearance of a ruby red colour, absorption maximum at 200–400 nm, and the presence of spherical nanoparticles with sizes of 50–150 nm. Also, the fenugreek seed extract produced selenium nanoparticles that exhibited cytotoxicity against human breast cancer cells. Another example of seed-mediated synthesis involves the use of cacao beans for the formation of 'chocolate'-derived silver nanoparticles.[171] Natural cacao beans are native to South America and Africa and used to make cacao powder, chocolate, and cocoa. At the same time, cacao beans provide numerous health benefits (*e.g.*, lower blood pressure, anti-inflammatory effects, improved brain function) by containing high amounts of polyphenols, flavonoids, and oxalic acid. In all, constituents of cacao are known to contribute to nanoparticle synthesis by acting as reducing and stabilising agents, plus it makes for a nice sweet treat. Consequently, cacao powder extract was added with $AgNO_3$ solution and incubated at room temperature or at 100 °C for 30 minutes to promote nanoparticle synthesis. During the course of incubation, the solution colour changed from light yellow to dark brown indicating the formation of silver nanoparticles. The resulting silver nanoparticles were spherical in morphology, 35–42 nm in size, and showed antibacterial properties against *Staphylococcus aureus*, *S. epidermidis*, *E. coli*, and *Pseudomonas aeruginosa*. Additionally, seed extracts of *Nigella sativa* (used for food spices, additives, preservatives, diuretics, anti-cancer and anti-inflammatory agents) were used for the synthesis of ZnO nanostructures.[172] Notably, ZnO nanoparticles are commercially used as additives in paints, cosmetics (*e.g.*, sunscreen), plastics and rubbers, and as broad-spectrum antimicrobial materials. Consequently, ZnO nanoparticles were synthesised by adding $Zn(NO_3)_2$ to *Nigella sativa* seed extract. In this case, the seed extract was reported to reduce Zn^{2+} ions to Zn metal. However, after seed extract-mediated reduction, the Zn metal was re-oxidised to form ZnO nanoparticles. By comparison, ZnO can be synthesised using a non-biological approach at high pH by adding KOH to $Zn(NO_3)_2$. At high pH, the precipitation of ZnO occurs under these conditions. ZnO nanoparticles formed by seed extracts showed a hexagonal wurtzite structure, sizes of ~24 nm, and inhibitory effects on quorum sensing in *Chromobacterium violaceum* and *P. aeruginosa*. Notably, these same particles were equally as effective at inhibiting biofilm formation of four common food pathogens.

To date, the plant-mediated synthesis of metallic nanoparticles has been heavily focused on the production of Au and Ag nanoparticles from the reduction of metal ion precursors ($HAuCl_4$ and $AgNO_3$) for reasons discussed above; whereas, reports of the plant-mediated formation of platinum, palladium, and alternative compositions of nanoparticles, such as metal oxides, are much more limited in scope. This is probably due to the inability of plants to actively reduce metals with low reduction potentials.[173] However, in one example, cinnamon bark extract was used for the bioreduction of Pd^{2+} ions and biosynthesis of palladium nanoparticles.[163] Initially, cinnamon bark extract was prepared by grinding sun-dried bark into a uniform powder, suspending in water, and boiling for 1 minute. After extraction of cinnamon bark, $PdCl_2$ was added to cinnamon extract and incubated at 30 °C for 72 h in order to reduce Pd^{2+} ions and form palladium nanoparticles. To achieve bioreduction, the presence of terpenoids in the cinnamon bark extract probably participated in the reduction of Pd^{2+} to Pd^0. The cinnamon extract-synthesised palladium nanoparticles were well dispersed, crystalline, and 15–20 nm in size by TEM.

2.3.2 Microbial-induced Nanoparticle Synthesis

Microorganisms play a critical role in maintaining, balancing, and regulating the Earth's diverse geochemistry. It is estimated there are 5×10^{30} microorganisms on Earth. In nature, microorganisms efficiently perform biotransformation reactions, metal remediation and speciation, pseudo-biomineralisation, production of metal-containing pigments, and precipitation and/or dissolution of minerals. In the latter, bacteria employ highly conserved redox processes in order to utilise metal ions available in the environment as an energy source for metabolism. As a result, the bacterial reduction of metal ions facilitates the precipitation of metal deposits, minerals, and nanoparticles. Similar to plants, microorganisms contain an equally high number of potential reducing agents amenable to the reduction of metal salts, as well as a variety of coloured pigments, such as melanin. Microbial pigmentation provides protection for microorganisms against oxidants, extreme temperatures, UV light, and heavy metals.[174] Additionally, pigmentation plays an important role in virulence of microorganisms and is believed to serve as a warning signal. In all, these microbial capabilities are perhaps the most useful for metal nanoparticle synthesis, as has been described above for plant extracts. Consequently, the ability to exploit and utilise these microbially controlled biochemical processes and complex systems for the *in vitro* production of nanoparticles (gold, silver, selenium, and silica) is highly appealing for green chemistry. As a result, the microbial-mediated synthesis of nanoparticles has been demonstrated using a large collection of biological microorganisms (*i.e.*, bacteria and fungi), but, by comparison, falls short of that shown for plant-mediated synthesis (see Figure 2.15). Additionally, the use of cell-free extracts has recently gained popularity and been applied to nanoparticle synthesis as an alternative to using living organisms, in much the same way as for plant extracts and for the same reasons. In this section, we will describe examples of microbial-mediated nanoparticle synthesis, compare nanoparticle properties from different organisms, and highlight the benefits of using cells *vs.* cell-free extracts.

In one example, the Gram-negative bacterial strain of *Acinetobacter* sp. SW30 was isolated from activated sewage sludge and shown to synthesise selenium nanoparticles.[175] Many different forms of *Acinetobacter* are ubiquitous in nature, are capable of forming biofilms, and display a high degree of antibiotic and metal resistance. In particular, resistance to heavy metal toxicity is an important property of live cells and ultimately determines

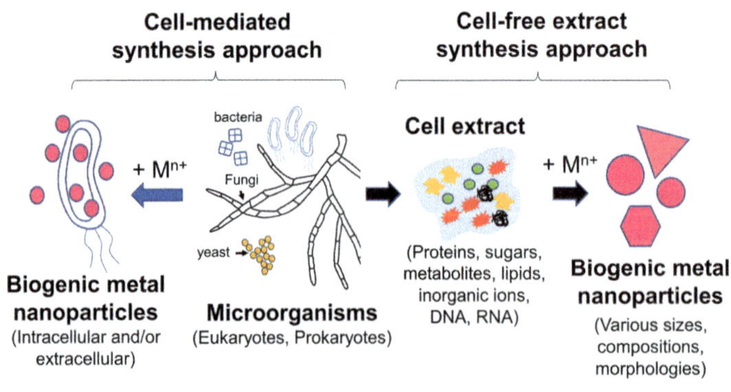

Figure 2.15 Microbial-mediated synthesis of nanoparticles.

their role and efficacy in nanoparticle synthesis. However, only a few cells are naturally resistant to metals by containing relevant genes, as exemplified by *Acinetobacter* and *Delftia acidovorans* (gold-associated bacterium), but, to some degree, most types of bacteria are capable of evolving metal resistance under increasing selection pressures. For example, certain strains of bacteria (*i.e. E. coli*) can be driven to evolve genes that confer resistance to silver under repeated exposure to silver nanoparticles over multiple generations, thereby rendering silver nanoparticles ineffective at killing these bacteria.[176] On the other hand, heavy metal toxicity has limited the use of live microorganisms and prompted the increased use of cell extracts for nanoparticle synthesis. As a result, the use of cell extracts is generally preferred for nanoparticle synthesis over the use of live cells due to metal toxicity and simplicity. However, as mentioned above, *Acinetobacter* exhibits a natural resistance to metals and, in particular, is tolerant to selenium. For microbial synthesis of selenium, *Acinetobacter* sp. SW30 was cultured until cells reached the stationary growth phase (2.7×10^9 CFU mL^{-1}), harvested by centrifugation, treated with sodium selenite (Na_2SeO_3) in sterile distilled water, and incubated at 30 °C to induce synthesis of selenium nanoparticles. After 24 h of incubation with Se, *Acinetobacter* sp. SW30 cells successfully reduced Se^{4+} ions to Se^0 nanoparticles intracellularly. The formation of selenium nanoparticles was observed by a colour change to bright orange and the presence of intense absorption peaks at 300 nm and 540 nm. In total, the cells produced amorphous selenium nanospheres that were 78 nm in size when exposed to 1.5 mM Na_2SeO_3 and crystalline selenium nanorods at concentrations of 2.0 mM Na_2SeO_3 and higher. In parallel, protein-only cell extracts of *Acinetobacter* sp. sW30 were compared against use of live cultures for selenium nanoparticle synthesis. Using a total cell protein content of 4 mg mL^{-1}, polygonal-shaped selenium nanoparticles were formed with sizes of 79 nm. This parallel study confirmed similarities of the selenium in addition to implications of heavy metal toxicity on cells, as mentioned above. Similarly, the silver-resistant bacterial strain of *Pseudomonas stutzeri* AG259 was one of the first organisms shown to accumulate large quantities of silver intracellularly upon exposure to silver salts.[177] After accumulation, silver became deposited as single crystals with distinct shapes (equilateral triangles and hexagons) and sizes (<200 nm) in the periplasmic space of *P. stutzeri*.

2.3.2.1 Extracellular Synthesis of Nanoparticles

As discussed above, a significant disadvantage of using live cells arises from the difficulty and challenges associated with extracting and retrieving synthesised nanoparticles from inside bacterial cells (*e.g.*, *Acinetobacter*, *P. stutzeri*). For example, extraction and isolation of intracellularly synthesised nanoparticles for application-based uses (*e.g.*, therapeutics, sensors) requires chemical (*e.g.*, lysozyme) and/or mechanical disruption (*e.g.*, sonication) of cell membranes and presents similar challenges to those from the use of living plants. An exception to this is the extracellular synthesis of nanoparticles by certain living microorganisms. In this case, bacteria secrete secondary metabolites, enzymes, and/or small signalling molecules that function in the extracellular synthesis of nanoparticles. This is exemplified by the synthesis of MgO nanoparticles using the cell filtrate of the endobacterium of *Burkholderia rinojensis*.[178] For synthesis, magnesium nitrate was added to bacterial cell cultures, incubated at 40 °C for 20 minutes, and then at room temperature for 10 h to induce precipitation of MgO nanoparticles. The precipitated MgO nanoparticles showed a cubic lattice structure, average sizes of 26.7 nm, and a zeta potential of −32.1 mV, indicative of a MgO surface stabilised by negatively charged biomolecules. Additionally, the bacterial-synthesised MgO nanoparticles showed good antifungal activity against the plant pathogen of *Fusarium oxysporum*. In a similar example, the common gold-associated bacterium of *Delftia acidovorans* was shown to biomineralise gold nanoparticles using highly specialised metallophores in nature.[179] For many years, it has been well known that bacterial communities reside within gold deposits, are linked to the production of bacterioform gold, and play a significant role in the accumulation and deposition of gold; however, until recently, the metabolic processes involved in gold biomineralisation by such organisms were relatively unknown. Population analysis of gold-resident bacteria revealed 90% of the community was composed of *Cupriavidus metallidurans* and *Delftia acidovorans*. However, only *D. acidovorans* was shown to produce darkened zones of gold precipitate around colonies when plated on agar and flooded with high concentrations of Au^{3+}. This result confirms that the *D. acidovorans* capability to form nanoparticles relies on an extracellular mechanism for the production of gold, whereas, by comparison, *C. metallidurans* produces gold intracellularly through bioaccumulation in its cytoplasm. Consequently, in these natural gold deposits, these bacteria are subjected to toxic concentrations of soluble Au^{3+} ions, but have adapted different detoxification mechanisms to deal with this gold toxicity. For reference, soluble Au^{3+} ions are toxic to bacteria at 10 μM, but become non-toxic when metallic Au^0 is precipitated. Accordingly, *D. acidovorans* showed secretion of a diffusible non-ribosomal peptide (delftibactin) that is capable of producing solid gold from Au^{3+} ions. Structurally, delftibactin is a linear polyketide metabolite that contains a single metal-binding site for gold (see Figure 2.16). The binding site is composed of an N^{δ}-hydroxy-N^{δ}-formylornithine and a cyclic N^{δ}-hydroxyornithine group and was confirmed by gallium NMR spectroscopy (see Figure 2.16). Notably, when delftibactin is depleted in the presence of gold, *D. acidovorans* produces more delftibactin peptide using a positive feedback mechanism to reach homeostasis. Mechanistically, delftibactin binds soluble Au^{3+} ions to form gold–peptide complexes, reduces complexed Au^{3+} ions through oxidative decarboxylation, and co-precipitates with solid gold metal deposits. By comparison to other gold templating and/or reducing molecules, delftibactin produces gold metal at room temperature, neutral pH, and at significantly faster rates than those of traditional agents (*i.e.*, citrate). Consequently, addition of Au^{3+} to a synthetic delftibactin

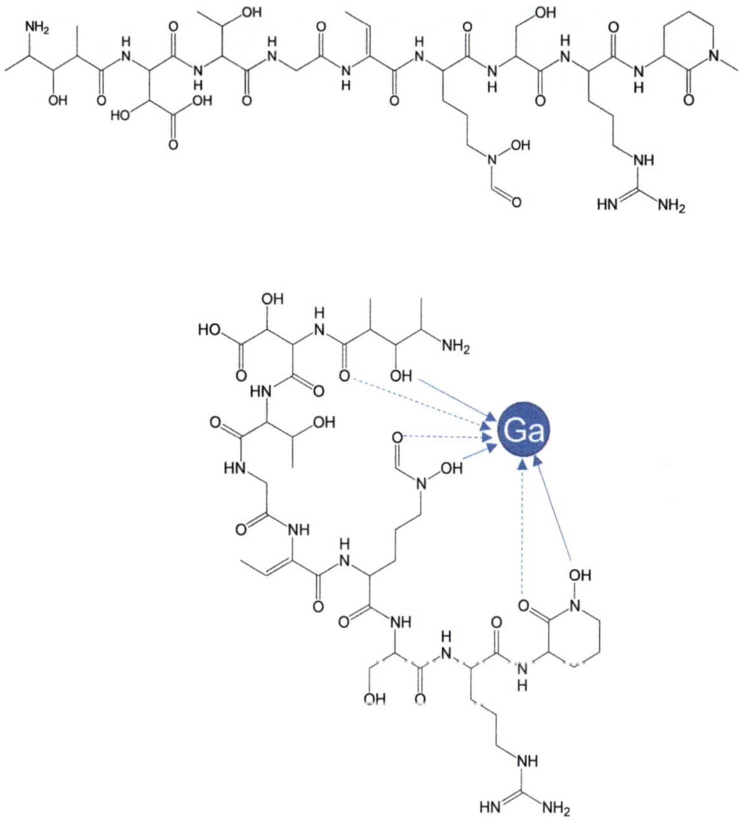

Figure 2.16 (Top) Chemical structure of polyketide non-ribosomal peptide delftibactin. (Bottom) Coordination complex of delftibactin-gallium showing single metal-binding site determined by gallium NMR spectroscopy.

analogue formed a delftibactin–gold coordination complex in a 2 : 1 stoichiometry, and, after 10 minutes, generated gold nanoparticles. Delftibactin produced gold nanoparticles with spherical and octahedral morphologies reminiscent of natural deposits of gold (see Figure 2.16).

2.3.2.2 Nanoparticle Synthesis Using Extremophilic Organisms

Extremophiles represent a unique class of Gram-positive bacteria for nanoparticle synthesis because of their tolerance for living in high temperatures and acidic conditions, and the ability to confer radiation resistance in some organisms (*e.g.*, *Deinococcus radiodurans*). This tolerance is achieved in part by possessing thermally stable proteins (γ-prefoldins) and cellular machinery, while also being able to adapt to harsh environmental conditions by utilising alternative metabolisms (anaerobic) for harvesting energy, which take advantage of available resources. Consequently, the ability to tolerate environmental stresses, such as high radiation exposure, heavy metal exposure, and high pH, provides an additional handle to tune nanoparticle synthesis and properties, by offering the ability to impose similar types of stresses on cells during synthesis. This concept was applied using the Gram-positive bacteria *D. radiodurans* for the cell-free synthesis of silver and silver/gold bimetallic nanoparticles.[180] Briefly, the approach developed for nanoparticle synthesis and production involved culturing *D. radiodurans* cells to stationary phase, adding environmental stressors (Ag^+ or Au^{3+} salts), incubating metal stressors with culture for 2 hours, isolating supernatant from metal-stressed cells (cell-free supernatant), adding metal target to cell-free supernatant (5 mM $AgNO_3$ or 1 mM $HAuCl_4$), and incubating for 48 hours to promote formation of nanoparticles (see Figure 2.17). In this approach, exposure of *D. radiodurans* cells to a metal stressor induces a defensive stress response mechanism, activates protein and gene expression, and secretes biologically active metabolites and/or enzymes in the supernatant for nanoparticle synthesis in order to reduce the concentration of toxic metals. By comparison, the use of *D. radiodurans* cells alone, instead of supernatant from metal-stressed cells, resulted in twofold lower nanoparticle yield, thus confirming the importance and value of imposing stress on the cells before nanoparticle synthesis. For example, when cells were stressed with Ag^+ ions, addition of Ag^+ target to cell-free supernatant led to the reduction and formation of Ag nanoparticles with sizes that ranged from 25 to 56 nm. Alternatively, addition of Au^{3+} target to silver-stressed cell-free supernatant led to the formation of Ag–Au bimetallic nanoparticles.

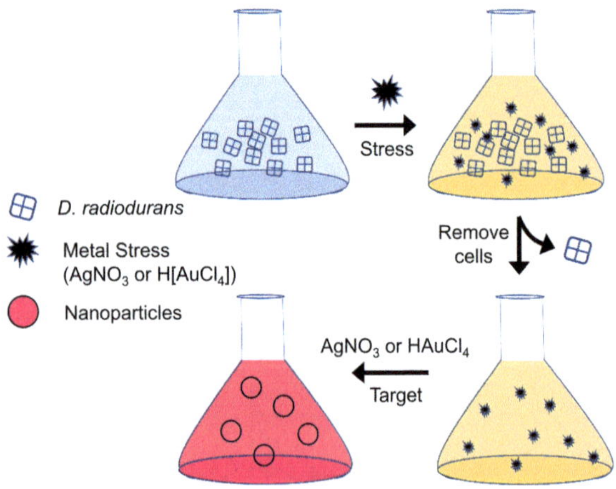

Figure 2.17 Nanoparticle synthesis scheme using *D. radiodurans* in the presence of metal stressors. Cells are cultured to stationary phase and then stressed with metal ions by addition of $AgNO_3$ or $HAuCl_4$ for 2 hours. After 2 hours, cells are removed from culture in order to isolate cell-free supernatant, which has been stressed by metals. Metal target ($AgNO_3$ or $HAuCl_4$) is added to cell-free supernatant and incubated for 48 hours. After 48 hours, Ag, Au, or Ag–Au bimetallic nanoparticles are formed by cell-free supernatant.

In this system, bimetallic nanoparticles were composed of alloys of 73% atomic gold and 27% atomic silver. However, the bimetallic composition could be tuned if cells were initially stressed with Au^{3+} and then the cell-free supernatant treated with Ag^{+} (28% atomic gold and 72% atomic silver). Mechanistically, bimetallic nanoparticles are probably formed by galvanic displacement reactions between the more noble metal (Au) and the less noble metal (Ag).

2.3.2.3 Nanoparticle Synthesis Using Fungi

Fungi account for a large and biologically diverse kingdom of microorganisms (includes yeast and mould) that are appealing for nanoparticle synthesis. Fungi possess unique features and functions that differ from bacteria. For instance, unlike bacteria, fungi are eukaryotic organisms that contain a nucleus, mitochondria and endoplasmic reticulum, vacuole, and exhibit higher order complexity, based on the hierarchal organisation of microorganisms. Additionally, fungi grow much more slowly than bacteria, with doubling times of ~90 minutes (*vs.* 20 minutes doubling time for bacteria), possess mechanically tough cell walls that are more difficult to lyse, and are much larger in size (0.5–5 μm for bacteria *vs.* 20–50 μm for yeast *vs.* ~100 μm for mould). Consequently, advantages of using fungi for synthesis of nanoparticles include the ability to grow on an industrial scale, cost effectiveness, and microbial diversity. This was demonstrated using multicellular organisms of filamentous fungi for the green synthesis of gold nanoparticles.[181] In this example, gold nanoparticles were synthesised using the extracellular fraction, autolysate, or intracellular extract from 29 different thermophilic filamentous fungi. As a result of fungal-mediated synthesis, gold nanoparticles were formed with sizes ranging from 6 nm to 40 nm and displaying varying degrees of quality. In all, nanoparticle sizes and quality were highly dependent on the type of strain, fungal fraction, and downstream processes (media, temperature). In a similar example, the yeast strain *Yarrowia lipolytica* was used for the synthesis of gold nanoparticles.[182] Notably, unlike other strains of yeast, *Yarrowia lipolytica* produces an exocellular pigment of pyomelanin in order to gain a survival advantage. Pyomelanin offers protection against oxidative stress, UV light, and other microorganisms. More importantly, however, it mediates energy transduction, promotes the chemical reduction of soluble and solid metal oxide materials, and mediates the synthesis of gold nanoparticles similarly to melanin. Pyomelanin is chemically and structurally similar to melanin, but, in contrast to melanin, is formed abiotically outside of the cell by L-tyrosine degradation, auto-oxidation of homogentisic acid into benzoquinoneacetic acid, and polymerisation into high molecular weight polymers of pyomelanin.[183] For reference, melanin is formed by tyrosinase-mediated oxidation of tyrosine or phenylalanine. Consequently, the extracellular formation of pyomelanin pigments by *Yarrowia lipolytica* offers a simple means for nanoparticle synthesis. This natural function of *Yarrowia* also eliminates the extra steps needed to obtain, purify, and prepare filtrates or extracts from microorganisms for nanoparticle synthesis. For example, pyomelanin was obtained and isolated from the culture supernatant of *Yarrowia* grown in YNBT medium for 120 hours. For gold nanoparticle synthesis, pyomelanin was added with $HAuCl_4$ at different concentrations, temperatures, and pH. Notably, the concentrations of pyomelanin and $HAuCl_4$ had the largest influence on nanoparticle size. When the gold salt concentration was high and pyomelanin concentration was low, large gold nanoparticles were formed (~104 nm). In contrast, small gold nanoparticles were obtained with high concentrations of pyomelanin and low gold salt concentration. This outcome is highly dependent on reagent concentrations and consistent with most other gold nanoparticle synthesis methods that involve the use of biomolecules and gold salts.

2.3.2.4 Microbial Manufacture of Cement

An overarching goal of green chemistry is to replace and/or augment current manufacturing processes that negatively impact and affect the environment with sustainable, inexpensive, and biologically friendly methods. In general, most current industrial-scale manufacturing processes are energy intensive, requiring high temperatures, involving the use of harsh chemicals, producing excessive amounts of waste by-products (CO_2), and/or resulting in irreversible damage to the environment. For example, concrete is arguably one of the most important technological achievements in human history and has revolutionised the construction industry and spurred rapid economic development worldwide; however, at the same time, it has caused significant environmental damage to the planet. In all, the manufacturing process of concrete utilises high temperatures and caustic reagents to convert calcite into Portland cement, while at the same time releasing massive amounts of CO_2 during curing/hardening. In this case, the development of a biologically equivalent process for concrete production would represent a major advancement in biomanufacturing and change the landscape of what is possible using biological approaches. Not surprisingly, nature is able to create intricate and complex cement-like structures of calcite and silica that occur in the form of sea sponges, spicules, diatoms, and shells with unprecedented quality, durability, and mechanical strength using highly orchestrated biomineralisation processes.[146] Unfortunately, the cost, complexity, and scalability of biologically driven approaches are major issues that have thus far prevented their practical implementation and/or use as substitutes for industrial grade cement and materials. Recently, however, bacteria have been used to grow bio-cement as a replacement material and/or to provide a means to augment industrially produced concrete. Bio-cement is a type of biologically controlled structural cement that is formed through bacterially induced hardening and precipitation of calcite/sand/soil aggregates under ambient conditions. In nature, many types of bacteria and fungi possess the ability to precipitate calcium carbonate (calcite) in the presence of calcium salts and urea. Natural sources of calcium in the environment are supplied in the form of calcium lactate, calcium nitrate, and/or calcium formate. In this process, Gram-positive bacteria (*Bacillus*) utilise a urease enzyme to facilitate the decomposition of urea into carbonate and ammonium ions. The carbonate ($[CO_3]^{2-}$) products then react with Ca^{2+} ions in the pre-cement mixture to promote calcium carbonate ($CaCO_3$) precipitation on or near the bacterial cell surface.[184] However, carbonate ions are not specific to only Ca^{2+} ions, but also readily react with other available metal ions present (Cu^{2+}, Cd^{2+}) to produce insoluble metal carbonate precipitates of copper carbonate ($CuCO_3$)[183] and otavite ($CdCO_3$).[184] In the presence of Ca^{2+}, the microbial-induced precipitation of calcium carbonate serves as a seed for nucleation and additional growth of calcium carbonate and as an adhesive to bind and cement loose sand or soil aggregates together. The process of microbial-mediated mineralisation enables control over mineral morphogenesis. In terms of practical applications, bacteria and mixtures of calcite aggregate can be applied to blank and undeveloped areas of land *via* spraying, in order to grow bio-cement structures on demand (*e.g.*, runways in austere environments), to seal fractures in oil and gas wells, for *in situ* concrete repair, to stabilise soil, and to create living building materials, such as structural bricks. Unlike the manufacture of Portland cement, bacterial-produced cement based on calcite precipitation results in no CO_2 released or the need for high temperatures.

In another example, urease-containing bacteria have also been used to create self-repairing concrete. In this case, calcite-forming bacterial spores are enshrined in concrete and cured into the final hardened material. The bacteria exist in a dormant sporulated state entombed in solidified concrete until cracks occur and expose the calcite-forming bacteria to moisture. The moisture activates the bacterial spore resulting in germination of vegetative and dividing cells localised within the concrete crack. This in turn results in bacteria-produced urease, enzymatic formation of carbonate, and precipitation of calcium carbonate within the micro-crack. In all, the microbial-produced cement results in an increase of 40% in compressive strength over control concrete, a decrease in water penetration and chloride ion permeability, and improved sealant properties.[184]

Reminiscent of eukaryotic organisms, bacteria–mineral interactions can impart a high level of biological control over the formation of complex mineral shapes and structures, as exemplified above by calcifying bacteria. In nature, only eukaryotes produce true biomineralised structures (*e.g.*, coral, bryozoan skeleton) by inducing mineralisation and precisely controlling the shape and structure of minerals independent of environmental conditions. By comparison, bacteria participate in pseudo-biomineralisation processes and exert a lot less control during mineralisation.[187] This level of biological control is evident in the final mineral shapes and structures, which are often ill-defined and irregular when compared to true biomineralised structures. In the case of pseudo-mineralisation, bacteria are able to partially control and shape the growing mineral structure, but are dependent upon available chemicals in the environment or require addition of mineral precursors to initiate and maintain mineralisation processes. For example, in the microbial-induced precipitation of calcite for bio-cement, as described above, sources of calcium and urea (converted to $[CO_3]^{2-}$) must be supplemented to the bacteria. This concept of pseudo-biomineralisation was tested by examining the dominant morphologies of minerals generated by exposing bacterial cells of *E. coli*, *P. fluorescens*, or *Bacillus coagulans* to calcifying conditions (see Figure 2.18). In a typical mineralisation experiment, bacteria in the form of planktonic cells or mature biofilms were exposed to Ca^{2+} ions and gaseous $(NH_4)_2CO_3$ to induce mineral formation. As a result, three general outcomes emerged from mineralisation in response to the calcifying conditions. In total, bacteria formed extensive biofilm networks (densely packed consortia of bacteria that stick to each other and are surrounded by a slimy extracellular matrix) to evade being encrusted by growing calcite mineral, were randomly and uncontrollably embedded inside mineral, as exemplified by *E. coli* in Figure 2.18, or showed an ability to control calcite mineralisation and encapsulation in mineral structures. The latter ability was demonstrated by *B. coagulans*, to produce cells surrounded by shell-like mineralised structures with clearly defined rimmed openings (see Figure 2.18). These distinct features support the idea that bacteria can participate in pseudo-mineralisation processes. Alternatively, bacteria can also influence mineral formation indirectly by altering physiochemical properties and environmental conditions. In this case, the organism does not impact the mineral shape and offers much less control than in pseudo-mineralisation processes.

The creation of engineered 'living' building materials represents a significant milestone in the development of microbial-based concrete materials.[188] Living building materials have been created by suspending photosynthetic cyanobacteria (*Synechococcus*) in an inert, biologically compatible scaffold consisting of sand and a physically crosslinked gelatin hydrogel. The gelatin hydrogel matrix provided structural support for the bacteria, a form factor

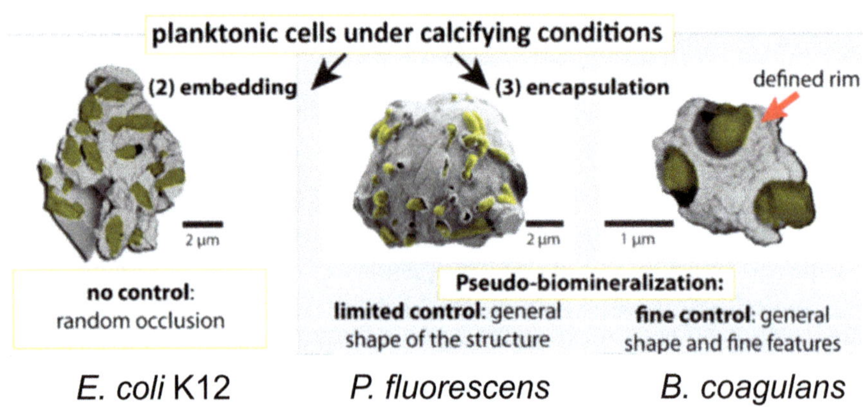

Figure 2.18 SEM images of *E. coli* K12, *P. fluorescens*, and *B. coagulans* bacterial strains exposed to calcifying conditions ($CaCl_2$ and $(NH_4)_2CO_3$ gas). Images show dominant morphologies of bacteria–calcite interactions and indicate pseudo-biomineralisation. Reproduced from ref. 187 with permission from American Chemical Society, Copyright, 2019.

for calcite biomineralisation used to cement the sand particles together, and increased biomineralisation rates of $CaCO_3$, which was otherwise slow when not contained in a hydrogel. Additionally, the use of photosynthetic bacteria offered significant benefits for microbial-induced precipitation of calcite. Notably, the benefits of using cyanobacteria over ureolytic Gram-positive bacteria, as described above to construct 'living' building materials, include the ability to capture energy through photosynthesis from light and water, absorption of CO_2, long-term viability, no production of ammonia by-products as for urea-based concrete, and ability to withstand high and low temperatures, pH, salinity, and humidity. For biomineralisation of calcite within the hydrogel matrix, cyanobacteria convert $[HCO_3]^-$ to intracellular CO_2 and OH^-. The generated hydroxyl ion is then exported outside the cell, leading to an increase in the local pH and immediate precipitation of $CaCO_3$ within the gelatin hydrogel. The precipitated $CaCO_3$ binds and cements sand to form the final hydrated 'living' building materials (see Figure 2.19). However, the hydration state of the precipitated calcite/sand/hydrogel material dictates its final mechanical properties and its ability to be regenerated and replicated into successive generations of newer 'living' building materials. For example, desiccation of the 'living' building material blocks resulted in material properties displaying greater structural integrity, compressive strength, and improved fracture energy; while exposure to increased humidity and elevated temperatures enabled three regenerations of new 'living' building material blocks to be created. After three generations, bacteria in the calcite/sand/hydrogel became less active and viable. In this latter case,

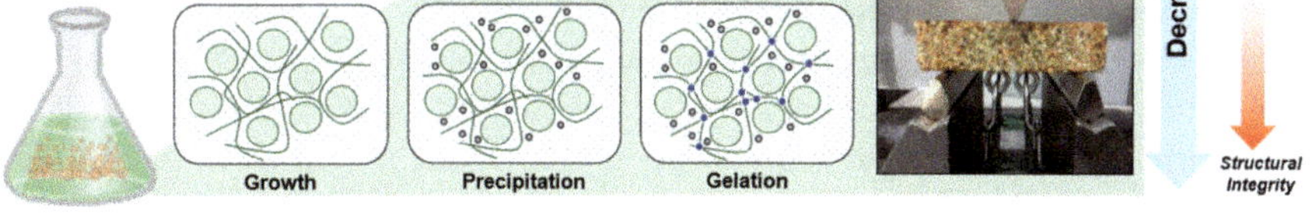

Figure 2.19 Formation of living building materials by mineralising calcite in the presence of a scaffold. Photosynthetic cyanobacteria, media, dissolved gelatin, and sand are combined and grown at 37 °C. During cell metabolism, the pH is increased, leading to calcite precipitation and reinforcement of material. After calcite precipitation, the living building material is cooled in the form of a block producing a three-dimensional hydrogel network. The block is then dehydrated, leading to increased physical crosslinks in the gelatin scaffold and improvement in strength and structural integrity. Reproduced from ref. 188 with permission from Elsevier, Copyright 2019.

humidity and temperature were used as switches to regulate the metabolic activity of the calcite-producing cyanobacteria in order to generate new materials from the original bio-cement block.

2.4 Synthesis of Nanoparticles by Mammalian Cells

In nature, there are numerous examples of animals carrying out true biomineralisation of inorganic nanoparticles as an essential part of their physiology and biological function. These biogenically derived nanoparticles provide and impart a non-biological function to organisms by augmenting or adding properties intrinsic to inorganic materials only (*i.e.*, magnetism). For example, mineralised forms of magnetic iron oxide particles (magnetite, maghemite) have been discovered in migratory animals, such as birds,[189] ants,[190] honey bees,[189] and salmon.[191] In all, these biogenic magnetic materials are used by biological systems for directional navigation, foraging, and/or geospatial orientation with the Earth's magnetic field. However, the *in vitro* and/or *in vivo* use of mammalian cells for inorganic nanoparticle synthesis is more limited, with only a few reported cases. Alternatively, mammalian cells are more applicable and versatile for the biomanufacture of therapeutic proteins and viral vaccines *via* genetic engineering or viral infection. For example, snake cells grown in laboratory cultures have been used to create venom-producing organoids capable of biosynthesising active snake venom.[192] The snake-free organoid produced venom that is biologically equivalent to venom milked and extracted from live snakes.

For nanoparticle synthesis using mammalian cells, crude protein extracts from weaver ant larvae were used for the production of silver particles.[193] Weaver ants are indigenous to Southern India, Australia, and Southeast Asia and recognised for their ability to build unique nests using leaves and larval silk. Due to this unique ability, weaver ant larvae extracts contain 13 mg of total protein per 1 g of larvae, with molecular weights ranging from 10 to 200 kDa, along with retinol, tocopherol, thiamine, niacin, riboflavin, and ascorbic acid. Notably, a major fraction of the protein content is attributed to the silk protein fibroin, which is used for nest construction. Consequently, $AgNO_3$ was added to the crude ant larvae extract at room temperature under fluorescent light exposure. Between 24 and 72 hours of light exposure, the formation of silver nanoparticles was detected by the appearance of a brown colour and plasmon resonance at ~430 nm. Also, silver nanoparticles synthesised in the presence of crude larvae extract yielded sizes of 7.87 + 2.54 nm. In another example, the condition of brain hypoxia (2% oxygen) was eloquently used to drive the biosynthesis of gold nanoparticles within living brain tissue, as a measure to restore the brain redox state and help protect it against hypoxia.[194] Hypoxia is a condition that occurs in response to low oxygen from being at high altitude and/or suffering from a pathological condition (*i.e.*, severe asthma attack). Fundamentally, induction of brain hypoxia results in increased levels of NADH cofactors, activation of enzymes (oxidases and synthases), and ultimately neuronal tissue breakdown and brain injury. However, conceptually, to restore the redox state of the hypoxic brain, Au^{3+} ions are added and effectively reduced by the strong electron-donating ability of NADH to form Au^0 metal nanoparticles. At the same time, NADH becomes oxidised to NAD^+ by Au^{3+}. This gold-mediated redox process in turn balances the ratio of $NADH/NAD^+$ to improve cellular respiration and restore normal brain function. This concept was tested *ex vivo* using coronal sections of mouse brain tissue that were either normally oxygenated (normoxia) or simulated to be hypoxic. In both conditions of hypoxia and normoxia, brain tissue was exposed to Au^{3+} precursor salts and incubated for 5 minutes. Interestingly, brain slices that suffered from hypoxia showed the *in situ* biosynthesis of Au nanoparticles with sizes of ~10 nm localised in the outer region of brain tissue; while normoxia-treated brain slices did not show the formation of gold (see Figure 2.20). Notably, the formation of gold nanoparticles was site specific and was deposited in the areas of brain tissue most affected by hypoxia. Additionally, the presence of gold nanoparticle labels could be used as a means to map specific hypoxic areas of the brain. These results confirmed the presence of biochemical differences between hypoxia and normoxia conditions and their effects on the formation of gold nanoparticles. In all, the biosynthesis of gold nanoparticles mediated by brain hypoxia is simply a necessary by-product of restoring redox processes in the brain, where the outcome is less concerned with gold nanoparticle properties (*i.e.*, size, shape, quality) and more with the biosynthesis pathway used to create the gold nanoparticles.

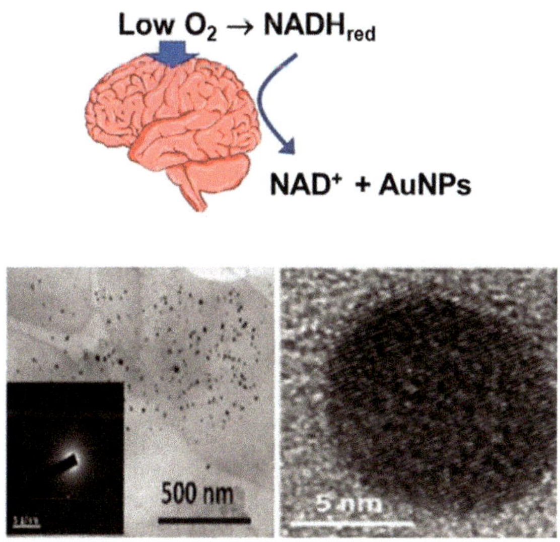

Figure 2.20 Formation of Au NPs by hypoxic brain. TEM micrographs of *in situ* biosynthesised gold nanoparticles extracted and isolated from brain slices treated with $AuCl_3$ under hypoxic conditions. Reproduced from ref. 194 with permission from the Royal Society of Chemistry.

2.5 Synthetic Biology for Nanoparticle Synthesis

The use of microorganisms for nanoparticle synthesis has led to new strategies, enhanced properties, and various nanostructures, as described in the previous sections in more detail. However, in all these examples, nanoparticle synthesis was achieved by exploiting the natural biochemical and metabolic processes of cells containing no genetic modifications. As a result, most cellular processes are aimed at cell survival and replication; and, by nature, are inherently inefficient at dealing with nanoparticle synthesis. Alternatively, synthetic biology offers a more directed approach to nanoparticle synthesis, with better control and a means to tune nanoparticle compositions and properties. In this regard, cellular resources and processes can be directed to nanoparticle synthesis through production of biomolecular templates and/or nanoparticle components *via* single or multiple genetic modifications to cells. In part, synthetic biology has been routinely used to genetically engineer microorganisms for the production of heterologous peptide/protein templates, modification of cell genotypes and phenotypes, ruggedisation of cells against extreme environmental conditions, and execution of post-translational modifications on biomolecules similar to eukaryotic organisms. This latter point is significant given that naturally occurring bacteria (prokaryotes) are unable to perform post-translational modifications on biomolecules, and in turn cannot produce true biomineralised structures, as discussed previously. Recently, this non-natural function was genetically engineered into *E. coli* for the purpose of post-translational modification of a silica-precipitating peptide from diatoms (silafin R5 peptide) for the synthesis of silica nanostructures.[16] In this example, post-translational modifications of the silafin R5 peptide were carried out using a library of 38 enzymes, and included phosphorylation of serine residues, methylation, acetylation, glycosylation, and biotinylation. Each post-translational modification of R5 resulted in a variety of different-sized silica spheres. In another example highlighting synthetic biology, intracellular compartments, called encapsulins, from *Thermotoga maritima* were engineered for the size-constrained, *in vivo* synthesis of silver nanoparticles.[195] In this case, encapsulins were designed to display silver-binding peptides (AG4) discovered previously.[196] By comparison, encapsulin nanocompartments are structurally similar to ferritin protein cages.

2.6 Conclusion

The use of biological approaches to create and synthesise inorganic nanomaterials presents opportunities for achieving new structures, functional materials, and sustainability that surpasses those obtained from traditional materials science methods. For example, the use of biological approaches enables production of materials under ambient conditions and in aqueous environments. Appealingly, biology also offers an unlimited set of biomolecules with differing levels of complexities (*e.g.*, amino acids, peptides, proteins, viruses), a variety of common and exotic plant species, and a diverse group of complex microorganisms that are capable of generating inorganic nanomaterials with superior properties and structures on demand. Notably, biological approaches are beginning to become recognised as competitive and viable methods for industrial and commercial uses due to improvements in cost and availability, and, in limited instances, even as a replacement or way to augment current industrial production of environmentally harsh materials. For example, microorganisms are being utilised for the formation of living bio-cement as an equivalent to Portland cement. The impact of using microorganisms to produce a bio-cement equivalent to Portland cement is tremendous and will undoubtedly lead to use of alternative biological methods for other industrial processes. Additionally, advances in synthetic biology provide an engineered approach to creating unique genetic circuits that can drive the production of proteins and peptides. These can control and template the synthesis of nanomaterials in various host organisms that are amenable to large-scale industrial production.

Acknowledgements

We acknowledge funding support from the Air Force Office of Scientific Research.

References

1. S. Lagziel-Simis, N. Cohen-Hadar, H. Moscovich-Dagan, Y. Wine and A. Freeman, *Curr. Opin. Biotechnol.*, 2006, **17**, 569.
2. S. Mann, *Nature*, 1988, **332**, 119.
3. K. Simkiss and K. M. Wilbur, *Biomineralization: Cell Biology and Mineral Deposition*, Academic Press, New York, 1989.
4. A. Sigel, H. Sigel and K. O. Sigel, *Biomineralization: From Nature to Application*, Metal Ions in Life Sciences, Wiley, Chichester, 2008.
5. W. Wang, C. F. Anderson, Z. Wang, W. Wu, H. Cui and C.-J. Liu, *Chem. Sci.*, 2017, **8**, 3310.
6. M. Holzinger, A. Le Goff and S. Cosnier, *Front. Chem.*, 2014, **2**, 63.
7. M. B. Dickerson, K. H. Sandage and R. R. Naik, *Chem. Rev.*, 2008, **108**, 4935.
8. R. de la Rica and H. Matsui, *Chem. Soc. Rev.*, 2010, **39**, 3499.
9. *Bioinspired Inorganic Materials: Structure and Function*, ed. S. Hall, *Series Inorganic Materials*, Series ed. D. W. Bruce, D. O'Hare and R. I. Walton, Royal Society of Chemistry, Cambridge, 2019.
10. *The Chemistry of Inorganic Biomaterials*, ed. C. Spicer, *Inorganic Materials Series*, Series ed. D. W. Bruce, D. O'Hare and R. I. Walton, Royal Society of Chemistry, Cambridge, 2021.
11. M. Uchida, M. T. Klem, M. Allen, P. Suci, M. Flenniken, E. Gillitzer, Z. Varpness, L. O. Liepold, M. Young and T. Douglas, *Adv. Mater.*, 2007, **19**, 1025.
12. Y. J. Jiang, S. M. Guan, Y. Zhang, G. H. Liu, X. B. Zheng and J. Gao, *J. Chem. Technol. Biotechnol.*, 2020, **95**, 2463.

13. M. Wakabayashi, S. Yokojima, T. Fukaminato, K.-i. Shiino, M. Irie and S. Nakamura, *J. Phys. Chem. A*, 2014, **118**, 5046.
14. Q. Y. Lin, J. A. Mason, Z. Y. Li, W. J. Zhou, M. N. O'Brien, K. A. Brown, M. R. Jones, S. Butun, B. Lee, V. P. Dravid, K. Aydin and C. A. Mirkin, *Science*, 2018, **359**, 669.
15. R. J. Macfarlane, B. Lee, M. R. Jones, N. Harris, G. C. Schatz and C. A. Mirkin, *Science*, 2011, **334**, 204.
16. A. K. Wallace, N. Chanut and C. A. Voigt, *Adv. Funct. Mater.*, 2020, 20008449.
17. J. M. Slocik and R. R. Naik, *Curr. Opin. Biotechnol.*, 2017, **46**, 1.
18. J. McCafferty and D. Schofield, *Curr. Opin. Chem. Biol.*, 2015, **26**, 16.
19. T. R. Walsh and M. R. Knecht, *Chem. Rev.*, 2017, **117**, 12641.
20. M. Y. Klem and T. Douglas, *J. Mater. Chem.*, 2010, **20**, 65.
21. T. Douglas, D. P. E. Dickson, S. Betteridge, J. Charnock, C. D. Garner and S. Mann, *Science*, 1995, **69**, 54.
22. D. He and J. Marles-Wright, *New Biotechnol.*, 2015, **32**, 651.
23. P. R. Selvakannan, S. Mandal, S. Phadtare, A. Gole, R. Pasricha, S. D. Adyanthaya and M. Sastry, *J. Colloid Interface Sci.*, 2004, **269**, 97.
24. P. R. Selvakannan, S. Mandal, S. Phadtare, R. Pasricha and M. Sastry, *Langmuir*, 2003, **19**, 3545.
25. P. R. Selvakannan, A. Swami, D. Srisathiyanarayanan, P. S. Shirude, R. Pasricha, A. B. Mandale and M. Sastry, *Langmuir*, 2004, **20**, 7825.
26. H. Joshi, P. S. Shirude, V. Bansal, K. N. Ganesh and M. Sastry, *J. Phys. Chem. B*, 2004, **108**, 1535.
27. A. Kumar, P. Mukherjee, A. Guha, S. D. Adyantaya, A. B. Mandale, R. Kumar and M. Sastry, *Langmuir*, 2000, **16**, 9775.
28. S. Mandal, A. Gole, N. Lala, R. Gonnade, V. Ganvir and M. Sastry, *Langmuir*, 2001, **17**, 6262.
29. N. Wangoo, K. K. Bhasin, S. K. Mehta and C. R. Suri, *J. Colloid Interface Sci.*, 2008, **323**, 247.
30. J. M. Slocik and D. W. Wright, *Biomacromolecules*, 2003, **4**, 1135.
31. J. Feng, R. B. Pandey, R. J. Berry, B. L. Farmer, R. R. Naik and H. Heinz, *Soft Matter*, 2011, **7**, 2113.
32. J. P. Palafox-Hernandez, Z. Tang, Z. E. Hughes, Y. Li, M. T. Swihart, P. N. Prasad, T. R. Walsh and M. R. Knecht, *Chem. Mater.*, 2014, **26**, 4960.
33. T. Ishida, T. Murayama, A. Taketoshi and M. Haruta, *Chem. Rev.*, 2020, **120**, 464.
34. X. Yang, M. Yangj, B. Pang, M. Vara and Y. Xia, *Chem. Rev.*, 2015, **115**, 10410.
35. T. Maruyama, Y. Fujimoto and T. Maekawa, *J. Colloid Interface Sci.*, 2015, **447**, 254.
36. S. Shankar, S. K. Soni, H.-K. Daima, P. R. Selvakannan, J. M. Khire, S. K. Bhargava and V. Bansal, *Phys. Chem. Chem. Phys.*, 2015, **17**, 21517.
37. H.-E. Lee, H.-Y. Ahn, J. Mun, Y. Y. Lee, M. Kim, N. H. Cho, K. Chang, W. S. Kim, J. Rho and K. T. Nam, *Nature*, 2018, **556**, 360.

38. L. Berquist and T. Hegmann, *ChemNanoMat*, 2017, **3**, 863.
39. X. Liu, J. Zhang, X. Guo, S. Wu and S. Wang, *Nanoscale*, 2010, **2**, 1178.
40. J. Zhang, X. Liu, X. Guo, S. Wu and S. Wang, *Chem. – Eur. J.*, 2010, **16**, 8108.
41. L. F. Drummy, S. E. Jones, R. B. Pandey, B. L. Farmer, R. A. Vaia and R. R. Naik, *ACS Appl. Mater. Interfaces*, 2010, **2**, 1492.
42. S. N. Kim, Z. Kuang, J. M. Slocik, S. E. Jones, Y. Cui, B. L. Farmer, M. C. McAlpine and R. R. Naik, *J. Am. Chem. Soc.*, 2011, **133**, 14480.
43. A. Velamakanni, C. W. Magnuson, K. J. Ganesh, Y. Zhu, J. An, P. J. Ferreira and R. S. Ruoff, *ACS Nano*, 2010, **4**, 540.
44. M. Sethi and M. R. Knecht, *ACS Appl. Mater. Interfaces*, 2009, **1**, 1270.
45. M. Sethi and M. R. Knecht, *Langmuir*, 2010, **26**, 9860.
46. L. B. Wright, N. A. Merrill, M. R. Knecht and T. R. Walsh, *ACS Appl. Mater. Interfaces*, 2014, **6**, 10524.
47. M. Sethi, W. C. Law, W. A. Fennell, P. N. Prasad and M. R. Knecht, *Soft Matter*, 2011, **7**, 6532.
48. A. Majzik, L. Fülöp, E. Csapó, F. Bogár, T. Martinek, B. Penke, G. Bíró and I. Dékány, *Colloids Surf., B*, 2010, **81**, 235.
49. M. Sethi, G. Joung and M. R. Knecht, *Langmuir*, 2009, **25**, 1572.
50. K. K. Caswell, J. N. Wilson, U. H. F. Bunz and C. J. Murphy, *J. Am. Chem. Soc.*, 2003, **125**, 13914.
51. S. Devi, B. Singh, A. K. Paul and S. Tyagi, *Anal. Methods*, 2016, **8**, 4398.
52. Y. Xianyu, Z. Wang, J. Sun, X. Wang and X. Jiang, *Small*, 2014, **10**, 4833.
53. C. Carnovale, G. Bryant, R. Shukla and V. Bansal, *ACS Omega*, 2019, **4**, 242.
54. B. Hu, S.-B. Wang, K. Wang, M. Zhang and S.-H. Yu, *J. Phys. Chem. C*, 2009, **112**, 1169.
55. J. Nanda, B. Adhikari, S. Basak and A. Banerjee, *J. Phys. Chem. B*, 2012, **116**, 12235.
56. S. Shankar and J.-W. Rhim, *Carbohydr. Polym.*, 2015, **130**, 353.
57. D. R. Bae, W. S. Han, J. M. Lim, S. Kang, J. Y. Lee, D. Kang and J. H. Jung, *Langmuir*, 2009, **26**, 2181.
58. J.-C. Yu, F.-G. Zhao, W. Shao, C.-W. Ge and W.-S. Li, *Nanoscale*, 2015, **7**, 8811.
59. C. J. Munro, E. C. Bell, M. O. Olagunju, J. L. Cohn, E. M. Zahran, L. G. Bachas and M. R. Knecht, *ACS Sustainable Chem. Eng.*, 2019, **7**, 17055.
60. A. Bhattacharjee and M. Ahmaruzzaman, *RSC Adv.*, 2016, **6**, 41348.
61. A. Samadi-maybodi and R. Reza Akhoondi, *Colloids Surf., A*, 2014, **447**, 111.
62. S. O. Oluwafemi, N. Revaprasadu and A. J. Ramirez, *J. Cryst. Growth*, 2008, **310**, 3230.
63. S. Tiwari, P. K. Gupta, Y. Bagbi, T. Sarkar and P. R. Solanki, *Biosens. Bioelectron.*, 2017, **89**, 1042.
64. G.-M. Han, H.-X. Jiang, Y.-F. Huo and D.-M. Kong, *J. Mater. Chem. B*, 2016, **4**, 3351.

65. A. Ghorbani-Choghamarani and G. Azadi, *Appl. Organomet. Chem.*, 2016, **30**, 247.
66. C. T. Dameron, B. R. Smith and D. R. Winge, *J. Biol. Chem.*, 1989, **264**, 17355.
67. C. T. Dameron and D. R. Winge, *Inorg. Chem.*, 1990, **29**, 1343.
68. W. Wang, C. F. Anderson, Z. Wang, W. Wu, H. Cui and C.-J. Liu, *Chem. Sci.*, 2017, **8**, 3310.
69. M. J. Limo, A. Sala-Rabada, E. Boix, V. Thota, Z. C. Westcott, V. Puddu and C. C. Perry, *Chem. Rev.*, 2018, **118**, 11118.
70. Z. E. Hughes, M. A. Nguyen, Y. Lie, M. T. Swihart, T. R. Walsh and M. R. Knecht, *Nanoscale*, 2017, **9**, 421.
71. D. B. Pacardo, M. Sethi, S. E. Jones, R. R. Naik and M. R. Knecht, *ACS Nano*, 2009, **3**, 1288.
72. Z. Tang, J. P. Palafox-Hernandez, W.-C. Law, Z. E. Hughes, M. T. Swihart, P. N. Prasad, M. R. Knecht and T. R. Walsh, *ACS Nano*, 2013, **7**, 9632.
73. M. A. Nguyen, Z. E. Hughes, Y. Liu, M. T. Swihart, M. R. Knecht and T. R. Walsh, *J. Phys. Chem. C*, 2018, **122**, 11532.
74. J. M. Slocik, M. O. Stone and R. R. Naik, *Small*, 2005, **1**, 1048.
75. M. Hnilova, E. E. Oren, U. O. S. Seker, B. R. Wilson, S. Collino, J. S. Evans, C. Tamerler and M. Sarikaya, *Langmuir*, 2008, **24**, 12440.
76. C. J. Munro and M. R. Knecht, *J. Phys. Chem. C*, 2018, **122**, 26686.
77. C. J. Munro, Z. E. Hughes, T. R. Walsh and M. R. Knecht, *J. Phys. Chem. C*, 2016, **120**, 18917.
78. C. J. Munro and M. R. Knecht, *Langmuir*, 2017, **33**, 13757.
79. M. Hnilova, X. Liu, E. Yuca, C. Jia, B. Wilson, A. Y. Karatas, C. Gresswell, F. Ohuchi, K. Kitamura and C. Tamerler, *ACS Appl. Mater. Interfaces*, 2012, **4**, 1865.
80. Y. Li, Z. Tang, P. N. Prasad, M. R. Knecht and M. T. Swihart, *Nanoscale*, 2014, **6**, 3165.
81. J. M. Slocik, D. B. Dennis, A. O. Govorov, N. M. Bedford, Y. Ren and R. R. Naik, *ACS Biomater. Sci. Eng.*, 2020, **6**, 2612.
82. C.-L. Chen and N. L. Rosi, *J. Am. Chem. Soc.*, 2010, **132**, 6902.
83. C.-L. Chen, P. Zhang and N. L. Rosi, *J. Am. Chem. Soc.*, 2008, **130**, 13555.
84. C. Zhang, Y. Zhou, A. Merg, C. Song, G. C. Schatz and N. L. Rosi, *Nanoscale*, 2014, **6**, 12328.
85. A. D. Merg, Y. Zhou, A. M. Smith, J. E. Millstone and N. L. Rosi, *ChemNanoMat*, 2017, **3**, 745.
86. A. D. Merg, J. C. Boatz, A. Mandal, G. Zhao, S. Punekar, C. Liu, Z. Wang, P. Zhang, P. C. A. van der Wel and N. L. Rosi, *J. Am. Chem. Soc.*, 2016, **138**, 13655.
87. S. W. Mokashi-Punekar, T. R. Walsh and N. L. Rosi, *J. Am. Chem. Soc.*, 2019, **141**, 15710.
88. S. Mokashi-Punekar and N. L. Rosi, *Part. Part. Syst. Charact.*, 2019, 1800504.
89. T. K. Sau and C. J. Murphy, *Langmuir*, 2004, **20**, 6416.

90. E. Zhu, X. Yan, S. Wang, M. Xu, C. Wang, H. Liu, J. Huang, W. Xue, J. Cai, H. Heinz, Y. Li and Y. Huang, *Nano Lett.*, 2019, **19**, 3730.
91. C.-Y. Chiu, Y. J. Li, L. Y. Ruan, X. C. Ye, C. B. Murray and Y. Huang, *Nat. Chem.*, 2011, **3**, 393.
92. Z. Zong, K. Xu, D. Li, Z. Tang, W. He, Z. Liu, X. Wang and Y. Tian, *J. Catal.*, 2018, **361**, 168.
93. J. M. Slocik and R. R. Naik, *Adv. Mater.*, 2006, **18**, 1988.
94. A. Parab, J. M. Slocik, R. R. Naik and M. R. Knecht, *ACS Appl. Nano Mater.*, 2018, **1**, 7149.
95. G. Huang, F. Li, X. Zhao, Y. Ma, Y. Li, M. Lin, G. Jin, T. J. Lu, G. M. Genin and F. Xu, *Chem. Rev.*, 2017, **117**, 12764.
96. S. Zhang, R. Geryak, J. Geldmeier, S. Kim and V. V. Tsukruk, *Chem. Rev.*, 2017, **117**, 12942.
97. A. D. Frankel and C. O. Pabo, *Cell*, 1988, **55**, 1189.
98. K. Bansal, M. Aqdas, M. Kumar, R. Bala, S. Singh, J. N. Agrewala, O. P. Katare, R. K. Sharma and N. Wangoo, *Bioconjugate Chem.*, 2018, **29**, 1102.
99. R. M. Kramer, C. Li, D. C. Carter, M. O. Stone and R. R. Naik, *J. Am. Chem. Soc.*, 2004, **126**, 13282.
100. J. Bain and S. S. Staniland, *Phys. Chem. Chem. Phys.*, 2015, **17**, 15508.
101. C. Yang and H. Yi, *ChemCatChem*, 2015, **7**, 2015.
102. M. Uchida, S. Kang, C. Reichhardt, K. Harlen and T. Douglas, *Biochim. Biophys. Acta*, 2010, **1800**, 834.
103. A. J. Love, V. Makarov, I. Yaminsky, N. O. Kalinina and M. E. Taliansky, *Virology*, 2014, **449**, 133.
104. W. Yang, W. Guo, J. Chang and B. Zhang, *J. Mater. Chem. B*, 2017, **5**, 401.
105. M. J. Parker, M. Allen, B. Ramsay, M. T. Klem, M. Young and T. Douglas, *Chem. Mater.*, 2008, **20**, 1541.
106. M. Uenuma, K. Kawano, B. Zheng, N. Okamoto, M. Horita, S. Yoshii, I. Yamashita and Y. Uraoka, *Nanotechnology*, 2011, **22**, 215201.
107. K. W. Pulsipher, S. Hong, S. Deng and I. J. Dmochowski, *J. Inorg. Biochem.*, 2017, **174**, 169.
108. D. Ensign, M. Young and T. Douglas, *Inorg. Chem.*, 2004, **43**, 3441.
109. T. Ueno, M. Suzuki, T. Goto, T. Matsumoto, K. Nagayama and Y. Watanabe, *Angew. Chem., Int. Ed.*, 2004, **43**, 2527.
110. M. T. Klem, J. Mosolf, M. Young and T. Douglas, *Inorg. Chem.*, 2008, **47**, 2237.
111. K. K. W. Wong, T. Douglas, S. Gider, D. D. Awschalom and S. Mann, *Chem. Mater.*, 1998, **10**, 279.
112. F. C. Meldrum, B. R. Heywood and S. Mann, *Science*, 1992, **257**, 522.
113. H.-A. Hosein, D. R. Strongin, M. Allen and T. Douglas, *Langmuir*, 2004, **20**, 10283.
114. P. Mackle, J. M. Charnock, C. D. Garner, F. C. Meldrum and S. Mann, *J. Am. Chem. Soc.*, 1993, **115**, 8471.
115. K. K. W. Wong and S. Mann, *Adv. Mater.*, 1996, **8**, 928.

116. K. Yoshizawa, K. Iwahori, K. Sugimoto and I. Yamashita, *Chem. Lett.*, 2006, **35**, 1192.
117. I. Yamashita, J. Hayashi and M. Hara, *Chem. Lett.*, 2004, **33**, 1158.
118. K. Iwahori, K. Yoshizawa, M. Muraoka and Y. Yamashita, *Inorg. Chem.*, 2005, **44**, 6393.
119. C. Jolley, V. Pool, Y. Idzerda and T. Douglas, *Chem. Mater.*, 2011, **23**, 3921.
120. C. C. Jolley, M. Uchida, C. Reichhardt, R. Harrington, S. Kang, M. T. Klem, J. B. Parise and T. Douglas, *Chem. Mater.*, 2010, **22**, 4612.
121. M. Uchida, M. L. Flenniken, M. Allen, D. A. Willits, B. E. Crowley, S. Brumfield, A. F. Willis, L. Jackiw, M. Jutila, M. J. Young and T. Douglas, *J. Am. Chem. Soc.*, 2006, **128**, 16626.
122. W. Arap, R. Pasqualini and E. Ruoslahti, *Science*, 1998, **279**, 377.
123. M. Okuda, Y. Kobayashi, K. Suzuki, K. Sonoda, T. Kondoh, A. Wagawa, A. Kondo and H. Yoshimura, *Nano Lett.*, 2005, **5**, 991.
124. T. Matsui, N. Matsukawa, K. Iwahori, K.-I. Sano, K. Shiba and I. Yamshita, *Langmuir*, 2007, **23**, 1615.
125. D. Kase, J. L. Kulp, M. Yudasaka, J. S. Evans, S. Lijima and K. Shiba, *Langmuir*, 2004, **20**, 8939.
126. R. M. Kramer, L. A. Sowards, M. J. Pender, M. O. Stone and R. R. Naik, *Langmuir*, 2005, **21**, 8466.
127. M. Künzle, T. Eckert and T. Beck, *J. Am. Chem. Soc.*, 2016, **138**, 12731.
128. M. Künzle, T. Eckert and T. Beck, *Inorg. Chem.*, 2018, **57**, 13431.
129. K. B. Narayannan and S. S. Han, *Bioinspiration Biomimetics*, 2017, **12**, 031001.
130. G. Stubbs, *Semin. Virol.*, 1990, **1**, 405.
131. W. Shenton, T. Douglas, M. Young, G. Stubbs and S. Mann, *Adv. Mater.*, 1999, **11**, 253.
132. E. Dujardin, C. Peet, G. Stubbs, J. N. Culver and S. Mann, *Nano Lett.*, 2003, **3**, 413.
133. K. M. Bromley, A. J. Patil, A. W. Perriman, G. Stubbs and S. Mann, *J. Mater. Chem.*, 2008, **18**, 4796.
134. M. L. Górzny, A. S. Walton and S. D. Evans, *Adv. Funct. Mater.*, 2010, **20**, 1295.
135. J.-S. Lim, S.-M. Kim, S.-Y. Lee, E. A. Stach, J. N. Culver and M. T. Harris, *Nano Lett.*, 2010, **10**, 3863.
136. M. Kobayashi, M. Seki, H. Tabata, Y. Watanabe and I. Yamashita, *Nano Lett.*, 2010, **10**, 773.
137. M. Knez, A. Kadri, C. Wege, U. Gösele, H. Jeske and K. Nielsch, *Nano Lett.*, 2006, **6**, 172.
138. W. K. Kegel and P. van der Schoot, *Biophys. J.*, 2006, **91**, 1501.
139. O. K. Zahr and A. S. Blum, *Nano Lett.*, 2012, **12**, 629.
140. Y. Pan, W. J. Paschoalino, S. S. Bayram, A. S. Blum and J. Mauzeroll, *Nanoscale*, 2019, **11**, 18595.
141. S. S. Bayram, O. K. Zahr, J. Del Re and A. S. Blum, *Nanotechnol*, 2016, **27**, 485603.

142. J. Zhang, K. Zhou, Y. Zhang, M. Du and Q. Wang, *Adv. Mater.*, 2019, **31**, 1901485.
143. T. Pavesi and J. C. Moreira, *J. Appl. Toxicol.*, 2020, **40**, 1183.
144. I. Sadeghi, E. Y. Liu, H. Yi and A. Asatekin, *ACS Appl. Nano Mater.*, 2019, **2**, 5233.
145. Y. Liu, Y. Xu, Y. Zhu, J. N. Culver, C. A. Lundgren, K. Xu and C. Wang, *ACS Nano*, 2013, **7**, 3627.
146. S. Mann, *Biomineralizaton Principles and Concepts in Bioinorganic Materials Chemistry*, Oxford Univ. Press, New York, 2002.
147. M. J. M. Christenhusz and J. W. Byng, *Phytotaxa*, 2016, **261**, 201.
148. Y. M. Bar-On, R. Phllips and R. Milo, *Proc. Natl. Acad. Sci. U. S. A.*, 2018, **115**, 6506.
149. M. S. Akhtar, J. Panwar and Y.-S. Yun, *ACS Sustainable Chem. Eng.*, 2013, **1**, 591.
150. P. Singh, Y.-J. Kim, D. Zhang and D.-C. Yang, *Trends Biotechnol.*, 2016, **34**, 588.
151. J. C. Steffens, *Annu. Rev. Plant Physiol. Plant Mol. Biol.*, 1990, **41**, 553.
152. J. L. Gardea-Torresdey, J. G. Parsons, E. Gomez, J. Peralta-Videa, H. E. Troiani, P. Santiago and M. Jose Yacaman, *Nano Lett.*, 2002, **2**, 397.
153. R. K. Sharma, S. Gulati and S. J. Mehta, *J. Chem. Educ.*, 2012, **89**, 1316.
154. A. N. Geraldes, A. A. da Silva, J. Leal, G. M. Estrada-Villegas, N. Lincopan, K. Katti and A. B. Lugao, *Adv. Nanopart.*, 2016, **5**, 176.
155. W. J. Keijok, R. H. A. Pereira, L. A. C. Alvarez, A. R. Prado, A. R. da Silva, J. Ribeiro, J. P. de Oliveira and M. C. C. Guimaraes, *Sci. Rep.*, 2019, **9**, 16019.
156. K. J. Rao and S. Paria, *ACS Sustainable Chem. Eng.*, 2015, **3**, 483.
157. F. Buazar, M. H. Baghlani-Nejazd, M. Badri, M. Kashisaz, N. A. Khaledi and F. Kroushawi, *Starch*, 2016, **68**, 796.
158. R. K. Sharma, S. Yadav, R. Gupta and G. Arora, *J. Chem. Educ.*, 2019, **96**, 3038.
159. S. Singh, P. K. More and S. M. Mohan, *Indian J. Sci. Res.*, 2014, **4**, 46.
160. M. N. Alam, S. Das, S. Batuta, N. Roy, A. Chatterjee, D. Mandal and N. A. Begum, *ACS Sustainable Chem. Eng.*, 2014, **2**, 652.
161. S. Rajeshkumar, C. Malarkodi, G. Gnanadhas, K. Paulkumar, M. Vanaja, C. Kannan and G. Annadurai, *J. Nanostruct. Chem.*, 2013, **3**, 44.
162. E. Burlacu, C. Tanase, N.-A. Coman and L. Berta, *Molecules*, 2019, **24**, 4354.
163. M. Sathishkumar, K. Sneha, I. S. Kwak, J. Mao, S. J. Tripathy and Y. S. Yun, *J. Hazard. Mater.*, 2009, **171**, 400.
164. N. Savithramma, M. L. Rao and P. S. Devi, *J. Biol. Sci.*, 2011, **11**, 39.
165. G. A. Kahrilas, L. M. Wally, S. J. Fredrick, M. Hiskey, A. L. Prieto and J. E. Owens, *ACS Sustainable Chem. Eng.*, 2014, **2**, 367.
166. Z. A. Ali, R. Yahya, S. D. Sekaran and R. Puteh, *Adv. Mater. Sci. Eng.*, 2016, 4102196.
167. P. Acharya, G. K. Jayaprakasha, K. M. Crosby, J. L. Jifon and B. S. Patil, *ACS Sustainable Chem. Eng.*, 2019, **7**, 14580.

168. K. Sneha, M. Sathishkumar, S. Y. Lee, M. A. Bae and Y.-S. Yun, *J. Nanosci. Nanotechnol.*, 2011, **11**, 1811.
169. C. H. Ramamurthy, K. S. Sampath, P. Arunkumar, M. S. Kumar, V. Sujatha, K. Premkumar and C. Thirunavukkarasu, *Bioprocess Biosyst. Eng.*, 2013, **36**, 1131.
170. M. Vahdati and T. T. Moghadam, *Sci. Rep.*, 2020, **10**, 510.
171. N. R. Chowdhury, M. MacGregor-Ramiasa, P. Zilm, P. Majewski and K. Vasilev, *J. Colloid Interface Sci.*, 2016, **482**, 151.
172. N. A. Al-Shabib, F. M. Husain, F. Ahmed, R. A. Khan, I. Ahmad, E. Alsharaeh, M. S. Khan, A. Hussain, M. T. Rehman, M. Yusuf, I. Hassan, J. M. Khan, G. M. Ashraf, A. Alsalme, M. F. Al-Ajmi, V. V. Tarasov and G. Aliev, *Sci. Rep.*, 2016, **6**, 36761.
173. R. K. Das, V. L. Pachapur, L. Lonappan, M. Naghdi, R. Pulicharla, S. Maiti, M. Cledon, L. M. A. Dalila, S. J. Sarma and S. K. Brar, *Nanotechnol. Environ. Eng.*, 2017, **2**, 18.
174. S. Tadepalli, J. M. Slocik, M. K. Gupta, R. R. Naik and S. Singamaneni, *Chem. Rev.*, 2017, **117**, 12705.
175. S. A. Wadhwan, M. Gorain, P. Banerjee, U. U. Shedbalkar, R. Singh, G. C. Kundu and B. A. Chopade, *Int. J. Nanomed.*, 2017, **12**, 6841.
176. J. L. Graves, M. Tajkarimi, Q. Cunningham, A. Campbell, H. Nonga, S. H. Harrison and J. E. Barrick, *Front. Genet.*, 2015, **6**, 42.
177. T. Klaus, R. Joerger, E. Olsson and C.-G. Granqvist, *Proc. Natl. Acad. Sci. U. S. A.*, 1999, **96**, 13611.
178. M. M. Abdel-Aziz, T. M. Emam and E. A. Elsherbiny, *Mater. Sci. Eng., C*, 2020, **109**, 110617.
179. C. W. Johnston, M. A. Wyatt, X. Li, A. Ibrahim, J. Shuster, G. Sutham and N. A. Magarvey, *Nat. Chem. Biol.*, 2013, **9**, 241.
180. A. Chen, L. M. Contreras and B. K. Keitz, *Appl. Environ. Microbiol.*, 2017, **83**, e00798-17.
181. Z. Molnar, V. Bodai, G. Szakacs, B. Erdelyi, Z. Fogarassy, G. Safran, T. Varga, Z. Konya, E. Toth-Szeles, R. Szucs and I. Lagzi, *Sci. Rep.*, 2018, **8**, 3943.
182. I. B. Tahar, P. Fickers, A. Dziedzic, D. Ploch and B. Skora, *Microb. Cell Fact.*, 2019, **18**, 210.
183. C. E. Turick, A. S. Knox, J. M. Becnel, A. E. Ekechukwu and C. E. Milliken, *Biopolymers*, 2010, **449**, 72.
184. K. Vijay, M. Murmu and S. V. Deo, *Constr. Build. Mater.*, 2017, **152**, 1008.
185. Q. Li and G. M. Gadd, *Appl. Microbiol. Biotechnol.*, 2017, **101**, 7397.
186. Q. Li, L. Csetenyi and G. M. Gadd, *Environ. Sci. Technol.*, 2014, **48**, 14409.
187. G. M. Durak, M. Laumann, S. L. P. Wolf, A. Pawar, D. Gebauer and T. Bottcher, *ACS Biomater. Sci. Eng.*, 2019, **5**, 5088.
188. C. M. Heveran, S. L. Williams, J. Qiu, J. Artier, M. H. Hubler, S. M. Cook, J. C. Cameron and W. V. Srubar III, *Matter*, 2020, **2**, 1.
189. J. Gould, *Am. Sci.*, 1980, **68**, 256.
190. D. Acosta-Avalos, E. Wajnberg, P. S. Oliveira, I. Leal, M. Farina and D. M. Esquivel, *J. Exp. Biol.*, 1999, **202**, 2687.

191. T. P. Quinn, *J. Comp. Physiol., A*, 1980, **137**, 243.
192. Y. Post, J. Puschhof, J. Beumer, H. M. Kerkkamp, M. A. G. de Bakker, J. Slagboom and B. de Barbanson, *et al.*, *Cell*, 2020, **180**, 233.
193. A. Khamhaengpol and S. Siri, *J. Photochem. Photobiol., B*, 2016, **163**, 337.
194. E. A. Rozhkova, B. Lee, J. A. Prasad, Y. Liu and E. V. Shevchenko, *Nanoscale*, 2019, **11**, 19285.
195. T. W. Giessen and P. A. Silver, *ACS Synth. Biol.*, 2016, **5**, 1497.
196. R. R. Naik, S. J. Stringer, G. Agarwal, S. E. Jones and M. O. Stone, *Nat. Mater.*, 2002, **1**, 169.

CHAPTER 3

The Role of Supercritical Carbon Dioxide and Water in the Synthesis of Metal and Metal Oxide Nanoparticles: Current State of the Art, Further Perspectives and Needs

M. TÜRK[a*], CH. SCHÜßLER[a], M. HOHM[a] AND M. CRONE[a]

[a]Karlsruhe Institute of Technology (KIT), Institute for Technical Thermodynamics and Refrigeration (ITTK), Engler-Bunte-Ring 21, Karlsruhe, 76131, Germany
*E-mail: michael.tuerk@kit.edu

3.1 Introduction

Design, development and synthesis of nanostructured materials, such as supported metal nanoparticles (NPs), are of particular interest for both scientific and industrial communities. The development of novel chemical processes that allow control of the particle size, size distribution and morphology of, for example, catalytic active nanostructures, discloses new possibilities for energy-related applications and automotive emission control.

Inorganic Materials Series No. 14
Unconventional Green Synthesis of Inorganic Nanomaterials
Edited by Silvia Gross

Published by the Royal Society of Chemistry, www.rsc.org

Noble metal nanoparticles (*e.g.*, gold, palladium, silver) and, in particular, platinum (Pt) nanoparticles have been demonstrated to be efficient catalysts for chemical reactions, such as hydrogenation, hydration and oxidation. The widespread conventional and established methods for the preparation of supported metal NPs, such as incipient wetness impregnation (IWI), co-precipitation, chemical vapour deposition (CVD) and flame spray pyrolysis (FSP) present some disadvantages and limitations. As an example, the synthesis of supported NPs prepared by wet impregnation of a substrate requires a subsequent reductive treatment and an energy-intensive drying step. Catalysts prepared by this procedure usually yield metal particles with a broad size distribution and involve large volumes of waste water.[1,2]

Ceria (CeO_2) is worth mentioning, since it is one of the most promising materials for environmental and energy-related applications. It can be used in various applications, such as catalysts, polishing agents, gas sensors and UV-shielding materials.[3–6] Further studies show that size- and shape-controlled CeO_2 synthesised in supercritical (sc) H_2O has high oxygen storage capacity, and that it is particularly useful for redox catalysis.[7] In addition, CeO_2 can be used as the active component of the catalyst or as the substrate of noble metal catalysts for CO and propane oxidation.[8] Reducing the CeO_2 particle size to the nanoscale has a significant influence on the catalytic behaviour.[9,10] Currently, many conventional formation processes for metal oxide nanoparticles, such as CVD, physical vapour deposition, liquid-phase precipitation, solid-state transformation or sol–gel methods rely on high temperatures or toxic reagents followed by time-consuming separation, purification and/or crystallisation processes.[11,12]

A promising environmentally friendly alternative to these conventional preparation methods is represented by the application of supercritical fluids (SCFs), such as supercritical (sc)CO_2 or scH_2O. Note that the supercritical state (see Figure 3.1) is defined as the region above the critical pressure and critical temperature of the fluid (*e.g.*, in the case of CO_2, 7.4 MPa and 304 K, while for H_2O significantly higher values of 22.1 MPa and 646 K are observed). Above the critical point of the fluid, only a single homogeneous phase exists due to the absence of phase boundaries. Furthermore, various fluid properties, such as density, viscosity, diffusion coefficient and surface tension, change significantly. The typical values of thermophysical properties of the gas, liquid and supercritical states are given in Table 3.1. It is obvious that SCFs are characterised by liquid-like densities and mass-transfer properties (viscosities and diffusivities) lying between those of gases and liquids.

The low critical temperature of CO_2 is the cause of its high compressibility at moderate temperatures. In this region, minor changes in the temperature or pressure result in large changes in the fluid density and hence in the solvent power. The influence of the density on the solubility of non-volatile organic substances, and thus on adsorption behaviour, is discussed in Section 3.2.1. Similar to density, the transport properties, such as viscosity, change with pressure and temperature. For example, at 313 K, the dynamic viscosity of CO_2 increases by a factor of nearly six as the pressure increases

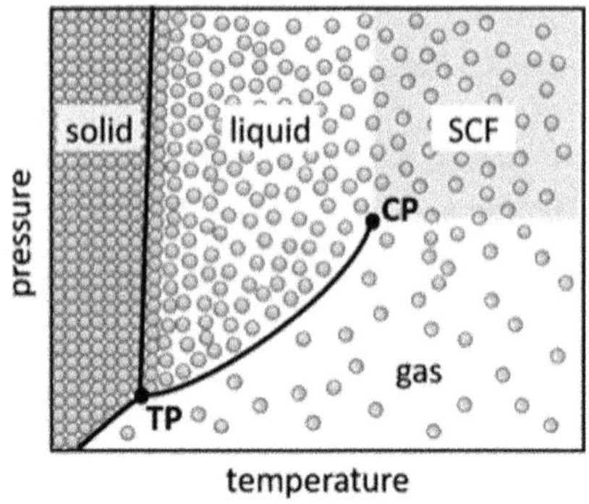

Figure 3.1 Schematic pressure–temperature diagram for a pure substance (TP is the triple point and CP is the critical point).

Table 3.1 Orders of magnitude for density, ρ, dynamic viscosity, η, diffusion coefficient, D, and surface tension, σ, in the gaseous, SCF, and liquid states.

	ρ/kg m^{-3}	η/Pa s	D/m^2 s^{-1}	σ/mN m^{-1}
Gas	0.6–2	10^{-5}	10^{-5}	0
SCF	$4–8 \times 10^2$	$10^{-4}–10^{-5}$	10^{-8}	0
Liquid	$8–12 \times 10^2$	10^{-3}	10^{-9}	72[a]

[a]Applies only for water.

from 0.1 to 30 MPa. By contrast, the kinematic viscosity decreases by two orders of magnitude, as the pressure increases from 0.1 to 7 MPa.[13] Thus, the kinematic viscosity is extremely low in the supercritical region, which is important for various applications in materials processing. In addition, the diffusivity in supercritical CO_2 is higher than that in liquid solvents. The high binary diffusion coefficient D_{12} and the absence of surface tension leads to enhanced mass-transfer rates and causes higher fluxes inside the pores of solid substrates, which is favourable for the impregnation of porous materials. Furthermore, since non-volatile organic substances can be easily dissolved in $scCO_2$, and due to the fact that $scCO_2$ is completely miscible with reacting gases such as hydrogen or oxygen as well as with the reaction products, it can act simultaneously as solvent, reactant and separation medium.

Compared to CO_2, H_2O is characterised not only by high critical data, but also by extraordinary physicochemical properties. The solvent behaviour of H_2O is controlled by its dielectric constant, which increases with increasing density and decreases with increasing temperature. Under ambient conditions, the dielectric constant of H_2O is 78, but it is significantly reduced to values around 2 to 10 at the supercritical state, which is a similar value to that of polar organic solvents. Thus, H_2O changes its character from a solvent for salts under ambient conditions to a solvent for non-polar compounds under supercritical conditions. As such, at low temperatures H_2O and salts are miscible, with H_2O and non-polar compounds being immiscible, while at higher temperatures organic compounds and light gases, such as H_2 or O_2, form a homogeneous phase with scH_2O. The change in the dielectric constant is proportional to the density and inversely proportional to the temperature. Furthermore, the influence of the temperature on the strength of hydrogen bonds is similar to that of the dielectric constant. The dielectric constant being a controlling factor of reaction rate, equilibrium and solubility of metal oxides is discussed in Section 3.2.4.

SCFs can dramatically influence the phase behaviour of mixtures. The high solubility of SCFs in liquids may cause the precipitation of previously dissolved solid compounds due to the anti-solvent effect. Adding SCFs to liquids significantly reduces their viscosity, making high-viscosity materials easier to handle. Furthermore, the increasing amount of SCF in the liquid phase causes a decrease in the surface tension. This enables mixtures containing SCFs to penetrate into small pores of complex porous substrates, and to deposit or remove substances in or out of porous structures without damaging the structure itself. In the following, the versatility and effectiveness of scCO_2- and scH_2O-based particle formation processes will be discussed with reference to various examples.

SCF-based particle formation processes are often classified as either physical (*e.g.*, rapid decompression, anti-solvent effect) or chemical (*e.g.*, hydrolysis, reduction) transformation processes.[1] Such transformations can be carried out completely in SCFs and usually involve the dissolution of a precursor in a suitable SCF followed by its chemical conversion to the desired material. In this chapter, the focus is on two different chemical transformation processes.[1,13]

The supercritical fluid reactive deposition (SFRD) process was first proposed by Watkins and McCarthy.[14] This process enables the deposition of mono- and bimetallic nanoparticles (<5 nm) on porous substrates (*e.g.*, Pt on MCM-41) and includes three major steps:

- dissolution of a metal precursor in scCO_2,
- adsorption of the metal precursor from scCO_2 onto a substrate, and
- conversion of the adsorbed precursor to its metal form.

The continuous hydrothermal synthesis (CHTS) process was pioneered by Adschiri *et al.*[11] This fast (<2 s) process enables the synthesis of small (<10 nm) metal oxide particles (*e.g.*, CeO_2) and can be subdivided into three steps:

- dissolution of a metal salt in cold compressed H_2O,
- mixing of the cold salt solution with hot compressed H_2O, and
- synthesis of metal oxide particles after mixing.

3.2 Fundamentals

3.2.1 Solubility of Metal Precursors in $scCO_2$

Knowledge of the phase equilibrium, especially the solubility of a pure solid component or a mixture of solids in SCFs, is essential for the design of particle formation processes and determination of the best operating conditions.[15] Thus, with regard to the SFRD process it would be extremely helpful to be able to answer the question: 'Is the substance of interest sufficiently soluble in $scCO_2$?', since insufficient solubility limits the practical applicability. A selected number of precursors, which have sufficient solubility in $scCO_2$, are summarised in Table 3.2.

Models based on equations of state, together with different mixing rules, are the most widely used to correlate and predict solubility in supercritical fluids. Therefore, accurate knowledge of the critical parameters (p_C, T_C, ρ_C), acentric factor, solid molar volume and sublimation pressure of the solids is essential. Due to the high critical temperatures and very low sublimation pressures, these data often cannot be determined by experiment. Common group-contribution methods, which are generally used to predict thermodynamic properties from molecular structures, are mostly empirical and often lead to inconsistent and unreliable results.[18] Further details on the numerous approaches used to predict the properties of pure substances can be found in the literature.[15,18] Thus, due to the lack of information on

Table 3.2 Molar mass, *M*, metal content and melting temperature, T_M, of selected precursors.

Precursor[a]	M/g mol^{-1}	Metal content/wt%	T_M/K	Reference
[Ag(acac)$_2$]	207	52.1	373	16
[Ag(hfac)(cod)]	422	25.6	395–397	16
[Au(acac)Me$_2$]	326	60.4	354–355	16
[Co(acac)$_3$]	356	16.5	483–486	16
[Cu(acac)$_2$]	262	24.3	557–561	16
[CuDI6]	610	10.4	387	17
[Cu(tmhd)$_2$]	430	14.8	471	16
[Fe(acac)$_3$]	353	15.8	453–455	16
[Fe(Cp)$_2$]	186	30.0	445–447	16
[Fe(tmhd)$_3$]	606	9.2	452–458	16
[Pd(acac)$_2$]	305	34.9	463	16
[Pd(hfac)$_2$]	521	20.4	373	16
[Pd(tmhd)$_2$]	473	22.5	511–513	16
[Pt(acac)$_2$]	394	49.5	522–525	16
[PtMe$_2$(cod)]	333	58.6	376–378	16
[Pt(hfac)$_2$]	609	32.0	416–418	16
[Rh(acac)$_3$]	400	25.7	536–537	16

[a]All acronyms used are explained in the List of Abbreviations.

the above-mentioned data, density-based models are often used for the correlation of experimental solubility data. One of the most commonly used density-based model, which correlates the solubility of a solid in an SCF to the density of the pure solvent, has been proposed by Stahl *et al.* and by Kumar and Johnston:[19,20]

$$\ln(b_2) = A \cdot \rho + B \tag{3.1}$$

In eqn (3.1), b_2 (mol kg^{-1}) is the solubility of the dissolved solid (using molality as the unit), ρ (mol dm^{-3}) is the density of the pure solvent, and A (dm^3 kg^{-1}) and B (mol kg^{-1}) are constants that are characteristic for the particular binary system and vary with temperature (see Table 3.3). As an example, the influence of the temperature and solvent density on the solubility of one common Pt precursor, dimethyl(1,5-cyclooctadiene)platinum(II) ([PtMe$_2$(cod)]) in scCO$_2$ is depicted in Figure 3.2.[21]

The experimental data show trends that are similar to those observed for a large number of solids in the supercritical region. The lines depicted in

Table 3.3 Constants A and B from eqn (3.1) for the solubility of [PtMe$_2$(cod)] in CO$_2$.

T/K	A/dm^3 kg^{-1}	B/mol kg^{-1}	Corr. R^2
313	0.248	−8.412	0.994
333	0.197	−6.769	0.994
353	0.213	−6.135	0.995

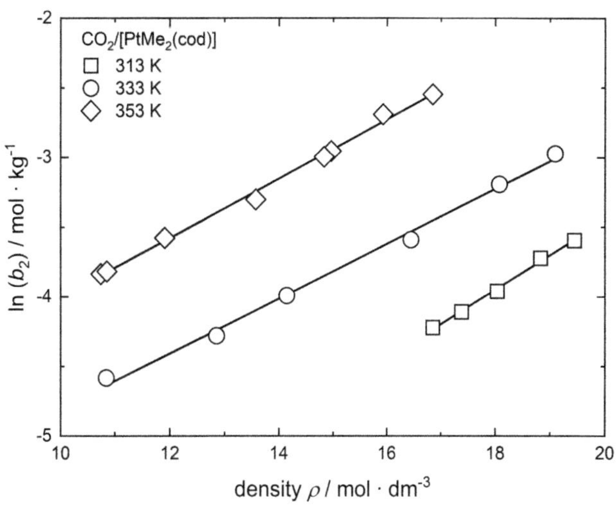

Figure 3.2 Solubility *vs.* solvent density for the CO$_2$/[PtMe$_2$(cod)] system. The solid lines represent the correlation eqn (3.1). Data from ref. 21.

Figure 3.2 are calculated with eqn (3.1) and confirm the linear relationship between the logarithm of solubility and the solvent density for all three isotherms. At constant temperature, the solubility of the dissolved solid increases almost linearly with the solvent density and thus the ability to dissolve solids. Figure 3.2 also shows the pronounced temperature effect on the solubility outside the retrograde region, *i.e.*, the solubility decreases as the temperature is increased isobarically, indicating exothermic dissolution of a pure solid into the SCF, whereas outside the retrograde region the solubility increases with temperature. In this region, the effect of the temperature on the solute sublimation pressure overlays the effect of solvent density.[18,21,22] More details about the different concepts used for the description of the solubility of solids in SCFs can be found in various extensive review articles.[19,20,23–32]

3.2.2 Adsorption Behaviour in Supercritical CO_2

The molecular picture of the adsorption behaviour in scCO_2 is complex, and parameters of adsorption isotherms derived from experimental data should always be interpreted with care.[33] The simplest form of an adsorption isotherm is a linear equation. In this approach, the uptake is proportional to the concentration of dissolved solid in the fluid phase. Usually, the linear adsorption behaviour will occur only in small ranges and the linear approach can be applied at low concentrations. For a non-linear adsorption behaviour, a number of models have been developed, *e.g.*, the Langmuir and the modified Langmuir equation, the Freundlich equation, the Sips equation, the ideal adsorbed solution theory or the Dubinin–Astakhov equation.[34–38] However, these widely used models can only be applied to adsorption isotherms with a single curvature, either convex or concave. For the description of isotherms that exhibit both curvatures, the more general isotherm of Hill should be used.[33,39] One of the most commonly used models for correlating adsorption isotherms is the Langmuir isotherm model and this is given by eqn (3.2):[34]

$$q = \frac{Q \cdot K \cdot c}{1 + K \cdot c} \tag{3.2}$$

In eqn (3.2), q (mol precursor per kg substrate) is the uptake at equilibrium conditions, Q (mol precursor per kg substrate) is the maximum monolayer adsorption capacity, K (kg CO_2 per mol precursor) is the Langmuir parameter related to free energy of adsorption and c (mol precursor per kg CO_2) is the concentration. An overview of adsorption isotherms for selected SFRD-relevant compounds is given in Table 3.4. Further details on the models mentioned in Table 3.4 for the description of adsorption isotherms can be found in the corresponding literature.

At this point it is worth mentioning that in batch processes the thermodynamics and kinetics of adsorption are only one step in the overall process, which involves dissolution of the precursor, its diffusion and adsorption to the surface and into the pores and, finally, reduction of the precursor to its

Table 3.4 Overview of adsorption isotherm data for metal organic compounds (precursors).

Precursor	Substrate[a]	*p*/MPa	*T*/K	Model	Reference
[Co(acac)$_3$]	MCM-41	12.5 15 20	333 313, 333, 353 333	Dubinin–Astakhov	41
[CuDI6]	CA	10.6	308	Langmuir Sips	17
[Ni(acac)$_2$]	CA	30	333	Linear	42
[Pd(acac)$_2$]	BP2000	20	333	Linear	43
[Pd(hfac)$_2$]	SBA-15	8.5	313	Langmuir Modified Langmuir Freundlich	44
[PtMe$_2$(cod)]	CA	10.6	308	Langmuir Sips	17
[PtMe$_2$(cod)]	BP2000	20	333	Langmuir	43
[PtMe$_2$(cod)]	(Al_2O_3 + CeO_2)	15.5, 25	353	Langmuir Modified Langmuir	45
[PtMe$_2$(cod)]	(Al_2O_3 + SnO_2)	15.5	353	Langmuir Modified Langmuir	45
[PtMe$_2$(cod)]	CA4	27.6	353	Langmuir	46
[PtMe$_2$(cod)]	CA4	27.7	353	Langmuir	47
[PtMe$_2$(cod)]	RFA	10.7, 20.7	308	Modified Langmuir	48
[Rh(acac)$_3$]	MCM-41 MSU-H HMS	15.0	313, 333, 353	Dubinin–Astakhov	49
[Ru(acac)$_3$]	SA	18.0	313, 353	Linear	50
[Ru(cod)(tmhd)$_2$]	CA	19.3	343	Langmuir Modified Langmuir Freundlich	51
[Ru(cod)(tmhd)$_2$]	CA4	27.6	353	Langmuir	52
[Ru(cod)(tmhd)$_2$]	CA4, CA21	27.6	353	Langmuir	53
[PtMe$_2$(cod)] and CuDI6	CA	10.6	308	Ideal adsorption solution theory	17

[a]All acronyms used are explained in the List of Abbreviations.

metal form. Thus, reliable data for transport properties, such as binary diffusion coefficients, are essential for the design of a large number of processes, such as particle synthesis, particle deposition on porous substrates or separation processes where mass transfer is involved as the dominant mechanism. For example, according to eqn (3.3), in a non-stirred high-pressure vessel with volume V, the time t to reach equilibrium strongly depends on, *i.e.*, is inversely proportional to, the binary diffusion coefficient D_{12}:

$$t \sim \frac{V^{\frac{2}{3}}}{D_{12}} \tag{3.3}$$

Typically, D_{12} is in the range of 0.5–1.5 × 10^{-8} m^2 s^{-1} and thus is more than one order of magnitude higher than in organic liquids under ambient conditions.[1] From eqn (3.3) it follows that the time to reach diffusion equilibrium is quite high. Assuming a non-stirred autoclave with V = 100 mL and D_{12} = 1 × 10^{-8} m^2 s^{-1}, nearly 60 hours are required to reach diffusion equilibrium. This is far longer than is typically applied in technical processes.[40] Therefore, in the case of batch processes, the use of a well-stirred autoclave is essential. With regard to process optimisation, the most important tendencies are that under isothermal conditions D_{12} decreases with pressure due to the density increase and increases only slightly with temperature at constant pressure. Furthermore, it is important to note that at isothermal–isobaric conditions $D_{12} \sim M^{-0.5}$.

Figure 3.3 shows the adsorption isotherms for [$PtMe_2$(cod)] at different pressures. It is obvious that increasing precursor concentration in $scCO_2$ leads to increased precursor uptake. In addition, the pressure (and temperature, not shown) has a significant effect on precursor uptake. An isothermal decrease in pressure leads to higher uptake, which can be explained by the solubility behaviour of the precursor. Note that, at a constant temperature, the solubility decreases with decreasing pressure while, at constant pressure, the solubility increases with temperature. The results shown in Figure 3.4 indicate that the interactions between [$PtMe_2$(cod)] and Al_2O_3 + CeO_2 are significantly lower than for [$PtMe_2$(cod)] and Al_2O_3 + SnO_2. Thus, the uptake of Pt precursor on Al_2O_3 + SnO_2 is about a factor of two higher than on Al_2O_3 + CeO_2.

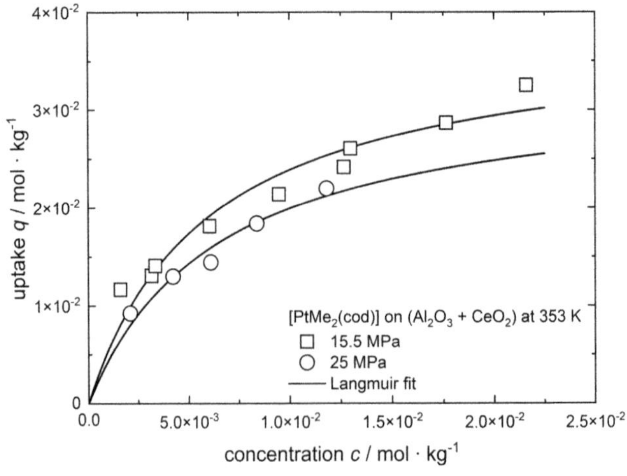

Figure 3.3 Influence of pressure on the [$PtMe_2$(cod)] uptake on Al_2O_3 + CeO_2. Data from ref. 45.

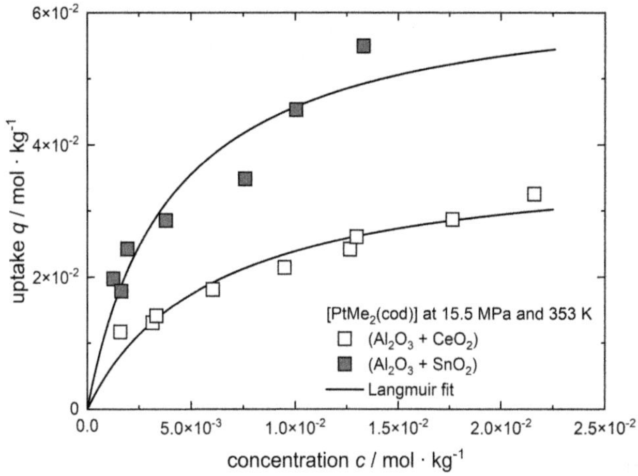

Figure 3.4 Influence of the chemical nature of the substrate surface on uptake. Data from ref. 45.

Table 3.5 Langmuir parameters from eqn (3.2) for the adsorption isotherms of [$PtMe_2$(cod)] on different substrates at 15.5 and 25 MPa and 353 K. Data from ref. 45

Substrate	*p*/MPa	*Q*/mol precursor per kg substrate	*K*/kg CO_2 per mol precursor	Corr. R^2
($Al_2O_3 + CeO_2$)	15.5	0.038	166.40	0.926
	25	0.033	154.04	0.941
($Al_2O_3 + SnO_2$)	15.5	0.064	248.63	0.873

The corresponding Langmuir parameters of eqn (3.2) are summarised in Table 3.5. From Table 3.5 it follows that the affinity parameter *K*, which is an indicator of how much of the substrate surface is covered by the adsorbed precursor, is about 1.5 times higher for the Al_2O_3 + SnO_2 system compared to the Al_2O_3 + CeO_2 system. Thus, the higher uptake is caused by the higher attraction of the [$PtMe_2$(cod)] molecules to the surface of the Al_2O_3 + SnO_2 system.

Note that the temperature and pressure dependence of the uptake can be explained by the $scCO_2$ density dependence of the precursor solubility. In the following, it is assumed that the solubility of the precursor in $scCO_2$ is below the equilibrium solubility under the given process conditions. Increasing pressure and decreasing temperature causes an increase in CO_2 density, which leads to an increase of the precursor amount in CO_2 (see Figure 3.2) and thus depletion of the adsorbed phase and thus lower precursor uptake.

Based on the results available in the literature, the following trends in the influence of pressure and temperature on the equilibrium uptake can be observed.[1]

- At constant pressure and fixed precursor concentration, a temperature increase leads to a decrease in the uptake.
- At constant temperature and fixed precursor concentration, a pressure increase leads to a decrease in the uptake (see Figure 3.3).
- At constant pressure and temperature and fixed precursor concentration, the uptake is strongly influenced by the chemical nature of the substrate surface (see Figure 3.4).
- Far above the critical point, the influence of pressure and/or temperature on the uptake is weak.

It is worth noting that opposite trends may be observed if an excess of the precursor exists, *i.e.*, adsorption is performed under saturated conditions. The uptake can be thermodynamically controlled and different adsorption isotherm models can be used to correlate the data.

Experiments performed by Bozbag *et al.* show a strong influence of uptake on the particle size.[48] The mean particle size of Pt nanoparticles increased with uptake from 2.0 nm at 10 wt% to 2.7 nm at 22 wt%, and up to 3.3 nm at 34 wt%. Thus, the mean particle size appears to increase monotonically with uptake. Such behaviour can be simply explained as follows: a higher concentration of Pt atoms leads to a more intensive particle coarsening and/or coalescence. More details about the relationship between uptake and particle size is given in Section 3.3.3 and is illustrated in Figure 3.9.

3.2.3 Solubility of Salts in H_2O

Knowledge of phase equilibria and the solubility of inorganic salts and metal oxides in H_2O is an important aspect for the development of particle synthesis processes. Among other factors, the solubility of metal oxides changes significantly around the critical point. First, the solubility increases with increase in temperature and then decreases under isobaric conditions. Furthermore, as discussed in more detail below, the solubility of salts decreases strongly in the near-supercritical and supercritical region with increasing temperature. This solubility behaviour is caused by the decrease in density and the dielectric constant. In addition, the strong Coulombic interactions between charged species lead to high critical temperatures for salt solutions. Therefore, H_2O is much more volatile than the salt, and the triple points of the salts are usually higher than the critical temperature of pure H_2O. Thus, the phase behaviour of a binary H_2O/salt mixture depends on whether the solid–liquid–gas three-phase line does not cut the critical line (type 1) or cuts it (type 2). As a rule, type 1 salts are highly soluble in dense scH_2O and this group includes most alkali and alkaline-earth halides and hydroxides, while the group of practically insoluble salts of type 2 include, among others, the sulfates.[54] Table 3.6 classifies some important salts according to their solubility behaviour.

Usually, in the literature, solubility data have been represented according to the water density, which considers pressure and temperature variations,

Table 3.6 Classification of salts according to their solubility in scH_2O. Reproduced from ref. 55 with permission from John Wiley & Sons, Copyright © 2018 WILEY-VCH Verlag GmbH & Co. KGaA, Weinheim.

Type 1 salts	Type 2 salts
$BaBr_2$, $BaCl_2$, $Ba(NO_3)_2$	BaF_2, CaF_2
KBr, KCl, K_2CO_3	K_2SO_4
LiBr, LiCl	Li_3PO_4, Li_2SO_4
NaBr, NaCl, $NaNO_3$, Na_2SeO_4	Na_2CO_3, Na_3PO_4, Na_2SO_4

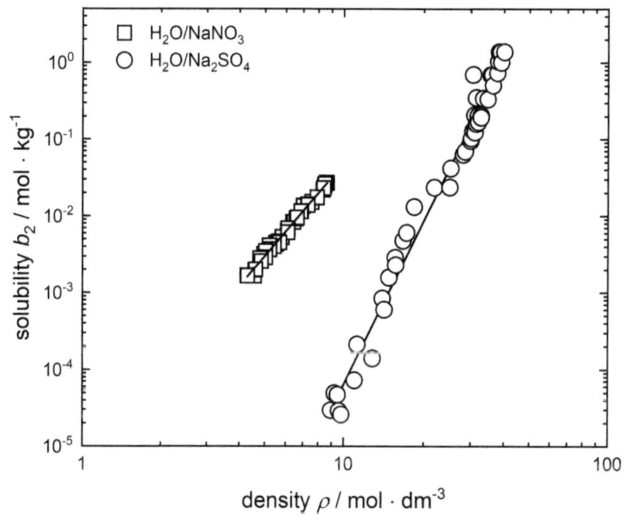

Figure 3.5 Solubility of $NaNO_3$ and Na_2SO_4 in sub and supercritical H_2O; lines are a guide to the eye. Data from ref. 56 and 57.

and is often the key parameter for salt solubility. Figure 3.5 shows the solubility of $NaNO_3$ (type 1 salt) and Na_2SO_4 (type 2 salt) in sub- and supercritical H_2O.[56,57] Similarly to a large number of other salts, a linear relationship between solubility and H_2O density is observed. It is clear that the solubility of $NaNO_3$ is up to three orders of magnitude higher than that of Na_2SO_4 in the identical sub- and supercritical water density ranges. Selected solubility data of salts in sub- and supercritical H_2O are summarised in Table 3.7.

More details about the complex phase behaviour of different H_2O/salt mixtures, experimental solubility data and the different modelling approaches can be found in various review articles.[54,57,58]

3.2.4 Kinetics of Supercritical Hydrothermal Synthesis

The basics of subcritical hydrothermal synthesis are thoroughly described in Chapter 6, and this approach can be applied to the preparation of several inorganic systems, ranging from oxides, to sulfides, to metal nanoparticles,

Table 3.7 Range of density, ρ, and solubility, b_2, data of inorganic compounds (using molality as the unit) in sub- and supercritical water[a]

Salt	p/MPa	T/K	ρ/mol dm^{-3}	b_2/mmol kg^{-1}	Reference
$Al_2(SO_4)_3$	15–29.2	619–675	3.5–36	0.89–1.63	59
$CaCl_2$	18.8–23.3	669–688	4.6–8.1	0.57–1.93	60
$CaHPO_4$	20.5–24.2	665–690	5.5–18	Data n.a.	61
$CaSO_4$	18.8–23.2	655–675	4.9–11	Data n.a.	61
CuO	7.6–41.4	330–824	1.2–56	0.0002–0.008	62
CuO	27.9–28.1	573–723	7.2–42	0.0013–0.008	63
KCl	17.9–24.2	667–680	4.5–9.3	1.77–22.78	56
KCl	9–12	643–673	1.9–2.9	0.013–0.084	64
KNO_3	20.1–24	675–682	5.3–8.1	2.48–12.4	56
KOH	22.1–30.4	696–798	4–11	1.09–10.6	65
K_2HPO_4	24.9–30.9	673–723	6–22	0.01–2.39	65
K_2SO_4	25	647–655	23–29	115–459	66
K_2SO_4	18.1–26.7	633–723	3.8–33	0.28–221.3	67
K_3PO_4	27–30	397–478	49–53	0.21–88	68
LiCl	19.6–23.6	661–692	4.8–9.8	4.72–15.45	56
$LiNO_3$	18.3–23.7	663–679	4.6–9.4	2.28–23.26	56
$MgCl_2$	19.4–22.8	667–668	5.5–7.9	1.04–3.70	60
$MgSO_4$	18.8–23.3	654–675	5.0–13	0.012–0.11	61
NaCl	17.1–23.5	654–685	4.1–13	1.44–87	56
NaCl	9–12	623–673	1.9–3.2	0.02–0.21	64
NaCl	10–25	723–823	1.6–6.1	0.015–1.73	69
NaCl	1.4–10.3	673–823	0.2–1.6	0.000019–0.019	70
NaCl	10–25	568–598	2.8–42	57–342	71
NaCl	11–15	553–613	3.0–42	3.7–208	72
$NaNO_3$	17.3–23.1	663–680	4.2–8.7	1.65–27.1	56
NaH_2PO_4	20.3–24.2	671–695	5.0–8.7	1.36–6.62	61
Na_2CO_3	25–30	371–492	48–54	0.19–111	68
Na_2CO_3	24.2–25	651–711	6.1–27	0.36–49.8	73
Na_2HPO_4	21–23.2	668–681	5.8–8.1	0.55–4.42	61
Na_2SO_4	25	616–636	32–36	211–704	66
Na_2SO_4	18–26.4	643–723	3.8–31	0.0071–100.8	67
Na_2SO_4	25	723	6.1	0.013	69
Na_2SO_4	11–15	553–613	3.0–42	2.3–124.5	72
Na_2SO_4	24.3–24.6	655–670	9–21	0.03–6	73
Na_2SO_4	19–30.6	626–648	6.3–36	65–209	74
Na_2SO_4	25	647–778	4.9–29	0.003–69.7	75
Na_2SO_4	20–25	594–638	28–39	352–1411	76
Na_2SO_4	25	673	9.3	0.028	77
PbO	25.9–34.3	523–774	5.2–46	0.351–8.246	63

[a]n.a. = not applicable.

as is extensively reported in Chapter 6. In the hydrothermal synthesis of oxides, an ionic metal salt is first hydrolysed to the metal hydroxide, which is then dehydrated to form metal oxide particles, which precipitate from the aqueous solution. Formally, the hydrothermal synthesis of metal oxide particles can be carried out at subcritical conditions in batch reactors. In contrast to batch reactors, which are used for long duration processes, flow reactors are used for short duration syntheses. In the early 1990s, Adschiri *et al.* showed that very small metal oxides could be synthesised under supercritical conditions *via* continuous hydrothermal synthesis within a very short time (<2 s).[11] Under such conditions, according to the theory of Born, the rate constant k (s^{-1}) can be expressed as:[78]

$$\ln\frac{k}{k_0}=\frac{\psi}{RT}\cdot\left(\frac{1}{\varepsilon}-\frac{1}{\varepsilon_0}\right)-\frac{E}{RT} \quad (3.4)$$

where k_0, ψ and E are constants that depend on the particular reaction system, R is the gas constant, T is the temperature and ε is the dielectric constant. Note that k_0, ψ and E are adjustable parameters that were fitted to experimental data in the range 473 K to 613 K. Therefore the authors chose ε_0 as the dielectric constant at 473 K.[79]

The key point here is that, in contrast to subcritical conditions, in the supercritical region the reaction rate increases dramatically due to the decreased dielectric constant of H_2O and the increased dissociation to H^+ and OH^-. The influence of the temperature on ε, and therefore on k, is clarified in Figure 3.6, for the formation of CeO_2 at 30 MPa. It is worth mentioning that the rate constants above 613 K were estimated using the parameters given by Aoki *et al.*[79] It is clear that, in the temperature range below the critical temperature, the rate constant k falls on a straight line, but increases significantly above the critical temperature. An opposite behaviour is observed for ε; when the temperature increases to the supercritical region, ε is decreases significantly from *ca.* 32 at 500 K to around 2 at 800 K.

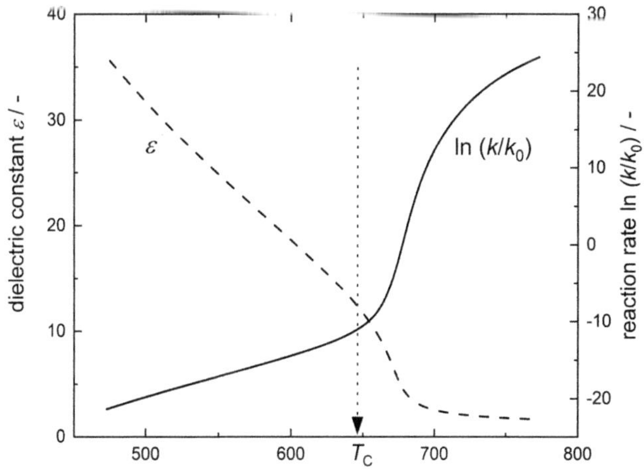

Figure 3.6 Influence of temperature on ε and Arrhenius plot of the first-order rate constant at 30 MPa for the synthesis of CeO_2. Values for ε were taken from Grigull.[80] The $\ln(k/k_0)$ curve was calculated using eqn (3.4) with the parameters given by Aoki *et al.*[79] Data from ref. 79 and 80.

A further major advantage of supercritical conditions is that both the salt and the metal oxide solubility decrease significantly in this region. This behaviour occurs because the density and the dielectric constant of H_2O decrease. The decrease in solubility leads to a high degree of supersaturation. According to nucleation theory, the nucleation rate depends on the degree of supersaturation and the surface energy.[81] Thus, if the solution is rapidly heated to the supercritical state, an extremely high nucleation rate in the range of 10^{23} to 10^{25} cm^{-3} s^{-1} can be expected, which leads to the formation of small nanoparticles of metal oxides.[13]

3.3 Synthesis of Supported Nanoparticles Using SFRD

As mentioned in the introduction, the established classical methods for the preparation of supported metal NPs, such as IWI, co-precipitation, CVD and FSP have some disadvantages and limitations. Among these, IWI is not suitable for aerogels due to the surface tension of the liquids used, which can cause collapse of the pores. In the case of co-precipitation, controlling the homogeneity of the particle size and its distribution is a major challenge. CVD requires relatively high temperatures and highly volatile metal precursors, which usually require complex methods of synthesis, and are expensive and/or toxic. FSP requires the use of liquid solutions, and the high temperatures required to reduce the precursors can cause undesirable by-products. Another drawback of these techniques is the fact that they cannot easily be applied to polymeric substrates.

A recent a review article by Xu *et al.* provides a detailed overview of the chemistry in SCFs for the synthesis of metal NPs.[82] Note that this review focuses exclusively on the application of supercritical fluids in the synthesis of non-supported metal NPs in both batch and flow reactors.

3.3.1 What Makes $scCO_2$ so Special as a Process Medium?

SCF-based deposition of metals on complex surfaces has been attracting increased attention due to the unique properties of SCFs. The advantages of using SCFs, especially $scCO_2$, as the process medium can be summarised as follows.

- The solvent power can be adjusted with small changes in pressure and temperature.
- Due to the liquid-like density of $scCO_2$, the dissolved metallic precursor concentration is significantly higher than in CVD (by up to three orders of magnitude).
- A solvent-free and dry product is obtained, because no liquid waste or by-products are generated.
- Due to the high binary diffusion coefficient D_{12}, mass-transfer rates are faster than in liquids.
- The absence of surface tension favours an effective nanostructuring of complicated pore architectures and complex geometries.

- Complete miscibility with reacting gases, such as hydrogen or oxygen, is observed.

3.3.2 Conversion of Adsorbed Metal Complex to Metal Nanoparticles

SFRD can be used to deposit metallic nanoparticles directly on a large number of solid substrates (*e.g.*, TiO_2, Al_2O_3, black pearls (BP2000), carbon aerogels (CA), carbon nanotubes). In this process, the SCF acts simultaneously as a solvent, and reaction and separation medium.[1,2,13] The SFRD technique involves the dissolution of an organometallic precursor in $scCO_2$ and exposure of the substrate to the solution. After sufficient time for adsorption, the precursor will be transformed to its metal form, either with a reducing agent such as H_2 or by high temperature. Thereby, four different methods can be used to convert the organometallic precursor to its metal form, as follows:

1. chemical reduction in $scCO_2$ with H_2,
2. chemical reduction at ambient pressure with H_2,
3. thermal reduction in $scCO_2$, and
4. thermal reduction at ambient pressure in an inert atmosphere.

Typically, method (1) is the usual way to synthesise metallic nanoparticles. More details about the different methods used for the conversion of the precursor to its metal form can be found in the literature.[1,13] In the case of the formation of binary NPs, two paths can be followed: the simultaneous deposition of both metal precursors or the subsequent deposition of the individual precursors.

It has been pointed out by Zhang and Erkey that transfer of the metallic precursor from the CO_2 phase to the surface of the substrate and the reduction of the precursor to its metal form by H_2 occur simultaneously.[83] Until now, only a few studies have been carried out to understand the reaction mechanism and kinetics. Experiments performed by Saquing *et al.* clarified that the reduction of [$PtMe_2$(cod)] according to method (1) can be described by the following chemical equation:

$$C_{10}H_{18}Pt + 3H_2 \rightarrow Pt + 2CH_4 + C_8H_{16}$$

and that the reaction is autocatalytic.[84]

In a typical SFRD experiment, the precursor and the substrate are placed in a high-pressure reactor; thereafter the system is treated in $scCO_2$ at process conditions for a certain time. During this time, the precursor is dissolved in CO_2 and adsorbed onto the substrate. Subsequently, a stoichiometric excess of pure H_2 is introduced into the reactor at isothermal and isobaric conditions and the precursor is converted to its metal form.

Figure 3.7 shows the influence of different substrates, and thus of different specific surface areas, on the particle size distribution (PSD) of Pt.[85] In these experiments [$PtMe_2$(cod)] was used as precursor and method

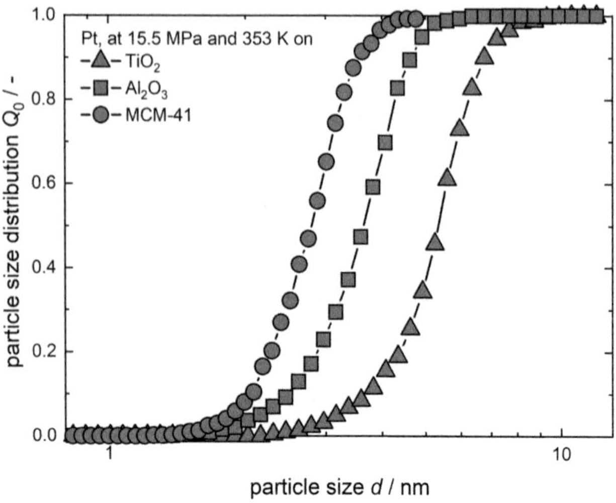

Figure 3.7 PSD of Pt NPs on TiO_2, on Al_2O_3 and on MCM-41. Particles were synthesised by method (1) using [PtMe$_2$(cod)] as precursor. Data from ref. 85.

(1) was used to convert the precursor into its metal form. Thereby, the amount of the [PtMe$_2$(cod)] was fixed at 50 mg, which corresponds to a concentration of 6.3×10^{-3} mol [PtMe$_2$(cod)] kg^{-1} CO_2. It is important to note that all experiments were performed below the solubility limit of [PtMe$_2$(cod)] in CO_2. It is clear that the Pt NPs are characterised by different particle sizes but by a similar PSD. The smallest particles are deposited on MCM-41 ($A_{BET} \approx 900$ m^2 g^{-1}) with a mean size of $d_{50} = 2.9$ nm, while the largest particles, with a mean size of $d_{50} = 5.3$ nm, are obtained for TiO_2 ($A_{BET} \approx 60$ m^2 g^{-1}). Thus, these results suggest that a higher surface area and thus less opportunity for particle aggregation and growth leads to smaller particles.

The mean particle size (PS) and PSD of Pt and Au particles synthesised from [PtMe$_2$(cod)] or [AuMe$_2$(acac)] (acac = acetylacetonate) are depicted in Figure 3.8. Clearly, Al_2O_3 ($A_{BET} \approx 155$ m^2 g^{-1}) supported Pt particles with a mean PS of $d_{50} = 3.7$ nm, which is significantly smaller than the Au particles ($d_{50} = 11.3$ nm).

From these results it follows that the mean PS and PSD are significantly influenced by the properties of the substrate (specific surface area and chemical nature), the type and the amount of the precursor in the system, and thus by the adsorption behaviour.

3.3.3 Control of Particle Size

Due to the large differences in the specific surface areas of the various substrates, it is more suitable to use the so-called 'number of molecular layers' (NML) instead of the metal uptake for purposes of comparison:[40]

$$\mathrm{NML} = \frac{N_\mathrm{A} \cdot n \cdot a}{A_\mathrm{BET}} \tag{3.5}$$

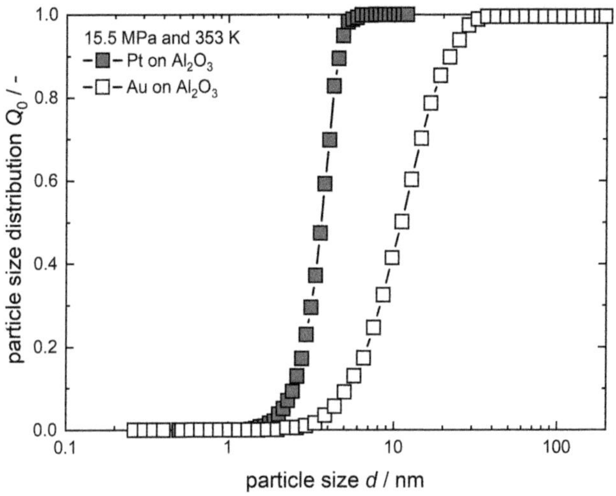

Figure 3.8 Influence of the type of precursor on PSD of Pt and of Au on Al_2O_3, particles were synthesised by method (1) using [$PtMe_2$(cod)] or [$AuMe_2$(acac)] as precursor. Data from ref. 85 and 86.

In eqn (3.5), N_A is the Avogadro number (6.02×10^{23} mol^{-1}), n (mol g^{-1}) is the metal uptake, a is the projection area of either Pt or Au atoms (9.8×10^{-20} m^2 or 9.5×10^{-20} m^2, respectively) and A_{BET} (m^2 g^{-1}) is the specific surface area of the substrate. This approach considers both the specific surface area of the substrate and the uptake, and therefore the intensity of the substrate–CO_2 interaction.[40,87] Figure 3.9 shows that the size of the supported NPs increases with increasing uptake and/or decreasing specific surface area of the substrate and thus with increasing NML. In the case of Pt, the PS increases from 2.1 nm at NML = 0.034 (BP2000, $A_{BET} \approx 1400$ m^2 g^{-1}) up to 6.7 nm at NML = 0.78 (TiO_2), while significantly larger Au particles are produced ranging from 3 nm on BP2000, up to 20 nm on TiO_2.

Thus these results again confirm that increasing surface area and thus less opportunity for particle aggregation and growth leads to smaller particles. Furthermore, the results also demonstrate that the PS of the supported particles depends strongly on both the surface area and the chemical nature of the substrate. This latter significantly influences the strength of the molecular interactions between the precursor and the substrate surface, which is confirmed by the results obtained from adsorption measurements (see Figure 3.4). Furthermore, one should consider that the coalescence rate of particles from high melting-point substances, such as Au (T_M = 1337 K) and Pt (T_M = 2041 K), is low. The reason therefore is that at ambient temperatures the solid-state diffusion coefficient, $D \propto \exp(-T_M/T)$, is low, but it increases significantly at temperatures higher than 0.75 T_M. Note that the melting temperature of Au clusters with a diameter of 4 nm is about 970 K. In a typical SFRD experiment the reduction of the precursor is performed at

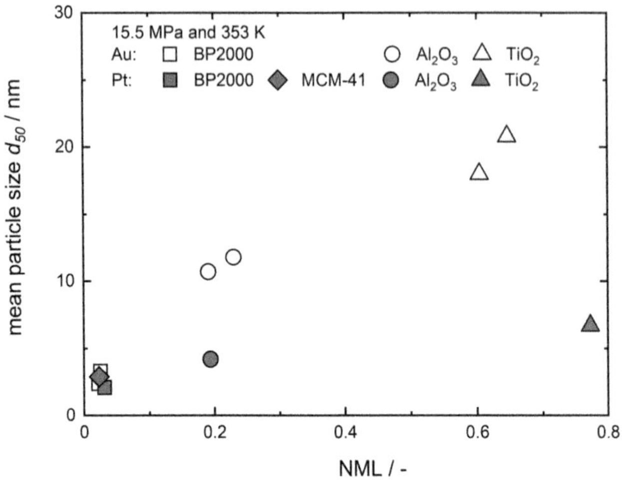

Figure 3.9 Influence of uptake and surface area on PS of Pt and Au NPs. The particles were synthesised by method (1), using [$PtMe_2$(cod)] or [$AuMe_2$(acac)] as precursor.

a temperature of 353 K, which is much lower than 0.75 T_M (727 K). Of course, this feature requires more detailed experimental and theoretical investigation of the effects of different CO_2–precursor–substrate interactions, and is part of current and future research activities.

Based on numerous experimental results published in the literature, one can summarise that the particle size of NPs synthesised by SFRD is influenced by pressure, temperature, specific surface area and precursor concentration in $scCO_2$ in the following way.[1,13]

- Increasing the reduction temperature leads to an increase in PS and a wider PSD, while increase of the reduction pressure results in a smaller PS and a narrower PSD.
- Increasing the specific surface area of the substrate results in highly dispersed and therefore smaller NPs.
- A higher precursor concentration in $scCO_2$, and hence higher uptake, leads to larger particles.

The SFRD process enables the formation of highly dispersed and uniform NPs with mean particle sizes generally smaller than 2.5 nm. It is even possible to synthesise core–shell particles or nanorods by selecting suitable process conditions. It is worth considering that the observed increase in the particle size with increasing precursor concentration is consistent with the previous experimental results and classical nucleation theory. Note that the nucleation rate is directly proportional to the square of the concentration of condensable molecules in $scCO_2$ and therefore with solute concentration.[81]

3.4 Synthesis of MO_x Nanoparticles Using Near- or Supercritical H_2O

Water has received increasing attention as a solvent and/or reaction medium because it is safe, non-toxic, readily available, inexpensive and environmentally benign. In the sub- and supercritical states, H_2O can be gradually compressed from gas-like to liquid-like densities. In this region, H_2O is characterised by extraordinary properties, which can be tuned by variation in pressure and/or temperature.

3.4.1 What Makes scH_2O so Special as a Process Medium?

The polar characteristics of H_2O at ambient temperature and pressure lead to high solvent power for salts or other polar compounds, but to less solvent power for non-polar molecules. However, the properties of H_2O change dramatically above its critical point. The dielectric constant of scH_2O varies in the range from around 2 at 800 K and 30 MPa, to 15 close to the critical point, and it is 78 at 298 K and 0.1 MPa.[80] These high values under ambient conditions result from preferred dipole orientations in the water structure, while the low values under supercritical conditions are the result of the breakdown of strong hydrogen bonding at these very high temperatures and pressures. This means that scH_2O becomes less polar and can dissolve non-polar substances, or shows a low solubility for salts and metal oxides. This feature, among others, leads to earlier supersaturation and to the precipitation of small metal oxide particles.

It has been pointed out by Weingärtner and Franck that the dissociation equilibrium of H_2O ($H_2O \rightleftharpoons H^+ + OH^-$) is a key property for a large number of applications.[54] Approaching the critical point, the self-dissociation of H_2O increases up to three orders of magnitude compared to ambient conditions, but is up to ten orders of magnitude lower at 800 K. These changes have an enormous influence on both hydrolysis and acid–base equilibria, and therefore on the solubility and chemical equilibrium of substances formed by metal ions in sub- and supercritical H_2O.[54,88]

Thus, changes in the dielectric constant and dissociation equilibrium of H_2O influence the solvent behaviour in the following ways.[54]

- Compared to ambient conditions, the solubility of inorganic salts is strongly reduced at sub- and supercritical conditions.
- Compared to ambient conditions, the solubility of non-polar compounds is high under near- and supercritical conditions.
- The chemical equilibrium of solvated metal ions shifts towards the corresponding metal oxide or hydroxide above the critical point.
- The solubility of gases, such as O_2, N_2, H_2 or CO_2, in scH_2O is unlimited.

Furthermore, diffusion rates are high and viscosities are low even at liquid-like densities, which enhances mass transfer and diffusion-controlled chemical reactions.

3.4.2 Synthesis of MO_x Particles from Aqueous Metal Salt Solutions

In simplified form, the synthesis of metal oxides occurs in two consecutive (equilibrium) reaction steps.[11,89]

1. Hydrolysis of the metal ions and formation of metal hydroxides.
2. Dehydration of the metal hydroxides to metal oxides and H_2O.

With increasing temperature, the equilibria of compressed aqueous metal salt solutions shift in the direction of metal oxides.[11,88,89] Furthermore, since the solubility of metal ions decreases by several orders of magnitude by switching from ambient to supercritical conditions, high supersaturations can be achieved. According to Sue *et al.* the driving force for precipitation, the supersaturation S, is defined as follows:[90]

$$S = \frac{c_0 - b}{b} \tag{3.6}$$

where c_0 (mol kg^{-1}) is the metal concentration of the starting solution and b (mol kg^{-1}) is the estimated metal oxide equilibrium solubility under the process conditions. With regard to the synthesis of NPs, achieving high supersaturations is absolutely necessary, since high supersaturation leads to a high nucleation rate and thus to small particles.[81] This can be easily realised by sudden mixing of the cold aqueous salt solution with hot compressed H_2O under near- or supercritical conditions. Note that batch reactors are usually not able to create fast supersaturations due to their lower heating rates, which leads to formation of particles in the micrometre range.

The reactions taking place during CHTS can be summarised as a combination of rapid hydrolysis (often several steps) and dehydration of the metal ions.

1. Hydrolysis	$M^{x+} + xOH^- \rightleftharpoons M(OH)_x$
2. Dehydration	$M(OH)_x \rightleftharpoons MO_{x/2}\ (s) + x/2\ H_2O$
3. Equilibrium	$MO_{x/2}\ (aq) \rightleftharpoons MO_{x/2}\ (s)$

According to reaction (1), the metal salt is first hydrolysed to metal hydroxide, which is then dehydrated, according to reaction (2), forming metal oxide particles in the nano to micro size range by precipitation from solution. In the chemical reactions (2) and (3), (aq) and (s) refer to the aqueous species and the crystalline solid oxide. It is worth noting that supercritical conditions favour rapid hydrolysis of metal salts and subsequent reaction to form hydroxides or oxides. The equilibria of these reactions shift to the formation of hydroxides and oxides with increasing temperature. Furthermore, in the supercritical region, due to the lower solubility of the reactants and lower dielectric constants, high reaction rates, and thus significantly shortened reaction times, are typical for hydrothermal reactions. Note that in such reactions H_2O acts as a reactant and as a solvent.[91]

The synthesis of CeO_2 from $Ce(NO_3)_3$ under hydrothermal conditions can be described by the following chemical equation:

$$2Ce(NO_3)_3 + 4H_2O \rightleftharpoons 2CeO_2 + 6HNO_3 + H_2$$

Among others, CHTS experiments have been performed with $Ce(NO_3)_3$ as precursor. In these experiments the concentration was varied between 0.005 mol dm^{-3} and 0.15 mol dm^{-3} and the mixing temperature was either 573 or 673 K, the total mass flow rate ranged from 35 g min^{-1} to 55 g min^{-1}, pressure was 25 MPa or 30 MPa, while the mixing ratio a (= cold/hot H_2O) was set to 0.1.[92]

Additional experiments were conducted with $Ce_2(SO_4)_3$ at 523 K and 573 K and 30 MPa. The results obtained are summarised in Figure 3.10, which shows the relationship between the mean particle size and the residence time, τ, which is inversely proportional to the mass flow rate. Clearly, the mean particle size increases with increasing residence time, which is in agreement with nucleation theory (see Section 3.2.4). A selected number of NPs made *via* the CHTS process are summarised in Table 3.8. Another summary of particles synthesised *via* CHTS can be found in the review by Hayashi and Hakuta.[93]

Clearly, the mean particle size decreases with decreasing residence time, from around 50 nm and τ = 1.5 s to 14 nm and τ < 0.5 s, due to improvement in mixing intensity and less time available for particle growth. Furthermore, particle size decreases with increasing temperature, *i.e.*, at 573 K large but uniform particles were synthesised, while at 673 K small nanoparticles less than 20 nm were produced. In addition, these results show that under

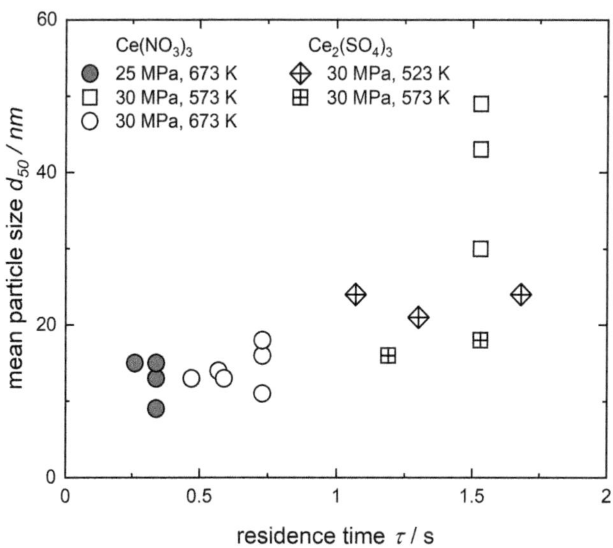

Figure 3.10 Influence of pressure, temperature and residence time as well as type of salt on the mean particle size of CeO_2. Data from ref. 92.

Table 3.8 Metal oxide nanoparticles synthesised by continuous hydrothermal synthesis.

Salt	Products	*p*/MPa	*T*/K	PS/nm	Reference
$Al(NO_3)_3$	AlOOH	35	673–763	600	11
$Al(NO_3)_3·9H_2O$	AlO(OH) Al_2O_3	30	673	5.6	90
$Al(NO_3)_3·9H_2O$	AlOOH	30	623–673	n.a.	94
$Al(NO_3)_3$	AlOOH	25–40	523–763	100–1500	89
$Al(NO_3)_3·9H_2O$ KOH, HNO_3	γ-Al_2O_3 AlOOH	30–40	673	2.8–18	95
$Ce(NO_3)_3·6H_2O$	CeO_2	30	523–673	20–220	96
$Ce(NO_3)_3·6H_2O$	CeO	30	573–653	10–50	79
$Co(NO_3)_2$	Co_3O_4	35	673	100	11
$Cu(NO_3)_2·3H_2O$ KOH	CuO	30	523	50–100	97
$Cu(NO_3)_2·3H_2O$	CuO	30	673	31	90
$Cu(NO_3)_2·3H_2O$ $C_2H_5NO_2$ (glycine) KOH	Cu_2O	20	523	5	97
$Fe(NO_3)_3$ $Fe_2(SO_4)_3$ $FeCl_2$	α-Fe_2O_3	35	673	50	11
$Fe(NO_3)_3·9H_2O$	Fe_2O_3	25–37.5	673	5.1–8.2	90
$Fe(NH_4)_2H(C_6H_5O_7)_2$	Fe_3O_4	35	673	50	11
$Gd(OH)_3$	$Gd(OH)_3$	30	773	30	98
$Ni(NO_3)_2$	NiO	35	673	200	11
$Ni(NO_3)_2·6H_2O$ KOH	NiO	30	673	3.4–16.7	90
$Ni(C_4H_6O_4)$ $C_6H_9N_3O_2$ (Histidine) $NiCO_3$	NiO	25	473–648	n. a.	99
$SnCl_2·5H_2O$ NaOH	SnO_2	25	523–673	3.5–9	100
$Ti(O^iPr)_4$ $Ba(O^iPr)_2$ (O^iPr = isopropoxide)	$BaTiO_3$	16	423–653	15–36	101
LiC_2H_5O $TiC_{12}H_{28}O_4$ (TTIP)	$Li_4Ti_5O_{12}$	25	623–773	2–20	102
$Ti(SO_4)_2$	TiO_2	35	673	20	11
$TiCl_4$	TiO_2	30	673–723	20	11
$Zn(NO_3)_2·6H_2O$	ZnO	30	663	n.a.	103
$Zn(NO_3)_2·6H_2O$ KOH	ZnO	30	657–673	27–57	104
$ZrOCl_2$	ZrO_2	30	673–763	10	11
$ZrO(NO_3)_2·2H_2O$	ZrO_2	30	673	4.3–4.7	90
$Al(NO_3)_3·9H_2O$ $Y(NO_3)_3·6H_2O$ $TbCl_3·nH_2O$ KOH	$Al_5Y_3O_{12}$:Tb (= YAG:Tb)	30	623–693	100	105
$Ba(NO_3)_2$ $ZrO(NO_3)_2·2H_2O$ KOH	$BaZrO_3$	30	673	7–14[a]	106

Table 3.8 (*Continued*)

Salt	Products	*p*/MPa	*T*/K	PS/nm	Reference
$Ba(NO_3)_2$ $Fe(NO_3)_3 \cdot 9H_2O$ KOH	$BaO \cdot 6Fe_2O_3$	30	723–773	100–200	107
$CoSO_4$ Al_2SO_4 NaOH	$CoAl_2O_4$	38	673	10	108
KOH Nb_2O_5 Nb_2O_5, PS 100 nm	$KNbO_3$	25–30	623–683	200–1000	109
LiOH $Co(NO_3)_2$ O_2	$LiCoO_2$ Co_3O_4	30 30	673 573, 623	500–1000 n.a.	88
$Co(NO_3)_2 \cdot 6H_2O$ $LiOH \cdot H_2O$ H_2O_2	$LiCoO_2$ Co_3O_4, CoO	30	573–684	550	110
LiOH $Co(NO_3)_2$ H_2O_2	$LiCoO_2$	30	673	600–1000	111
LiOH $Fe(SO_4)_2(NH_4)_2 \cdot 6H_2O$ o-H_3PO_4	$LiFePO_4$	25	523–673	n.a.	112
LiOH $Mn(NO_3)_2$ H_2O_2	$LiMn_2O_4$	30	673	100–500	111
$Mg(NO_3)_2 \cdot 6H_2O$ $Fe(NO_3)_3 \cdot 9H_2O$ KOH	$MgFe_2O_4$	30	673–733	20	113
TiO_2 $Ba(OH)_2 \cdot 8H_2O$	$BaTiO_3$	30	673	18–35	114

[a]Estimated from Figure 4 in ref. 106.

constant process conditions, *i.e.*, constant residence time, the mean particle size increases with increasing reactant concentration, from around 12 nm at the lowest salt concentrations, up to 20 nm at higher concentrations, which is in accordance with classical nucleation theory.[81]

3.4.3 Control of Particle Size

One of the main goals of basic research in chemical engineering is the systematic development of high-throughput processes for industrial production. Supercritical hydrothermal syntheses of nanoparticles can be performed either by using batch-type or flow-type reactors. More details about the different reactor concepts can be found in review articles by Hayashi and Hakuta, Ruiz-Jorge *et al.* as well as Darr *et al.*, which discuss specific features of supercritical hydrothermal synthesis and summarise the particle sizes of various materials obtained either from hydrothermal batch reactors or from continuous hydrothermal supercritical synthesis.[93,115,118]

Conventional batch reactors are easy to handle and are suitable for laboratory-scale production. However, the fast, precise and reliable control of process and reaction conditions, as well as the development and implementation of large-scale production, are problematic. In contrast, flow-type reactors enable the precise control of reaction conditions and high-throughput production. Thus, in order to enhance the effectiveness of flow-type reactors, the development of a design method is required. Therefore, the influence of different process parameters (flow rate, reaction temperature and mixer channel size) on the supercritical nanoparticle synthesis in a flow-type reactor has been examined.[79] The results of these experiments showed that rapid mixing and higher process temperatures favour the formation of small CeO_2 nanoparticles. The obtained results were correlated with the Reynolds number (Re) and showed that the size of the synthesised CeO_2 particles is independent of the Re number at values higher than a certain value (*e.g.*, Re = 2.5×10^4 at 653 K). This threshold value for the Re number for reaction-controlled conditions depends on the reaction temperature. Since the Re number does not consider the reaction rate, the results were also correlated using the Damköhler (Da) number, which is defined as the ratio of reaction to mixing rate. Typically, the Da number increases with increasing temperature at a fixed mean flow velocity, since k increases sharply at supercritical temperatures (see Figure 3.6 and eqn (3.4)), while the PS decreases with decreasing Da number and becomes constant at Da $<10^{-1}$. Based on these results, the authors concluded that, in contrast to the Re number, a threshold value of the Da number of 10^{-1} is valid for a fixed reaction system and is independent of the reaction temperature. Thus, for process conditions where the particle size is independent of the flow rate, the effect of mixing rate can be predicted from the threshold value of the Da number. These results suggest that this approach enables the prediction of reaction-controlled conditions under which the particle size is independent of the flow/mixing rate and reaction temperature. Nevertheless, this finding requires more detailed experimental and theoretical investigation so that these first results, based on the hydrothermal synthesis of CeO_2, can also be transferred to other, more complex chemical reactions.

The most important results of the various experimental investigations published in the literature can be summarised as follows.[13,79,91]

- The dielectric constant is a controlling factor of salt solubility, reaction rate and solubility of the resulting metal oxides.
- The residence time is a key value: increasing the total mass flow rate and/or the mixing ratio between the aqueous metal salt solution and the scH_2O stream leads to smaller particles and agglomerates.
- As predicted from (classical) nucleation theory, the size of the synthesised metal oxide particles decreases with increasing supersaturation.
- A higher reaction temperature and decreasing reactant concentration result in smaller particles.
- Generally, an increasing reactant concentration leads to larger particles.

Further information regarding the materials investigated, the process and the reaction conditions, as well as the size and morphology of the synthesised particles and more details about the different concepts currently employed in the synthesis of metal oxide nanoparticles using sub- and/or supercritical H_2O, can be found in various articles (without any claim that this is a complete list of such articles).[12,13,116–122]

3.5 Applications

SFRD has been widely used to deposit various supported mono- (*e.g.*, Pt, Pd, Ru) and bimetallic (*e.g.*, Pt/Pd, Pt/Ru, Pt/Cu) NPs on different substrates. Such supported metal NPs are frequently used in catalysis, electronics, sensing and electrochemical energy conversion. The catalytic properties of these deposited NPs have been mainly investigated in hydrogenation and oxidation reactions. This section provides a survey (that can never pretend to be exhaustive) of published knowledge about the properties of catalytically active NPs.

The great potential of the SFRD process for the preparation of highly active catalysts was verified by Incera Garrido *et al.* for the first time.[123] The results of these experiments demonstrated that SFRD leads to highly dispersed Pt NP with a mean diameter of 3.1 nm and a Pt content of 0.71 wt%. The catalysts were characterised by superior activity towards preferential CO oxidation compared with a catalyst prepared by conventional aqueous impregnation of Pt. Furthermore, the results showed that a reduction temperature higher than the melting temperature at the given pressure led to relatively large particles associated with a lower catalytic activity. On the basis of these promising results, PtCuO NPs were deposited on sponges of α-Al_2O_3 + CeO_2 by simultaneous deposition of bis(2,2,6,6-tetramethyl-3,5-heptanedionato)copper(II) ([Cu(tmhd)$_2$]) and [PtMe$_2$(cod)] from $scCO_2$.[124] Again, these catalysts were tested for preferential oxidation of CO and showed high activity and selectivity. The activity of PtCuO NPs was found to be twice as high as the activity of a reference catalyst prepared by conventional wet impregnation.

Pascu *et al.* have shown that the kinetically controlled synthesis of supported metal NPs enables the production of active catalysts.[125] The results of these experiments showed that the properties (*i.e.*, PS, PSD, metal loading and morphology) of the NPs can be controlled by changing the metal precursor type, its concentration and the composition of the reaction media. Furthermore, the authors showed that the catalytic activity and the reaction yield are directly correlated with the properties of the final product, mainly the size of the particles.

In an extensive research study performed by Casapu *et al.*, the influence of different synthesis methods, *i.e.*, IWI, FSP and SFRD, on PS and PSD and the hysteresis behaviour during CO oxidation over Pt/γ-Al_2O_3 catalysts was systematically investigated.[126] The lowest temperature for 50% CO conversion was observed for the catalyst prepared by SFRD with a mean particle size of 2.3 nm and a Pt content of 1.7 wt%. Both smaller and larger particles produced with the above-mentioned methods were less active under these reaction conditions.

Morere *et al.* used reduced graphene oxide (RGO) as substrate for the synthesis of Pt/RGO, Ru/RGO and Ni/RGO catalysts *via* SFRD.[127] Depending on the metal, deposition method and metal loading, the PS of the various metal NPs on RGO ranged from 2 to 15 nm and these nanocomposites were successfully used as selective catalysts in the partial hydrogenation of limonene.

Qiao *et al.* demonstrated that the catalytic properties of SBA-15-supported Pt/Pd bimetallic catalysts are strongly influenced by the choice of the solvent used for impregnation of the substrate.[128] The comparison between the catalysts synthesised by SFRD and those prepared using the common liquid-phase impregnation method showed that SFRD leads to a superior activity for the hydrogenation of levulinic acid to γ-valerolactone, with almost perfect selectivity.

Sanchez-Miguel *et al.* investigated the hydrogenation of furfural to furfuryl alcohol in $scCO_2$.[129] In this study PtRu and PtCu NPs were deposited *via* SFRD on mesoporous SBA-15 by the reduction of Pt, Ru and Cu metal organic precursors in either $scCO_2/H_2$ (method (1)) or N_2/H_2 (method (3)) atmosphere. The catalysts were tested in the hydrogenation of furfural in $scCO_2$ at 80 °C and the PtRu catalysts showed a high activity and selectivity towards furfuryl alcohol.

Hu *et al.* applied SFRD to deposit mono-Pt and -Pd and bimetallic PtPd NPs on porous carbon by reduction in a $scCO_2/H_2$ atmosphere.[130] The synthesised NPs displayed superior performance in the selective conversion of biomass-based furfural into furfuryl alcohol and tetrahydrofurfuryl alcohol. The catalyst containing 3 wt% Pd or Pt exhibited a higher performance than commercial catalysts with 5 wt% Pd or Pt. In addition, the yield of the 3 wt% Pd catalyst was approximately seven times higher than that of the commercial 5 wt% Pd catalyst.

Ogel *et al.* systematically investigated the influence of the preparation method and the resulting PSD on the catalytic performance during CO/NO oxidation.[131] Three different categories of Pt/Al_2O_3 catalysts were obtained: (1) catalysts containing homogeneously distributed Pt nanoparticles with $d < 2$ nm, prepared with a chlorine-containing precursor; (2) catalysts with homogeneously distributed but slightly larger Pt NPs with $d \approx 2$–4 nm, prepared by SFRD and FSP; and (3) catalysts with heterogeneous PSD containing either a combination of small and large Pt particles or an insufficient distribution of the noble metal component prepared by pulse laser ablation. Further investigations showed that an optimal PS of about 2–3 nm was found for the CO oxidation, as obtained by SFRD, whereas slightly larger Pt particles with diameters of 3–4 nm led to the highest NO conversion at lower temperatures.

A summary of a large number of mono-, bi- and tri-metallic NPs synthesised using $scCO_2$ can be found in a recent review.[2] More details on promising applications of NPs synthesised by SFRD with regard to catalysis, electronics, sensing and electrochemical energy conversion can be found elsewhere in the literature.[1,13,132,133]

CHTS was first explored by Adschiri *et al.* in the early 1990s.[89] Based on this fundamental work, a large number of experimental studies have shown that scH_2O enables the controlled formation of various types of materials that are synthesised in both conventional hydrothermal batch autoclave systems and in flow systems. In addition to the fast and controlled synthesis of different kinds of simple and multiple metal oxide NPs, the *in situ* functionalisation of metal oxide NPs with clickable anchors has also been investigated in detail.[134,135] Below we present a short survey summarising the results of recently published studies on the production of multiple metal oxide particles for different applications.

Recently, Auxemery *et al.* synthesised nanostructured $Ce_xZr_{(1-x)}O_2$ particles that were used as supports for Ni NPs (10 wt%) to obtain catalysts for the dry reforming of methane (DRM) reaction.[136] Additional experiments were performed to compare the activity, the stability as well as the selectivity of coke formation in the presence of the prepared catalytic materials. The results showed that the industrial catalyst led to heavy coking with a rapid loss of activity. By contrast, the nanostructured $Ce_{0.50}Zr_{0.50}O_2$ loaded with Ni NPs showed the highest turnover frequency and the smallest selectivity for coke formation in DRM amongst the catalysts.

Mamakhel *et al.* prepared SnO_2 NPs by a continuous-flow hydrothermal method using $SnCl_4{\cdot}5H_2O$ and aqueous sodium hydroxide as precursor solutions.[100] The experiments were performed at four different temperatures (523, 573, 623, 673 K) and four different precursor solution pH values (1, 3, 7, 11) and the obtained results showed that the shape and size of the SnO_2 NPs can be controlled by adjusting the synthesis temperature and precursor pH value. At pH values of 3 and 7, spherical NPs in the size range from 3 to 10 nm were synthesised, while at pH values of either 1 or 11 nanorods were achieved. Furthermore, selected samples were checked for their suitability as anode materials in Li-ion batteries and an increased first cycle capacity was observed for the smaller NPs.

Zhu *et al.* demonstrated that a supercritical hydrothermal process enables the synthesis of highly Cr-substituted CeO_2 nanoparticles (Cr-CeO_2) for the low-temperature bitumen upgrading method.[137] In these experiments, a Cr-substitution concentration of 22.7 mol% was achieved, which strongly affected the rise in the oxygen storage capacity of Cr-CeO_2. Further results showed that, in the presence of highly Cr-substituted CeO_2 nanoparticles, a high conversion of asphaltene (up to 47.9% at 623 K) and a high selectivity of syngas (H_2 and CO, up to 58.3% at 623 K) were achieved through low-temperature catalytic bitumen upgrading.

Silva *et al.* used supercritical flow technology for the synthesis of highly dispersed PtPd and (PtPd)Fe NPs on γ-Al_2O_3 with high surface area.[138] The results of their catalytic tests showed that the addition of a low amount of Fe (0.12 wt%) enhanced the C_3H_6 oxidation activity of the (PtPd)1/γ-Al_2O_3 catalyst. At a temperature of 393 K, the turnover frequency increased from 0.02 s^{-1} to 0.06 s^{-1}. Furthermore, the (PtPd)Fe/γ-Al_2O_3 catalyst showed a significantly lower T_{50} (temperature of half conversion) for C_3H_6 oxidation than the commercial reference catalyst.

In terms of process intensification and optimisation, Hellstern *et al.* demonstrated that a hydrothermal continuous-flow, dual-stage reactor enables the surfactant-free synthesis of core–shell NPs.[139] Based on these promising results, Aymonier *et al.* proposed a one-pot multistep process for the synthesis of advanced materials.[140] The key idea behind this concept is that the core material is synthesised in the first reactor stage, while the inorganic shell material is deposited in the second reactor. Finally, the core–shell particles can be overcoated with an organic shell in a third reactor. Of course, this very promising concept needs to be investigated in more depth because control of the synthesis depends strongly on mastering the flow characteristics of the reactors and the reaction kinetics, and on controlling the nucleation and growth phenomena.

Further examples of successful applications for metal oxides can be found in extensive review articles by Ruiz-Jorge *et al.*, Darr *et al.*, Adschiri *et al.* and Staerz *et al.*[115,118,141,142]

3.6 Summary and Future Directions

The discovery of novel energy sources, energy generation and storage methods, as well as energy saving techniques, are some of the greatest challenges, both currently and in the future. One important aspect is that new processes must be environmentally friendly when employed at large scale. This section summarises what is still missing to achieve further progress in SCF-based particle formation processes, provides an overview of the requirements for future developments and points out the need for a better understanding of the fundamentals to pave the way for new products.

Reliable thermophysical data and transport properties for mixtures in which SCFs are involved are of great importance for process design and for the development of optimised operating conditions for various technical processes, such as extraction, impregnation and particle formation. With particular emphasis on the synthesis of supported nanostructured materials by SFRD, information about the pure and impregnated substrate characteristics, ranging from, for example, pore volume, pore diameter, porosity, specific surface area, pore size distribution to surface properties, should be known.

Thus, to obtain improved understanding of particle synthesis processes in which the SCF acts either as a solvent and/or reactant and/or separation medium, fundamental knowledge of the adsorption/desorption behaviour in dense fluids, as well as the complex multiphase behaviour and flow, is essential for a deeper understanding of the relationship between process conditions and product properties. In addition, such an improved understanding is a prerequisite for the transformation of laboratory-scale findings to large-scale operations. Thus, closer cooperation between fundamental and experimental research is essential for progress in SCF precipitation processes.

For all SCF-assisted particle formation processes, especially those based on fast chemical transformations, fast online and *in situ* measurement techniques are urgently needed for real-time process analysis and/or control. However, the high pressures and temperatures that occur under supercritical conditions complicate the design of *in situ* setups. Despite these challenges, several designs have been successfully developed by various groups that allow *in situ* studies in batch and flow reactors.[140,143–149]

Nowadays, (synchrotron) X-ray methods, such as small-/wide-angle X-ray scattering, total scattering, tomography and energy-dispersive diffraction imaging, are used widely as the radiation source in most developed *in situ* techniques.[82] Further *in situ* techniques based on Raman spectroscopy, infrared radiation and visible light have also been developed to study chemical and physical phenomena in SCFs.[150] Furthermore, such real-time monitoring of precipitation can be compared with numerical simulations, which helps in understanding the nucleation and growth phenomena and the evolution of crystallites and/or particle size. These techniques can provide a lot of useful information about the kinetics and thermodynamics of particle formation, and the growth process in the very early stages of particle generation. Furthermore, such data will be helpful for identifying optimal process conditions, and can enable a deeper insight into the relationship between process conditions and particle size, and hence product properties. In addition, these measurement methods should be complemented by modelling of nucleation, cluster formation, and particle growth by either condensation or coagulation on a molecular scale.

The simulation of particle synthesis in SCFs can be achieved by coupling computational fluid dynamics (CFD), which considers thermodynamic and hydrodynamic models, with population balance equations (PBEs), which can be solved by the standard method of moments.[13] However, the results of numerous modelling works illustrate the lack of knowledge about the determination of nucleation and growth parameters, such as the fluid–solid interfacial tension.[140] An interesting approach to obtain nucleation and growth parameters under supercritical conditions was proposed by Voisin *et al.*[143] The authors coupled *in situ* synchrotron wide-angle X-ray scattering with CFD and PBE simulation to determine the mass-transfer coefficient for the growth rate.

When developing new processes, it is important to always consider the advice given by Val Krukonis in 1977: "There is no point in doing something in a SCF just because it is neat. Using the fluid must have some real advantage". Following this strategy, it is important to verify explicitly the significant improvement of the obtained product. For example, metallic nanoparticles that are deposited on porous solid substrates in $scCO_2$ should exhibit at least a similar or superior catalytic activity compared to a catalyst prepared by classical methods, such as IWI, co-precipitation, CVD and FSP. Such design, realisation, verification and confirmation of the improved product properties requires a few (at least two) process steps and interdisciplinary cooperation between researchers from different but complementary disciplines.

Thus, in summary, the synthesis of metallic and metal oxide nanoparticles by SCF-based particle formation processes is a large unexplored area that merits intense attention and needs to be explored by both fundamental and experimental researchers. In order to enlarge the feasibility of the process and its optimisation, it is important to understand the influence of the different operating parameters and processing conditions on the obtained product properties.

So, as things stand, we can conclude that the following are needed.

- Comparison between catalysts synthesised by SCF-based processes and those produced by conventional methods in terms of activity, selectivity, light-off and light-on temperature, reaction rate and long-term stability.
- Determination of experimental multiphase behaviour and thermophysical data in the near- and supercritical state and development of predictive methods.
- Understanding the mechanisms for nucleation and growth of nanoparticles in dense fluids using both modelling and fast online and *in situ* measurement techniques, such as X-ray diffraction, extended X-ray absorption fine structure, Raman and infrared spectroscopy under high pressure.
- Improved understanding of the interfacial phenomena in dense fluids.
- Extension of the limited knowledge on chemical reactions and their rate constants in dense fluids, as well as at solid–SCF interfaces.
- In the case of batch processes, the development of a continuous process.
- Transformation of laboratory-scale findings or pilot plant studies to large-scale operations.
- Studies on the handling of nanomaterials with regard to technological and safety aspects.

List of Abbreviations Used for Precursors and Substrates

acac	acetylacetonate
cod	1,5-cyclooctadiene
Cp	cyclopentadienyl
DI6	bis(1,1,1,3,5,5,6,6,6-nonafluorohexane-2,4-diiminate)
hfac	hexafluoroacetylacetonate
me	methyl
tmhd	2,2,6,6-tetramethyl-3,5-heptanedionato
BP2000	black pearls
CA	carbon aerogel
SA	silica aerogel
RFA	resorcinol-formaldehyde aerogel
MCM-41	mesoporous silica
SBA-15	mesoporous silica
MSU-H	mesoporous silica
HMS	mesoporous silica

References

1. M. Türk and C. Erkey, *J. Supercrit. Fluids*, 2018, **134**, 176.
2. P. F. Siril and M. Türk, *Small*, 2020, **16**, 2001972.
3. R. Si and M. Flytzani-Stephanopoulos, *Angew. Chem.*, 2008, **120**, 2926.
4. J. Murata, K. Yodogawa and K. Ban, *Int. J. Mach. Tools Manuf.*, 2017, **114**, 1.
5. D. E. Motaung, G. H. Mhlongo, P. R. Makgwane, B. P. Dhonge, F. R. Cummings, H. C. Swart and S. S. Ray, *Sens. Actuators B*, 2018, **254**, 984.
6. F. Caputo, M. De Nicola, A. Sienkiewicz, A. Giovanetti, I. Bejarano, S. Licoccia, E. Traversa and L. Ghibelli, *Nanoscale*, 2015, **7**, 15643.
7. J. Zhang, H. Kumagai, K. Yamamura, S. Ohara, S. Takami, A. Morikawa, H. Shinjoh, K. Kaneko, T. Adschiri and A. Suda, *Nano Lett.*, 2011, **11**, 361.
8. M. Melchionna and P. Fornasiero, *Mater. Today*, 2014, **17**(7), 349.
9. Z. Hu, X. Liu, D. Meng, Y. Guo, Y. Guo and G. Lu, *ACS Catal.*, 2016, **6**, 2265.
10. A. Trovarelli and J. Llorca, *ACS Catal.*, 2017, **7**, 4716.
11. T. Adschiri, K. Kanazawa and K. Arai, *J. Am. Ceram. Soc.*, 1992, **75**(4), 1019.
12. M. K. M. Lane and J. B. Zimmerman, *Green Chem.*, 2019, **21**, 3769.
13. C. Erkey and M. Türk, *Synthesis of Nanostructured Materials in Near and/or Supercritical Fluids*, Supercritical Fluid Science and Technology Series, Elsevier, Amsterdam, 2021, vol. 8.
14. J. J. Watkins and Th. J. McCarthy, *Chem. Mater.*, 1995, **7**(11), 1991.
15. I. Kikic, A. Lora and M. Bertucco, *Ind. Eng. Chem. Res.*, 1997, **36**, 5507.
16. https://www.chemicalbook.com/ChemicalProductProperty_US_CB8722377.aspx, last accessed May 2023.
17. S. E. Bozbag, U. Unal, M. A. Kurykin, C. J. Ayala, M. Aindow and C. Erkey, *J. Phys. Chem. C*, 2013, **117**, 6777.
18. M. Türk, M. Crone and T. Kraska, *J. Supercrit. Fluids*, 2010, **55**, 462.
19. E. Stahl, W. Schilz, E. Schütz and E. Willing, *Angew. Chem., Int. Ed.*, 1978, **17**, 731.
20. S. K. Kumar and K. P. Johnston, *J. Supercrit. Fluids*, 1988, **1**, 15.
21. M. Türk, M. Crone and G. Upper, *J. Supercrit. Fluids*, 2011, **58**, 1.
22. M. Türk and T. Kraska, *J. Chem. Eng. Data*, 2009, **54**, 1592.
23. R. T. Kurnik and R. C. Reid, *AIChE J.*, 1981, **27**(5), 861.
24. C. A. Eckert, B. L. Knutson and P. G. Debenedetti, *Nature*, 1996, **383**, 313.
25. J. Mendez-Santiago and A. S. Teja, *Fluid Phase Equilib.*, 1999, **158–160**, 501.
26. F. P. Lucien and N. R. Foster, *J. Supercrit. Fluids*, 2000, **17**, 111.
27. G. I. Burgos-Solorzano, J. F. Brennecke and M. A. Stadtherr, *Fluid Phase Equilib.*, 2004, **220**, 57.
28. P. Coimbra, C. M. M. Duarte and H. C. de Sousa, *Fluid Phase Equilib.*, 2006, **239**, 188.
29. R. B. Gupta and J.-J. Shim, *Solubility in Supercritical Carbon Dioxide*, CRC Press, Boca Raton, 2007.

30. M. Skerget, Z. Knez and M. Knez-Hrncic, *J. Chem. Eng. Data*, 2011, **56**, 694.
31. W. H. Teoh, R. Mammucari and N. R. Foster, *J. Organomet. Chem.*, 2013, **724**, 102.
32. Z. Knez, D. Cör and M. Knez-Hrncic, *J. Chem. Eng. Data*, 2018, **63**, 860.
33. G. Brunner and M. Johannsen, *J. Supercrit. Fluids*, 2006, **38**, 181.
34. I. Langmuir, *J. Am. Chem. Soc.*, 1918, **40**(9), 1361.
35. H. M. F. Freundlich, *J. Phys. Chem.*, 1906, **57**, 385.
36. R. Sips, *J. Chem. Phys.*, 1948, **16**(5), 490.
37. C. J. Radke and J. M. Prausnitz, *AIChE J.*, 1972, **18**(4), 761.
38. M. M. Dubinin and V. A. Astakhov, Molecular Sieve Zeolites – II, *Advances in Chemistry*, American Chemical Society, Washington D.C., 1971, ch. 44, 69, vol. 102.
39. T. L. Hill, *An Introduction into Statistical Thermodynamics*, Dover Publications Inc., New York, 1986.
40. P. Gurikov and I. Smirnova, *J. Supercrit. Fluids*, 2018, **132**, 105.
41. I. Ushiki, M. Koike, T. Shimizu, Y. Sato, S. Takishima and H. Inomata, *J. Supercrit. Fluids*, 2018, **140**, 329.
42. S. E. Bozbag, L. C. Zhang, M. Aindow and C. Erkey, *J. Supercrit. Fluids*, 2012, **66**, 265.
43. B. Cangül, L. C. Zhang, M. Aindow and C. Erkey, *J. Supercrit. Fluids*, 2009, **50**, 82.
44. M. J. Tenorio, C. Pando, J. A. R. Renuncio, J. G. Stevens, R. A. Bourne, M. Poliakoff and A. Cabanas, *J. Supercrit. Fluids*, 2012, **69**, 21.
45. R. Metzner, Bachelor Thesis, Karlsruhe Institute for Technology (KIT), 2010.
46. Y. Zhang, D. Kang, C. Saquing, M. Aindow and C. Erkey, *Ind. Eng. Chem. Res.*, 2005, **44**, 4161.
47. C. D. Saquing, T.-T. Cheng, M. Aindow and C. Erkey, *J. Phys. Chem. B*, 2004, **108**, 7716.
48. S. E. Bozbag, N. S. Yasar, L. C. Zhang, M. Aindow and C. Erkey, *J. Supercrit. Fluids*, 2011, **56**, 105.
49. I. Ushiki, N. Takahashi, T. Shimizu, Y. Sato, M. Ota, R. L. Smith Jr. and H. Inomata, *J. Supercrit. Fluids*, 2017, **120**, 240.
50. G. Caputo, I. De Marco and E. Reverchon, *J. Supercrit. Fluids*, 2010, **54**, 243.
51. Y. Zhang, B. Cangul, Y. Garrabos and C. Erkey, *J. Supercrit. Fluids*, 2008, **44**, 71.
52. C. Erkey, *Supercritical Fluids and Organometallic Compounds*, Elsevier, Amsterdam, 1st edn, 2011.
53. Y. Zhang, D. Kang, M. Aindow and C. Erkey, *J. Phys. Chem. B*, 2005, **109**, 2617.
54. H. Weingärtner and E. U. Franck, *Angew. Chem., Int. Ed.*, 2005, **44**, 2672.
55. M. Türk, *Chem. Ing. Tech.*, 2018, **90**(4), 436.
56. I. Leusbrock, S. J. Metz, G. Rexwinkel and G. F. Versteeg, *J. Chem. Eng. Data*, 2009, **54**, 3215.

57. Q. Liu, X. Ding, B. Du and T. Fang, *Crit. Rev. Anal. Chem.*, 2017, **47**(6), 513.
58. T. Voisin, A. Erriguible, D. Ballenghien, D. Mateos, A. Kunegel, F. Cansell and C. Aymonier, *J. Supercrit. Fluids*, 2017, **120**, 18.
59. J. Rincon, R. Camarillo and A. Martin, *J. Chem. Eng. Data*, 2012, **57**, 2084.
60. I. Leusbrock, S. J. Metz, G. Rexwinkel and G. F. Versteeg, *J. Supercrit. Fluids*, 2010, **53**, 17.
61. I. Leusbrock, S. J. Metz, G. Rexwinkel and G. F. Versteeg, *J. Supercrit. Fluids*, 2010, **54**, 1.
62. B. Hearn, M. R. Hunt and A. Hayward, *J. Chem. Eng. Data*, 1969, **14**(4), 442.
63. K. Sue, Y. Hakuta, R. L. Smith, T. Adschiri and K. Arai, *J. Chem. Eng. Data*, 1999, **44**, 1422.
64. H. Higashi, Y. Iwai, K. Matsumoto, Y. Kitani, F. Okazaki, Y. Shimoyama and Y. Arai, *Fluid Phase Equilib.*, 2005, **228–229**, 547.
65. W. T. Wofford, P. C. Dell'Orco and E. F. Gloyna, *J. Chem. Eng. Data*, 1995, **40**, 968.
66. M. Hodes, P. Griffith, K. A. Smith, W. S. Hurst, W. J. Bowers Jr. and K. Sako, *AIChE J.*, 2004, **50**, 2038.
67. X. Ding, T. Zhang, S. Zhang, Z. Shen and T. Fang, *Fluid Phase Equilib.*, 2019, **483**, 31.
68. G. Lemoine, H.-A. Turc, A. Leybros, J.-C. Ruiz, Y. S. de Gelicourt and H. Muhr, *J. Supercrit. Fluids*, 2017, **130**, 91.
69. F. J. Armellini and J. W. Tester, *Fluid Phase Equilib.*, 1993, **84**, 123.
70. J. F. Galobardes, D. R. van Hare and L. B. Rogers, *J. Chem. Eng. Data*, 1981, **26**, 363.
71. X. Ding, Y. Lei, Z. Shen, Y. Yu, Q. Zhou, J. Wei and T. Fang, *J. Chem. Eng. Data*, 2017, **62**, 3374.
72. B. Liu, X. Ding, Z. Jiang, B. Wang and T. Fang, *J. Solution Chem.*, 2020, **49**, 1186.
73. M. S. Khan and S. N. Rogak, *J. Supercrit. Fluids*, 2004, **30**, 359.
74. D. Shvedov and P. R. Tremaine, *J. Solution Chem.*, 2000, **29**(10), 889.
75. S. N. Rogak and P. Teshima, *AIChE J.*, 1999, **45**(2), 240.
76. M. M. DiPippo, K. Sako and J. W. Tester, *Fluid Phase Equilib.*, 1999, **157**, 229.
77. T. Voisin, A. Erriguible, G. Aubert and C. Aymonier, *Ind. Eng. Chem. Res.*, 2018, **57**, 2376.
78. T. Adschiri, Y. Hakuta, K. Sue and K. Arai, *J. Nanopart. Res.*, 2001, **3**, 227.
79. N. Aoki, A. Sato, H. Sasaki, A.-A. Litwinowicz, G. Seong, T. Aida, D. Hojo, S. Takami and T. Adschiri, *J. Supercrit. Fluids*, 2016, **110**, 161.
80. U. Grigull, *BKW*, 1983, **35**(6), 1.
81. M. Türk, *Particle Formation with Supercritical Fluids: Challenges and Limitations*, Elsevier, Amsterdam, 2014.
82. Y. Xu, V. Musumeci and C. Aymonier, *React. Chem. Eng.*, 2019, **4**, 2030.
83. Y. Zhang and C. Erkey, *J. Supercrit. Fluids*, 2006, **38**, 252.

84. C. D. Saquing, D. Kang, M. Aindow and C. Erkey, *Microporous Mesoporous Mater.*, 2005, **80**, 11.
85. S. Wolff, M. Crone, T. Muller, M. Enders, S. Bräse and M. Türk, *J. Supercrit. Fluids*, 2014, **95**, 588.
86. S. Müller and M. Türk, *J. Supercrit. Fluids*, 2015, **96**, 287.
87. M. Crone and M. Türk, *J. Supercrit. Fluids*, 2022, **182**, 105461.
88. T. Adschiri, Y. Hakuta, K. Kanamura and K. Arai, *Int. J. High Pressure Res.*, 2001, **20**(1–6), 373.
89. T. Adschiri, K. Kanazawa and K. Arai, *J. Am. Ceram. Soc.*, 1992, **75**(9), 2615.
90. K. Sue, M. Suzuki, K. Arai, T. Ohashi, H. Ura, K. Matsui, Y. Hakuta, H. Hayashi, M. Watanabe and T. Hiaki, *Green Chem.*, 2006, **8**, 634.
91. A. Yoko, G. Seong, T. Tomai and T. Adschiri, *KONA Powder Part. J.*, 2020, **37**, 28.
92. Ch. Schüßler, PhD Thesis, Karlsruhe Institute of Technology (KIT), 2023.
93. H. Hayashi and Y. Hakuta, *Materials*, 2010, **3**, 3794.
94. Y. Hakuta, T. Adschiri, H. Hirakoso and K. Arai, *Fluid Phase Equilib.*, 1999, **158–160**, 733.
95. T. Sato, K. Sue, Y. Akiyama, K. Shibata, S.-i. Kawasaki, S. Tanaka, K. Saitoh, A. Kawai-Nakamura, K. Aida and T. Hiaki, *Chem. Lett.*, 2008, **37**(3), 242.
96. Y. Hakuta, S. Onai, H. Terayama, T. Adschiri and K. Arai, *J. Mater. Sci. Lett.*, 1998, **17**, 1211.
97. T. Togashi, H. Hitaka, S. Ohara, T. Naka, S. Takami and T. Adschiri, *Mater. Lett.*, 2010, **64**, 1049.
98. T. Sasaki, S. Ohara, M. Umetsu, T. Naka, T. Adschiri, C. Nabeta, H. Ichikawa and Y. Fukumori, *J. Soc. Powder Technol., Jpn.*, 2006, **43**, 440.
99. M. Le Clercq, T. Adschiri and K. Arai, *Biomass Bioenergy*, 2001, **21**, 73.
100. A. Mamakhel, M. Sondergaard, K. Borup and B. B. Iversen, *J. Supercrit. Fluids*, 2020, **166**, 105029.
101. H. Reveron, C. Aymonier, A. Loppinet-Serani, C. Elissalde, M. Maglione and F. Cansell, *Nanotechnology*, 2005, **16**, 1137.
102. Y. Shen, J. R. Eltzholtz and B. B. Iversen, *Chem. Mater.*, 2013, **25**, 5023.
103. S. Ohara, T. Mousavand, M. Umetsu, S. Takami, T. Adschiri, Y. Kuroki and M. Takata, *Solid State Ionics*, 2004, **172**, 261.
104. K. Sue, K. Murata, K. Kimura and K. Arai, *Green Chem.*, 2003, **5**, 659.
105. Y. Hakuta, K. Seino, H. Ura, T. Adschiri, H. Takizawa and K. Arai, *J. Mater. Chem.*, 1999, **9**, 2671.
106. A. Yoko, M. Akizuki, N. Hirao, S. Kohara, M. Kumar, N. Umezawa, T. Ohno and Y. Oshima, *J. Supercrit. Fluids*, 2016, **107**, 746.
107. Y. Hakuta, T. Adschiri, T. Suzuki, T. Chida, K. Seino and K. Arai, *J. Am. Ceram. Soc.*, 1998, **81**(9), 2461.
108. D. Rangappa, T. Naka, A. Kondo, M. Ishii, T. Kobayashi and T. Adschiri, *J. Am. Chem. Soc.*, 2007, **129**, 11061.
109. K. Kaseda, M. Takesue, T. M. Aida, M. Watanabe, H. Hayashi and R. L. Smith, *J. Supercrit. Fluids*, 2011, **58**, 279.

110. Y. H. Shin, S.-M. Koo, D. S. Kim, Y.-H. Lee, B. Veriansyah, J. Kim and Y.-W. Lee, *J. Supercrit. Fluids*, 2009, **50**, 250.
111. K. Kanamura, T. Umegaki, K. Toyoshima, K.-I. Okada, Y. Hakuta, T. Adschiri and K. Arai, *Key Eng. Mater.*, 2000, **181–182**, 147.
112. A. Aimable, D. Aymes, F. Bernhard and F. Le Cras, *Solid State Ionics*, 2009, **180**, 861.
113. T. Sasaki, S. Ohara, T. Naka, J. Vejpravova, V. Sechovsky, M. Umetsu, S. Takami, B. Jeyadevan and T. Adschiri, *J. Supercrit. Fluids*, 2010, **53**, 92.
114. M. Atashfaraz, M. Shariaty-Niassar, S. Ohara, K. Minami, M. Umetsu, T. Naka and T. Adschiri, *Fluid Phase Equilib.*, 2007, **257**, 233.
115. F. Ruiz-Jorge, J. R. Portela, J. Sanchez-Oneto and E. J. M. de la Ossa, *Appl. Sci.*, 2020, **10**, 5508.
116. A. S. Teja and P.-Y. Koh, *Prog. Cryst. Growth Charact. Mater.*, 2009, **55**, 22.
117. O. Pascu, S. Marre and C. Aymonier, *Nanotechnol. Rev.*, 2015, **4**(6), 487.
118. J. A. Darr, J. Zhang, N. M. Makwana and X. Weng, *Chem. Rev.*, 2017, **117**, 11125.
119. E. Lester, P. Blood, J. Denyer, D. Giddings, B. Azzopardi and M. Poliakoff, *J. Supercrit. Fluids*, 2006, **37**, 209.
120. Y. Marcus, *Supercritical Water: A Green Solvent, Properties and Uses*, John Wiley & Sons, Inc., 2012.
121. P. W. Dunne, A. S. Munn, C. L. Starkey, T. A. Huddle and E. H. Lester, *Philos. Trans. R. Soc., A*, 2015, **373**, 20150015.
122. T. Adschiri and A. Yoko, *J. Supercrit. Fluids*, 2018, **134**, 167.
123. G. I. Garrido, F. C. Patcas, G. Upper, M. Türk, S. Yilmaz and B. Kraushaar-Czarnetzki, *Appl. Catal., A*, 2008, **338**, 58.
124. S. Lang, M. Türk and B. Kraushaar-Czarnetzki, *J. Catal.*, 2012, **286**, 78.
125. O. Pascu, B. Cacciuttolo, S. Marre, M. Pucheault and C. Aymonier, *J. Supercrit. Fluids*, 2015, **105**, 84.
126. M. Casapu, A. Fischer, A. M. Gänzler, R. Popescu, M. Crone, D. Gerthsen, M. Türk and J.-D. Grunwaldt, *ACS Catal.*, 2017, **7**, 343.
127. J. Morere, E. Sanchez-Miguel, M. J. Tenorio, C. Pando and A. Cabanas, *J. Supercrit. Fluids*, 2017, **120**, 7.
128. Y. Qiao, N. Said, M. Rauser, K. Yan, F. Qin, N. Theyssen and W. Leitner, *Green Chem.*, 2017, **19**, 977.
129. E. Sanchez-Miguel, M. J. Tenorio, J. Morere and A. Cabanas, *J. CO2 Util.*, 2017, **22**, 382.
130. D. Hu, H. Xu, Z. Yi, Z. Chen, C. Ye, Z. Wu, H. F. Garces and K. Yan, *ACS Sustainable Chem. Eng.*, 2019, **7**, 15339.
131. E. Ogel, M. Casapu, D. E. Doronkin, R. Popescu, H. Störmer, C. Mechler, G. Marzun, S. Barcikowski, M. Türk and J.-D. Grunwaldt, *J. Phys. Chem. C*, 2019, **123**, 5433.
132. S. A. Müller, D. Degler, C. Feldmann, M. Türk, R. Moos, K. Fink, F. Studt, D. Gerthsen, N. Barsan and J.-D. Grunwaldt, *ChemCatChem*, 2018, **10**, 864.
133. S. B. Barim, E. Uzunlar, S. E. Bozbag and C. Erkey, *J. Electrochem. Soc.*, 2020, **167**, 054510.

134. M. D. de Tercero, M. Bruns, I. G. Martinez, M. Türk, U. Fehrenbacher, S. Jennewein and L. Barner, *Part. Part. Syst. Charact.*, 2013, **30**, 229.
135. M. D. de Tercero, I. G. Martínez, M. Herrmann, M. Bruns, C. Kübel, S. Jennewein, U. Fehrenbacher, L. Barner and M. Türk, *J. Supercrit. Fluids*, 2013, **82**, 83.
136. A. Auxemery, B. B. Frias, E. Smal, K. Dziadek, G. Phillippot, P. Legutko, M. Simonov, S. Thomas, A. Adamski, V. Sadykov, K. Parkhomenko, A.-C. Roger and C. Aymonier, *J. Supercrit. Fluids*, 2020, **162**, 104855.
137. Y. Zhu, G. Seong, T. Noguchi, A. Yoko, T. Tomai, S. Takami and T. Adschiri, *ACS Appl. Energy Mater.*, 2020, **3**, 4305.
138. H. Silva, P. Hernandez-Fernandez, A. K. Baden, H. L. Hellstern, A. Kovyakh, E. Wisaeus, T. Smitshuysen, I. Chorkendorff, L. H. Christensen, D. Chakraborty and C. Kallesoe, *Catal. Sci. Technol.*, 2019, **9**, 6691.
139. H. L. Hellstern, A. Mamakhel, M. Bremholm and B. B. Iversen, *Chem. Commun.*, 2016, **52**, 3434.
140. C. Aymonier, G. Philippot, A. Erriguible and S. Marre, *J. Supercrit. Fluids*, 2018, **134**, 184.
141. T. Adschiri, Y.-W. Lee, M. Goto and S. Takami, *Green Chem.*, 2011, **13**, 1380.
142. A. Staerz, U. Weimar and N. Barsan, *Sens. Actuators, B*, 2022, **358**, 131531.
143. T. Voisin, A. Erriguible, G. Philippot, D. Ballenghien, D. Mateos, F. Cansell, B. B. Iversen and C. Aymonier, *Chem. Eng. Sci.*, 2017, **174**, 268.
144. S. Marre, Y. Roig and C. Aymonier, *J. Supercrit. Fluids*, 2012, **66**, 251.
145. G. Philippot, K. M. O. Jensen, M. Christensen, C. Elissalde, M. Maglione, B. B. Iversen and C. Aymonier, *J. Supercrit. Fluids*, 2014, **87**, 111.
146. T. Fujita, H. Kasai and E. Nishibori, *J. Supercrit. Fluids*, 2019, **147**, 172.
147. M. Bremholm, H. Jensen, S. B. Iversen and B. B. Iversen, *J. Supercrit. Fluids*, 2008, **44**, 385.
148. K. M. O. Jensen, M. Christensen, P. Juhas, C. Tyrsted, E. D. Bojesen, N. Lock, S. J. L. Billinge and B. B. Iversen, *J. Am. Chem. Soc.*, 2012, **134**, 6785.
149. J.-L. Mi, T. N. Jensen, M. Christensen, C. Tyrsted, J. E. Jorgensen and B. B. Iversen, *Chem. Mater.*, 2011, **23**, 1158.
150. C. Quilfen, T. Tassaing, D. Uzio and C. Aymonier, *J. Supercrit. Fluids*, 2018, **141**, 104.

CHAPTER 4

Highly Efficient Rapid Preparation of Inorganic Nanostructured Materials by Microwave Heating

FEI-FEI CHEN[a,b] AND YING-JIE ZHU[a*]

[a]State Key Laboratory of High Performance Ceramics and Superfine Microstructure, Shanghai Institute of Ceramics, Chinese Academy of Sciences, Shanghai 200050, China; [b]College of Materials Science and Engineering, Fuzhou University, Fuzhou 350108, China
*E-mail: y.j.zhu@mail.sic.ac.cn

4.1 Introduction

Wet-chemistry synthetic methods have been regarded as the most efficient and versatile routes for the preparation of inorganic compounds and materials.[1] The chemical composition and structure of inorganic compounds and materials can be well tailored by wet-chemistry synthetic methods. In recent years, synthetic techniques displaying advantages such as energy saving, sustainability, and low cost have become increasingly popular. However, conventional wet-chemistry synthetic methods usually suffer from being time consuming, high cost, or yielding low-quality products. In this regard, unconventional wet-chemistry synthetic methods provide a possible solution.

Inorganic Materials Series No. 14
Unconventional Green Synthesis of Inorganic Nanomaterials
Edited by Silvia Gross

Published by the Royal Society of Chemistry, www.rsc.org

In recent years, the rapid microwave heating method has been recognised as a highly valuable and efficient alternative for the synthesis of both inorganic and organic compounds and materials.[2,3] Microwave heating can accelerate the synthesis and processing of materials, with high energy efficiency and time savings, but can also yield novel materials.

The discovery of the microwave heating effect involves an interesting story. In 1945, an American engineer, Percy LeBaron Spencer, performed some tentative tests on radar microwaves. One day, he was surprised to find that a piece of chocolate in his pocket had melted. Initially, he thought that this phenomenon had originated from his own body temperature. Finally, the microwave heating effect was found after careful and repetitive experiments.

Electromagnetic waves with frequencies of 0.3–300 GHz, corresponding to wavelengths of 1 mm–1 m, are defined as microwaves. In the electromagnetic spectrum, microwaves are situated between the infrared and radio wavelengths. Different microwave bands are used for various applications. In order to avoid interference with radar or telecommunications, microwave bands for daily use should be regulated at national levels. Microwaves with a frequency of 2.45 GHz (wavelength 12.24 cm) are commonly used for industrial and domestic equipment.

However, it was only in the 1980s that the application of microwave heating for materials synthesis was initiated. The milestones for this were established with two pioneering studies: Komarneni and Roy[4] prepared inorganic titania gel spheres in liquid phase with the help of microwave heating; and Gedye *et al.*[5] studied the use of microwave heating for rapid organic synthesis and observed considerable increase in reaction rates. These were the first two studies to report microwave-assisted inorganic and organic synthesis, respectively. Since then, studies on materials synthesis and processing by microwave heating technology have become one of the hottest research fields in materials chemistry.

The most attractive feature of microwave-assisted preparation is the incredible increase in reaction rate.[6,7] It usually takes many hours or even days for conventional heating methods to accomplish chemical reactions or materials synthesis, whereas similar processes can be completed in minutes using microwave heating. Therefore, microwave-assisted preparation is highly efficient, rapid, energy saving, low cost, and environmentally friendly, and, over the past few decades, fast-growing progress in microwave-assisted synthesis of organic compounds and materials has been achieved, from laboratories to homes to industries. In addition, the synergistic combination of microwave chemistry with well-established wet-chemistry synthesis methods shows promising applications and prospects. Consequently, it is rationally expected that research activities towards microwave chemistry and microwave-assisted synthesis will continue to expand in the coming decades. As shown in Figure 4.1, yearly publications for microwave-assisted synthesis of nanostructured materials have increased rapidly over the past two decades.

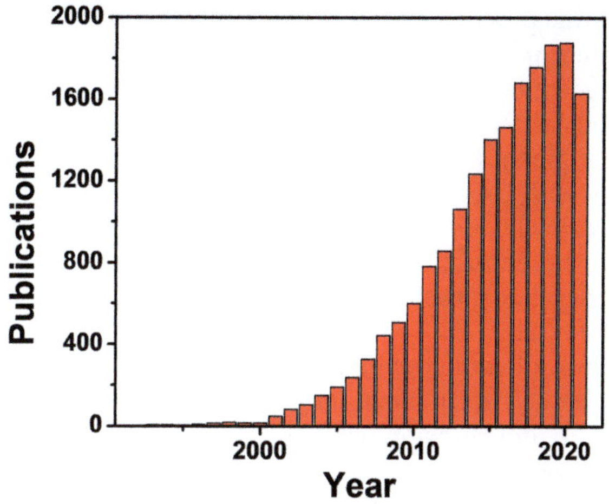

Figure 4.1 Yearly publications for microwave-assisted synthesis of nanostructured materials. The data were collected from Web of Science, and key words used for searching were 'microwave, nano*, and synthesis'.

It this chapter, the microwave-assisted preparation of inorganic nanostructures in the liquid phase is highlighted. The basic concepts of microwave chemistry are introduced, including the microwave heating mechanism, microwave effects, and related microwave heating apparatus. The microwave-assisted preparation of various inorganic nanostructures is reviewed and discussed, including metals, metal oxides, metal chalcogenides, metal fluorides, metal phosphates, graphene, metal–organic frameworks, and nanocomposites. In particular, the comparison between microwave heating and conventional heating for the preparation of inorganic nanostructures in the liquid phase is discussed.

4.2 Microwave Chemistry

4.2.1 Microwave Heating Mechanism

Since microwave heating as a non-classical energy source can significantly shorten the reaction time and save energy, it is necessary to investigate the microwave heating mechanism. Some previous papers have provided thorough and comprehensive reviews on the basic principles and mechanism of microwave heating.[8–12] Briefly, microwave heating involves two main mechanisms: (1) dipolar polarisation and (2) ionic conduction. Polar molecules and conducting ions, respectively, in the reaction systems account for the two heating mechanisms.

4.2.1.1 Dipolar Polarisation and Ionic Conduction

During microwave heating by dipolar polarisation, the polar molecules, such as water, align along the direction of the electric field. As the electric field oscillates, dipolar molecules attempt to realign with the alternating electric field. In this process, energy is lost by the rotation, friction, or collision of polar molecules, and heat is generated and thus the temperature is increased. It is worth noting that the microwave frequency has a great effect on the amount of heat generated by the dipolar polarisation. Under irradiation with high-frequency microwaves, there is not enough time for polar molecules to realign with the alternating electric field. In contrast, the dipoles will reorient too slowly under irradiation with low-frequency microwaves.[11] Under these two situations, microwave heating will hardly occur. Based on the above considerations, the frequency of commercial microwave systems is allocated at 2.45 GHz.

During microwave heating by ionic conduction, the ions in the reaction system move back and forth constantly under the influence of the microwave field. Friction and collision will occur between neighbouring charged ions, causing a rise in temperature. The ionic conduction heating effect is extremely obvious when reactions are performed in ionic liquids under microwave irradiation.[12]

4.2.1.2 Features of Microwave Heating

Starting from the above two mechanisms of microwave heating, it can be reasonably deduced that microwave heating technology is a non-classical, unconventional heating method, which is much more rapid and energy and time efficient than conventional heating techniques. The following advantages of microwave heating are unique and very different from conventional heating methods.[3] (1) Direct heating: the microwave irradiation can directly interact with reactants, minimising the effects of energy loss in the apparatus and environment. Direct heating can avoid the lengthy heat convection and conduction processes that are usually involved in conventional heating. Consequently, there are no evident temperature gradients in the microwave-heated reaction system. (2) Volumetric heating: since microwave heating can directly occur within the reactants, volumetric heating is realised. The reaction system therefore experiences a uniform thermal history by volumetric heating. (3) Instantaneous heating: the direct and volumetric heating result in instantaneous heating. The temperature of the reaction mixture will increase immediately once the power is switched on, which can significantly reduce the reaction time. (4) Selective heating: different substances have different microwave absorption abilities. The selective heating of different reactants occurs in a heterogeneous reaction system, resulting in inhomogeneous energy dissipation and temperature gradient.[13,14]

Schanche published a review article on microwave synthesis that focused on microwave equipment. The principle of microwave synthesis provides three major benefits for chemistry development: speed, reproducibility, and a scalable approach to synthesis. Incorporating planning systems, automation, distributed system access, searchable databases, and reaction optimisation kits, the technology has proved itself to be able to significantly improve the quality of output for medicinal chemistry, leading to optimisation by major pharmaceutical and biotechnology companies. In addition, Schanche investigated the temperature profile after 60 s using two different heating methods: microwave heating and oil bath. The whole reaction mixture was heated homogeneously under microwave irradiation, whereas for the oil bath heating the reaction mixture in contact with the vessel wall was heated first and the temperature at the centre was evidently lower than that in regions close to the vessel wall.[15]

4.2.2 Microwave Effects

Compared with conventional heating methods, the reaction rate can be enhanced by several orders of magnitude using microwave heating. In addition, some chemical reactions, which cannot be driven by conventional heating, can successfully occur under microwave irradiation. Gaining deep insight into the intrinsic microwave effects is indispensable. In addition to the thermal/kinetic effects, the current debate on rate acceleration by microwave heating focuses on two aspects: "specific microwave effects" and "non-thermal microwave effects", which are described hereafter.

4.2.2.1 Specific Microwave Effects

The 'specific microwave effects' are defined as those in which the rate acceleration cannot be achieved or reproduced by conventional heating.[11,16] The specific microwave effects include the following: (1) The superheating effect. The boiling point of a liquid under microwave heating is higher than that at atmospheric pressure. (2) The selective heating of different reactants in a heterogeneous reaction system. The selective heating creates hot spots with higher temperatures than other zones. The hot spots have sizes of ~100 μm.[17] Hence, the chemical reactions can be accelerated in these hot spots. (3) The direct coupling of microwave energy with specific reactants to generate 'molecular radiators'. (4) The elimination of vessel wall effects caused by inverted temperature gradients.[18] The much shorter reaction time in the microwave synthesis is just induced by the rapid heating rate and by the superheating/hot-spot effects.

4.2.2.2 Non-thermal Microwave Effects

The 'non-thermal microwave effects' are defined as those in which the rate acceleration cannot be rationalised by thermal/kinetic and specific microwave effects.[11,16] Instead of the macroscopic temperature effect, non-thermal microwave effects may result from direct interaction of microwaves with specific molecules. The non-thermal microwave effects include: (1) increased collision efficiency; (2) possible excitation of rotational or vibrational transitions; and (3) varied activation energy. For example, Godinho *et al.*[19] reported that the preparation time of gadolinium-doped ceria nanorods was drastically shortened with the use of microwave heating during hydrothermal treatment. They postulated that microwave heating increased the effective collision rate.

4.2.2.3 *Specific Microwave Effects* vs. *Non-thermal Microwave Effects*

Researchers have performed several experiments to understand the intrinsic microwave effects, and different interpretations have been presented. Jacob *et al.*[20] have provided a thorough discussion and comparison between thermal and non-thermal effects of microwave irradiation. They demonstrated the following: (1) Microwave dielectric heating causes thermally beneficial effects, such as hot spots, superheating effect, and rapid heating rate. (2) The dipoles shift rapidly under microwave irradiation. The intermolecular bonds hinder such dipole movement, resulting in a lag in the dipoles following microwave irradiation. The whole process is regarded as molecular agitation, which is responsible for the non-thermal microwave effects. (3) The rate-limiting step in solid-state reactions is the diffusion of reactants. Compared with conventional heating, microwave irradiation can promote the diffusion of reactants, which then accelerates the chemical reactions.

How to discriminate between specific and non-thermal microwave effects is a challenge due to the following issues: (1) the temperature profiles created by microwave heating cannot be reproduced by conventional heating, which makes comparison between them very difficult under the same conditions; (2) the inability to accurately measure reaction temperature under microwave irradiation may cause misinterpretation of the experimental results and induce controversies over the 'non-thermal microwave effects'.[11] To overcome these obstacles, Obermayer *et al.*[21] reported an innovative approach to separate thermal and non-thermal effects. Their key point was to use a silicon carbide (SiC) reaction vessel. The temperature was measured by external infrared and internal fibre optic probes. The SiC used in the vessel is distinct from other materials because of its: (1) strong microwave-absorbing capability; (2) chemical inertness; (3) high melting point (~2700 °C); and (4) low thermal expansion coefficient.[22] Under microwave irradiation, the SiC vessel could be efficiently heated by the resistance heating mechanism. In particular, the reactants in the SiC vessel could be effectively shielded from the electromagnetic field due to the strong microwave-absorbing ability of the SiC. In addition, microwave-transparent Pyrex® vials were used for comparison. The experimental results showed that there were no significant differences in conversion, purity, and yield using SiC or Pyrex® vessels. The authors proposed that thermal effects were the dominant heating mechanism for the rate acceleration and direct influence of the electromagnetic field on the reactions was not observed.

Conner and Tompsett[23] studied the microwave effects from a different standpoint. They introduced the concept of 'interfaces' to understand the rate acceleration with the help of Monte Carlo simulation. They demonstrated that microwaves could change the energies and/or the 'effective temperature' of individual species at the interfaces, causing the observed enhancements in reaction rate or selectivity. The greatest effect of microwaves on the chemical reactions was attributed to the immediate and periodic changes in the reaction profile (or relative temperature). These changes were the result of differing absorption of microwaves by molecules at the interfaces. Therefore, microwave irradiation had the ability to periodically provide a more preferable reaction coordinate, which could markedly promote the reaction rate.

Although many innovative research works have been reported, and various viewpoints have been demonstrated, there is still ongoing debate about the actual mechanisms underlying microwave heating. It is recognised by many researchers that the rate acceleration under microwave irradiation is a result of thermal/kinetics effects (rapid heating rate).[11] However, the controversies surrounding 'specific microwave effects' and 'non-thermal microwave effects' are expected to continue for years because the intrinsic and unequivocal mechanism has not yet been identified. The biggest obstacle to overcome is the discrepancies in the experimental conditions used amongst the various studies. Comparison of results obtained under totally different synthetic parameters is unreasonable, and makes the conclusions inaccurate and speculative. Therefore, identical synthetic parameters should be set up for both microwave heating and conventional heating, in order to compare the experimental results. However, there are some inherent challenges; for example, the temperature profiles of microwave heating cannot be well reproduced using conventional heating.

4.2.3 Role of Synthetic Parameters

In this section, we will discuss the effects of different synthetic parameters on the products under microwave irradiation. These synthetic parameters include the solvent, the temperature, the microwave frequency, the microwave power, the heating type, the reaction vessel, and the reaction system.

4.2.3.1 Solvent

The use of a solvent is inevitable for wet-chemistry synthesis. The polarity of the solvent plays a critical role in microwave-assisted preparation, which represents the coupling capability with microwaves. In order to quantitatively determine the effect of different solvents, the loss tangent ($\tan\delta$) is defined as $\tan\delta = \varepsilon''/\varepsilon'$, which is the conversion efficiency of materials from microwave energy to heat energy. ε'' and ε' are the dielectric loss and dielectric constant of the solvent, respectively. The loss tangents and boiling points of common solvents under microwave irradiation are listed in Table 4.1.

Table 4.1 Loss tangents (tan *δ*) and boiling points of various solvents at 2.45 GHz and 20 °C. Reproduced from ref. 11 and 24 with permission from John Wiley & Sons, Copyright 2011 WILEY-VCH Verlag GmbH & Co. KGaA, Weinheim, and the Royal Society of Chemistry.

Solvent	Boiling point/°C	tan *δ*
Hexane	68–69	0.020
Toluene	111	0.040
Dichloromethane	38.8	0.042
Tetrahydrofuran	66	0.047
Acetone	56–57	0.054
Chlorobenzene	131	0.101
Water	100	0.123
1,2-Dichloroethane	84	0.127
N,*N*-Dimethylformamide	153	0.161
Acetic acid	118–119	0.174
N-Methyl-2-pyrrolidone	202	0.275
1,2-Dichlorobenzene	180.5	0.280
Methanol	65	0.659
2-Propanol	82	0.779
Ethanol	78	0.941
Ethylene glycol	~198	1.350

The solvents with tan $\delta > 0.5$, 0.1–0.5, and <0.1 are classified as high, medium, and low microwave-absorbing solvents, respectively. Among these solvents, water and alcohols are ideal solvents for microwave-assisted preparation of inorganic nanostructures owing to their good microwave-absorbing ability and moderate boiling points. The loss tangent is inversely proportional to the microwave penetration depth, that is, the solvents with higher loss tangent have shorter penetration depth. For example, the penetration depth of water is only a few centimetres.

In some cases, two or more different solvents are used, which allows adjustment of synthetic parameters, such as the volume ratio, and a mixture of a polar solvent and a non-polar solvent. As an example, a binary system consisting of a polar solvent and a non-polar solvent (such as water/chloroform) can be used for phase-transfer reactions under microwave irradiation. The temperature of water can reach 100 °C, whereas non-polar chloroform maintains a low temperature below its boiling point (~61 °C), and the reactants can transfer from one phase to the other.

A relevant point to address is: what should we do if the chemical reactions must be performed in the non-polar solvent? Leadbeater and Torenius[25] provided an efficient solution by introducing a small quantity of an ionic liquid. They added 1,3-dialkylimidazolium iodide into hexane, and the temperature of the reaction system reached 217 °C under microwave irradiation (200 W for 10 s). As expected, the temperature of the reaction system maintained a low temperature (46 °C) without ionic liquid under the same conditions.

4.2.3.2 Temperature and Microwave Frequency

The temperature and microwave frequency significantly affect the loss tangent and penetration depth. For example, ethanol has a high loss tangent ($\tan\delta = 0.941$) at room temperature. However, its loss tangent is only 0.27 and 0.08 at 100 and 200 °C, respectively.[24] In contrast, the loss tangent of the ionic liquid [BMIM]PF_6 (BMIM, 1-*n*-butyl-3-methylimidazolium) is 0.185, 1.804, and 3.592 at 20, 100, and 200 °C, respectively.[24] The big difference in loss tangent of various solvents at different temperatures originates from the intrinsic heating mechanism. Ethanol is dominantly heated by the dipolar polarisation mechanism. At a higher temperature, the viscosity of ethanol decreases, thus the friction and collision of molecules are reduced, and the microwave absorption ability becomes weaker. For [BMIM]PF_6, the ionic conduction mechanism is involved in the heating process, and its ability to absorb microwaves is stronger at a higher temperature.

The effect of microwave frequency on the loss tangent and penetration depth was studied by Horikoshi *et al.*[26] It was found that: (1) the dielectric loss factor (ε'') of water decreased with temperature under microwave irradiation at a frequency of either 2.45 or 5.8 GHz; and (2) the penetration depth is inversely proportional to frequency. For example, the penetration depth in water decreased from ~1.8 cm with 2.45 GHz microwaves to 0.34 cm using 5.8 GHz microwaves. Horikoshi *et al.*[27] studied the effect of microwave frequency on the synthesis of nanoparticles. The nanoparticles could not form in non-polar solvents because they were not microwave absorbers. However, gold nanoparticles were successfully prepared in oleylamine at a microwave frequency of 5.8 GHz. In contrast, the architecture of gold nanoparticles displayed no obvious variation using either 2.45 or 5.8 GHz microwaves in the presence of polar ethylene glycol as the solvent.

4.2.3.3 Variable Frequency and Fixed Frequency

In contrast to fixed-frequency microwave heating, variable-frequency microwave heating involves a selected bandwidth, which is swept around a central frequency over a certain time. It is desirable that arcing and localised heating that occur in fixed-frequency microwave heating can be avoided by varying the microwave frequencies.[28] Previous studies reported the application of variable-frequency microwave heating on the synthesis of silver nanoparticles, manganese oxides, $BaTiO_3$, and so on.[28–30] As a representative example, Nyutu *et al.*[30] investigated the effect of microwave frequency on the hydrothermal synthesis of nanocrystalline $BaTiO_3$ morphologies. They found that: (1) $BaTiO_3$ particles prepared with 2.45 GHz microwaves had a cubic morphology (Figure 4.2a); (2) $BaTiO_3$ particles prepared *via* conventional heating under similar conditions had an irregular morphology with a bimodal particle size distribution (Figure 4.2b); (3) samples prepared by high-frequency and variable-frequency microwaves consisted of regularly shaped particles with a unimodal size distribution (Figure 4.2c–f); (4) the specific surface

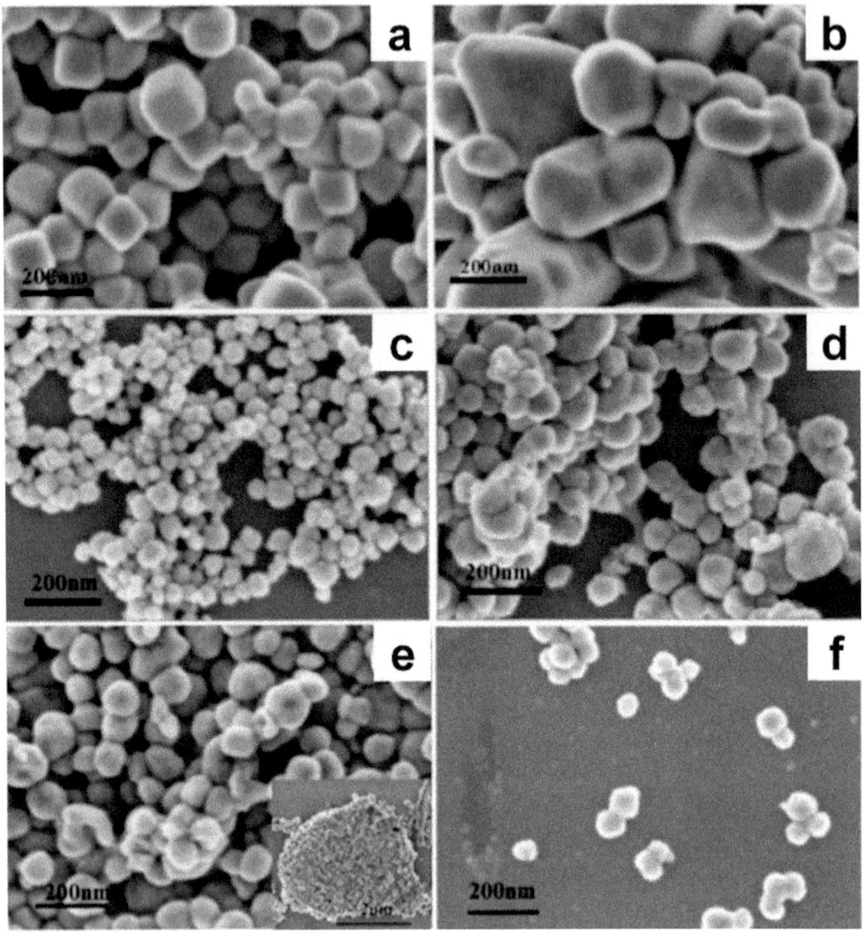

Figure 4.2 Scanning electron microscopy (SEM) images of the samples prepared for 20 h at different microwave frequencies: (a) 2.45 GHz; (b) conventional heating; (c) 5.5 GHz; (d) 4.0 GHz; (e) 3–5.5 GHz (5 s); (f) 3–5.5 GHz (1 s). Reproduced from ref. 30 with permission from American Chemical Society, Copyright 2008.

area of the sample increased with extended reaction time under variable-frequency microwave irradiation, but decreased with increasing reaction time under fixed-frequency microwave irradiation.

4.2.3.4 *Microwave Power and Heating Type*

Microwave power has an effect on the formation of materials. Seol *et al.*[31] and Yang *et al.*[32] studied the effect of applied microwave power on the morphology of gold and silver nanostructures, respectively. In the microwave-assisted preparation of gold nanostructures,[31] the higher microwave power increased

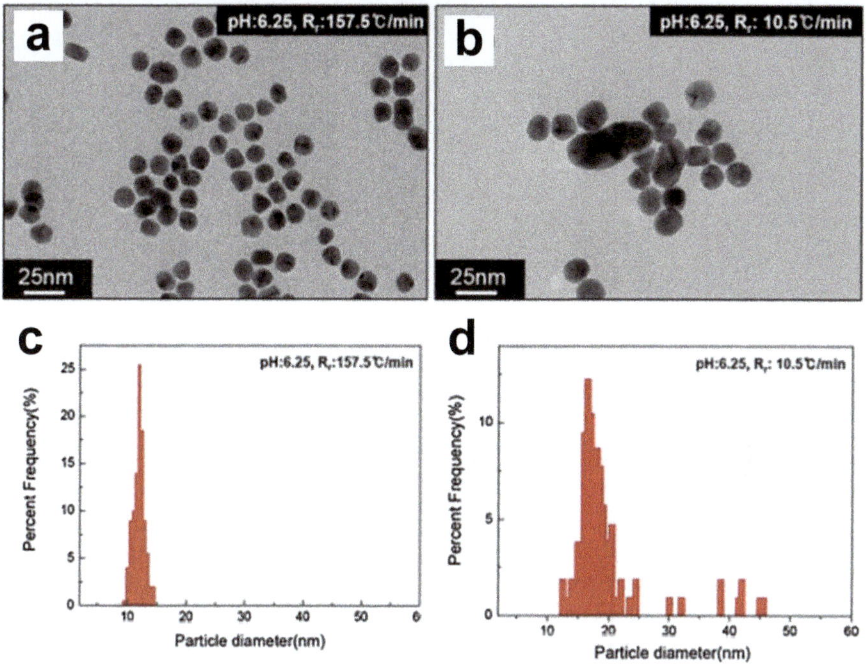

Figure 4.3 Transmission electron microscopy (TEM) images and size distributions of gold nanoparticles prepared under different conditions: (a and c) pH 6.25, heating rate = 157.5 °C min^{-1}; (b and d) pH 6.25, heating rate = 10.5 °C min^{-1}. The heating rate was controlled by microwave power. Reproduced from ref. 31 with permission from Elsevier, Copyright 2011.

the temperature ramping rate, and the as-prepared gold nanoparticles experienced homogeneous nucleation, which resulted in smaller average size and narrow size distribution (Figure 4.3).

The type of microwave heating has also an effect on the product. Pulsed and continuous microwave heating are two heating types. Shen *et al.*[33] prepared tungsten carbide (WC) nanoparticles with an average size of 24.1 nm under pulsed microwave heating. The pulsed microwaves were set as 15 s on and 15 s off for 20 cycles. However, larger WC nanoparticles, with an average size of 35.7 nm, were obtained under continuous microwave heating for 5 min.

4.2.3.5 *Reaction Vessel*

Because the penetration depth of microwaves is limited, the volume and shape of the reaction vessel have a great impact on the product. For example, Bonaccorsi and Proverbio[34] prepared zeolite A under microwave irradiation. The experimental equipment is shown in Figure 4.4a. They found that the product had a lower content of zeolite A and higher level of impurities if the volume

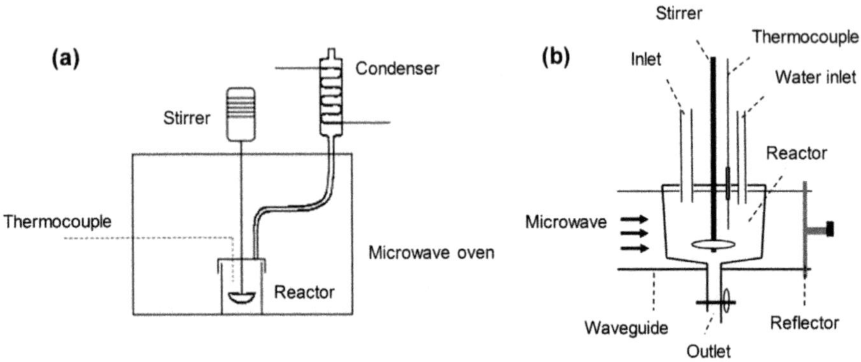

Figure 4.4 Schematic illustration of experimental equipment for microwave-assisted synthesis of: (a) zeolite A and (b) $BaTiO_3$. (a) Reproduced from ref. 34 with permission from Elsevier, Copyright 2003. (b) Reproduced from ref. 36 with permission from Elsevier, Copyright 2012.

of the reactor increased from 100 to 500 mL. The difference was attributed to the penetration depth of the microwaves. If the thickness of the bulk materials was much larger than the penetration depth of the microwaves, only the outer regions were efficiently heated. In a spherical glass reactor with a volume of 500 mL, the same amount of material had a lower surface-to-volume ratio compared with the 100 mL reactor. The effect of reactor shape on the product was also investigated by Conner *et al.*[35] Reactors with different diameters had different distributions of microwave energy and, consequently, the preparation of silicate crystals in a larger reactor resulted in larger sizes.

4.2.3.6 *Mechanical Stirring*

It is well known that mechanical stirring can produce a stable and homogeneous reaction environment in wet-chemistry synthesis, which can affect the nucleation and growth of nanostructures. In microwave-assisted synthesis, *in situ* stirring also plays an important role in achieving uniform microwave irradiation and energy distribution. For example, Komarneni and Katsuki[37] demonstrated that mechanical stirring could not only improve the yield of $BaTiO_3$ nanoparticles, but also was favourable for the formation of smaller and more uniform nanoparticles. Katsuki *et al.*[36] prepared $BaTiO_3$ nanoparticles using a home-built semi-continuous microwave reactor (Figure 4.4b). With the help of mechanical stirring, the conversion rate was enhanced from 62–65% (without stirring) to 100%.

4.2.3.7 *Type of Reactor*

There are two types of microwave reactors: open and closed reactors. The open microwave reactor is simple, safe, and cost effective. However, reactions in the open microwave reactor are performed under atmospheric pressure and the reaction temperature is limited by the boiling point of the solvent used. Consequently, solvents with relatively high boiling points, such as water and ethylene glycol, are usually adopted in microwave-assisted synthesis using open reactors. As a typical example, although methanol is an excellent microwave-absorbing solvent, with $\tan\delta = 0.659$ (Table 4.1), the temperatures below its boiling point (65 °C) are too low to enable many chemical reactions to occur in open microwave reactors.

In contrast, the reaction temperature can be higher than the boiling point of the solvent in closed microwave reactors. In addition, because of the autogeneous pressure built up, a high-pressure environment is formed spontaneously with vaporisation of the solvent. This high-temperature and high-pressure environment is desirable for chemical reactions and materials synthesis. Such processes are called 'hydrothermal' or 'solvothermal' synthesis (see Chapter 6), depending on whether the solvent is water or other solvents. Much attention and caution are needed for 'hydrothermal' or 'solvothermal' synthesis methods because explosions may occur at overly high temperatures and pressures. In recent years, the microwave-assisted hydrothermal/solvothermal method has been extensively applied for the rapid synthesis of inorganic nanostructured materials, owing to the significant reaction rate acceleration.[12]

4.2.3.8 *Other Factors*

It is worth noting that many other factors affect the microwave-assisted synthesis of nanostructured materials. The optimisation of synthetic parameters is essential to obtain high-quality materials.[38] Additional factors that influence the synthesis include the nature of the reactants, their concentrations, the nature and amounts of additives, the volume of the reaction system, the shape of the reactor, the microwave cavity design, the reaction time, and the microwave power.

4.2.4 Apparatus

In this section, three major components of the microwave reactor will be discussed, including the reaction vessel, the cavity, and the temperature sensor.

4.2.4.1 *Reaction Vessel*

Quartz, Pyrex® glass, and Teflon® are microwave-transparent materials and they are widely used as building materials for microwave reaction vessels.[2] A more specialised microwave reaction vessel can be made from SiC with strong microwave-absorbing ability, and its features and shielding ability have been discussed in Section 4.2.2.3. The choice of microwave reaction vessel depends on the research purpose.

4.2.4.2 *Multimode Cavity*

In the design of the microwave apparatus, maximum conversion efficiency from microwave energy to heat is pursued. In the early stages, domestic microwave ovens were usually used for microwave-assisted chemical reactions and materials synthesis. They were easy to operate, commercially available, and small, which greatly contributed to the burgeoning development of this research field.

The cavities of domestic microwave ovens have dimensions that are much larger than the wavelength of the microwaves used. Consequently, the cavities in domestic microwave ovens can support different configurations of the microwave field (modes).[2] This kind of cavity is the so-called 'multimode cavity'. Parallel microwave synthesis can be realised in a multimode microwave reactor. Ferguson published an overview of the typical reactors used for microwave synthesis.[39] The multimode microwave reactor has some shortcomings, for example, the multimode cavity induces random microwave field patterns, causing inhomogeneous temperature distributions, and the repeatability of experiments is consequently poor.

4.2.4.3 *Single-mode Cavity*

A microwave apparatus with a single-mode cavity is an ideal alternative. In contrast to multimode cavity, the microwave field in a single-mode cavity is well defined in space. Taking a cylinder single-mode cavity as a typical example, there is only one configuration of the microwave field. Therefore, uniform temperature and repeatability of experiments can be achieved based on the known configuration of the microwave field. The synthetic parameters, including temperature, pressure, power, and time, can be well controlled by these advanced microwave reactors.[3] Depending on the requirements of the microwave synthesis, accessories for temperature monitoring or mechanical stirring are also available (Figure 4.4).[34,36]

In microwave-assisted preparation, the penetration depth of microwaves is usually a few centimetres for most solvents, which are unable to penetrate the whole reaction system if the reactors are large in size. To overcome this problem, Nishioka *et al.*[40] reported a flow-type, single-mode microwave reactor, which could enable continuous-flow reactions under elevated pressure. The microwave reactor consisted of a variable-frequency microwave generator, a pressure pump, a single-mode cavity, and a quartz tube. The reaction fluid was introduced into the quartz tube inlet with the help of a pressure pump (Figure 4.5). During the synthetic process, the temperature of the liquid flow was detected and controlled automatically using a temperature feedback module. In this way, the instantaneous heating of different flow solvents, including water, ethylene glycol, and ethanol, could be accomplished. They also demonstrated the successful application of such a flow-type microwave reactor in the continuous synthesis of Cu nanoparticles.

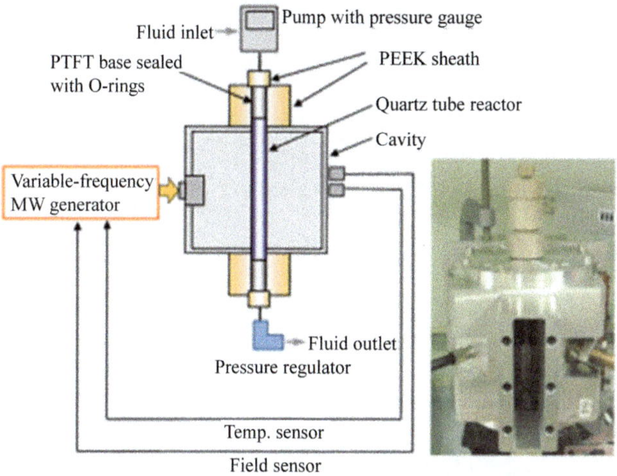

Figure 4.5 Experimental setup of a flow-type microwave (MW) reactor. Inset: an image of the microwave cavity part. Reproduced from ref. 40 with permission from American Chemical Society, Copyright 2013.

4.2.4.4 *Temperature Measurement*

In microwave-assisted synthesis, one challenge is how to accurately measure the reaction temperature. There are many factors influencing the temperature measurement, such as viscosity, microwave absorptivity, stirring rate, and so on. Hayden *et al.*[41] studied accurate internal temperature measurement during microwave heating. They combined an external infrared sensor with an internal fibre optic sensor to enable simultaneous temperature monitoring (Figure 4.6). In addition, a built-in camera was used for real-time observation of the stirring efficiency (inset of Figure 4.6). Currently, many sensors have been developed for synergistic temperature measurement that can provide reliable and reproducible results.

4.2.5 Microwave Heating *vs.* Conventional Heating

Microwave heating is distinct from conventional heating in its rate acceleration, selectivity, and yield of the product. As mentioned, rational comparison between microwave heating and conventional heating depends on accurate temperature measurements and consistent experimental conditions. Unfortunately, in many cases, comparisons have been made under different synthetic parameters and it should be noted that comparison studies between microwave heating and conventional heating should be performed under the same experimental conditions. Some comparison studies from different standpoints, which focus on energy consumption, conversion ratio, structure, size distribution, morphology, crystal phase, crystallinity, and kinetics and thermodynamics, are reviewed in this section.

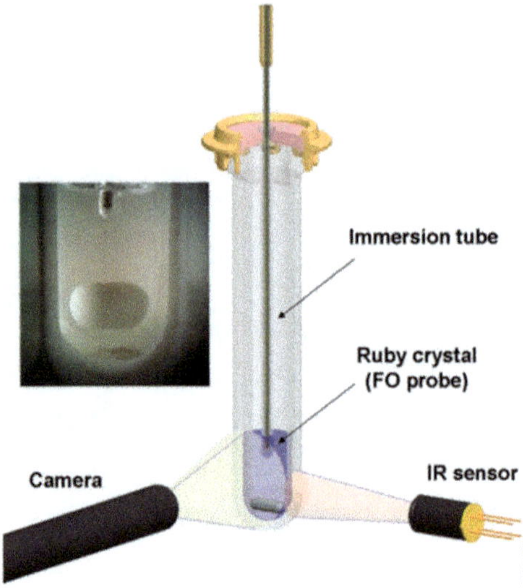

Figure 4.6 Schematic illustration of temperature measurement with the help of dual external infrared (IR) and internal fibre optic (FO) sensors and the location of the built-in camera. Inset: image of 2 mL liquid and a magnetic stir bar. Reproduced from ref. 41 with permission from John Wiley & Sons, Copyright 2013 WILEY-VCH Verlag GmbH & Co. KGaA, Weinheim.

4.2.5.1 *Energy Consumption*

It is well known that the energy consumption of chemical reactions can be considerably reduced under microwave irradiation. It is necessary to quantitatively determine the energy consumption by microwave heating or conventional heating. Katsuki *et al.*[36] determined the energy consumption by calculation from the output power. They synthesised $BaTiO_3$ nanoparticles (Figure 4.7a) using a home-built, single-mode, semi-continuous microwave reactor. The approximate energy consumption from room temperature to 90 °C was 89 kJ, and 29 kJ for heating at 90 °C for 5 min, in the microwave-assisted hydrothermal synthesis, giving a total energy consumption of 118 kJ. For the energy consumed under conventional heating, the approximate energy consumption from room temperature to 90 °C was 648 kJ, and 144 kJ for heating at 90 °C for 5 min, giving a total energy consumption of 792 kJ. The energy consumption of conventional heating is in this case 6.7 times that of microwave heating. It can be concluded that the energy consumed in microwave heating is much lower than for conventional heating under the same experimental conditions.

4.2.5.2 *Conversion Ratio*

Katsuki and Komarneni[43] prepared $BaTiO_3$ nanoparticles by microwave heating and conventional heating using commercial TiO_2 and $Ba(OH)_2$ as the raw materials to determine the conversion ratio. The conversion ratio of TiO_2

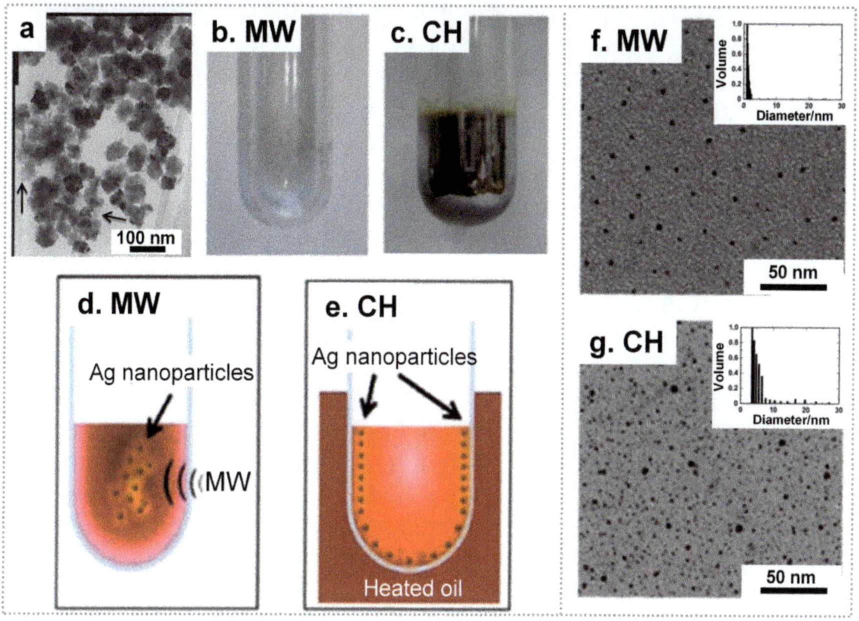

Figure 4.7 (a) TEM image of $BaTiO_3$ nanoparticles prepared at 90 °C for 5 min by microwave heating. (b and c) Photographs of samples obtained by microwave (MW) heating for 4 min and conventional heating (CH) for 5 min. (d and e) Schematic illustration of the temperature distributions in the reactors heated by microwaves and oil bath. (f and g) TEM images of Ag nanoparticles prepared by microwave heating and conventional heating. Inset: size distributions measured by light scattering. (a) Reproduced from ref. 36 with permission from Elsevier, Copyright 2012. (b–g) Reproduced from ref. 42 with permission from the Royal Society of Chemistry.

and $Ba(OH)_2$ to $BaTiO_3$ nanoparticles at reaction time $t = 1$ min was 73.7% and 0% for microwave heating and conventional heating, respectively, and reached 95.8% for microwave heating and remained at 0% for conventional heating at $t = 5$ min. It took 10 min for microwave heating and 1 h for conventional heating to achieve 100% conversion, demonstrating the benefits of microwave heating. Similar results were reported for the microwave-assisted synthesis of $LaF_3/LaPO_4$:Ce,Tb nanocrystals and metal nanoparticles on a mesoporous material.[44,45]

4.2.5.3 *Size Distribution*

Compared with conventional heating, the products prepared by microwave heating usually show more uniform and narrow size distributions.[44] Horikoshi *et al.*[42] compared Ag nanoparticles prepared by microwave heating and conventional heating at 100 °C. It was found that continuous 64 W microwave irradiation had a ramping rate of 0.51 °C s^{-1}, whereas oil bath heating with a power of 400 W had a heating rate of 0.09 °C s^{-1}. The colour of the reaction suspension under microwave irradiation for 4 min became slightly yellow (Figure 4.7b). On the other hand, a black Ag film was observed using conventional heating for 4 min (Figure 4.7c). The difference resulted from the initial differing temperature distributions for the two heating methods. As shown in Figure 4.7d, the reaction temperature increased because of volumetric heating under microwave irradiation. In contrast, heat was transferred from the oil bath to the reactor walls using conventional heating, and then to the solution *via* thermal conduction and convection (Figure 4.7e). The transmission electron microscopy (TEM) images of Ag nanoparticles prepared by microwave heating and conventional heating are shown in Figure 4.7f and g. Ag nanoparticles prepared by microwave heating had a uniform size distribution (1.8–3.6 nm, average size ~3 nm), whereas a broader size distribution (1–5 nm) was observed for Ag nanoparticles prepared by conventional heating. The size distributions measured by light scattering were located in the range of 1–3 nm and 3–5.7 nm for microwave heating and conventional heating, respectively (inset of Figure 4.7f and g). Similar results were also reported for the rapid microwave-assisted synthesis of polyacrylamide–metal (Ag, Pt, Cu) nanocomposites and ZnO nanoparticles.[46,47]

4.2.5.4 Crystal Phase and Crystallinity

Under the same experimental parameters, the crystal phase of the product obtained by microwave heating and conventional heating may be different. Katsuki and Komarneni[48] prepared α-Fe_2O_3 particles using $FeCl_3$ as a precursor in the temperature range 100–160 °C. The as-prepared product consisted of acicular β-FeOOH particles with a length of 300 nm and a thickness of 40 nm by the conventional hydrothermal method at 100 °C for 2–3 h. In contrast, α-Fe_2O_3 spherical nanoparticles with diameters of 30–66 nm were obtained using the microwave–hydrothermal method at 100 °C for 2 h.

Microwave irradiation is able to promote the crystallisation of materials in a much shorter time.[49] Jhung *et al.*[50] prepared silicate-1 molecular sieves under microwave irradiation (MW) and conventional electric heating (CE). Two-step methods were used, including MW–MW, MW–CE, CE–MW, and CE–CE. The crystallisation curves of silicate-1 under different heating modes are shown in Figure 4.8. It took 3 h to complete crystallisation under microwave irradiation (MW–MW), but the time was ~90 h by conventional heating (CE–CE). The crystallisation curve using microwave heating had a steeper slope, indicating the faster crystallisation under microwave irradiation.

4.2.5.5 Morphology

Heating type greatly affects the nucleation and growth of crystals, causing significant differences in the morphology of the product. Dahal *et al.*[51] studied the beneficial effects of microwave heating *vs.* conventional heating in the preparation of noble metal (Rh, Pd, and Pt) nanoparticles. Compared with conventional heating, the beneficial effects of microwave heating included

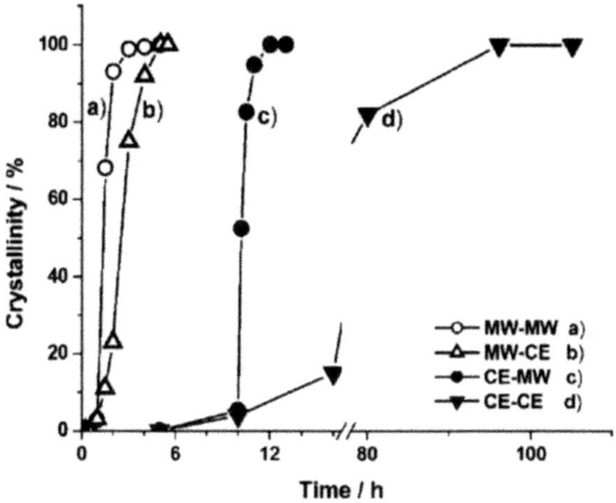

Figure 4.8 Crystallisation curves of silicate-1 synthesised under microwave irradiation and conventional electric heating. Reproduced from ref. 50 with permission from John Wiley & Sons, Copyright 2007 WILEY-VCH Verlag GmbH & Co. KGaA, Weinheim.

improved monodispersion, morphological control, and higher crystallinity. In the nucleation stage, differences in Rh nanoparticle formation were observed using different heating methods. Microwave heating enabled faster nucleation and the formation of more uniform-sized nanoparticles. In addition, more homogeneous seeds could be obtained by microwave heating, which were used for further crystal growth. The experimental results showed that Rh nanoparticles prepared by conventional heating exhibited mixed morphologies (spheres, tetrahedra, square plates). In contrast, microwave synthesis produced a majority of tetrahedral Rh nanoparticles with sizes of 5–7 nm or larger nanocubes (~10 nm) upon further growth (Figure 4.9). Microwave-prepared Rh nanoparticles exhibited approximately twice the catalytic activity of similar-sized conventionally prepared nanoparticles, as demonstrated in the vapour-phase hydrogenation of cyclohexene. Haque *et al.*[49] analysed the acceleration for microwave-assisted synthesis of metal–organic frameworks and aluminophosphates using the Arrhenius and Eyring equations. They found that, irrespective of the type of materials and metal ions in the isotypic metal–organic frameworks, microwaves could accelerate the synthesis by about 5–33 times for both nucleation and crystal growth. The accelerated synthesis using microwaves was mainly due to the increased pre-exponential factor of the Arrhenius equation and the decreased activation free energy of the Eyring equation, respectively. The decreased activation free energy under microwave irradiation was mainly ascribed to the high activation entropy of microwave-assisted synthesis compared with the entropy of synthesis using conventional heating. They proposed that the microwave-accelerated synthesis resulted from a more favourable reaction coordinate or

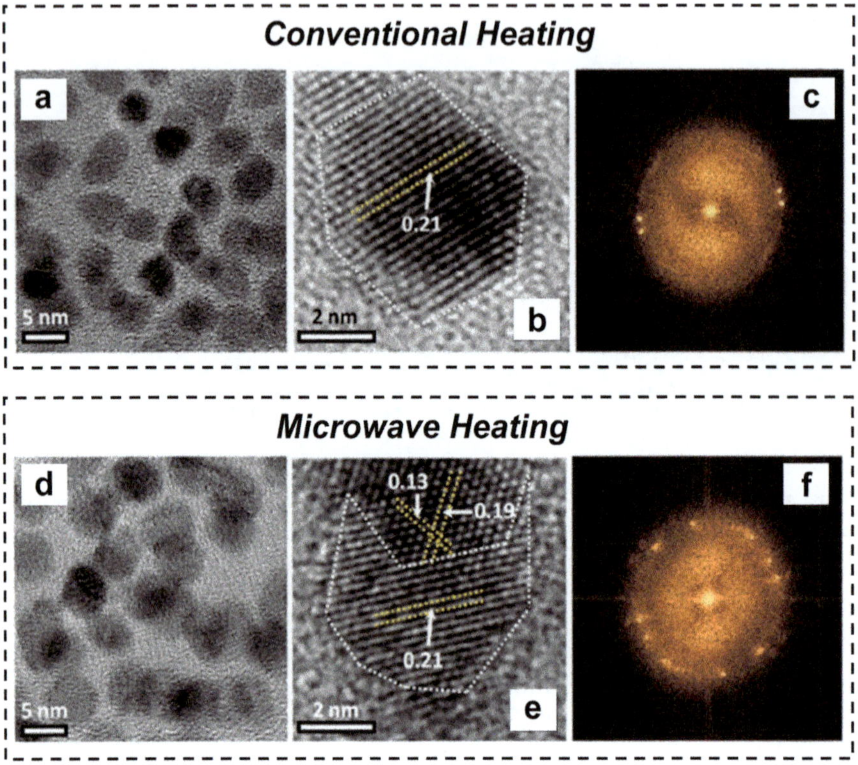

Figure 4.9 Rh nanocrystals: (a and d) TEM images; (b and e) high-resolution TEM images; (c and f) fast Fourier transform maps. Reproduced from ref. 51 with permission from American Chemical Society, Copyright 2012.

changes of relative energies of intermediates for high activation entropy with increased randomness.

4.2.5.6 Nucleation and Crystal Growth

Nucleation and crystal growth are two key stages for the formation of inorganic nanostructured materials. The acceleration of these two stages by microwave irradiation has been proved. In work reported by Bilecka *et al.*, a non-aqueous sol–gel approach was used to prepare ZnO nanoparticles from zinc acetate and benzyl alcohol (Figure 4.10).[47] They performed similar experiments using conventional oil bath heating for comparison. As shown in Figure 4.10a, the concentration of zinc acetate over time was significantly different using microwave heating and conventional heating. The dissolution of zinc acetate was slower using conventional heating, resulting in a smaller rate constant (0.0142 min^{-1}; Figure 4.10c). In contrast, the same rate constant reached 0.124 min^{-1} with microwave heating. Hence, the nucleation of ZnO by microwave heating was faster. The difference in the growth of ZnO

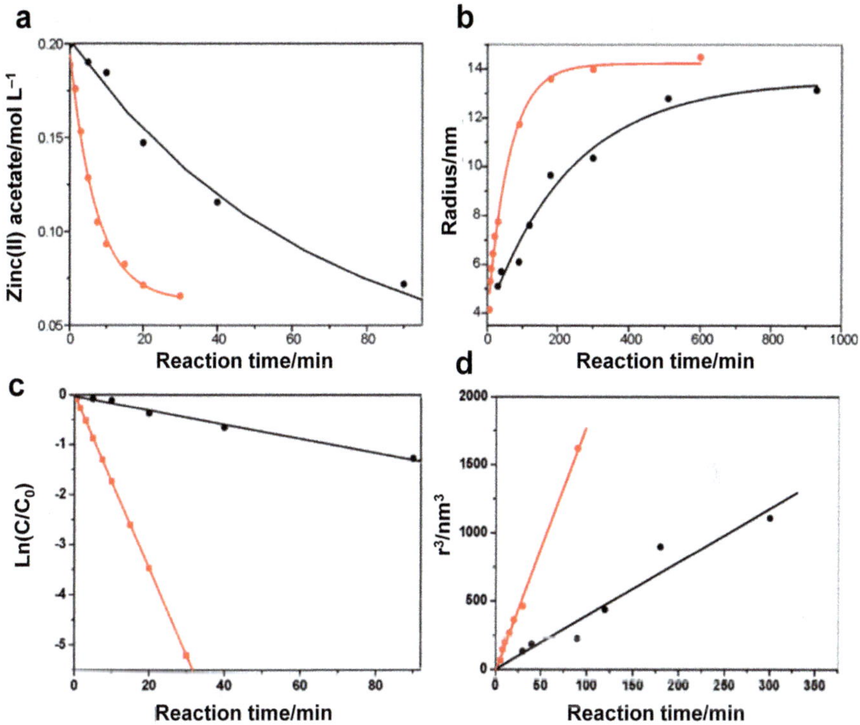

Figure 4.10 Comparison of benzyl acetate (a and c) and ZnO formation (b and d) using microwave heating (red) and conventional heating (black) at 120 °C. (a) Zinc acetate concentration *vs.* reaction time; (c) corresponding first-order kinetic plot. (b) Radius of ZnO nanocrystals *vs.* reaction time. (d) r^3 *vs.* reaction time according to the Lifshitz–Slyozov–Wagner (LSW) model. Reproduced from ref. 47 with permission from American Chemical Society, Copyright 2009.

using different heating methods was also studied (Figure 4.10b and d). The rate constant k_{growth} was 3.9 and 15.36 nm^3 min^{-1} for conventional heating and microwave heating, respectively. Comparing microwave heating and conventional heating, the authors proposed that: (1) microwave heating could facilitate dissolution of the precursor in the solvent; (2) microwave heating could increase the rate constant, resulting in faster production of monomer and, consequently, in an earlier nucleation event; and (3) microwave heating could promote crystal growth. Similar results were also reported for the fast microwave-assisted synthesis of microporous materials such as silicalite-1 and VSB-5 molecular sieves.[50]

4.2.5.7 *Kinetics and Thermodynamics*

Understanding the kinetics and thermodynamics under microwave irradiation provides the possibility to tune the physical and chemical properties of the materials. Studies on the activation energy and free energy using microwave heating have been reported. However, there are controversies surrounding the thermodynamics of microwave chemistry. Some papers report a decrease in activation energy of chemical reactions under microwave irradiation.[52–54] For example, Brosnan *et al.*[53] reported the microwave sintering of alumina at 2.45 GHz, and the activation energy was determined as 85 ± 10 and 520 ± 11 kJ mol^{-1} for microwave sintering and conventional sintering, respectively. However, opposite results were obtained in some other studies.[49,55] For example, Haque *et al.*[55] reported the synthesis of Cu_2O using three types of heating, including ultrasound, microwave, and conventional electric heating. They found that the calculated activation energies of both nucleation and crystal growth using conventional heating were the lowest. As shown in Table 4.2, the activation energy of nucleation was 61.5 and 44.1 kJ mol^{-1} for microwave and conventional heating, respectively. Although the activation energy increased using microwave heating, rate acceleration under microwave irradiation was still observed. It was found that the pre-exponential factor (A) of the Arrhenius equation was higher, and the free energy (ΔG) of the chemical reaction was lower, for microwave heating, which contributed to the rapid synthesis under microwave irradiation. Based on these experimental results, they attributed the rate acceleration of microwave heating to physical effects, such as hot spots, rather than to chemical effects. Similar results were also reported in further reports.[49] In conclusion, the effects of microwave irradiation on the kinetics and thermodynamics have not been well understood and more detailed studies are necessary to clarify the relevant issues.

4.2.5.8 Some Studies Highlighting Negligible Differences Between Microwave Heating and Conventional Heating

Although rate acceleration of chemical reactions with the help of microwaves has been extensively reported, there are some papers that report negligible differences between microwave heating and conventional heating.[56,57] For example, Pein *et al.*[57] prepared three kinds of $CuInS_2$ nanoparticles at 220 °C using 15 min microwave heating in a Pyrex® vessel, microwave heating

Table 4.2 Calculated pre-exponential factor (A) and activation energy (E_a) for the synthesis of Cu_2O. Reproduced from ref. 55 with permission from the Royal Society of Chemistry.

Heating method	A for nucleation/min^{-1}	E_a for nucleation/kJ mol^{-1}	A for crystal growth/min^{-1}	E_a for crystal growth/kJ mol^{-1}
Ultrasound	2.72×10^{11}	72.7	5.32×10^{12}	84.7
Microwave heating	1.36×10^{9}	61.5	2.92×10^{10}	75.1
Conventional heating	9.23×10^{4}	44.1	2.81×10^{6}	58.6

in a SiC vessel, and heating in an oil bath, respectively (Figure 4.11). TEM images showed similar morphology and size distribution in the range 5–10 nm for the three samples (Figure 4.11a–c). The heating profiles were carefully controlled to minimise differences resulting from a thermal effect (Figure 4.11d). XRD patterns of the three samples prepared under different conditions showed similar diffraction peaks and intensities (Figure 4.11e). The difference between microwave heating and conventional heating could not be identified from the analysis results by XRD, TEM and UV–visible (UV–vis) absorption spectroscopy (Figure 4.11f). A small difference was observed from the thermogravimetric (TG) analysis (Figure 4.11g), with the sample prepared by microwave heating in the Pyrex® vessel exhibited a larger weight loss during heating, indicating a higher amount of capping ligand on the surface.

4.3 Microwave-assisted Preparation of Inorganic Nanostructures

The preparation of various inorganic nanostructures under microwave irradiation has been extensively investigated. Herein, we focus on reviewing the microwave-assisted preparation of nanostructures of metals, metal oxides, metal chalcogenides, metal fluorides, metal phosphates, carbon materials, metal–organic frameworks, and nanocomposites. Considering the critical roles of the solvent in microwave chemistry, the contents will be classified by the type of solvent used.

4.3.1 Metals

The preparation of high-quality metal nanostructures has been a hot research topic owing to their unique physical, optoelectronic, and chemical properties, and promising applications in many fields.[58,59] Microwave-assisted rapid preparation can provide the possibility for designing and producing novel metal nanostructures. Hence there are a large number of papers reporting the microwave-assisted rapid preparation of metal nanostructures. We recommend some review articles about microwave-assisted preparation of metal nanostuctures,[11,12,60,61] and discuss some typical examples in this section.

4.3.1.1 Preparation in Aqueous Solution

Preparation Strategy. Preparation of metal nanostructures in aqueous solution has been the most widely studied because water is easily accessible, environmentally friendly, and low cost. In addition, water has a relatively low vapour pressure and good microwave-absorbing ability, making it an ideal solvent for microwave synthesis. The loss tangent of water at 2.45 GHz and 20 °C is 0.123 (Table 4.1), indicating that water is an effective medium

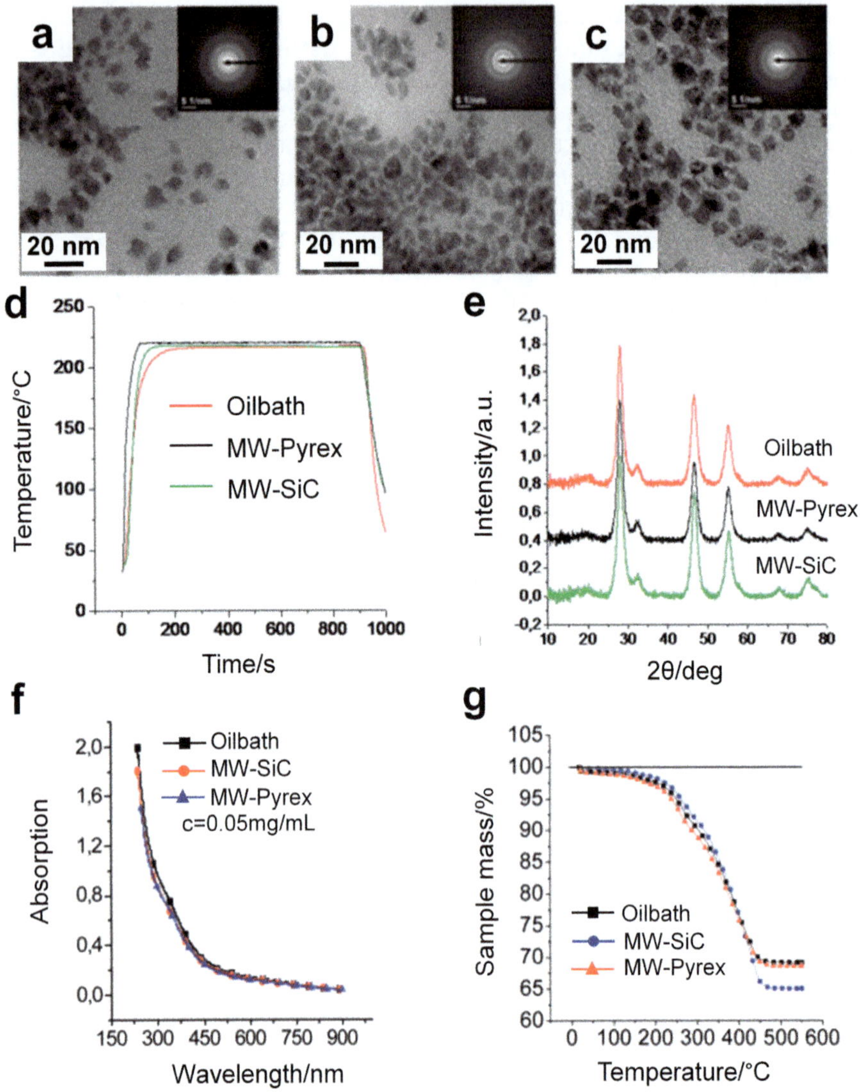

Figure 4.11 (a–c) TEM images and selected-area electron diffraction patterns of $CuInS_2$ nanoparticles prepared at 220 °C under different conditions: (a) microwave-Pyrex®; (b) microwave-SiC; (c) oil bath. (d) Heating profiles. (e) XRD patterns. (f) UV–vis absorption spectra. (g) Thermogravimetric curves. Reproduced from ref. 57 with permission from American Chemical Society, Copyright 2011.

to absorb microwave energy. The microwave heating mechanism in aqueous solution is illustrated in Figure 4.12. As already mentioned, polar water molecules under microwave irradiation are induced to align with the alternating electric field, during which process heat is generated due to friction

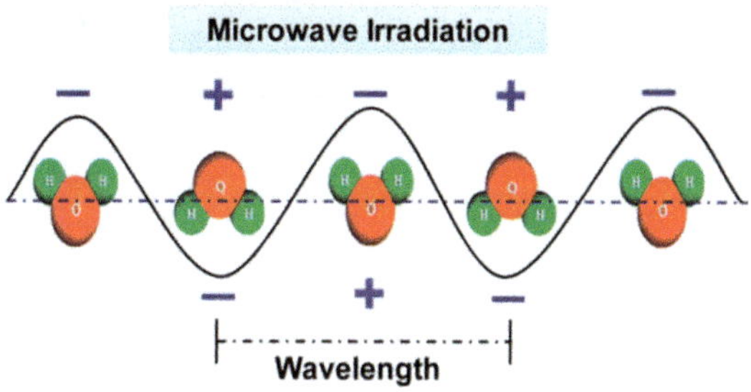

Figure 4.12 Schematic illustration of the microwave heating mechanism in aqueous solution. Reproduced from ref. 12 with permission from American Chemical Society, Copyright 2014.

and collision. Hence, the temperature of the reaction solution increases rapidly.

The reduction reaction of metal ions is employed to produce metal nanostructures in aqueous solution. In a typical synthetic process, a soluble metal salt and a reducing agent are needed. To avoid aggregation and control the morphology and size of the metal nanostructures, additives or surfactants are usually added into the aqueous solution. Frequently used reducing agents include $NaBH_4$,[62] hydrazine hydrate,[63] sodium citrate,[64] sugars (α-D-glucose, sucrose, maltose),[65,66] amino acid,[67] glutathione,[68] 2,7-dihydroxy naphthalene,[69] wine pomace,[70] beet juice,[71] ascorbic acid,[72,73] carboxymethyl cellulose sodium,[74] homocellulose,[75] and so on. The representative works on microwave-assisted preparation of metal nanostructures in aqueous solution are listed in Table 4.3.

4.3.1.1.1 Single Metal Nanostructures. *Gold.* The fascinating properties of Au nanostructures are very attractive for optical, biomedical, and catalytic applications.[80–84] The properties of Au nanostructures depend on their size and morphology,[82] which can be adjusted by means of the synthesis parameters. High-quality Au nanostructures can be obtained by microwave-assisted preparation. The reducing agent, surfactant, and their concentration and molar ratio have significant effects on the size and morphology of Au nanostructures. Kundu *et al.*[69] reported the preparation of multiple-shaped Au nanoparticles in aqueous solution under microwave irradiation. They used 2,7-dihydroxy naphthalene (2,7-DHN) as a reducing agent and cetyltrimethylammnonium bromide (CTAB) as a surfactant. After addition of CTAB, the colour of the aqueous solution changed from light yellow to deep orange (Figure 4.13a). Under microwave heating in the presence of 2,7-DHN, the colour of the suspension became pink and a new peak at 520 nm appeared in the UV–vis absorption spectrum, which resulted from surface plasmon resonance (SPR) of spherical Au nanoparticles (curve C in Figure 4.13a). It was found that the concentrations of 2,7-DHN and NaOH had an important influence on the colour and morphology of the Au nanoparticles.

Table 4.3 Some examples of metal nanostructures prepared in aqueous solution by microwave heating[a]

Metal	Precursor	Solvent	Additive	Reducing agent	Nanostructure	Reference
Au	$HAuCl_4$	Water	CTAB	2,7-DHN	Spheres, polyhedrons, rods, triangles, prisms	69
Ag	$AgNO_3$	Water	Starch	Amino acids	Particles	67
Ag	$AgNO_3$	Water		Orange peel extract	Particles	76
Cu	$CuCl_2$	Water	Hexadecylamine	Ascorbic acid	Nanowires	77
Ag, Cu, Ag/Cu	$AgNO_3$, $Cu(NO_3)_2$	Water	Chitosan	Ascorbic acid	Particles	73
Pt	$HPtCl_4$	Water		Wine pomace	Particles	70
Pd	$PdCl_2$	Water	PEG	Glucose	Particles	66
Rh	$RhCl_3$	Water	CTAB	CTAB	Nanowire network	78
Ag@Au	$HAuCl_4$, $AgNO_3$	Water	NaOH	Curdlan	Core–shell particles	79
Pd@Pt	K_2PtCl_4, $PdCl_2$	Water	CTAB	Ascorbic acid	Core–shell particles	72

[a]CTAB: cetyltrimethylammonium bromide; 2,7-DHN: 2,7-dihydroxy naphthalene; PEG: poly(ethylene glycol).

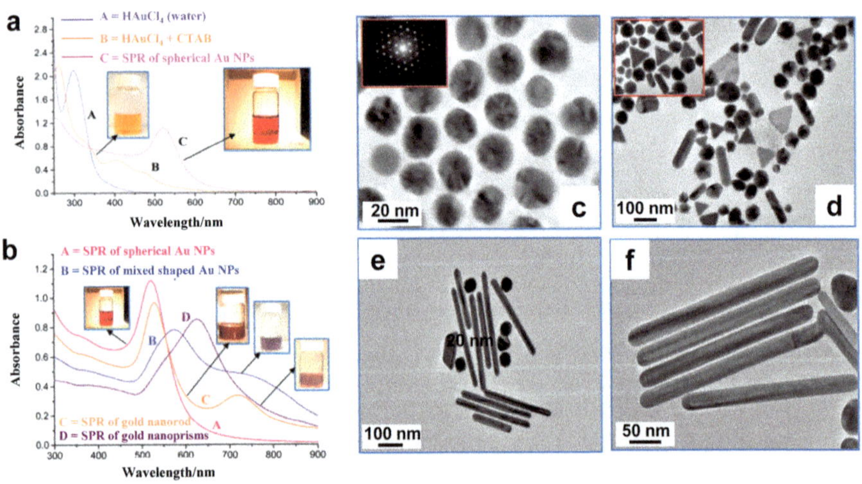

Figure 4.13 (a) UV–vis absorption spectra of Au nanoparticles prepared by microwave heating. (b) UV–vis absorption spectra of four different Au nanoparticle solutions with different shapes. (c–f) TEM images of different-shaped Au nanoparticles. Reproduced from ref. 69 with permission from American Chemical Society, Copyright 2008.

As shown in Figure 4.13b, the difference in the morphology of the Au nanoparticles was responsible for different optical properties. TEM images of various CTAB-capped Au nanoparticles are shown in Figure 4.13c–f. Various shapes of Au nanoparticles, including spheres (Figure 4.13c), triangles and polyhedra (Figure 4.13d), and nanorods (Figure 4.13e and f) were obtained. Kundu *et al.*[85] reported Au nanoparticles prepared using $HAuCl_4{\cdot}3H_2O$ and poly(*N*-vinyl-2-pyrrolidone) (PVP) under microwave heating for just 1 min in aqueous solution. The size of the Au nanoparticles could be controlled by adjusting the molar ratio of PVP/Au(III) and by using PVP molecules with different molecular weights. The Au nanoparticles were stable for more than three months in ambient conditions in the dark. The synthesised Au nanoparticles exhibited excellent catalytic activity for the reduction of aromatic nitro compounds by $NaBH_4$ in aqueous solution.

Silver. Hu *et al.*[67] reported the microwave-assisted preparation of Ag nanoparticles in aqueous solution containing $AgNO_3$, starch, and basic amino acid (L-lysine or L-arginine). The authors proposed that the amino acid with two amino groups in each molecule contributed to the formation of uniform silver nanoparticles. Recently, the preparation of Ag nanoparticles using plant extracts as reducing agents was studied. Kahrilas *et al.*[76] used orange peel extract as the reducing and capping agent to prepare Ag nanoparticles, and Ag nanoparticles with a mean diameter of 7.36 ± 8.06 nm were obtained in 15 min under microwave irradiation. In addition, the use of *Biophytum sensitivum* leaf extract and *Tamarindus indica* natural fruit were also reported for the microwave-assisted preparation of Ag nanoparticles.[86,87]

Copper. Liu *et al.*[77] reported the microwave-assisted preparation of Cu nanowires in aqueous solution containing $CuCl_2$, hexadecylamine and ascorbic acid. The as-prepared Cu nanowires had an average diameter of 50 ± 10 nm and lengths of >10 μm, and aspect ratios of >200. Both ascorbic acid and hexadecylamine had an influence on the morphology and aspect ratio. The as-prepared Cu nanowires exhibited high conductivity.

4.3.1.1.2 Bimetallic Nanostructures. Bimetallic nanostructures consist of two kinds of metals, which can combine their respective advantages to improve the properties of the resulting nanostructures.

Ag@Au. El-Naggar *et al.*[79] adopted sodium curdlan as a reducing and stabilising agent for the formation of Ag@Au nanoparticles under microwave irradiation, and they also proposed the formation mechanism (Figure 4.14a). Curdlan was insoluble in water but soluble in the presence of NaOH, which played a dual role: the reduction of Ag^+/Au^{3+} ions by primary alcoholic groups and the coordination interaction between O^- and Ag^+/Au^{3+} ions. The TEM image of the as-prepared Ag@Au nanoparticles showed the difference in contrast between the Ag core and the Au shell (Figure 4.14b).

Pd@Pt. Zhang *et al.*[72] reported Pd@Pt core–shell nanostructures and their enhanced electrocatalytic activity towards oxygen reduction and methanol oxidation. They used ascorbic acid as a reducing agent and CTAB as

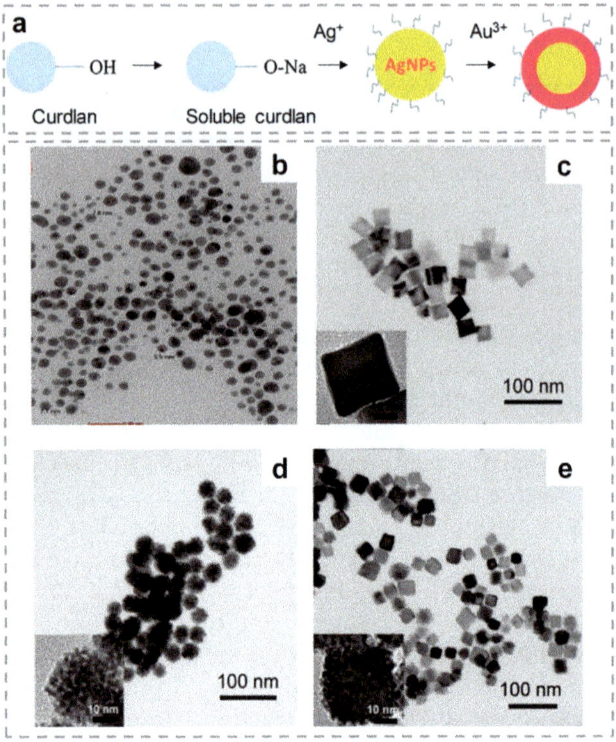

Figure 4.14 (a) Schematic illustration for the formation of Ag@Au core–shell nanoparticles. (b) TEM image of Ag@Au core–shell nanoparticles. (c–e) TEM images of Pd (c), Pt (d), and Pd@Pt (e) nanostructures. (a and b) Reproduced from ref. 79 with permission from Elsevier, Copyright 2015. (c–e) Reproduced from ref. 72 with permission from American Chemical Society, Copyright 2010.

a shape-control agent (or capping ligand). The long alkyl chains of CTAB anchored to the Pt or Pd surface formed a capping shell to tailor the structure and improve the dispersion of nanoparticles. Three kinds of metal nanostructures, including Pd, Pt, and Pd@Pt, were prepared by microwave heating in aqueous solution. The as-prepared Pd nanocubes had an average edge length of 25 nm (Figure 4.14c), and spherical Pt nanostructures with an average size of 33 nm were formed by self-assembly of Pt primary nanoparticles with an average diameter of 3–4 nm (Figure 4.14d). When the reaction was carried out with a Pd/Pt molar ratio of 3 : 1, Pd@Pt nanocubes with a rough surface and sizes of about 25–30 nm were obtained (Figure 4.14e). The formation of well-defined Pd@Pt core–shell nanostructures resulted from the growth of small Pt nanoparticles on the surface of the Pd nanocubes. The authors also demonstrated the electrocatalytic performance of Pd@Pt nanostructures for the oxygen reduction reaction and methanol electro-oxidation reaction. The experimental results showed that Pd@Pt nanostructures had a higher catalytic activity than those of pure Pd and Pt nanostructures for both reactions, indicating the synergistic effect of the bimetallic nanostructures. The microwave-assisted preparation of bimetallic alloys such as Cu–Ag and Au–Ag have also been reported in the literature.[63,88]

4.3.1.2 *Preparation in Polyols*

Preparation Strategy. Polyols usually have strong microwave-absorbing ability (Table 4.1), and they are ideal solvents for microwave-assisted preparation of inorganic nanostructures. For example, the frequently used ethylene glycol (EG) has a high loss tangent tan δ of 1.350 at 2.45 GHz and 20 °C, and a high boiling point of 198 °C, which are desirable for microwave heating. There are abundant hydrogen bonds formed amongst polyol molecules, resulting in high viscosities. One of the most attractive features of polyols is their reduction ability. Therefore, the use of additional reducing agents can be avoided. The reduction mechanism of polyols towards metallic ions was proposed to be as follows:[89]

$$CH_2OH-CH_2OH \rightarrow CH_3CHO + H_2O$$

$$CH_3CHO + M^{n+} \rightarrow M + CH_3COCOCH_3 + H_2O$$

The frequently used polyols include ethylene glycol, diethylene glycol,[90] triethylene glycol,[90] tetraethylene glycol,[91] glycerol,[92,93] 1,3-propanediol,[92] and so on. The microwave-assisted preparation of metal nanostructures in polyols has been widely reported. The representative reports are listed in Table 4.4.

4.3.1.2.1 Single Metal Nanostructures. *Gold*. Tsuji *et al.*[94] reported shape- and size-controlled preparation of gold nanocrystals by the microwave–polyol process. When $HAuCl_4$ was reduced in EG in the presence of PVP at a low $HAuCl_4$ concentration, a mixture of spherical, triangular, octahedral, and decahedral particles was obtained in a few minutes (Figure 4.15). They found that the addition of MCl (M = Na, K, H) was effective to enhance the shape-selective oxidative etching by $[AuCl_4]^-$ + Cl^- anions. With increasing concentrations of $[AuCl_4]^-$ and Cl^-, the yields of large plate-type and ribbon-type single twin particles increased, while those of spherical and octahedral particles decreased in most cases.

Silver. Zhao *et al.*[95] reported on the microwave-assisted polyol reduction of $AgNO_3$ in EG by adding different amounts of sodium sulfide into the solution. The key role of Na_2S in tuning the morphology of silver nanoparticles was highlighted. When the amount of Na_2S was 1 mmol and the reaction time was 20 s, silver nanocubes with sizes of 80 ± 5 nm were obtained. Interestingly, when the amount of Na_2S was up to 60 mmol, well-defined silver

Table 4.4 Some examples of metal nanostructures prepared in polyols by microwave heating[a]

Metal	Precursor	Solvent	Additive	Nanostructure	Reference
Au	$HAuCl_4$	EG	PVP, Cl^- ions	Spheres, octahedra, plates, rods, decahedra, icosahedra	94
Ag	$AgNO_3$	EG	PVP, Na_2S	Nanocubes, nanowires	95
Cu	$CuCl_2$	EG		Particles	96
Pt	H_2PtCl_6	EG, glycerol, 1,3-propanediol	PVP	Particles	92
Ru	$RuCl_3$	EG	PVP	Particles	93
Fe	$FeCl_3$	EG	Tetrabutylammonium, urea	Flowers	97
Co	$Co(CH_3COO)_2$	EG	PVP, $PdCl_2$	Particles	98
Ni	Nickel acetylacetonate	EG	Trioctylphosphine, sodium formate	Icosahedra	99
Bi	$Bi(NO_3)_3$	EG		Nanowires	100
$FeNi_3$	Iron acetylacetonate, nickel acetylacetonate	EG	Hydrazine monohydrate	Nanochains	101
Ag	Ag_2O	1,2-Ethanedithiol		Nanowires	102

[a]PVP: poly(vinylpyrrolidone).

nanowires with an average diameter of ~80 nm could be obtained in 1.5–3 min.

Copper. Zhu *et al.*[103] used hydrazine to reduce $CuSO_4$ in EG under microwave irradiation. They studied the influences of experimental parameters such as molar ratio of $N_2H_4{\cdot}H_2O/CuSO_4$ and $NaOH/CuSO_4$, heating method, and reaction temperature on the product. Well-dispersed Cu nanoparticles with a diameter of ~15 nm were obtained in the absence of a protective polymer.

Iron. Jung *et al.*[97] reported the preparation of flower-like Fe particles under microwave irradiation for 30 min combined with thermal treatment at 400 °C for 2 h in 95% nitrogen and 5% hydrogen gas. The reaction system consisted of $FeCl_3$, tetrabutylammonium, and urea in EG. During the reaction process, Fe-EG (alkoxide), α-Fe_2O_3, and FeOOH nanoparticles were formed first, and subsequently transformed to three-dimensional (3D) hierarchical flower-like particles by the self-assembly of primary nanoparticles (Figure 4.16a). They proposed a three-step growth mechanism: (1) fast nucleation of amorphous

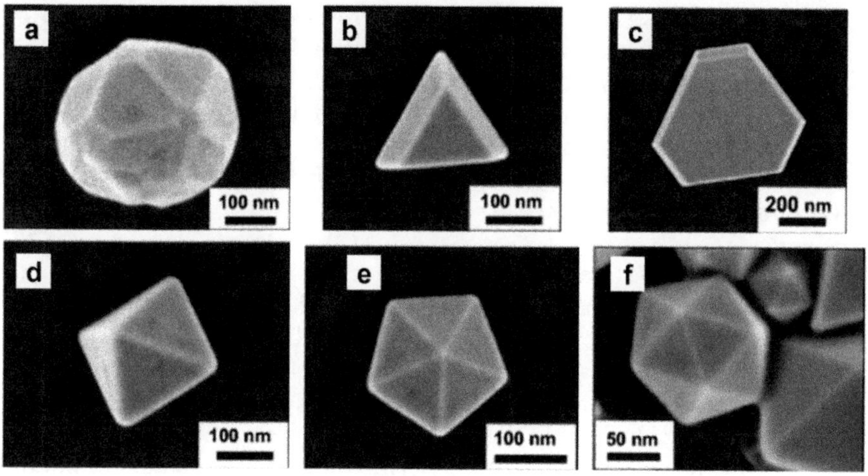

Figure 4.15 TEM images of gold particles with different shapes: (a) quasi-spherical; (b) triangular; (c) hexagonal; (d) octahedral; (e) decahedral; (f) icosahedral. Reproduced from ref. 94 with permission from Elsevier, Copyright 2007.

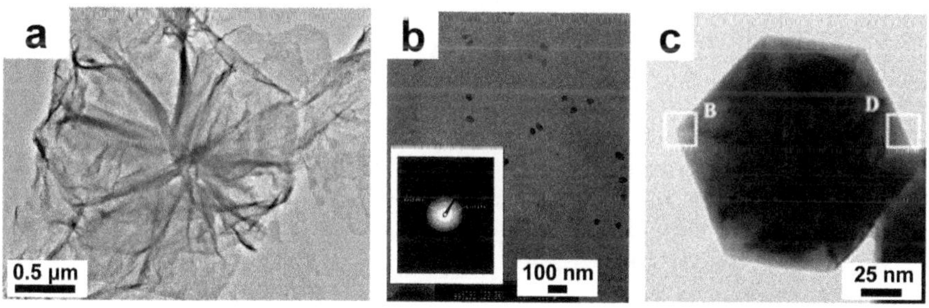

Figure 4.16 TEM images of metal nanostructures prepared by microwave heating: (a) iron flowers; (b) cobalt nanoparticles; (c) nickel icosahedra. (a) Reproduced from ref. 97 with permission from the Royal Society of Chemistry. (b) Reproduced from ref. 98 with permission from John Wiley & Sons, Copyright 2009 American Institute of Chemical Engineers (AIChE). (c) Reproduced from ref. 99 with permission from the Royal Society of Chemistry.

primary nanoparticles; (2) aggregation of nanoparticles into the microspheres with a rough surface; and (3) self-assembly of thin nanoplates onto the microspheres.

Cobalt. Cheng and Cheng[98] used cobalt acetate as a precursor, EG as a solvent, and PVP as a capping agent to prepare face-centred cubic cobalt nanoparticles under microwave irradiation (Figure 4.16b). The yields of cobalt nanoparticles with an average diameter of 40 and 81 nm were 71 and 85%, respectively.

Nickel. Donegan *et al.*[99] reported the microwave-assisted preparation of nickel nanocrystals with an icosahedral morphology (Figure 4.16c). They used nickel acetylacetonate as the metal source, and sodium formate and trioctylphosphine oxide were employed as the reducing agent and capping ligand, respectively. The as-prepared nanocrystals with a mean diameter of 237 ± 43 nm exhibited enhanced ferromagnetic behaviour at room temperature compared with bulk nickel.

4.3.1.2.2 Bimetallic Nanostructures. *$FeNi_3$.* Jia *et al.*[101] reported the microwave-assisted preparation of magnetic $FeNi_3$ nanochains by reducing iron(III) acetylacetonate and nickel(II) acetylacetonate with hydrazine in EG under microwave irradiation. They proposed the formation mechanism shown in Figure 4.17a. The complexes of $[Fe(N_2H_4)_3]^{3+}$ and $[Ni(N_2H_4)_3]^{2+}$ were formed in a hydrazine solution, and the excess hydrazine reduces $[Fe(N_2H_4)_3]^{3+}$ and $[Ni(N_2H_4)_3]^{2+}$ complexes into Fe and Ni nuclei under microwave irradiation. Because of the excellent microwave-absorbing characteristics of polar EG molecules and metallic $FeNi_3$, 'hot spots' in the bulk solution and 'hot surfaces' on the metallic $FeNi_3$ were created, which speeded up the mass transfer and crystal growth. Microwave irradiation also induced localised ionic currents on the 'hot surfaces' of metallic $FeNi_3$ in an alternating electromagnetic field, providing an additional driving force for directional crystallographic fusion of nuclei into nanocrystals. The nanocrystallites attached to each other, probably driven by the 'oriented attachment' mechanism associated with the Ostwald ripening, and grew into a chain because of the attractive magnetic force between magnetic nanoparticles as well as the van der Waals interactions. Figure 4.17b–e shows the typical high-angle annular dark-field scanning TEM and elemental energy-dispersive X-ray spectroscopy (EDS) mapping images of an individual nanochain, exhibiting the homogeneous distribution of Fe and Ni elements.

4.3.1.2.3 Microwave–Polythiol Method. The most common polyol used for microwave-assisted preparation is EG, which has a high boiling point of 198 °C. Zhu's research group developed an alternative and innovative microwave–polythiol method using 1,2-ethanedithiol as a reducing agent, which has a much lower boiling point (~147 °C) than its polyol counterpart EG.[104] In order to lower the preparation temperature for nanomaterials, a polythiol (1,2-ethanedithiol) was adopted as a reducing reagent, with a stronger reducing property than EG. For example, silver nanowires were prepared by the microwave-assisted polythiol method using 1,2-ethanedithiol as both the reducing reagent and the solvent from solid Ag_2O through a solid–liquid reaction mechanism at temperatures between 80 and 140 °C for a short processing time (10 min).[102] They also prepared Ag samples by the microwave-assisted polyol method for comparison. The experiments indicated that Ag samples prepared by microwave heating of Ag_2O in EG at temperatures between 80 and 190 °C were composed of only spherical particles. No Ag nanowires or nanorods were observed in all

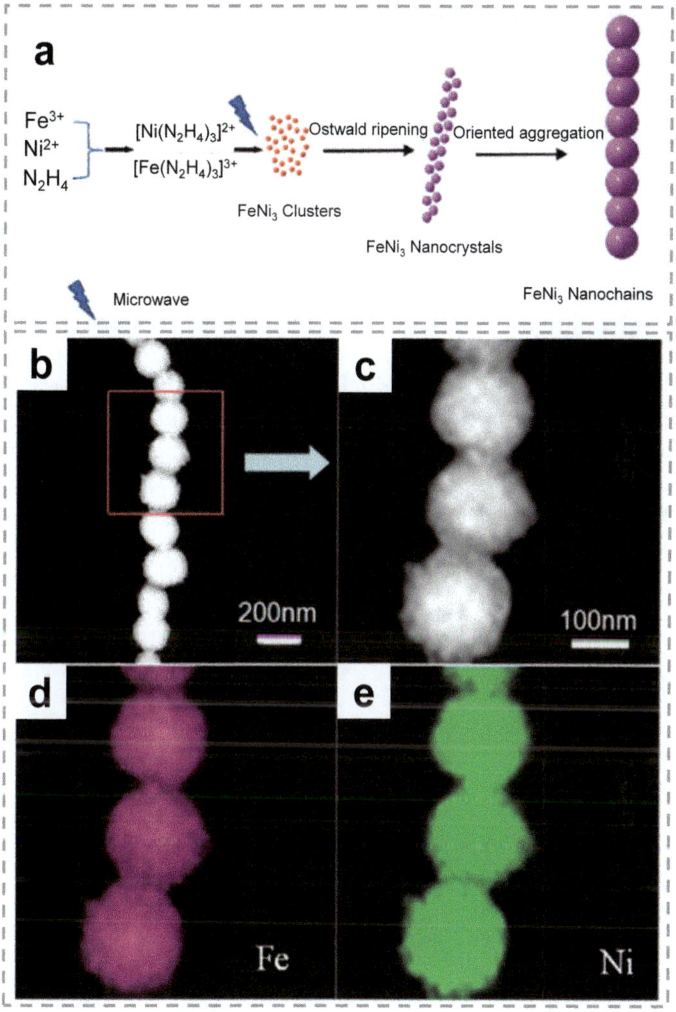

Figure 4.17 (a) Schematic illustration of a proposed mechanism for the formation of $FeNi_3$ nanochains. (b and c) High-angle annular dark-field scanning TEM and (d and e) elemental mapping images of a single $FeNi_3$ nanochain. Reproduced from ref. 101 with permission from American Chemical Society, Copyright 2010.

of these samples. Zhu's research group also prepared tellurium with various morphologies using solid TeO_2 as the reactant and 1,2-ethanedithiol as both the reducing reagent and the solvent by microwave heating at 140 °C for 30 min. The morphology of Te could be adjusted by changing the experimental parameters, such as the microwave heating temperature. It is expected that this method could also be used to prepare many other kinds of nanostructures.[104]

4.3.1.3 Preparation in Ionic Liquids

Preparation Strategy. Ionic liquids represent a class of organic salts that melt below about 100 °C and have a wide liquid temperature range. Ionic liquids are usually composed of a combination of organic cations and organic/inorganic anions. The most commonly studied ionic liquids contain ammonium, phosphonium, pyridinium, or imidazolium cations. Common anions of ionic liquids include hexafluorophosphate $[PF_6]^-$, tetrafluoroborate $[BF_4]^-$, and chloride Cl^-. Ionic liquids are a new family of environmentally friendly solvents, and they have desirable properties for microwave-assisted synthesis, including high fluidity, low toxicity, negligible vapour pressure, wide temperature range for the liquid state, air and water stability, non-flammability, low interfacial tension, high ionic conductivity, and high microwave absorption ability because of their ionic character and high polarisability. In addition, ionic liquids can be used as stabilisers or structure-directing agents.[105,106] The frequently used ionic liquids include: $[BMIM][BF_4]$,[107,108] [BMIM][OTf],[107] $[BtMA][NTf_2]$,[107] $[EmimCO_2H][BF_4]$,[105] and $[BuPy][BF_4]$[109,110] ([BMIM] = *n*-butylmethylimidazolium; [BtMA] = *n*-butyltrimethylammonium; [OTf] = $[O_3SCF_3]^-$; $[NTf_2] = [N(O_2SCF_3)_2]^-$; $[EmimCO_2H]$ = 1-methyl-3-(3-carboxyethyl)-imidazolium; [BuPy] = *N*-butylpyridinium). The chemical structures of some typical ionic liquids are shown in Figure 4.18.

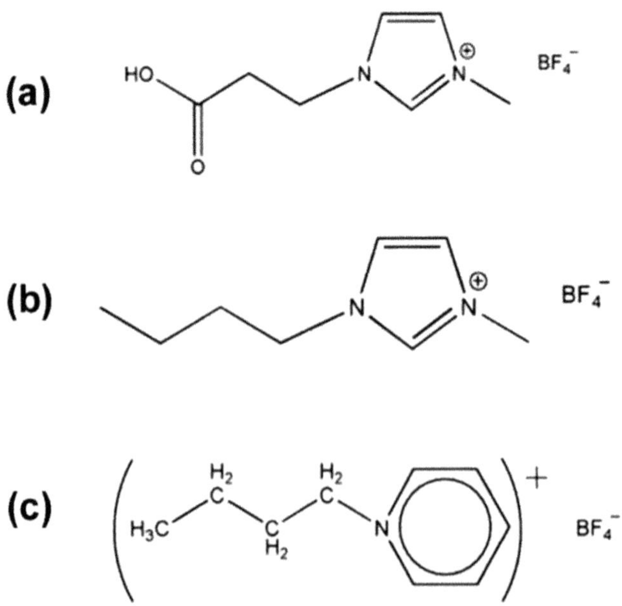

Figure 4.18 Chemical structures of typical ionic liquids: (a) 1-methyl-3-(3-carboxyethyl)-imidazolium tetrafluoroborate $[EmimCO_2H][BF_4]$; (b) 1-*n*-butyl-3-methylimidazolium tetrafluoroborate $[BMIM][BF_4]$; (c) *N*-butylpyridinium tetrafluoroborate. Reproduced from ref. 12 with permission from American Chemical Society, Copyright 2014.

In 2004, Zhu's research group reported pioneering work on the application of microwave chemistry in the preparation of nanostructured materials using ionic liquids, and developed the microwave-assisted ionic liquid method.[110] Tellurium (Te) nanorods and nanowires were rapidly prepared by microwave heating using TeO_2, PVP, $NaBH_4$, and the ionic liquid [BuPy][BF_4]. Figure 4.19a and b shows TEM images of the as-prepared Te nanorods, with diameters ranging from ~15 to ~40 nm and lengths up to ~700 nm. Figure 4.19c shows a single Te nanorod with a diameter of 32 nm and a length of 610 nm. An electron diffraction pattern is shown in Figure 4.19d, indicating a single-crystalline and hexagonal structure. The high-resolution transmission electron microscopy (HRTEM) image shows that the Te nanorods have a periodic fringe spacing of 5.927 Å along the longitudinal axis (Figure 4.19e), which corresponds to the interplanar spacing of (001) planes. In addition, Te nanowires could be obtained by adjusting the experimental parameters (Figure 4.19f). Most of the

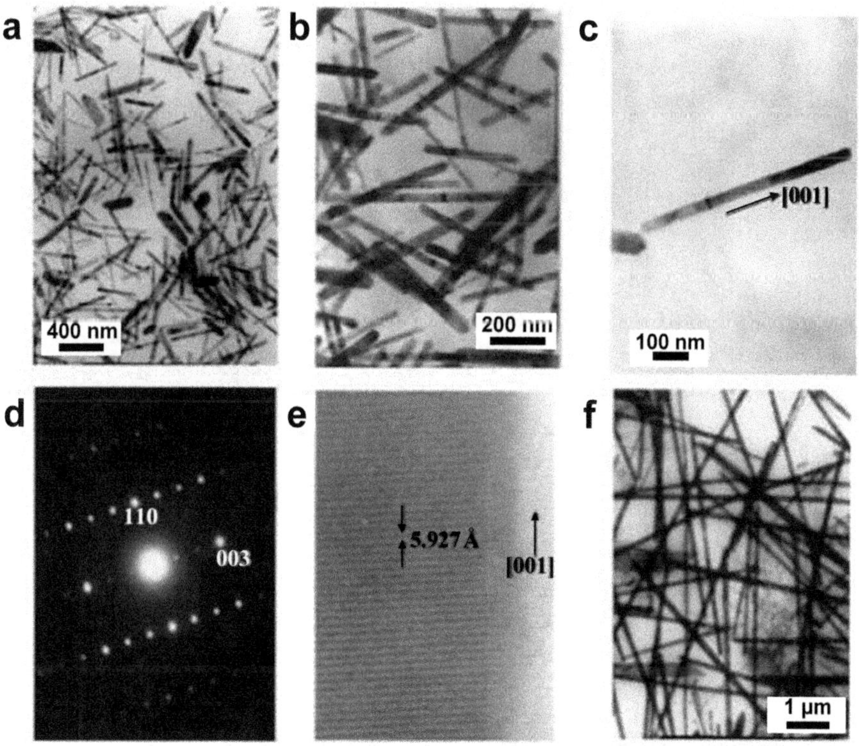

Figure 4.19 Characterisation of Te nanorods and nanowires prepared by the microwave-assisted ionic liquid method. (a–c) TEM images of Te nanorods. (d) Electron diffraction pattern of a single Te nanorod. (e) HRTEM image of a single Te nanorod. (f) Te nanowires. Reproduced from ref. 110 with permission from John Wiley & Sons, Copyright 2004 WILEY-VCH Verlag GmbH & Co. KGaA, Weinheim.

nanowires had diameters ranging from ~20 to ~100 nm and lengths up to tens of micrometres. This work emphasised the critical role of the ionic liquid. The high ionic conductivity and polarisability of $BuPy^+$ with excellent microwave-absorbing ability induced a high heating rate and could shorten the reaction time. The movement and polarisation of ions under the rapidly changing electric field of the microwave reactor resulted in transient, anisotropic microdomains for the reaction system, facilitating the anisotropic growth of Te nanorods and nanowires. Furthermore, the ionic liquid could act as a capping reagent or a surfactant. They proposed that $BuPy^+$ acted as a capping reagent and attached to the {001} facets less strongly than to other facets, leading to easier addition of Te atoms to the {001} facets and thus to a preferential growth along the *c* axis of the crystal lattice.

The microwave-assisted rapid preparation of metal nanostructures in ionic liquids has been widely reported, and representative examples are listed in Table 4.5.

Li *et al.*[106] reported the microwave-assisted preparation of gold nanosheets using $HAuCl_4$ in the ionic liquid [BMIM][BF_4]. They found that the ionic liquid as a template could direct the formation of Au nanosheets (Figure 4.20a). Ren *et al.*[109] used various ionic liquids to prepare gold nanostructures without the addition of any capping agent or additional reducing agent under microwave irradiation. The morphologies of the as-prepared gold nanostructures included polyhedral crystals, single-crystalline nanoplates, hollow trapeziform crystals, holey polyhedra, and dendrites. They also studied the effects on the product of cations and anions of the ionic liquids and the experimental parameters, including the precursor concentration, reaction temperature, and heating method.

Marquardt *et al.*[105] used metal carbonyls $Co_2(CO)_8$ and $Mn_2(CO)_{10}$ as precursors and functionalised ionic liquid [$EmimCO_2H$][BF_4] as the solvent to prepare Co and Mn nanoparticles under microwave irradiation. The as-prepared Co and Mn nanoparticles had sizes of 1.6 ± 0.3 and 4.3 ± 1.0 nm, respectively. When the non-functionalised ionic liquid [BMIM][BF_4] was used, the as-prepared Co and Mn nanoparticles had larger sizes of 5.1 ± 0.9 and 28.6 ± 11.5 nm, respectively. They also demonstrated the high stability of the as-prepared metal nanoparticles in the absence of the capping agent.

Table 4.5 Some examples of metal nanostructures prepared in ionic liquids by microwave heating[a]

Metal	Precursor	Solvent	Nanostructure	Reference
Au	$HAuCl_4$	[BMIM][BF_4]	Nanosheets	106
Co	$Co_2(CO)_8$	[$EmimCO_2H$][BF_4]	Nanoparticles	105
Mn	$Mn_2(CO)_{10}$	[$EmimCO_2H$][BF_4]	Nanoparticles	105

[a][$EmimCO_2H$] = 1-methyl-3-(3-carboxyethyl)-imidazolium.

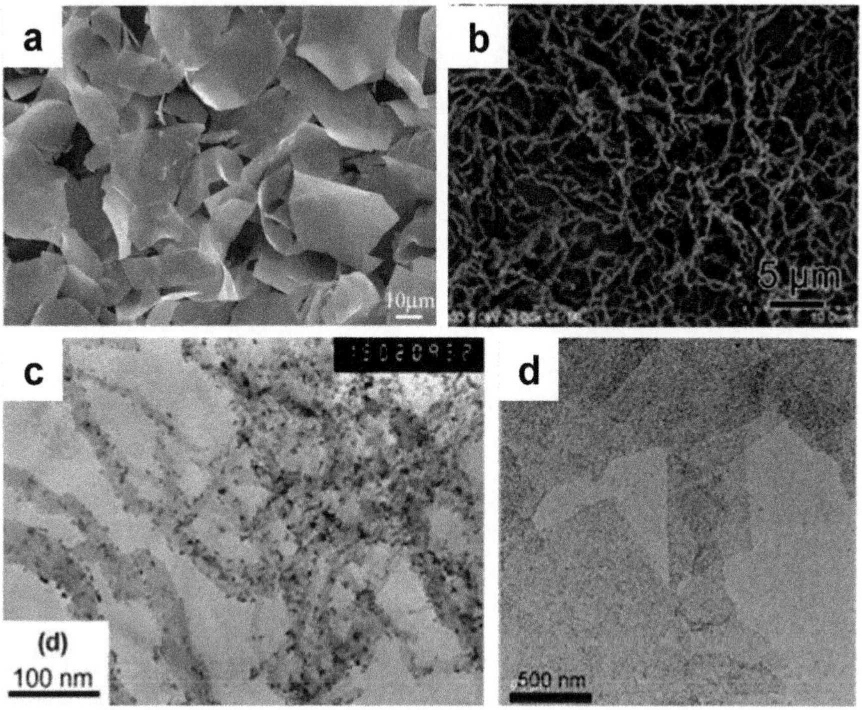

Figure 4.20 SEM images: (a) gold nanosheets; (b) Ni nanowires. TEM images: (c) Pt/carbon nanotubes; (d) Pt/graphene. (a) Reproduced from ref. 106 with permission from American Chemical Society, Copyright 2005. (b) Reproduced from ref. 111 with permission from the Royal Society of Chemistry. (c) Reproduced from ref. 113 with permission from Elsevier, Copyright 2005. (d) Reproduced from ref. 112 with permission from American Chemical Society, Copyright 2010.

4.3.1.4 Preparation in Mixed Solvents

Preparation Strategy. As mentioned previously, the solvent has significant effects on the microwave-assisted preparation of inorganic nanostructures. The interaction between the solvent and the microwaves depends on the polarity, dielectricity, loss tangent, and other properties of the solvent. The appropriate selection and use of mixed solvents will provide more possibilities for designing and preparing inorganic nanostructures. The microwave-assisted preparation of inorganic nanostructures in mixed solvents can combine the respective advantages of the different solvents. For example, polyols with high loss tangent ($\tan\delta$) values are excellent solvents for microwave heating, and their ability to absorb microwaves is generally much better than that of water. However, polyols with two or more –OH functional groups usually have high viscosities. The solubilities of some reactants in polyols are relatively low. The introduction of water into polyols may offset such shortcomings, and at the same time the microwave absorption of water can

Table 4.6 Some examples of metal nanostructures prepared in mixed solvents by microwave heating[a]

Metal	Precursor	Solvent	Additive	Nanostructure	Reference
Pt	H_2PtCl_6	Water, EG		Particles	120
Cu	$Cu(CH_3COO)_2$	Water, EG	DMG	Particles	121
Ni	$NiCl_2$	Water, EG	PVP, hydrazine	Nanowires	111
Pt–Ru	H_2PtCl_6, $RuCl_3$	Water, EG		Particles	120
Pt/CNT	H_2PtCl_6, CNTs	Water, EG		Particles@ nanotubes	113
Pt/GO	GO/H_2PtCl_6	Water, EG	PMMA	Particles@ nanosheets	112
Au	$HAuCl_4$	Water, oleyl-amine, oleic acid		Particles, hexagons	118
Ag	$AgNO_3$	EG, toluene, dodecylthiol		Rectangles	119

[a]CNTs: carbon nanotubes; GO: graphene oxide or graphite oxide; PMMA: poly(methyl methacrylate); DMG: dimethylglycoxime.

be enhanced by polyols. Fortunately, polyols are highly miscible with water. Frequently used mixed solvents include water/polyol, water/alcohol,[114] water/sodium bis(2-ethylhexyl)sulfosuccinate (AOT)-heptane,[115] methyl isobutyl ketone/butanol,[116] EG/methyl isobutyl ketone,[117] water/oleylamine/oleic acid,[118] and EG/toluene/dodecylthiol.[119] The microwave-assisted preparation of metal nanostructures in mixed solvents has been widely reported in the literature, and some representative examples are listed in Table 4.6.

4.3.1.4.1 Water/Polyol. *Nickel.* Amongst various water/polyol reaction systems, water/EG mixed solvents have been the most extensively investigated. For example, Tang *et al.*[111] reported the preparation of polycrystalline nickel nanowires under microwave irradiation using $NiCl_2$, PVP, hydrazine, and EG. The as-prepared nickel nanowires had diameters of 70–380 nm (Figure 4.20b), which could be adjusted by means of the concentrations of the reactants. It was also found that the nickel nanowires had a rough surface at low concentrations of reactants, but became smooth at high concentrations of reactants.

Pt/carbon. In water/EG mixed solvents, the microwave-assisted preparation of carbon-supported platinum nanoparticles is described here, due to their excellent catalytic performance. The carbon materials used as supports include commercial carbon,[122,123] mesoporous carbon,[124] carbon nanotubes (Figure 4.20c),[113] graphene oxide (Figure 4.20d),[112,125] and so on. For example, Guo *et al.*[112] reported the preparation of platinum nanoparticle/graphene hybrid nanosheets (Figure 4.20d). Graphene oxide, H_2PtCl_6, and poly(methyl methacrylate) were mixed in water/EG mixed solvents, followed by microwave irradiation. The as-prepared platinum/graphene hybrid nanosheets were used to modify the glassy carbon electrode, which was further employed to detect H_2O_2 and trinitrotoluene.

4.3.1.4.2 Multinary Solvents. *Gold.* Mohamed *et al.*[118] reported the microwave-assisted preparation of gold nanostructures with different morphologies capped with a mixture of oleylamine and oleic acid in ternary water/oleylamine/oleic acid solvents. The size and morphology of the nanostructures could be tailored by varying the ratio of oleylamine to oleic acid, the microwave heating time, and the concentration of the gold compound. Exclusive use of oleylamine resulted in the formation of spherical nanoparticles only. The addition of oleic acid increased the crystal growth rate and promoted the formation of anisotropic shapes.

Silver. Gao *et al.*[119] reported the microwave-assisted solvothermal preparation of hexagonally arranged spherical silver nanoparticles in EG/toluene/dodecylthiol ternary solvents. The ratio of EG to dodecylthiol had an effect on the nucleation, crystal growth, and self-assembly of the silver nanostructures. At a low volume ratio of EG to dodecylthiol, silver nanostructures with a rectangular shape, which were arranged in orthogonal arrays, were observed.

4.3.2 Metal Oxides

The tailorable properties and high stability of metal oxide nanostructures enable their wide application in chemical industry, biomedical, environmental, and energy-related areas.[126] Wet-chemistry synthetic methods are desirable for the preparation of metal oxide nanostructures because their chemical composition, structure, size, morphology, and self-assembly can be well tuned by changing the synthesis parameters. As mentioned in previous sections, microwave heating outperforms conventional heating in terms of rate acceleration, high efficiency, short reaction time, and energy saving. As a result, microwave-assisted rapid preparation of metal oxide nanostructured materials has been intensively studied. Mirzaei and Neri[127] published a review article on the microwave-assisted rapid synthesis of metal oxide nanostructures and their gas sensing applications. Faraji and Ani[128] reviewed the microwave-assisted rapid synthesis of metal oxide/hydroxide composite electrodes for high-power supercapacitors. Some representative examples will be reviewed in this section.

4.3.2.1 Preparation in Aqueous Solution

The microwave-assisted rapid synthesis of metal oxide nanostructures in aqueous solution has been widely reported in the literature, and some representative examples are listed in Table 4.7.

4.3.2.1.1 Transition Metal Oxide Nanostructures. Transition metal oxides are very important functional materials, and have promising applications in many fields, due to their variable properties. For example, ZnO and TiO_2 are typical semiconductors and Fe_3O_4 has magnetic properties.

Table 4.7 Some examples for the synthesis of metal oxide nanostructures by microwave heating in aqueous solution[a]

Metal oxide	Precursor	Solvent	Additive	Nanostructure	Reference
SnO_2	$SnCl_2$	Water	Hydrazine hydrate	Particles	129
SnO_2	$SnCl_2$	Water	CTAB, HMTA	Nanosheets	130
ZnO	$Zn(Ac)_2$	Water	Ammonia	Nanorod array	131
Fe_2O_3	$FeCl_3$	Water	$NH_4H_2PO_4$	Nanorings	132
Iron oxide	$FeCl_2$, $FeCl_3$	Water	Na_2CO_3, citric acid	Multicore particles	133
Co_3O_4	$Co(NO_3)_2$	Water	P123, urea	Nanowires	134
MnO_2	$KMnO_4$	Water	HCl	Nanosheets, nanowires, nanotubes	135
MnO_2	$KMnO_4$, $MnSO_4$	Water		Nanorods	136
TiO_2	$TiCl_3$	Water		Nanowires	137
TiO_2	$Ti(SO_4)_2$	Water	Urea	Nanoparticle-assembled microspheres	138
WO_3	Na_2WO_4	Water	$(NH_4)_2SO_4$	Nanowires	139
CuO	Copper acetate hydroxide	Water	NaOH	Nanosheets	140
CuO	$CuSO_4$	Water	PEG, HMTA	Flowers, nanosheets	141
ZrO_2	Zirconyl nitrate	Water		Particles	142
$NiCo_2O_4$	$CoCl_2$, $NiCl_2$	Water	Urea	Nanopetal-assembled flowers	143
Titanate	TiO_2	Water	NaOH	Nanotubes	144
$ZnWO_4$	$ZnCl_2$, Na_2WO_4	Water		Particles	145
Bi_2WO_6	$Bi(NO_3)_3$, Na_2WO_4	Water	HMTA, HNO_3	Nanosheet-assembled flowers	146
MFe_2O_4 (M = Zn, Ni, Mn, Co)	$M(NO_3)_2$, $Fe(NO_3)_3$	Water		Particles	147
$ZnAl_2O_4$	$Al_2(OH)_{6-x}(NO_3)_x$, $Zn(CH_3COO)_2$	Water		Particles	148
$BaTiO_3$	$TiCl_4$, $BaCl_2$	Water	KOH	Particles	149
$BiVO_4$	$Bi(NO_3)_3$, NH_4VO_3	Water	EDTA	Nanorod-assembled microspheres	150
$Zn_3V_2O_7(OH)\cdot 2H_2O$	$Zn(NO_3)_2$, NH_4VO_3	Water		Nanowires	151
$LiNi_{0.5}Mn_{1.5}O_4$	$MnSO_4$, $NiSO_4$, $KMnO_4$, LiOH	Water		Particles, octahedra	152

[a]HMTA: hexamethylenetetramine; P123: poly(ethylene glycol)-*block*-poly(propylene glycol)-*block*-poly(ethylene glycol); EDTA: ethylenediamine tetraacetic acid.

ZnO. Cho *et al.*[131] reported density-controlled synthesis of ZnO nanorod arrays on a flexible substrate using inductively coupled plasma etching with anodic aluminium oxide membranes and microwave irradiation without a catalyst at 90 °C. During the microwave irradiation, ZnO nanorods grew in the pores (Figure 4.21a). Thus, the density of ZnO nanorods could be controlled by changing the interpore distance of the anodic aluminium oxide membrane mask.

Fe_2O_3. Hu *et al.*[132] prepared α-Fe_2O_3 nanorings using $FeCl_3$ and $NH_4H_2PO_4$ in aqueous solution by the microwave–hydrothermal process (Figure 4.21b). α-Fe_2O_3 nanocrystals formed at the early stage were inclined to aggregate to reduce the high surface energy, resulting in the formation of monodisperse hematite nanodiscs. α-Fe_2O_3 had a strong microwave-absorbing ability, hence 'hot surfaces' on the hematite nanodiscs were created under microwave irradiation, promoting crystal growth. Furthermore, the core of the nanodiscs consisted of dense primary phosphate ion-stabilised nanocrystals. The phosphate ions at the centres of 'etched' α-Fe_2O_3 nanodiscs finally produced small holes at the centres of nanodiscs.

Co_3O_4. Meher *et al.*[134] studied the effect of microwave heating on the nanowire morphology, optical, magnetic, and pseudocapacitance behaviour

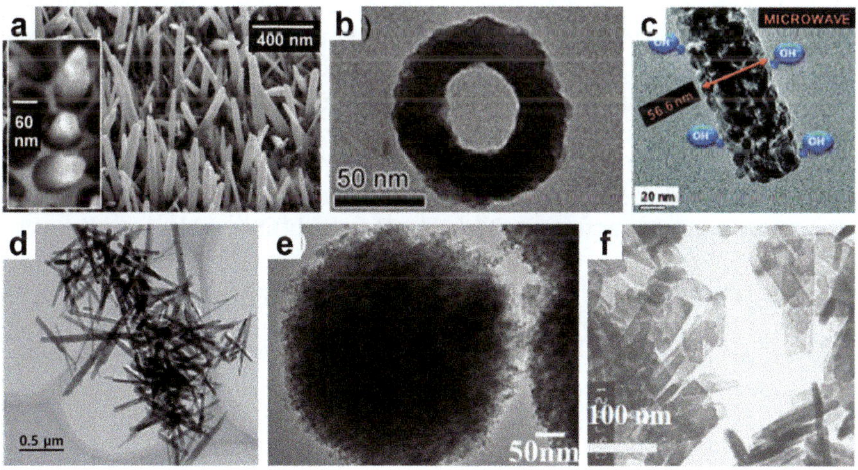

Figure 4.21 SEM (a) and TEM (b–f) images of different transition metal oxide nanostructures with various morphologies prepared by microwave heating: (a) ZnO; (b) Fe_2O_3; (c) Co_3O_4; (d) MnO_2; (e) TiO_2; (f) CuO. (a) Reproduced from ref. 131 with permission from American Chemical Society, Copyright 2008. (b) Reproduced from ref. 132 with permission from John Wiley & Sons, Copyright 2007 WILEY-VCH Verlag GmbH & Co. KGaA, Weinheim. (c) Reproduced from ref. 134 with permission from American Chemical Society, Copyright 2011. (d) Reproduced from ref. 136 with permission from the Royal Society of Chemistry. (e) Reproduced from ref. 138 with permission from American Chemical Society, Copyright 2014. (f) Reproduced from ref. 140 with permission from Oxford University Press, Copyright 2005 The Chemical Society of Japan.

of Co_3O_4. The microwave synthesis was performed in an aqueous solution containing $Co(NO_3)_2$, urea, and poly(ethylene glycol)-*block*-poly(propylene glycol)-*block*-poly(ethylene glycol) (P123) as the surfactant. They compared two kinds of Co_3O_4 nanowires obtained by the conventional refluxing method and by the microwave refluxing method. The conventional refluxing method produced randomly distributed thin nanowires, while the microwave refluxing method generated higher dimensional and organised Co_3O_4 nanowires. Because of the smaller crystallite size and more accessible surface sites, Co_3O_4 nanowires synthesised by the conventional reflux method exhibited better charge storage, high Coulombic efficiency, and enhanced rate response during the pseudocapacitance studies. By contrast, Co_3O_4 nanowires obtained by microwave heating showed better high-rate cyclic stability due to the more rigid, oriented nanowire structure (Figure 4.21c).

MnO_2. Truong *et al.*[135] studied the morphological and crystallisation evolution of MnO_2 nanostructures prepared by microwave-assisted hydrothermal treatment of potassium permanganate in the presence of hydrochloric acid. Three kinds of MnO_2 nanostructures were obtained, including α-MnO_2 nanotubes, α-MnO_2 nanowires, and δ-MnO_2 nanosheet-assembled microflowers. The as-prepared MnO_2 nanostructures were incorporated into air cathodes of lithium–air batteries as electrocatalysts for the oxygen reduction and evolution reactions. The experiments indicated that the electrodes made of single-crystalline α-MnO_2 nanotubes had much higher stability than those made of α-MnO_2 nanowires and δ-MnO_2 nanosheet-assembled microflowers, in both charge and discharge processes. Islam *et al.*[136] reported microwave-assisted hydrothermal synthesis of tunnel-type β-MnO_2 nanorods with exposed (101) planes within only 10 min (Figure 4.21d). They used β-MnO_2 nanorods as a high-performance cathode for aqueous Zn-ion batteries, which exhibited a high discharge capacity, high rate capability, and long cycling stability over 200 cycles.

TiO_2. Zhang *et al.*[137] reported the microwave-assisted hydrothermal synthesis of rutile TiO_2 nanostructures using $TiCl_3$ as the only reactant, and the resulting rutile nanowires could assemble into a 3D hierarchical architecture. Yang *et al.*[138] reported the microwave-assisted hydrothermal synthesis of TiO_2 microspheres with abundant mesopores and high specific surface area (124 m^2 g^{-1}). The as-prepared TiO_2 microspheres were uniform in size and had an average diameter of ~0.5 μm. It was found that the TiO_2 microspheres were constructed by well-crystallised anatase nanoparticles with sizes of 5–10 nm (Figure 4.21e).

CuO. It is challenging to prepare CuO nanosheets because CuO has no layered structure. Zhu's research group developed a layered precursor conversion strategy to synthesise CuO nanosheets under microwave irradiation.[140] Two processes are involved in the synthesis of CuO nanosheets. First, the layer-structured precursor, copper acetate hydroxide, was obtained by microwave heating of an EG solution containing $Cu(CH_3COO)_2$ and NaOH. Second, the as-prepared layer-structured precursor was dispersed in NaOH aqueous solution, and heated by microwaves at 100 °C. As shown in Figure 4.21f, the product consisted of quasi-rectangular CuO nanosheets. The layered precursor conversion method is promising for the preparation of metal oxide nanosheets whose structures are not layered.

4.3.2.1.2 Bimetallic Oxide Nanostructures. *$NiCo_2O_4$*. Lei *et al.*[143] prepared the precursor for $NiCo_2O_4$ using an aqueous solution containing $CoCl_2{\cdot}6H_2O$, $NiCl_2{\cdot}6H_2O$, and urea by microwave-assisted reflux for 15 min, and the precursor was pyrolysed in air at 300 °C for 2 h to obtain 3D hierarchical flower-shaped $NiCo_2O_4$ microspheres consisting of ultrathin nanopetals with thickness of ~15 nm, a large specific surface area (148.5 $m^2\ g^{-1}$), and a narrow pore size distribution (5–10 nm). They also demonstrated the application of $NiCo_2O_4$ microspheres in a high-performance supercapacitor. Khalid *et al.*[153] used a similar method to prepare porous $NiCo_2O_4$ microspheres constructed by self-assembly of nanosheets.

Bi_2WO_6. Cao *et al.*[146] reported the microwave-assisted synthesis of Bi_2WO_6 nanosheet-assembled flowers using an aqueous solution containing $Bi(NO_3)_3$, Na_2WO_4, hexamethylenetetramine (HMTA), and HNO_3. The experimental results showed that HMTA and HNO_3 played important roles in the formation of nanosheet-assembled flowers, and the Ostwald ripening process was proposed for the mechanism of formation.

MFe_2O_4. Komarneni *et al.*[147] reported the microwave–hydrothermal synthesis of nanostructured ferrites MFe_2O_4 (M = Zn, Ni, Mn, Co). Nanostructured ferrites with high specific surface area in the range 72–247 $m^2\ g^{-1}$ were synthesised in a few minutes at a temperature of 164 °C using nitrates of zinc, nickel, manganese, or cobalt, and ferric nitrate in aqueous solution.

$ZnAl_2O_4$. Zawadzki[148] reported the microwave-assisted hydrothermal synthesis of spinel $ZnAl_2O_4$ nanoparticles with a mean particle size of 2.4 nm and with a narrow size distribution. They used aluminium hydroxy nitrate with an empirical formula $Al_2(OH)_{6-x}(NO_3)_x$, where x is equal or close to 1, and hydrated zinc acetate in the synthesis.

$BaTiO_3$. Jhung *et al.*[149] studied the effect of experimental parameters, such as H_2O/Ti, Ba/Ti, and OH^-/Ti ratios, on the synthesis of $BaTiO_3$ nanoparticles. $BaTiO_3$ nanoparticles with sizes of less than 30 nm were synthesised under microwave irradiation using $TiCl_4$, $BaCl_2$, and KOH. It was found that the crystal size decreased with increasing of concentrations of hydroxide and barium ions.

$BiVO_4$. Zhu *et al.*[150] reported the microwave–hydrothermal synthesis of $BiVO_4$ nanorod-assembled hollow microspheres in aqueous solution containing $Bi(NO_3)_3$ and NH_4VO_3 in the presence of EDTA. Nanocage structures were obtained at a higher amount of EDTA. The evolution process of $BiVO_4$ with different microstructures indicated that EDTA played a critical role in determining the shape. In addition, it was found that EDTA was vital for mediating the crystal growth, and the hydrothermal time and temperature were key synthetic parameters in determining the morphology.

$Zn_3V_2O_7(OH)_2{\cdot}2H_2O$. Xia *et al.*[151] reported the microwave-assisted synthesis of ultralong zinc pyrovanadate ($Zn_3V_2O_7(OH)_2{\cdot}2H_2O$) nanowires with a

porous crystal framework. The as-prepared zinc pyrovanadate nanowires exhibited significantly enhanced electrochemical performance as the intercalation cathode for the aqueous zinc-ion battery. It is expected that this synthetic strategy may also be extended to prepare other nanostructured metal pyrovanadates.

4.3.2.2 Preparation in Polyols

The microwave-assisted synthesis of metal oxide nanostructures in polyols has been widely reported. Some representative examples are listed in Table 4.8.

4.3.2.2.1 Transition Metal Oxide Nanostructures. *ZnO.* Hu *et al.*[154] reported the microwave–polyol synthesis of monodisperse ZnO colloidal nanocrystal clusters. They used diethylene glycol with a boiling point of ~245 °C as the solvent. The size of the clusters could be controlled from 57 to 274 nm by varying the experimental conditions. They also demonstrated that sensors made of ZnO colloidal nanocrystal clusters showed high sensitivity for humidity measurements at room temperature.

Table 4.8 Some examples for the synthesis of metal oxide nanostructures by microwave heating in polyols[a]

Metal oxide	Precursor	Solvent	Additive	Nanostructure	Reference
ZnO	$Zn(CH_3COO)_2$	DEG		Particles	154
Fe_2O_3	$FeCl_3$	EG	SDBS	Nanosheet-assembled hollow spheres	155
Fe_3O_4	$FeCl_3$	EG	CH_3COONa, P123	Nanosheet-assembled nanoroses	156
MnO	$Mn(acac)_2$	TEG	PVP	Particles	157
$ZnAl_2O_4$	$Zn(CH_3COO)_2$, aluminium isopropoxide	1,4-Butanediol		Particles	158
MFe_2O_4 (M = Zn, Ni, Mn, Co, Cu)	Acetylacetonate metal	TEG		Particles	159
Bi_2WO_6	$Bi(NO_3)_3$, $(NH_4)_{10}W_{12}O_{41}$,	EG		Nanosheets	160
$CaMoO_4$	$Ca(NO_3)_2$, Na_2MoO_4	EG		Particles	161
Nd-Doped $SrTiO_3$	Tetre-*n*-butyl titanate, $Sr(NO_3)_2$, $Nd(NO_3)_2$	EG	NaOH	Spheres, plates	162

[a]SDBS: sodium dodecyl benzenesulfonate; acac: 2,4-pentanedionate; DEG: diethylene glycol; TEG: triethylene glycol.

Fe_2O_3. Cao *et al.*[155] reported the microwave–polyol synthesis of hierarchically nanostructured α-Fe_2O_3 hollow spheres in EG containing $FeCl_3$, NaOH, and SDBS. The hierarchically nanostructured α-Fe_2O_3 hollow spheres consisted of organised nanosheets. They also investigated the photocatalytic properties and potential application in water treatment of the as-prepared, hierarchically nanostructured α-Fe_2O_3 hollow spheres.

Fe_3O_4. Ai *et al.*[156] reported the microwave–polyol synthesis of Fe_3O_4 nanoroses in EG containing $FeCl_3$, CH_3COONa, and P123. In addition to size and morphology-dependent magnetic properties, the as-prepared Fe_3O_4 nanoroses showed high sensitivity and good reversibility for gas sensing of ethanol vapour at room temperature, and are promising for magnetic and sensing applications.

MnO. Lu *et al.*[157] reported the microwave–polyol synthesis of hydrophilic and paramagnetic MnO nanoparticles in triethylene glycol containing $Mn(acac)_2$ and PVP. The as-prepared MnO nanoparticles could be used as a magnetic resonance imaging contrast agent with r_1 value of 0.81 mM^{-1} s^{-1}, and are promising for biomedical applications.

4.3.2.2.2 Bimetallic Oxide Nanostructures. *$ZnAl_2O_4$*. Staszak *et al.*[158] reported the microwave–polyol synthesis of $ZnAl_2O_4$ nanoparticles with an average size of 3 nm in 1,4-butanediol containing zinc acetate and aluminium isopropoxide as the reactants at 200 °C for 30 min. The as-prepared $ZnAl_2O_4$ nanoparticles had superior thermal stability (up to 900 °C). Pt catalysts supported on the as-prepared $ZnAl_2O_4$ nanoparticles showed good activity in isobutane combustion.

MFe_2O_4. Solano *et al.*[159] reported the microwave–polyol synthesis of MFe_2O_4 (M = Mn, Fe, Co, Ni, Zn, and Cu) spinel magnetic nanoparticles with diameters of <10 nm. The microwave–polyol synthesis was performed in triethylene glycol containing metal acetylacetonate. A narrow nanoparticle size distribution and high crystallinity could be achieved through selected nucleation and suitable growth conditions.

Bi_2WO_6. Wu *et al.*[160] reported the microwave–polyol synthesis of Bi_2WO_6 nanosheets. The microwave–polyol synthesis was performed in EG containing $Bi(NO_3)_3{\cdot}5H_2O$ and $(NH_4)_{10}W_{12}O_{41}{\cdot}5H_2O$ at 160 °C for different heating times.

$CaMoO_4$. Phuruangrat *et al.*[161] reported the microwave–polyol synthesis of $CaMoO_4$ nanoparticles. The microwave–polyol synthesis was performed in EG containing $Ca(NO_3)_2$ and Na_2MoO_4. Body-centred tetragonal structured $CaMoO_4$ nanoparticles with a narrow size distribution (16–44 nm) were obtained.

4.3.2.3 Preparation in Ionic Liquids

The microwave-assisted synthesis of metal oxide nanostructures in ionic liquids has been reported in the literature, and some representative examples are listed in Table 4.9.

Table 4.9 Some examples of the synthesis of metal oxide nanostructures by microwave heating in ionic liquids.

Metal oxide	Precursor	Solvent	Additive	Nanostructure	Reference
ZnO	$Zn(NO_3)_2$	$[BMIM][BF_4]$, water	NaOH	Flowers, needles	163
TiO_2	H_2TiO_3	$[BMIM][BF_4]$		Particles	164
TiO_2	Tetra-butyl titanate	$[BMIM][BF_4]$, water, acetic acid		Hollow cube-like nanocrystals	165
CuO	$CuCl_2$	$[BMIM][BF_4]$, water	NaOH	Nanowhiskers	166
MnO_2	$MnCl_2$, $KMnO_4$	$[BMIM][BF_4]$, water		Nanoneedles, nanorods	167
WO_3	WCl_4	$[BMIM][BF_4]$		Nanoneedles	168
SnO_2	$Sn(O\mathit{t}Bu)_4$	$[BMIM][BF_4]$		Quantum dots	169
$ZnFe_2O_4$	$Zn(CH_3COO)_2$, $Fe(NO_3)_3$	$[BMIM][BF_4]$, water	$CO(NH_2)_2$	Nanoparticles	170

4.3.2.3.1 Metal Oxide Nanostructures. *ZnO*. Wang and Zhu[163] reported the microwave-assisted ionic liquid method for the synthesis of ZnO nanostructures. They found that both the ionic liquid ($[BMIM]BF_4$) and microwave heating had significant effects on the formation of ZnO nanostructures. By controlling the synthetic parameters, ZnO needles or flowers (Figure 4.22a) could be obtained in a short period of time (5 or 10 min). Each ZnO flower consisted of a number of petals with sharp tips. The single ZnO flower was not a simple aggregation of small crystals, but was instead composed of single-crystalline petals grown concentrically. The ZnO needles had widths ranging from 60 to 450 nm and lengths up to several micrometres. Most ZnO needles had a sharp tip and one flat end, while a few had two tips or two flat ends. The preferential growth direction of ZnO needles was along the *c* axis of the crystal lattice.

TiO_2. Chen *et al.*[165] reported a green strategy for microwave-assisted ionic liquid synthesis, followed by a vacuum-activation process, for the synthesis of highly active Ti^{3+} self-doped TiO_2 nanocrystals with a hollow structure. The as-prepared TiO_2 nanocrystals were uniform and most of them had cube-like or truncated octahedral bipyramidal shapes. The high-resolution TEM image shows *d*-spacings of 0.23 and 0.35 nm, corresponding to the (001) and (101) crystal planes of anatase TiO_2 (Figure 4.22b).

CuO. Wang *et al.*[166] reported the microwave synthesis of CuO nanowhiskers using $CuCl_2$ and NaOH in an ionic liquid. It was found that the ionic liquid $[BMIM][BF_4]$ had a significant influence on the morphology of the CuO. Interestingly, CuO nanowhiskers and nanosheets were obtained in the presence and absence of $[BMIM][BF_4]$, respectively. The formation of CuO nanowhiskers in the presence of $[BMIM][BF_4]$ was probably caused by selective adsorption of $[BMIM][BF_4]$ onto the crystal surfaces of CuO, and this was

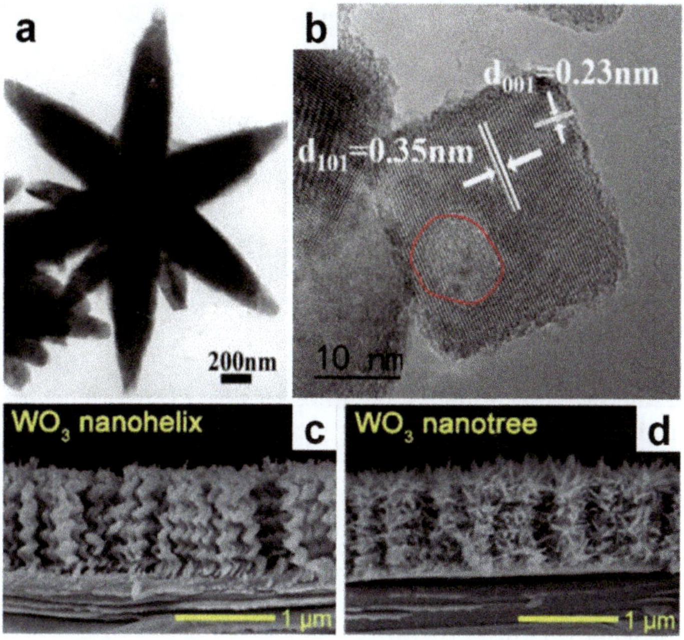

Figure 4.22 TEM (a and b) and SEM (c and d) images of different transition metal oxide nanostructures with various morphologies synthesised by microwave heating in an ionic liquid: (a) ZnO; (b) TiO_2; (c and d) WO_3. (a) Reproduced from ref. 163 with permission from Elsevier, Copyright 2004. (b) Reproduced from ref. 165 with permission from Elsevier, Copyright 2016. (c and d) Reproduced from ref. 168 with permission from the Royal Society of Chemistry.

supported by the thermogravimetric/differential scanning calorimetry and Fourier-transform infrared (FTIR) analysis.

MnO_2. Yang *et al.*[167] reported the microwave-assisted ionic liquid method for the rapid synthesis of single-crystalline cryptomelane-type manganese oxide octahedral molecular sieve (OMS-2) nanoneedles and nanorods, which was performed in a mixture of $MnCl_2$, $KMnO_4$, [BMIM][BF_4], and deionised water at a relatively low temperature (90 °C) for a short period of time. [BMIM][BF_4] acted as a co-solvent, structure-directing agent, and reducing reagent in the reaction system. Based on the redox reaction of $MnCl_2$ and $KMnO_4$ in the mixed solvents of water and [BMIM][BF_4], the formation of OMS-2 nanoneedles followed the rolling mechanism, with the lamellae as an intermediate. However, direct reaction of $KMnO_4$ with [BMIM][BF_4] resulted in the formation of OMS-2 nanorods with diameters of 3–6 nm. This method is simple, fast, and surfactant-free. It is expected that this method may be extended to rapidly synthesise nanostructures of other metal oxides.

WO_3. Lee *et al.*[168] reported the microwave-assisted synthesis of branched WO_3 nanotrees consisting of [001]-oriented WO_3 needle-like structures growing radially from the surface of WO_3 nanohelices in the presence of an ionic

liquid [BMIM][BF_4]. First, WO_3 nanohelices (Figure 4.22c) were prepared as the template using an electron-beam evaporator. Then, WO_3 nanohelices were immersed in a precursor solution containing WCl_4 in [BMIM][BF_4] under magnetic stirring, followed by microwave irradiation. They observed densely arranged and high-crystalline WO_3 nanoneedles grown perpendicular to the surface of the WO_3 nanohelices, resulting in branched nanotrees (Figure 4.22d). Furthermore, the WO_3 nanotrees could be transformed into WS_2/WO_3 hierarchical nanotrees by a simple sulfurisation process.

$ZnFe_2O_4$. Cao *et al.*[170] reported the microwave-assisted hydrothermal ionic liquid synthesis of $ZnFe_2O_4$ nanoparticles and photocatalytic properties over phenol, which was performed in a mixture of $Zn(CH_3COO)_2$, $Fe(NO_3)_3$, $CO(NH_2)_2$, and [BMIM][BF_4] at 160 °C for 30 min. The ionic liquid [BMIM][BF_4] and microwave heating temperature greatly affected the crystal phase of the product. Different amounts of [BMIM][BF_4] or different microwave heating temperatures could produce different products, such as $ZnFe_2O_4$ and β-FeOOH. The $ZnFe_2O_4$ nanoparticles prepared by the microwave–hydrothermal method showed high photocatalytic activity for the degradation of phenol.

4.3.2.4 *Preparation in Mixed Solvents*

The microwave-assisted synthesis of metal oxide nanostructures in mixed solvents has been widely reported in the literature, and some examples are listed in Table 4.10.

4.3.2.4.1 Water/Polyol. *ZnO*. Zhu *et al.*[172] reported the microwave synthesis of ZnO nanostructures in water/EG solution containing $Zn(CH_3COO)_2$. By only changing the microwave heating parameters, ZnO hierarchical nanostructures with straw bundle-like, wide chrysanthemum-like, and oat arista-like morphologies were obtained.

Co_3O_4. Chen *et al.*[178] synthesised mesoporous Co_3O_4 nanoflakes with an interconnected architecture using the microwave-assisted solvothermal method combined with thermal conversion. First, the layered-structure intermediate cobalt carbonate hydroxide hydrate ($Co(CO_3)_{0.5}(OH)_{0.11}\cdot H_2O$) interconnected architectures were obtained by microwave heating in mixed solvents of ethylene glycol and water at 180 °C for 10 min. Then, the layered $Co(CO_3)_{0.5}(OH)_{0.11}\cdot H_2O$ interconnected architectures were treated at 250 °C for 3 h in air. Finally, mesoporous Co_3O_4 nanoflakes with the interconnected architecture were obtained. The as-prepared Co_3O_4 nanoflakes exhibited high electrochemical performance as anode material in lithium-ion batteries and as the catalyst in the oxygen evolution reaction.

Indium Tin Oxide. Hammarberg *et al.*[182] reported the microwave synthesis of indium tin oxide (ITO) nanoparticles with an average size of 15 nm in water/diethylene glycol solution containing $InCl_3$, $SnCl_4$, and $N\text{-}(CH_3)_4OH\cdot 5H_2O$. The as-prepared ITO suspension was colloidally stable within weeks and showed a transparent appearance with a bright blue colour.

Table 4.10 Some examples of microwave-assisted synthesis of metal oxide nanostructures in mixed solvents[a]

Metal oxide	Precursor	Solvent	Additive	Nanostructure	Reference
SnO_2	$SnCl_2$	Water, ethanol		Quantum dots	171
ZnO	$Zn(CH_3COO)_2$	Water, EG		Straw bundle, chrysanthemum, oat arista nanostructures	172
Mn_3O_4	$MnCl_2$	Water, EG	NaOH	Particles	173
CeO_2	$Ce(NO_3)_3$	Water, DEG	Hexamine	Particles	174
TiO_2	Titanium isopropoxide	Water, isopropanol	HNO_3	Particles	175
Fe_2O_3	$Fe(NO_3)_3$	Water, ethylenediamine		Spindle	176
Fe_2O_3	$FeCl_3$	EG, ethanolamine		Nanosheet-assembled microspheres	177
Co_3O_4	$CoCO_3$	Water, EG	HMTA, $(NH_4)_2CO_3$	Nanoflakes	178
Co_3O_4	$Co(NO_3)_2$	Water, ethanol, hexane	Sodium oleate	Nanocubes	179
TiO_2	TiF_4	Water, cyclohexane, 1-pentanol	Urea/CTAB	Hierarchical nanostructures	180
CeO_2	$Ce(NO_3)_3$	Water, EG, oleic acid, *tert*-butylamine		Nanoparticles, nanocubes, nanorods	181
Indium tin oxide (ITO)	$InCl_3$, $SnCl_4$	Water, DEG	$N\text{-}(CH_3)_4OH{\cdot}5H_2O$	Particles	182
$NiCo_2O_4$	$Ni(NO_3)_2$, $Co(NO_3)_2$	Water, EG	Urea	Nanosheets	183
$MgCo_2O_4$	$Mg(NO_3)_2$, $Co(NO_3)_2$	Water, EG	Urea	Nanosheets	184
$CaMoO_4$	$Ca(CH_3COO)_2$, H_2MoO_4	Water, EG		Corn cob and dumbbell	185
$CoFe_2O_4$	$Co(CH_3COO)_2$, $FeCl_3$, CH_3COONa	Water, 1,2-propanediol		Particles	186
$PbTiO_3$	$Pb(CH_3COO)_2$, $Ti(OC_4H_9)_4$	Water, ethanol	KOH	Nanowires	187
$Y_4Al_2O_9$	$Y(NO_3)_3$, $Al(NO_3)_3$	Water, DMF		Nanosheet-assembled hierarchical microspheres	188

[a]DMF: *N,N*-dimethylformamide.

$NiCo_2O_4$. Mondal *et al.*[183] employed two steps to prepare $NiCo_2O_4$ nanosheets. First, the mixed metal (Ni, Co) hydroxide precursor was produced in the water/EG solution under microwave irradiation. In the second step, the mixed metal hydroxide precursor was thermally converted into $NiCo_2O_4$ nanosheets with a large specific surface area (143 $m^2\ g^{-1}$). The as-prepared $NiCo_2O_4$ nanosheets were used as the electrode material for a lithium-ion battery and a supercapacitor and exhibited high electrochemical performance and good cycling stability.

4.3.2.4.2 Multinary Solvents. *Co_3O_4*. Sun *et al.*[179] reported the microwave-assisted solvothermal synthesis of Co_3O_4 nanocubes with a lateral size of ~20 nm, which was performed in water/ethanol/hexane solution containing $Co(NO_3)_2$ at 200 °C for 20 min. The as-prepared Co_3O_4 nanocubes exhibited good gas sensing performance towards xylene and ethanol vapour.

TiO_2. Rahal *et al.*[180] reported the microwave synthesis of TiO_2 hierarchical nanostructures, which was performed in water/cyclohexane/1-pentanol solution containing TiF_4, urea, and CTAB. The TiO_2 hierarchical nanostructures prepared under microwave irradiation showed a high photocatalytic activity.

CeO_2. Tao *et al.*[181] reported the microwave-assisted synthesis of CeO_2 nanoparticles, nanocubes, and nanorods in water/EG/oleic acid/*tert*-butylamine solution containing $Ce(NO_3)_3$. It was found that the bandgap of the CeO_2 nanostructures depended on the morphology, and the absorption peaks of the CeO_2 nanorods shifted to a higher wavelength compared with those of CeO_2 nanoparticles.

4.3.3 Metal Chalcogenides

In the past decades, metal chalcogenide nanostructured materials have attracted much attention due to the rich choice of materials with tuneable electronic, optical, physical, and chemical properties. Hence, the applications of metal chalcogenide nanomaterials have expanded, from environmental remediation to photocatalytic conversion.[189] Controlled synthesis of nanostructured metal chalcogenides is desirable in order to tune their properties. Microwave chemistry offers a unique synthetic technique to obtain high-quality metal chalcogenide nanostructured materials. Consequently, the microwave-assisted rapid synthesis of metal chalcogenide nanostructures has been extensively studied, and some examples will be reviewed in this section.

4.3.3.1 Preparation in Aqueous Solution

The microwave-assisted synthesis of metal chalcogenide nanostructures in aqueous solution has been widely reported, and some examples are listed in Table 4.11.

Table 4.11 Some examples of metal chalcogenide nanostructures synthesised by microwave heating in aqueous solution[a]

Metal chalcogenide	Precursor	Solvent	Additive	Nanostructure	Reference
PbS	$PbCl_2$, MPA, thioacetamide	Water	NaOH	Quantum dots	190
Ag:PbS	$Pb(NO_3)_2$, $AgNO_3$, Na_2S	Water	CTAB	Particles	191
Bi_2S_3	$Bi(NO_3)_3$, thiourea	Water	PVP	Nanorod-assembled flowers	192
CuS	$CuCl_2$, thiourea	Water	NaOH/EDTA	Nanosheet-assembled flowers	193
ZnS	$Zn(CH_3COO)_2$, thioacetamide	Water	PEG	Particles	194
NiS_2	$Ni(NO_3)_2$, sulfourea	Water	PVP, NaOH	Nanocubes, nanospheres, nanoparticles	195
CdS	$CdCl_2$, $Na_2S_2O_3$	Water		Particles	196
CdS	Cd Fe Prussian blue analogues, Na_2S	Water		Yolk shell microcubes	197
$CdIn_2S_4$	$Cd(NO_3)_2$, $In_2(NO_3)_2$, thiourea	Water	Triton X, SDS or SLS	Flowers, spheres, pyramids	198
CdSe	$CdSO_4$, Na_2SeSO_3	Water	Potassium nitrilotriacetate	Particles	199
PbSe	$Pb(CH_3COO)_2$	Water	Potassium nitrilotriacetate	Particles	199
ZnSe	$ZnCl_2$, NaHSe,	Water	MPA	Particles	200
$CuInSe_2$	$CuCl_2$, $InCl_3$, Na_2Se	Water	Mercapto-acetic acid	Particles	201
CdTe	$CdCl_2$, NaHTe	Water	NaOH, MPA	Particles	202
CdTe	$CdCl_2$, Na_2TeO_3, MPA	Water	Trisodium citrate/ $NaBH_4$	Quantum dots	203

[a]SDS: sodium dodecyl sulfate; SLS: sodium lauryl sulfate; MPA: 3-mercaptopropionic acid.

4.3.3.1.1 Metal Sulfide Nanostructures. *PbS*. He *et al.*[190] reported the microwave-assisted synthesis of PbS quantum dots (Figure 4.23a), which was performed in an aqueous solution containing $PbCl_2$, 3-mercaptopropanoic acid, thioacetamide, and NaOH. After microwave irradiation, a light grey-green suspension was obtained. A further ageing process was needed for the growth of well-crystallised PbS nanostructures. After 10 days of ageing, the

Figure 4.23 TEM (a and d) and SEM (b, c, e and f) images of different metal sulfide nanostructures with various morphologies synthesised by microwave heating in aqueous solution: (a) PbS; (b) Bi_2S_3; (c) CuS; (d) ZnS; (e) NiS_2; (f) $CdIn_2S_4$. (a) Reproduced from ref. 190 with permission from American Chemical Society, Copyright 2012. (b) Reproduced from ref. 192 with permission from Elsevier, Copyright 2010. (c) Reproduced from ref. 193 with permission from Elsevier, Copyright 2010. (d) Reproduced from ref. 194 with permission from Elsevier, Copyright 2011. (e) Reproduced from ref. 195 with permission from Springer Nature, Copyright 2014. (f) Reproduced from ref. 198 with permission from the Royal Society of Chemistry.

primary PbS colloid separated into two layers, which contained PbS quantum dots in the top clear solution layer and PbS nanocrystals in a dark suspension at the bottom. After separation and centrifugation of the dark bottom and clear top solutions, PbS nanocrystals (~50 nm) and PbS quantum dots (~5 nm) were obtained, respectively.

Bi_2S_3. Thongtem *et al.*[192] reported the microwave-assisted synthesis of Bi_2S_3 nanorod-assembled flowers, which was performed in aqueous solution containing $Bi(NO_3)_3$, thiourea, and PVP (Figure 4.23b). They also proposed the formation mechanism of the Bi_2S_3 nanorod-assembled flowers as follows: (1) interaction between Bi^{3+} ions and thiourea molecules to form the Bi–thiourea complex; (2) Bi–thiourea complex decomposes to form the Bi_2S_3 nuclei under microwave-assisted hydrothermal conditions; and (3) Bi_2S_3 nanorods grow on Bi_2S_3 nuclei to form the Bi_2S_3 nanorod-assembled flowers.

CuS. Mu *et al.*[193] reported the microwave-assisted synthesis of CuS nanosheet-assembled flowers in aqueous solution containing $CuCl_2$, thiourea, NaOH, and EDTA (Figure 4.23c). The driving force for the self-assembly of CuS nanosheets was attributed to the interfacial tension and the interaction between the hydrophilic surfaces of the CuS nanosheets.

EDTA could act as both a chelating reagent and a surface-capping agent in the synthetic process.

ZnS. Mehta *et al.*[194] reported the microwave-assisted preparation of PEG-coated ZnS nanoparticles in aqueous solution containing $Zn(CH_3COO)_2$, thioacetamide, and PEG. A thin layer of PEG coating on the surface of the ZnS nanoparticles could be clearly observed from the TEM image (Figure 4.23d). They proposed the formation mechanism of ZnS nanoparticles as follows: (1) decomposition of thioacetamide was accelerated by microwave heating, and S^{2-} ions were produced; and (2) nucleation of ZnS occurred due to the reaction between Zn^{2+} and S^{2-} ions.

NiS_2. Pang *et al.*[195] reported the microwave-assisted preparation of NiS_2 nanostructures in an aqueous solution containing $Ni(NO_3)_2$, sulfourea, PVP, and NaOH. NiS_2 nanocubes, nanospheres, and nanoparticles could be obtained by controlling the microwave heating time. A scanning electron microscopy (SEM) image of the NiS_2 nanocubes is shown in Figure 4.23e. NiS_2 nanostructures with different morphologies exhibited different electrochemical and H_2 production catalytic activities.

$CdIn_2S_4$. Apte *et al.*[198] reported the microwave-assisted preparation of $CdIn_2S_4$ nanostructures with mixed morphologies of flowers, spheres, and pyramids in an aqueous solution containing $Cd(NO_3)_2$, $In_2(NO_3)_2$, and thiourea (Figure 4.23f). Various capping agents, including sodium dodecyl sulfate (SDS) and sodium lauryl sulfate (SLS), were used to tune the morphology of the $CdIn_2S_4$ nanostructures.

4.3.3.1.2 Metal Selenide Nanostructures. *CdSe (or PbSe)*. Zhu *et al.*[199] reported the microwave-assisted synthesis of CdSe nanoparticles in aqueous solution containing $CdSO_4$ (or $Pb(CH_3COO)_2$), Na_2SeSO_3, and potassium nitrilotriacetate. It was found that the crystalline phase of CdSe was dependent on the microwave heating time. CdSe with a cubic structure was obtained by microwave heating for 10 min; however, longer microwave heating times led to a crystal phase transition. When the microwave heating time was prolonged to 30 min, CdSe with a hexagonal structure was obtained.

ZnSe. Qian *et al.*[200] reported the microwave-assisted synthesis of ZnSe nanocrystals in an aqueous solution containing $ZnCl_2$, NaHSe, and 3-mercaptopropionic acid (MPA). It was found that there were significant effects of the pH value, molar ratio of precursors, and microwave heating time on the optical properties of the as-prepared ZnSe nanocrystals.

$CuInSe_2$. Bensebaa *et al.*[201] reported the microwave-assisted synthesis of $CuInSe_2$ nanoparticles in aqueous solution containing $CuCl_2$, $InCl_3$, Na_2Se, and mercaptoacetic acid.

4.3.3.1.3 Metal Telluride Nanostructures. *CdTe.* The synthesis of CdTe nanostructures has attracted much interest because CdTe, although highly toxic, is an important semiconductor and is widely used in various fields. The Te sources for the synthesis of CdTe nanostructures include NaHTe[202,204] and Na_2TeO_3.[203] For example, Li *et al.*[202] reported the microwave-assisted synthesis of CdTe nanoparticles in aqueous solution containing $CdCl_2$, NaHTe, MPA, and NaOH. At $T \leq 160$ °C, CdTe nanocrystals with different sizes and high quantum yield were synthesised by changing the microwave heating temperature and time. Duan *et al.*[203] reported the microwave-assisted synthesis of CdTe quantum dots in aqueous solution containing $CdCl_2$, Na_2TeO_3, MPA, trisodium citrate, and $NaHB_4$. The experimental parameters affected the crystal size and photoluminescence quantum yield of the CdTe quantum dots. The as-prepared photoluminescence quantum yield of the CdTe quantum dots reached 40–60%.

4.3.3.2 *Preparation in Polyols*

The microwave-assisted synthesis of metal chalcogenide nanostructures in polyols has been widely reported, and some examples are listed in Table 4.12.

4.3.3.2.1 Metal Sulfide Nanostructures. *CoS*. Luo *et al.*[205] reported the preparation of 3D CoS flower-like hierarchical structures constructed by nanopetals under microwave irradiation. They also used the as-prepared CoS flowers as the electrode for a supercapacitor, which showed high cycling stability, high rate capability, and enhanced energy density.

CdS. Oliveira *et al.*[206] reported the microwave-assisted synthesis of CdS nanosheet-assembled flowers in EG (or ethylenediamine) containing $CdCl_2$ and thiourea. It was found that the morphology of the product depended on the solvent used. The nanosheet-assembled flowers were obtained in EG, while aggregates of nanoplates were prepared in ethylenediamine.

CuS. Wang *et al.*[207] reported the microwave-assisted synthesis of CuS hierarchical nanostructures in EG containing $CuCl_2$, thiourea, and histidine. The histidine biomolecules had a great effect on the formation of the CuS hierarchical nanostructures. It was found that the morphology and size of the CuS nanostructures could be controlled by changing the molar ratio of Cu^{2+}/histidine and Cu^{2+}/thiourea.

NiS_2. Tian *et al.*[208] prepared NiS_2 hollow microspheres by template conversion under microwave irradiation. A Ni-based (Ni-BTC, BTC = benzene-1,3,5-tricarboxylic acid) metal–organic framework (MOF) was prepared to form a 3D architecture, and was employed as a sacrificial template to produce NiS_2 hollow microspheres by direct sulfidation under microwave irradiation. The resulting NiS_2 hollow microspheres exhibited high performance in electrocatalysing the hydrogen evolution reaction.

Cu_3BiS_3. Aup-Ngoen *et al.*[211] reported the microwave-assisted synthesis of Cu_3BiS_3 dendrites in EG containing CuCl, $BiCl_3$, and L-cysteine. It was found that L-cysteine could act as both a sulfur source and a complexing agent.

4.3.3.2.2 Metal Selenide/Telluride Nanostructures. *Sb_2Se_3*. Antimony selenide is a promising thermoelectric material with a high Seebeck coefficient, but its figure of merit is limited by its low electrical conductivity. Mehta *et al.*[213] reported the microwave-assisted synthesis of Sb_2Se_3 nanowires and nanotubes

Table 4.12 Some examples of metal chalcogenide nanostructures synthesised by microwave heating in polyols[a]

Metal chalcogenide	Precursor	Solvent	Additive	Nanostructure	Reference
CoS	$Co(NO_3)_2$, thioacetamide	EG	Citric acid	Flowers	205
CdS	$CdCl_2$, thiourea	EG		Nanosheet-assembled flowers	206
CuS	$CuCl_2$, thiourea	EG	Histidine	Hierarchical nanostructures	207
NiS_2	Ni-BTC-MOF, thioacetamide	EG		Hollow microspheres	208
FeS_2	$FeSO_4$, S powder	EG	PVP	Microspheres	209
Bi_2S_3	CdS, $Bi(NO_3)_3$	EG	CTAB	Nest-shaped nanostructures	210
Cu_3BiS_3	CuCl, $BiCl_3$, L-cysteine	EG		Dendrites	211
CdSe	$Cd(CH_3COO)_2$, Se powder	EG or TEG		Nanoballs	212
Sb_2Se_3	Tri-*n*-octylphosphine-ligated selenium, thioglycolic acid-ligated antimony salt	1,5-Pentanediol		Nanowires, nanotubes	213
Sb_2Te_3	$Na(SbO)C_4H_6O_6$, Te powder	EG		Nanoplates	214
Sb_2Te_3	$SbCl_3$, TeO_2	EG	PEG	Nanorods, nanosheets, nanotubes	215
Bi_2Te_3	$Bi(NO_3)_3$, Te powder	EG		Nanorods, nanoflakes	216
$Cu_{(2-x)}Te$	$Cu(NO_3)_2$, Te nanotubes	EG	$NaBH_4$	Nanowires	217
$CuInTe_2$, $CuInSe_2$	CuCl, In, Te or Se	Tetraethylene glycol or TEG		Particles	218

[a]Ni-BTC-MOF: Ni-benzene-1,3,5-tricarboxylic acid metal–organic framework.

in 1,5-pentanediol containing tri-*n*-octylphosphine-ligated selenium and thioglycolic acid-ligated antimony salt. The as-prepared one-dimensional nanocrystals of sulfurised antimony selenide exhibited significantly enhanced electrical conductivity. As the nanocrystal size increased, the nanowires could transform into nanotubes through void formation and coalescence.

$CuInTe_2$ and $CuInSe_2$. Grisaru *et al.*[218] reported the microwave-assisted polyol synthesis of body-centred tetragonal $CuInTe_2$ nanoparticles (100 nm) and $CuInSe_2$ nanoparticles (85 nm) in tetraethylene glycol or triethylene

glycol containing CuCl, In, and Te or Se. The polyol acted as both the solvent and the reducing agent.

4.3.3.3 Preparation in Ionic Liquids

The microwave-assisted synthesis of metal chalcogenide nanostructures in ionic liquids has been reported in the literature, and some examples are listed in Table 4.13.

Bi_2S_3 (or Sb_2S_3). Jiang *et al.*[220] reported the microwave-assisted ionic liquid synthesis of Bi_2S_3 (or Sb_2S_3) nanorods using Bi_2O_3 (or Sb_2O_3), $Na_2S_2O_3$, HCl, EG (or ethanolamine), and [BMIM][BF_4]. The experimental results showed that the ionic liquid had a great effect on the morphology of the product. Single-crystalline Bi_2S_3 nanorods were prepared in the presence of [BMIM][BF_4]; however, urchin-like Bi_2S_3 structures consisting of nanorods were obtained in the absence of [BMIM][BF_4]. In addition, single-crystalline Sb_2S_3 nanorods were obtained in the presence of [BMIM][BF_4]; but single-crystalline Sb_2S_3 nanosheets were prepared in the absence of [BMIM][BF_4].

Bi_2Se_3. Jiang *et al.*[221] reported the microwave-assisted ionic liquid method for the rapid synthesis of Bi_2Se_3 nanosheets with thicknesses of 50–100 nm using $Bi(NO_3)_3$, Se powder, HNO_3, EG (or ethanolamine), and [BMIM][BF_4]. It was found that both the solvent and the ionic liquid played key roles in

Table 4.13 Some examples of metal chalcogenide nanostructures synthesised by microwave heating in ionic liquids.

Metal chalcogenide	Precursor	Solvent	Additive	Nanostructure	Ref.
ZnS	$Zn(CH_3COO)_2$, Na_2S	[BMIM][BF_4]		Particles	219
CdS	$CdCl_2$, Na_2S	[BMIM][BF_4]		Particles	219
Bi_2S_3	Bi_2O_3, $Na_2S_2O_3$	[BMIM][BF_4], EG (or ethanolamine)		Nanorods	220
Sb_2S_3	Sb_2O_3, $Na_2S_2O_3$	[BMIM][BF_4], EG (or ethanolamine)		Nanorods	220
Bi_2Se_3	$Bi(NO_3)_3$, Se powder	[BMIM][BF_4], EG		Nanosheets	221
Bi_2Te_3	$Bi(NO_3)_3$, Te powder	1-Butyl-3-methylimidazolium bromide, EG		Nanosheets	222
CdTe	$CdCl_2$, H_2Te gas	1-Methoxymethyl-1-methylpyrrolidinium bis(trifluoromethanesulfonyl)amide	2-(Dimethylamino)ethanethiol hydrochloride	Particles	223

the morphology of the product. When ethylenediamine (EDA) was used as the solvent, both nanoparticles and irregularly shaped nanosheets were obtained in the absence of [BMIM]BF_4, while irregularly shaped nanosheets formed exclusively in the presence of [BMIM]BF_4. When using EG as the solvent, irregularly shaped nanosheets could be obtained in the absence of [BMIM]BF_4, while hexagonally shaped nanosheets were prepared in the presence of [BMIM]BF_4.

CdTe. Hayakawa *et al.*[223] reported the microwave-assisted synthesis of thiol-capped CdTe nanocrystals with a zinc blende crystal structure using $CdCl_2$, H_2Te gas, 2-(dimethylamino)ethanethiol hydrochloride, and an ionic liquid, 1-methoxymethyl-1-methylpyrrolidinium bis(trifluoromethanesulfonyl)amide.

4.3.3.4 Preparation in Other Solvents and Mixed Solvents

Some examples of the microwave-assisted synthesis of metal chalcogenide nanostructures in other solvents and in mixed solvents are listed in Table 4.14.

CoS. You *et al.*[224] reported a two-step microwave-assisted preparation of hollow CoS nanoprisms. The synthetic procedures are illustrated in Figure 4.24a. The cobalt acetate hydroxide nanoprisms were first prepared under microwave irradiation at 85 °C for 10 min, then they were subsequently converted to CoS hollow structures in the presence of a sulfur source at 120 °C within 5 min. Ethanol was used as the solvent in both steps. The obtained CoS nanoprisms showed a hollow structure with a shell thickness of ~30 nm (Figure 4.24b). It was demonstrated that CoS nanoprisms prepared by microwave heating exhibited higher catalytic activity for hydrogen evolution and better specific capacitance compared with their counterpart prepared by the traditional solvothermal method.

MoS_2. Gao *et al.*[225] reported a microwave-assisted synthesis of edge-terminated layered MoS_2 nanosheets with a 9.4 Å interlayer spacing. The synthetic procedures are illustrated in Figure 4.24c. The *N*,*N*-dimethylformamide (DMF) solution containing $(NH_4)_2MoS_4$ was subjected to microwave irradiation, and flower-like MoS_2 nanosheets were obtained (Figure 4.24d). In addition to the edge-terminated structure, which could provide more active edge sites, the expanded interlayer spacing could change the electronic structure and enhance the catalytic activity of the MoS_2 nanosheets.

$CoNi_2S_4$. Rafai *et al.*[227] reported the microwave–anion-exchange route to ultrathin cobalt nickel sulfide ($CoNi_2S_4$) nanosheets for hybrid supercapacitors. As shown in Figure 4.25a, a homogeneous hydrolysis of urea to form OH^- and CO_3^{2-} anions occurred in the first stage, leading to alkalisation of Co^{2+} and Ni^{2+} cations to produce hydrated inorganic oligomers under microwave irradiation. The rapid microwave heating supersaturated the hydrated inorganic oligomers, to instantaneously form ultrafine nanocrystals, which self-assembled by the intrinsic driving forces of the lamellar transition metal hydroxide ($Ni_{0.75}Co_{0.25}(OH)_2(CO_3)_{0.125}$) for two-dimensional (2D) anisotropic growth. After thermal treatment, $Ni_{0.75}Co_{0.25}(OH)_2(CO_3)_{0.125}$ nanosheets were

Table 4.14 Some examples of metal chalcogenide nanostructures synthesised by microwave heating in other solvents and mixed solvents[a]

Metal chalcogenide	Precursor	Solvent	Additive	Nanostructure	Reference
CoS	Cobalt acetate hydroxide, thioacetamide	Ethanol	PVP	Nanoprisms	224
MoS_2	$(NH_4)_2MoS_4$	DMF		Nanosheets	225
Bi_2S_3	Bi_2O_3, $Na_2S_2O_3$	Water, EG		Urchin-/rod-like	226
$CoNi_2S_4$	$Co(NO_3)_2$, $Ni(NO_3)_2$, Na_2S	Water, EG	Urea	Nanosheets	227
Ru_xSe_y	$RuCl_3$, Na_2SeO_3	Water, EG	KOH	Particles	228
Bi_2Te_3	TeO_2, $Bi(NO_3)_3$	Water, EG		Hollow nanospheres	229
$Bi_2Te_{3-x}Se_x$	$Bi(NO_3)_3$, Na_2TeO_3, Na_2SeO_3	Water, EG	NaOH	Nanoplates	230
Sb_2Te_3	$SbCl_3$, Na_2TeO_3	Hydrazine hydrate, EG		Nanosheets	231
CuS	$CuCl_2$, thioacetamide	Water, ethanol		Hierarchical nanosheets	232
CdSe	$Cd(CH_3COO)_2$, $NaSeSO_3$	Water, ethanol		Quantum dots	233
ZnS	$ZnCl_2$, thiourea	Water, EDA		Nanorods	234
$Zn_xCd_{1-x}S$	$Cd(CH_3COO)_2$, $Zn(CH_3COO)_2$/$(C_2H_5)_2NCS_2Na$	Water, EDA		Nanorods	235
CdS	$Cd(CH_3COO)_2$, thiourea	Water, diethylenetriamine		Nanorods, nanosticks, nanoflowers	236
Cu_2S	$CuSO_4$, dodecanethiol	Water, toluene		Particles	237
WS_2	Tungstic acid, sulfur	Water, monoethanolamine		Nanowires	238
ZnSe	$Zn(CH_3COO)_2$, Se	Tri-*n*-butylphosphine, octylamine (or octyldecylamine)		Rice-like particles, rods	239
CdSe	CdO, Se powder	Oleic acid, DMSO, TOPO, HDA		Cubic and hexagonal particles	240

[a]DMSO: dimethyl sulfoxide; TOPO: trioctylphosphine oxide; HDA: 1-hexadecylamine; EDA: ethylenediamine.

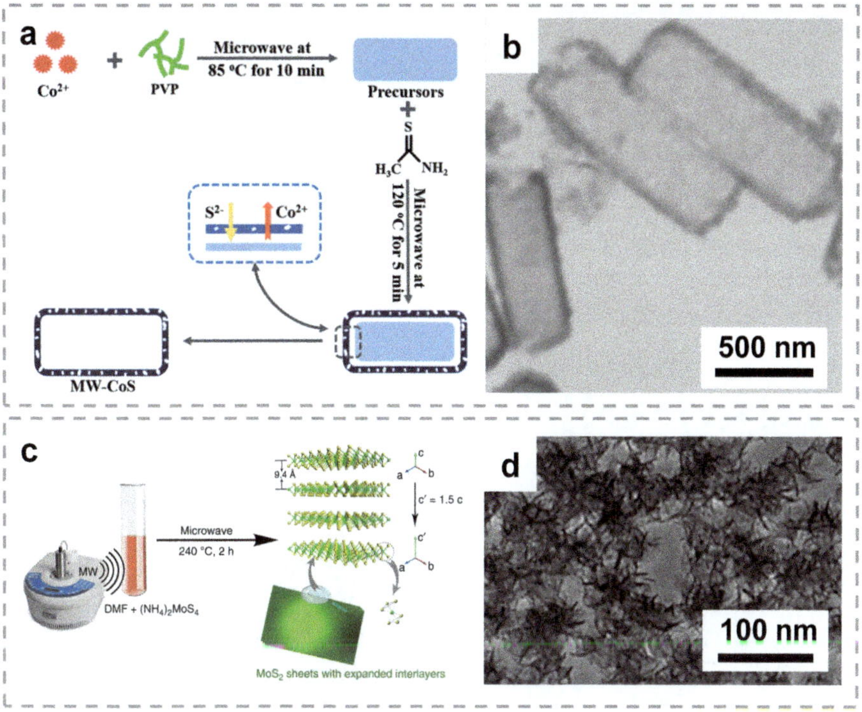

Figure 4.24 (a) Schematic illustration of synthetic procedures for CoS nanoprisms. (b) TEM image of CoS nanoprisms. (c) Schematic illustration of synthetic procedures for MoS_2 nanosheets. (d) TEM image of MoS_2 nanosheets. (a and b) Reproduced from ref. 224 with permission from the Royal Society of Chemistry. (c and d) Reproduced from ref. 225 with permission from Springer Nature, Copyright 2015.

transformed into $CoNi_3O_4$. Finally, the ultrathin $CoNi_2S_4$ nanosheets were obtained by sulfurisation of $CoNi_3O_4$ (Figure 4.25b). The $CoNi_2S_4$ nanosheets had a micrometre-sized planar area, ultrathin thickness (~2 nm), and mesoporous structure. As a positive electrode material for a hybrid supercapacitor, $CoNi_2S_4$ nanosheets exhibited excellent electrochemical performance with high energy density.

Ru_xSe_y. Nekooi *et al.*[228] reported the microwave-assisted synthesis of Ru_xSe_y nanoparticles in water/EG containing $RuCl_3$, Na_2SeO_3, and KOH. The as-prepared Ru_xSe_y nanoparticles were loaded onto the carbon substrate and investigated as a catalyst for the electrochemical oxygen reduction reaction. Increasing the Se concentration up to 15 mol% could enhance the catalytic activity for the oxygen reduction reaction, and the catalytic activity was considerably higher than that observed for the pure Ru catalyst synthesised under exactly the same conditions.

$Bi_2Te_{3-x}Se_x$. Hong *et al.*[230] reported the synthesis of high-quality $Bi_2Te_{3-x}Se_x$ nanoplates by the microwave-assisted solvothermal method in

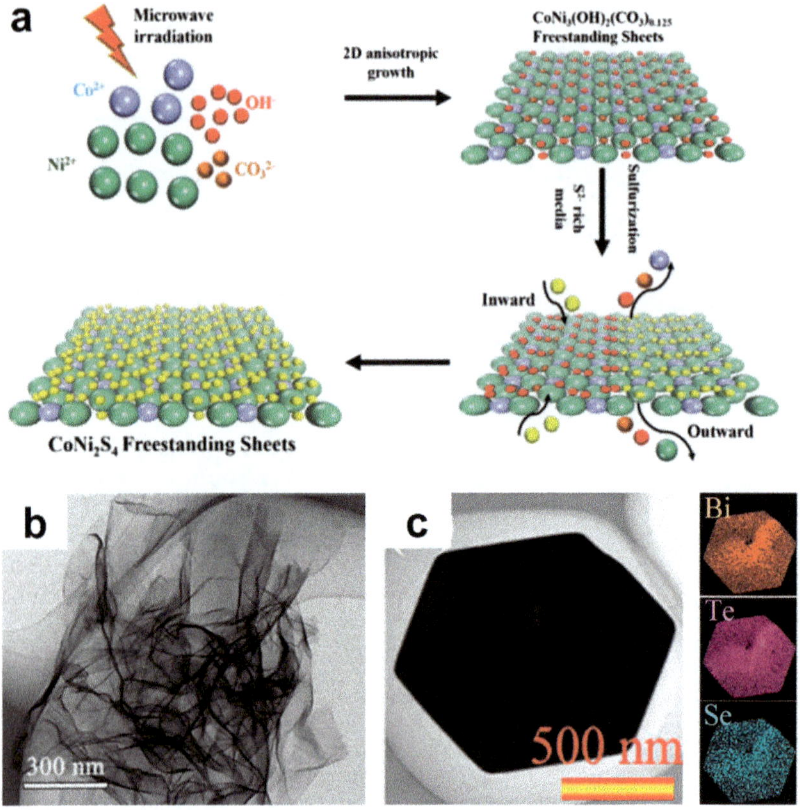

Figure 4.25 (a) Schematic illustration of synthetic procedures for $CoNi_2S_4$ nanosheets. (b) TEM image of $CoNi_2S_4$ nanosheets. (c) TEM image of a single $Bi_2Te_{2.7}Se_{0.3}$ nanoplate, and EDS mapping of Bi, Te, and Se of $Bi_2Te_{2.7}Se_{0.3}$ nanoplates. (a and b) Reproduced from ref. 227 with permission from Elsevier, Copyright 2019. (c) Reproduced from ref. 230 with permission from American Chemical Society, Copyright 2016.

EG and water. The morphology of the as-synthesised $Bi_2Te_{2.7}Se_{0.3}$ nanosheet is shown in Figure 4.25c, and one can see that a single nanoplate had a lateral size of ~1 μm. From energy-dispersive X-ray spectroscopy (EDS), Bi, Te, and Se signals could be detected, and EDS mapping showed the uniform distribution of the three elements in the nanoplate (Figure 4.25c). The obtained n-type $Bi_2Te_{2.7}Se_{0.3}$ nanostructures exhibited a relatively high figure of merit (ZT) of 1.23 at 480 K measured from the corresponding sintered pellets.

Sb_2Te_3. Dong *et al.*[231] reported the microwave-assisted preparation of Sb_2Te_3 single-crystalline nanosheets with thicknesses of 50–70 nm, which was performed in hydrazine hydrate/EG containing $SbCl_3$ and Na_2TeO_3. The as-prepared Sb_2Te_3 nanosheets were used as the raw material for the preparation of the bulk sample by spark plasma sintering. High electrical conductivity ($2.49 \times 10^4\ \Omega^{-1}\ m^{-1}$), high Seebeck coefficient (210 $\mu V\ K^{-1}$), and low thermal

conductivity (0.76 W m^{-1} K^{-1}) at 420 K were achieved. It is difficult to achieve high electrical conductivity without obvious growth of nanocrystals after sintering, and here a successful attempt was made using microwave-assisted synthesis.

4.3.4 Other Inorganic Nanostructures

In this section we will review the microwave-assisted preparation of other inorganic nanostructures in addition to metals, metal oxides, and metal chalcogenides.

4.3.4.1 Non-metals

Some examples of the microwave-assisted preparation of non-metal nanostructures are listed in Table 4.15.

He *et al.*[243] reported the microwave-assisted preparation of Si quantum dots in aqueous solution containing Si nanowires and glutaric acid. The as-prepared Si quantum dots (Figure 4.26a and b) exhibited good aqueous dispersibility, high stability, and strong fluorescence (Figure 4.26c). The as-prepared Si quantum dots were investigated for immunofluorescent cellular imaging.

Liu *et al.*[246] reported the microwave-assisted preparation of Te nanowires in aqueous solution containing Na_2TeO_3, hydrazine hydrate, PVP, and

Table 4.15 Some examples of non-metal nanostructures prepared by microwave heating[a]

Material	Precursor	Solvent	Additive	Nanostructure	Reference
Si	APTMS	Water	Trisodium citrate	Particles	241
Si	APTMS	Water	Trisodium citrate	Particles	242
Si	Si nanowires	Water	Glutaric acid	Quantum dots	243
Si	Si nanowires	Water	Immunoglobulin G	Particles	244
Si	Na_4Si_4	DMF	NH_4Br, C_3H_5N, C_3H_7N, $C_{11}H_{23}N$	Particles	245
Te	Na_2TeO_3	Water	Hydrazine hydrate, PVP, ammonia	Nanowires	246
Te	TeO_2	EG		Nanorods, nanowires	247
Te	TeO_2	BuPy[BF_4]	PVP/$NaBH_4$	Nanorods, nanowires	110
Se	SeO_2	EG		Nanorods, nanowires	248
Se	Na_2SeO_3	Ethylenediamine	$NaBH_4$	Nanorods	249

[a]APTMS: (3-aminopropyl)trimethoxysilane.

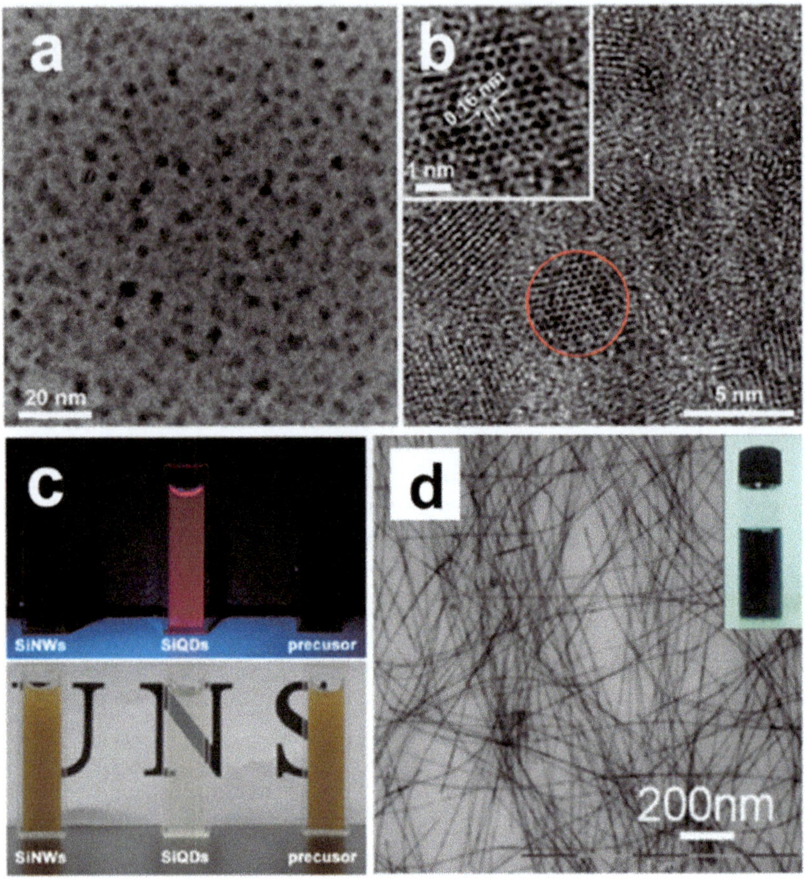

Figure 4.26 (a and b) TEM and high-resolution TEM images of Si quantum dots. (c) Photographs of aqueous solutions dispersed with Si nanowires (left), Si quantum dots (middle), and reaction precursors (right) under 365 nm irradiation (up) or ambient light (bottom). (d) TEM image of Te nanowires. (a–c) Reproduced from ref. 243 with permission from American Chemical Society, Copyright 2011. (d) Reproduced from ref. 246 with permission from American Chemical Society, Copyright 2010.

ammonia. Single-crystalline Te nanowires with a diameter of 20 nm and lengths of tens of micrometres were obtained (Figure 4.26d). Zhu *et al.*[110] had reported pioneering work on the microwave-assisted ionic liquid method for the preparation of Te nanorods and nanowires in ionic liquid *N*-butylpyridinium tetrafluoroborate containing TeO_2, $NaBH_4$, and PVP. Exclusive formation of Te nanorods or nanowires could be obtained by tuning the synthetic parameters (Figure 4.19). It was found that both the ionic liquid and microwave heating greatly affected the formation of Te nanorods and nanowires.

4.3.4.2 Halides

There are many reports on the microwave-assisted synthesis of halide nanostructures in the literature, and some examples are listed in Table 4.16.

4.3.4.2.1 Fluorides. *CeF_3.* Ma *et al.*[251] reported the microwave-assisted synthesis of CeF_3 nanoflowers in aqueous solution containing $Ce(NO_3)_3$, KF, HNO_3, and EDTA disodium as a complexing reagent. The as-prepared CeF_3 nanoflowers with a mean diameter of 190 nm were formed by self-assembly of nanodiscs.

GdF_3. Wang *et al.*[252] reported the microwave-assisted synthesis of GdF_3 raisin-like nanostructures in aqueous solution containing $Gd(NO_3)_3$ and $NaBF_4$. They doped the lanthanide ions into GdF_3 nanocrystals, which showed multicolour upconversion photoluminescence and paramagnetism.

LaF_3:Ce,Tb and LaF_3:Eu. Mi *et al.*[254] reported the microwave-assisted synthesis of polyethyleneimine (PEI)-functionalised luminescent rare earth-doped LaF_3 nanoparticles in EG containing $La(NO_3)_3$, NaF, and PEI (rare earth sources were $CeCl_3$, $Tb(NO_3)_3$, Eu_2O_3). The as-prepared nanoparticles had a hexagonal structure with an average size of about 12 nm. The morphology and structure of the nanoparticles could be preserved after doping with different ions, whereas the optical properties changed with different doped ions and their molar ratio. Therefore, four different colours (green, yellow, orange, and red) were obtained by simply adjusting the type of doping ion (Eu^{3+} and Tb^{3+}) and the molar ratio of the two doping ions.

CaF_2. Zhao *et al.*[258] reported the rapid microwave-assisted synthesis of water-dispersible CaF_2 and Yb^{3+}/Er^{3+} co-doped CaF_2 (CaF_2:Yb^{3+}/Er^{3+}) nanocrystals with different sizes and different Yb^{3+} and Er^{3+} dopant concentrations. The ionic liquid [BMIM][BF_4] was used as a fluorine source. It was found that the morphology, size, and crystallinity of the CaF_2:Yb^{3+}/Er^{3+} nanocrystals could be adjusted by using adenosine 5′-triphosphate (ATP). Yb^{3+} and Er^{3+} ions were doped into CaF_2 nanocrystals to enable upconversion luminescence, and the as-prepared CaF_2:Yb^{3+}/Er^{3+} nanocrystals exhibited upconversion luminescence emission under excitation at 980 nm.

4.3.4.2.2 Bismuth Oxyhalides. Bismuth oxyhalides (BiOX) are ternary semiconductors that are composed of group V-VI-VII family elements. They have a tetragonal structure, where [X-Bi-O-Bi-X] layers are stacked one above the other by van der Waals interaction through the halogen atoms along the *x* axis. Such a structure of BiOX is desirable because: (1) the internal electric field formed between the Bi–O and X^- layers can promote separation of photogenerated electrons and holes and thus improve the photocatalytic activity; and (2) it is inclined to form nanostructures with 2D plate-like morphology and high specific surface area.

BiOBr. Zhang *et al.*[260] reported the microwave-assisted synthesis of BiOBr nanosheet-assembled microspheres in diethylene glycol containing $Bi(NO_3)_3$ and CTAB, with or without oleic acid. The as-prepared BiOBr microspheres

Table 4.16 Some examples of halide nanostructures synthesised by microwave heating[a]

Material	Precursor	Solvent	Additive	Nanostructure	Reference
PrF_3	Praseodymium oxide powder, NaF	Water	HNO_3/ammonia	Hollow particles	250
CeF_3	$Ce(NO_3)_3$, KF	Water	Na_2H_2EDTA/HNO_3	Nanodisc-assembled flowers	251
GdF_3	$Cd(NO_3)_3$, $NaBF_4$	Water		Raisin-like nanostructures	252
Yb/Er-RbY_2F_7	YCl_3, $YbCl_3$, $ErCl_3$, NH_4F, RbF	Water		Particles	253
LaF_3:Ce,Tb, LaF_3:Eu	$La(NO_3)_3$, $CeCl_3$, $Tb(NO_3)_3$, Eu_2O_3, NaF	EG	Polyethyleneimine	Particles	254
MF_2 (M = Ca, Mg, Sr)	$Ca(NO_3)_2$, $Mg(NO_3)_2$, $Sr(NO_3)_2$	[BMIM][BF_4], water	NaH_2PO_4	Hollow microspheres	255
REF_3 (RE = La–Lu, Y)	Metal nitrate	[BMIM][BF_4], water		Nanodiscs, nanoparticles	256
$NaYF_4$	$Na(CF_3COO)$, $Y(CF_3COO)_3$	[BMIM][BF_4]		Spheres	257
CaF_2:Yb/Er	$CaCl_2$, $Yb(NO_3)_3$, $Er(NO_3)_3$	[BMIM][BF_4], EG	ATP	Nanocrystals	258
Nd-PbI_2	$Pb(CH_3COO)_2$, NaI, $Nd(NO_3)_3$	Water	SDS, PVA	Nanorods, nanosheets	259
BiOBr	$Bi(NO_3)_2$, CTAB	Diethylene glycol	Oleic acid	Nanosheet-assembled microspheres	260
$Bi_4O_5I_2$	$Bi(NO_3)_3$, KI	Water, EG		Nanosheet-assembled flowers	261
BiOX:Eu (X = Cl, Br, I)	$Eu(NO_3)_3$, $Bi(NO_3)_3$, KX (X = Cl, Br, I)	Water	PVA	Nanoflakes	262
$CsPbX_3$ (X = Cl, Br, I)	CH_3COOCs, PbX_2 (X = Cl, Br, I)	ODE, TMPPA, OLA		Nanocrystals	263
$CsPbX_3$ (X = Cl, Br, I)	CH_3COOCs, PbX_2 (X = Cl, Br, I)	ODE, OA, OLA, TOP	TOPO	Nanocubes, nanoplates, nanorods	264
$CsPbBr_3$	Cs_2CO_3, $PbBr_2$	ODE, OA, OLA		Nanocrystals	265

[a]Na_2H_2EDTA: ethylenediaminetetraacetic acid disodium; ATP: adenosine triphosphate disodium salt; PVA: poly(vinyl alcohol); ODE: 1-octadecene; TMPPA: bis(2,4,4-trimethylpentyl)phosphinic acid; OLA: oleylamine; OA: oleic acid; TOP: trioctylphosphine; TOPO: trioctylphosphine oxide.

exhibited excellent photocatalytic activities for methyl orange, rhodamine B, and phenol, and are promising for applications as adsorbents and photocatalysts for the treatment of organic pollutants.

$Bi_4O_5I_2$. He *et al.*[261] studied the effect of the preparation method on the photocatalytic performance of $Bi_4O_5I_2$ hierarchical nanosheet-assembled flowers. The preparation methods compared encompassed the microwave-assisted solvothermal method, oil bath synthesis, and direct hydrolysis. It was found that $Bi_4O_5I_2$ nanosheet-assembled flowers obtained by microwave treatment exhibited the highest photocatalytic activity towards phenol degradation under visible light irradiation.

BiOX. Dash *et al.*[262] reported the microwave-assisted synthesis of PVA-coated Eu^{3+}-doped BiOX (X = Cl, Br, I) nanoflakes and studied the optical properties and photocatalytic activity. It was found that both Eu^{3+}-doped BiOCl and BiOBr nanoflakes exhibited strong optical properties, which were quenched for Eu^{3+}-doped BiOI nanoflakes. The estimated bandgaps of Eu^{3+}-doped BiOX (X = Cl, Br, I) nanoflakes were 3.36, 2.74, and 1.67 eV, respectively. The nanoflakes exhibited high photocatalytic activity for dye degradation under visible light irradiation.

4.3.4.2.3 Caesium Lead Halides ($CsPbX_3$). Caesium lead halides ($CsPbX_3$) are an emerging class of materials with excellent optoelectronic properties and wide applications in solar cells, light-emitting diodes, and photodetectors. In particular, $CsPbX_3$ (X = Cl, Br, I) nanocrystals exhibit enhanced optical properties, which are comparable and even superior to metal chalcogenide quantum dots.[263]

One of the most widely used methods to prepare $CsPbX_3$ nanostructures is the hot-injection method, during which nanocrystals can be quickly obtained. The rapid nucleation and crystal growth makes it difficult to study the formation mechanism. A 'slowed down' wet-chemical synthesis may provide a solution (Figure 4.27a and b). Li *et al.*[265] reported the microwave-assisted synthesis of $CsPbBr_3$ nanocrystals, with the aim of determining the formation mechanism of the nanoparticles. Aiming to slow down the reaction rate and, specifically, to be able to obtain intermediate products, $PbBr_2$ and Cs_2CO_3 were separately placed at the two ends of a quartz boat, and a mixture of 1-octadecene (ODE), oleic acid (OA), and oleylamine (OLA) was carefully transferred into the container without disturbing the powders. The obtained $CsPbBr_3$ nanocrystals were uniform and monodisperse, and possessed a cubic shape with an average edge size of 10 nm (Figure 4.27c).

Pan *et al.*[264] reported the microwave-assisted synthesis of $CsPbX_3$ (Cl, Br, I) nanocrystals with different morphologies (nanocubes, nanoplates, and nanorods). It was found that the reaction temperature greatly affected the morphology and optical properties of the $CsPbX_3$ nanocrystals. When the synthesis was carried out at 160 °C, nanocubes with an average edge length of ~10.8 nm were obtained. When the reaction temperature was 80 °C, plate-like nanocrystals were obtained. Figure 4.27d shows the as-prepared $CsPbBr_3$ nanoplates with a thickness of ~3.3 nm. The as-prepared $CsPbBr_3$ nanoplates

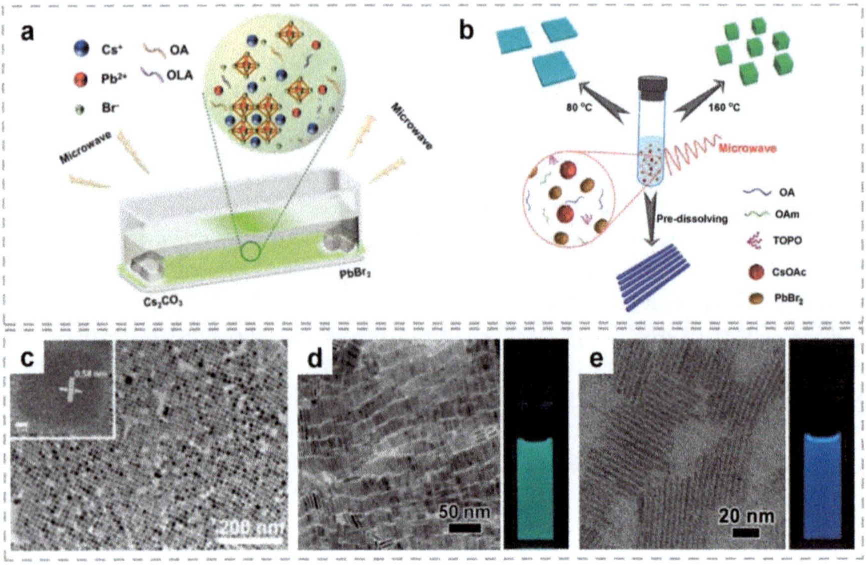

Figure 4.27 (a) Schematic illustration of synthetic procedures for $CsPbBr_3$ nanocrystals. (b) Schematic illustration of synthetic procedures for $CsPbX_3$ (X = Cl, Br, I) nanocrystals. (c) TEM image of $CsPbBr_3$ nanocrystals. (d and e) TEM images of plate- and rod-like $CsPbBr_3$ nanocrystals and photographs of nanocrystals dispersed in hexane under 365 nm light irradiation. (a and c) Reproduced from ref. 265 with permission from John Wiley & Sons, Copyright 2018 Wiley-VCH Verlag GmbH & Co. KGaA, Weinheim. (b, d and e) Reproduced from ref. 264 with permission from the Royal Society of Chemistry.

emitted a bright greenish blue colour. In addition, rod-like $CsPbBr_3$ nanocrystals could be obtained when the precursors were predissolved. As shown in Figure 4.27e, rod-like $CsPbBr_3$ nanocrystals were prepared with a diameter of ~2.7 nm, and a bright blue emission could be observed under UV light irradiation.

4.3.4.3 *Phosphates*

Some examples of the microwave-assisted synthesis of metal phosphate nanostructures are listed in Table 4.17.

4.3.4.3.1 Calcium Phosphate Nanostructures. Calcium phosphates are typical biomaterials with excellent biocompatibility and biodegradability and have been extensively investigated for biomedical applications.[279] Hydroxyapatite, $Ca_{10}(PO_4)_6(OH)_2$, is a member of the calcium phosphate family and is the main inorganic component of bones and teeth. Zhao *et al.*[266] reported the microwave-assisted synthesis of hydroxyapatite (HAP) nanosheet-assembled flowers in aqueous solution containing $Ca(CH_3COO)_2$ and NaH_2PO_4 in

Table 4.17 Some examples of metal phosphate nanostructures synthesised by microwave heating[a]

Material	Precursor	Solvent	Additive	Nanostructure	Reference
HAP	$Ca(CH_3COO)_2$, NaH_2PO_4	Water		Nanosheet-assembled flowers	266
HAP	$CaCl_2$, Na_2HPO_4	Water	PLA-mPEG	Nanorods	267
HAP	$CaCl_2$, creatine phosphate	Water		Nanorod arrays	268
W-HAP	$Ca(NO_3)_2$, $(NH_4)_2HPO_4$, $H_4Na_2O_6W$	Water	CTAB	Nanorods	269
Amorphous calcium phosphate	$CaCl_2$, ATP	Water		Porous nanospheres	270
Amorphous magnesium phosphate	$MgCl_2$, fructose 1,6-bisphosphate	Water		Nanosheet-assembled flowers	271
$Mg_3(PO_4)_2$	$MgCl_2$, creatine phosphate	Water		Nanosheets	272
$Mg_3(PO_4)_2$	$MgCl_2$, NaH_2PO_4	Water		Nanosheets	273
$BiPO_4$	$Bi(NO_3)_3$/NaH_3PO_4	Water		Nanoneedles	274
$LiFePO_4$	$LiOH/FeSO_4/H_3PO_4$	Water		Particles	275
$LnPO_4$ (Ln = La, Ce, Nd, Sm, Eu, Gd, Tb)	$Ln(NO_3)_3$/$NH_4H_2PO_4$	Water		Nanowires, nanorods	276
$LiFePO_4$	$Fe(CH_3COO)_2$, H_3PO_4, CH_3COOLi	Tetraethylene glycol		Nanorods	277
$LaPO_4$:Ce,Tb	$LaCl_3$, $CeCl_3$, $TbCl_3$	$[MeBu_3N][(SO_2CF_3)_2N]$		Particles	278

[a]HAP: hydroxyapatite; PLA-mPEG: polylactide-*block*-monomethoxy(polyethyleneglycol); $[MeBu_3N][(SO_2CF_3)_2N]$: tributylmethylammonium triflylimide.

the absence of any additional surfactant or additive, and the morphology of the product could be controlled by adjusting the synthetic parameters, such as the microwave heating temperature. HAP nanosheet-assembled flowers were obtained at 100 °C, while HAP polyhedra were obtained at 200 °C. They used HAP nanosheet-assembled flowers as a drug nanocarrier to load bovine serum albumin, haemoglobin (Hb), and fish sperm DNA. The protein release process was conducted at different pH values (pH 7.2, 5.5, and 4.8), and a pH-responsive protein release behaviour was found. Chen at al.[267] reported the microwave-assisted preparation of Eu^{3+}/Gd^{3+} dual-doped HAP nanorods in aqueous solution containing $CaCl_2$, Na_2HPO_4, europium nitrate, gadolinium nitrate, NaOH, and PLA-mPEG. The dual doping of Eu^{3+}/Gd^{3+} into HAP nanorods could provide the material with photoluminescent and magnetic properties, enabling multiple-model imaging capabilities, such as magnetic resonance imaging, photoluminescence imaging, and computed tomography imaging. They also demonstrated high biocompatibility, high drug loading capacity, and sustained drug release for the as-prepared Eu^{3+}/Gd^{3+} dual-doped HAP nanorods. In addition, Zhu's research group has synthesised various HAP nanostructured materials by microwave heating; for example, nanosheet-assembled microspheres,[280–282] nanowires,[283] nanorod-assembled microflowers.[284]

4.3.4.3.2 Biomolecule-assisted Microwave Rapid Synthesis of Calcium Phosphate Nanostructures. Recently, Zhu's research group has developed an innovative biomolecule-assisted green method for the synthesis of metal phosphate nanostructured materials.[279] The conventional synthesis of phosphate nanostructures uses inorganic phosphate salts as the phosphorus source, which leads to rapid nucleation and disordered growth of phosphate materials. To overcome this problem, they used phosphorus-containing biomolecules as the organic phosphorus sources based on the following considerations: (1) Phosphate groups are in phosphorus-containing biomolecules, and do not exist as free $[PO_4]^{3-}$ ions in the reaction solution, which can avoid rapid nucleation and disordered growth of the nanomaterial. (2) In order to release $[PO_4]^{3-}$ ions from phosphorus-containing biomolecules, heating or other experimental conditions are needed for hydrolysis of the phosphorus-containing biomolecules in aqueous solution. Hence, the hydrolysis rate of phosphorus-containing biomolecules can be controlled using the synthetic parameters, which finally determines the structure, size, morphology, and self-assembly of the phosphate nanostructures. (3) The hydrolysis of phosphorus-containing biomolecules proceeds progressively, and the rapid nucleation and disordered growth of nanomaterials can be inhibited. (4) The crystal growth process can be regulated by biomolecules, which can stabilise the initially formed amorphous calcium phosphate. (5) Biomolecules are highly biocompatible, and the as-prepared calcium phosphate nanostructures are promising for biomedical applications.

Therefore, phosphorus-containing biomolecules have been used as novel organic phosphorus sources for the microwave-assisted rapid synthesis of calcium phosphate nanostructures. As shown in Figure 4.28a, the phosphorus-containing biomolecules used for microwave-assisted synthesis include adenosine 5′-triphosphate (ATP),[270,285–291] adenosine 5′-diphosphate (ADP),[292] fructose 1,6-bisphosphate (FBP),[271,293–295] creatine phosphate (CP),[268,272,296,297] cytidine 5′-triphosphate (CTP),[298] pyridoxal-5′-phosphate (PP),[299] riboflavin-5′-phosphate (RP),[300] and glycerophosphate (GP).[301] HAP and amorphous calcium phosphate (ACP) with various nanostructures have been prepared by biomolecule-assisted microwave synthesis. Figures 4.28b–e shows HAP nanowires,[290] HAP nanorods,[300] HAP nanosheet-assembled microspheres,[297] and HAP hierarchically nanostructured porous hollow microspheres.[268] For example, Yu *et al.*[297] prepared mesoporous HAP microspheres in the

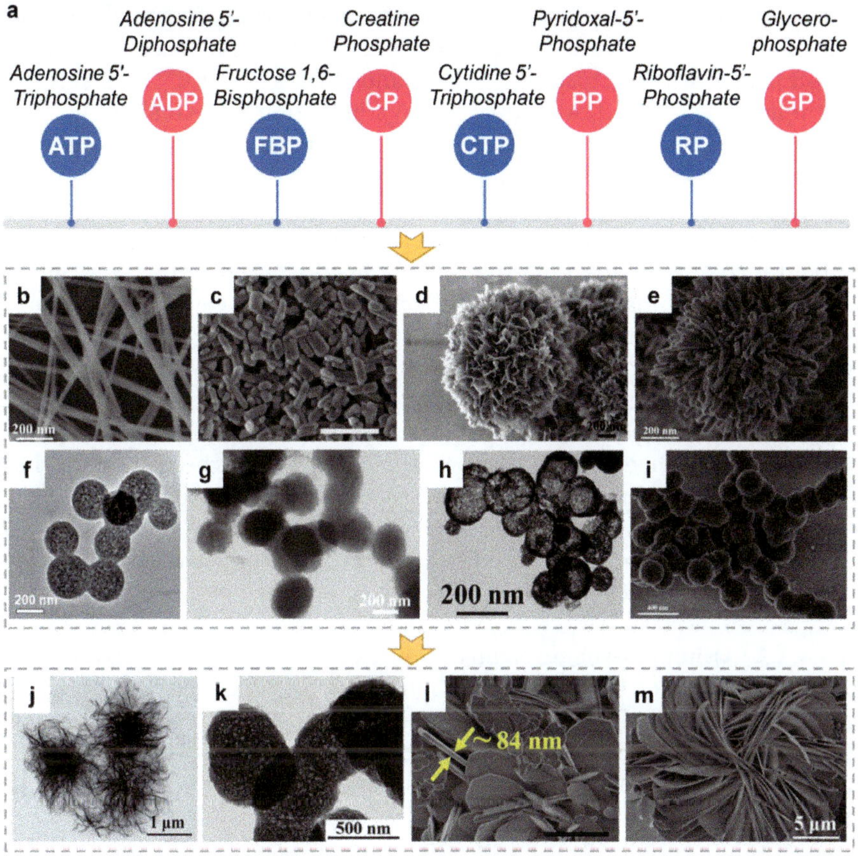

Figure 4.28 Phosphate nanostructures synthesised by microwave heating. (a) Phosphorus-containing biomolecules used for microwave-assisted rapid synthesis of phosphate nanostructures. (b–e) SEM and TEM images of HAP nanostructures. (f–i) TEM and SEM images of amorphous calcium phosphate (ACP) microspheres and nanospheres. (j) TEM image of amorphous magnesium phosphate nanosheet-assembled flowers. (k) TEM image of amorphous calcium magnesium phosphate. (l and m) SEM images of magnesium phosphate nanosheets. (b) Reproduced from ref. 290 with permission from Elsevier, Copyright 2012. (c) Reproduced from ref. 300 with permission from the Royal Society of Chemistry. (d) Reproduced from ref. 297 with permission from the Royal Society of Chemistry. (e) Reproduced from ref. 268 with permission from John Wiley & Sons, Copyright 2013 WILEY-VCH Verlag GmbH & Co. KGaA, Weinheim. (f) Reproduced from ref. 291 with permission from the Royal Society of Chemistry. (g) Reproduced from ref. 294 with permission from American Chemical Society, Copyright 2017. (h) Reproduced from ref. 287 with permission from the Royal Society of Chemistry. (i) Reproduced from ref. 270 with permission from John Wiley & Sons, Copyright 2013 WILEY-VCH Verlag GmbH & Co. KGaA, Weinheim. (j) Reproduced from ref. 271 with permission from the Royal Society of Chemistry. (k) Reproduced from ref. 296 with permission from the Royal Society of Chemistry. (l) Reproduced from ref. 272 with permission from Elsevier, Copyright 2015. (m) Reproduced from ref. 288 with permission from Elsevier, Copyright 2014.

presence of creatine phosphate, and the as-prepared microspheres consisted of nanosheets and nanorods (Figure 4.28d). Figure 4.28f–i shows the as-prepared ACP microspheres and nanospheres.[270,287,291,294] As a typical example, Qi *et al.*[270] reported the rapid synthesis of highly stable ACP porous nanospheres (Figure 4.28i) with an average pore diameter of ~10 nm by the microwave-assisted hydrothermal method using ATP as a phosphorus source and stabiliser.

The biomolecule-assisted microwave rapid synthesis method is also applicable for the preparation of other metal phosphate nanostructures, including amorphous magnesium phosphate, amorphous calcium magnesium phosphate, and magnesium phosphate (Figure 4.28j–m).[271,272,288,296] For example, Qi *et al.*[271] reported the microwave-assisted hydrothermal rapid synthesis of amorphous magnesium phosphate hierarchical nanostructures (Figure 4.28j) using FBP as an organic phosphorus source. They also reported the microwave–hydrothermal rapid synthesis of magnesium phosphate nanosheets using creatine phosphate as an organic phosphorus source.[272] The as-prepared nanosheets had an obvious pH-dependent dissolution performance, which could be used for an ideal pH-responsive nanocarrier for drug and gene delivery.

4.3.4.3.3 Other Phosphate Nanostructures. *$BiPO_4$*. Li *et al.*[274] reported the microwave-assisted synthesis of $BiPO_4$ nanoneedles in aqueous solution containing $Bi(NO_3)_3$ and NaH_2PO_4. By varying the type of solvent, $BiPO_4$ nanoparticles and rod-like, needle-like, and rice-like $BiPO_4$ nanostructures could be prepared. They studied the relationship between the morphology and the photocatalytic performance of the product. It was found that different $BiPO_4$ nanostructures showed different optical properties, specific surface areas, and photocatalytic activities for the degradation of methyl orange under UV and visible light irradiation, and that hexagonal $BiPO_4$ nanoparticles exhibited the best photocatalytic performance.

$LiFePO_4$. Yang *et al.*[275] reported the microwave-assisted hydrothermal synthesis of $LiFePO_4$ nanoparticles in aqueous solution containing LiOH, $FeSO_4$, and H_3PO_4 at 200 °C for 5 min. It was found that the particle size of $LiFePO_4$ was dependent on the microwave-assisted hydrothermal conditions.

$LnPO_4$. Patra *et al.*[276] reported the microwave-assisted synthesis of $LnPO_4 \cdot nH_2O$ (Ln = La, Ce, Nd, Sm, Ru, Gd, and Tb, n = 0–0.6) nanowires/nanorods in aqueous solution containing $Ln(NO_3)_3$ and $NH_4H_2PO_4$ in the pH range 1.8–2.2 for 20 min. The length and width of the as-prepared lanthanide orthophosphate nanowires/nanorods were in the range 70–2200 and 6–130 nm, respectively.

Aluminophosphate (AlPO-18). Ng *et al.*[302] reported the synthesis of nanosized aluminophosphate molecular sieves by microwave heating. They investigated the environmental issue concerning the re-use and disposal of non-reacted chemicals. Nanosized AlPO-18 was prepared by a multicycle synthesis approach *via* re-use of non-reacted compounds from precursor suspensions, with minimal requirement for chemical compensation after recovering crystalline nanoparticles from each step. The use of microwave heating resulted in the synthesis of nanocrystalline material within 5 min, instead of 3 days for conventional heating.

4.3.4.4 *Hydroxides*

Some examples of metal hydroxide nanostructures synthesised by microwave heating are listed in Table 4.18.

4.3.4.4.1 Single Metal Hydroxides. *$Mg(OH)_2$*. Umar *et al.*[303] reported the microwave-assisted synthesis of $Mg(OH)_2$ hexagonal nanodiscs in aqueous solution containing $Mg(CH_3COO)_2$ and HMTA. The as-prepared $Mg(OH)_2$ hexagonal nanodiscs were used as an efficient ethanol chemical sensor.

$Ni(OH)_2$. Ran *et al.*[304] reported the microwave-assisted synthesis of $Ni(OH)_2$ nanosheet-assembled flowers in aqueous solution containing $Ni(NO_3)_2$, urea, and SDS. Flower-like β-$Ni(OH)_2$ with monolayer petals, and flower-like α-$Ni(OH)_2$ with bilayer petals, were obtained at 120 °C and 150 °C, respectively. They used $Ni(OH)_2$ nanosheet-assembled flowers for application in water treatment. Zhu *et al.*[306] reported the α-$Ni(OH)_2$ precursor conversion method to prepare non-layer-structured NiO nanosheets. First, homogeneous

Table 4.18 Some examples for microwave-assisted synthesis of metal hydroxide nanostructures[a]

Material	Precursor	Solvent	Additive	Nanostructure	Reference
$Mg(OH)_2$	$Mg(CH_3COO)_2$	Water	HMTA	Nanodiscs	303
$Ni(OH)_2$	$Ni(NO_3)_2$	Water	Urea/SDS	Nanosheet-assembled flowers	304
$Ni(OH)_2$	$Ni(NO_3)_2$	EG	HMTA	Nanoballs	305
$Ni(OH)_2$	$Ni(NO_3)_2$	Water, EG	Urea	Nanosheets	306
$Co(OH)_2$	$CoCl_2$	Water	HMTA/SDS	Nanocones	307
$Co(OH)_2$	$Co(NO_3)_2$	Water, EG	Urea	Nanosheets	308
$Co(OH)_2$, $Ni(OH)_2$, $Cd(OH)_2$	Metal nitrate	TBAH, TEAH		Plates, networks	309
$In(OH)_3$	$InCl_3$	Water	Urea, trisodium citrate	Nanospheres	310
$Nd(OH)_3$	$Nd(NO_3)_3$	Water	Ammonia	Nanorods	311
Zn,Co-LDH	$Zn(NO_3)_2$, $Co(NO_3)_2$	Water, EG	Urea	Nanosheets	312
Ni,Al-LDH	Al_2O_3, $Ni(NO_3)_2$	Water	Urea	Nanosheet-assembled hollow microspheres	313

[a]TBAH: tetrabutylammonium hydroxide; TEAH: tetraethylammonium hydroxide; LDH: layered double hydroxide.

alkalisation of nickel nitrate by urea hydrolysis under microwave irradiation could produce α-$Ni(OH)_2$ nanosheets with an average thickness of ~1.52 nm. After thermal treatment, non-layered NiO nanosheets were obtained. The as-prepared ultrathin α-$Ni(OH)_2$ and NiO nanosheets exhibited enhanced supercapacitor performance.

$Co(OH)_2$. Liu *et al.*[307] reported the microwave-assisted synthesis of $Co(OH)_2$ nanocones in aqueous solution containing $CoCl_2$, HMTA, and SDS. They proposed that the conical nanostructures with hollow interiors might form through a rolling process of lamellar hydroxide formation. Unilamellar cobalt hydroxide nanosheets could be obtained by direct exfoliation of layered cobalt hydroxide nanocones in formamide. In addition, layered cobalt hydroxide nanocones could be transformed into CoOOH and Co_3O_4 nanocones by oxidation in alkaline solution and thermal decomposition, respectively. Souleymen *et al.*[308] reported the $Co(OH)_2$ precursor conversion method to prepare non-layered CoS_x nanosheets. The first step was a homogeneous alkalisation of cobalt nitrate solution by urea hydrolysis to form hydrated inorganic oligomers through microwave irradiation. Then, the anion exchange reaction was adopted to chemically transform the transition metal hydroxide to its corresponding sulfide nanosheets.

4.3.4.4.2 Layered Double Hydroxides. Layered double hydroxides (LDHs) have an intrinsic layered structure. The positively charged layers contain alternately arranged metal cations, while the charge-compensating anions are positioned within the interlayer space. In addition, the flexible choice of metal cations and unique intercalation characteristics enable facile physical and chemical engineering.[314]

Zn,Co-LDH. Qiao *et al.*[312] reported the one-step microwave-assisted synthesis of Zn,Co-LDH nanosheets with a high specific surface area of 116.4 $m^2 g^{-1}$, and high activity for the oxygen evolution reaction.

Ni,Al-LDH. Wang *et al.*[313] reported the microwave-assisted synthesis of Ni,Al-LDH hollow microspheres using Al_2O_3 as the template. Al_2O_3 porous microspheres composed of nanosheets with diameters of 1.5–2 μm and high specific surface area (>280 $m^2 g^{-1}$) were prepared. Then, the microwave synthesis of Ni,Al-LDH was carried out in aqueous solution containing $Ni(NO_3)_2$, urea, and Al_2O_3 porous microspheres. The as-prepared Ni,Al-LDH hollow microspheres exhibited high specific capacitance and good cycling performance as the supercapacitor electrode.

4.3.4.5 *Carbon Materials*

Carbon materials have become a hot and rapidly evolving research field in the past two decades owing to their intriguing physical and chemical properties.[315] These superior properties enable the use of carbon materials in many applications, including energy storage and conversion, catalysis, sensors, flexible electronics, and environmental protection. Schwenke *et al.*[316] published a comprehensive review on the microwave-assisted preparation of carbon nanomaterials, including the synthesis and processing of carbon nanotubes, carbon nanofibres, carbon nanoparticles, and graphene. Some carbon nanostructured materials prepared by microwave heating are reviewed in this section and summarised in Table 4.19.

4.3.4.5.1 Carbon Dots. Carbon dots (CDs) are an emerging luminescent material with superior properties that have attracted much interest in recent years. CDs are promising for various applications, such as sensors,[321,333] fluorescent inks,[334] and bioimaging.[335,336] Microwave heating is an attractive approach for rapid preparation of high-quality CDs. Commonly used precursors for the preparation of CDs by microwave heating include sugars, citric acid and citrates,[321,334,335] PEG,[319] eggshell membrane ash,[337] ascorbic acid,[336] and amino acid.[333,338,339] In order to tune the fluorescent emissions or improve the optical properties of CDs, different elements are usually doped into the CDs such as N-doped CDs,[334,336] N,P-doped CDs,[338,340] N,S-doped CSs,[333] and N-doped hydroxyl-functionalised CDs.[335]

Wang *et al.*[317] reported the one-step microwave-assisted preparation of photoluminescent CDs. They used a carbohydrate (glycerol, glycol, glucose, sucrose) and a tiny amount of an inorganic ion, and the preparation could be completed in just a few minutes. The as-prepared CDs exhibited blue, yellow, and red photoluminescence under ultraviolet (330–385 nm), blue (450–480 nm), and green (510–550 nm) light excitation, respectively. Wang *et al.*[318] used three different heating conditions to prepare CDs: conventional heating under reflux in nitric acid, microwave-assisted heating under reflux, and microwave–hydrothermal treatment. Compared with conventional heating under reflux, the microwave heating under reflux and microwave–hydrothermal method were both able to shorten the reaction time. It was found that CDs prepared by microwave heating exhibited increased absorption, higher quantum yield, and longer fluorescence lifetime compared with those prepared by conventional heating under reflux.

Tian *et al.*[323] used three kinds of solvents, including water, glycerol, and DMF, to prepare CDs. They proposed the possible formation mechanism of the CDs, as shown in Figure 4.29a. Citric acid and urea first assembled into a nanoplate structure through intermolecular hydrogen bonding. Under microwave–solvothermal conditions, the dehydration process occurred between citric acid and urea to form nitrogen-rich CDs with abundant carboxyl, hydroxyl, and amide groups on the surface. It was found that the solvent had a great effect on the extent of dehydration and carbonisation between citric acid and urea. DMF could provide a higher degree of dehydration and carbonisation of the CD precursor than water. They synthesised full-colour emissive CDs simply by tuning the volume ratio of water, glycerol, and DMF employed as solvents in a reaction mixture. Under UV light, the respective solutions showed bright luminescence covering the whole blue to red spectral range (Figure 4.29b), with the emission peaks gradually shifting from 448 nm to 638 nm (Figure 4.29c). In addition, by using a sodium silicate aqueous solution under microwave heating, CDs

Table 4.19 Some examples of carbon nanostructured materials prepared by microwave heating[a]

Material	Precursor	Solvent	Additive	Nanostructure	Reference
CDs	Inorganic phosphate salt or chloride salt, glycerol, glycol, glucose, sucrose	Water		Dots	317
CDs	Graphite oxide	Nitric acid		Dots	318
CDs	PEG	Water		Dots	319
CDs	Urea, citric acid	Water		Dots	320
CDs	Urea, thiourea, citric acid	Water		Dots	321
CDs	Sucrose	Diethylene glycol	Concentrated H_2SO_4	Dots	322
CDs	Urea, citric acid	Water or glycerol or DMF		Dots	323
Carbon nanoparticles	Glucose, PEG	Water		Particles	324
Carbon nanoparticles	Sucrose	H_3PO_4		Particles	325
Carbon nanoparticles	Ammonium citrate dibasic	Water		Particles	326
Carbon spheres	Glucose	Water		Nanospheres	327
Carbon microspheres	Sucrose	Water, EG	Concentrated H_2SO_4	Microspheres	328
Graphene	Graphite oxide	Water	Ascorbic acid	Nanosheets	329
Graphene	Graphite oxide, acrylamide, azodiisobutyronitrile	Water	Hydrazine hydrate	Nanosheets	330
Graphene	Expanded graphite	Aqueous ammonia		Nanosheets	331
Graphene	Graphite oxide	Water	H_2O_2, ammonium peroxydisulfate, azodicarbonamide	Nanosheets	332

[a]CDs: carbon dots.

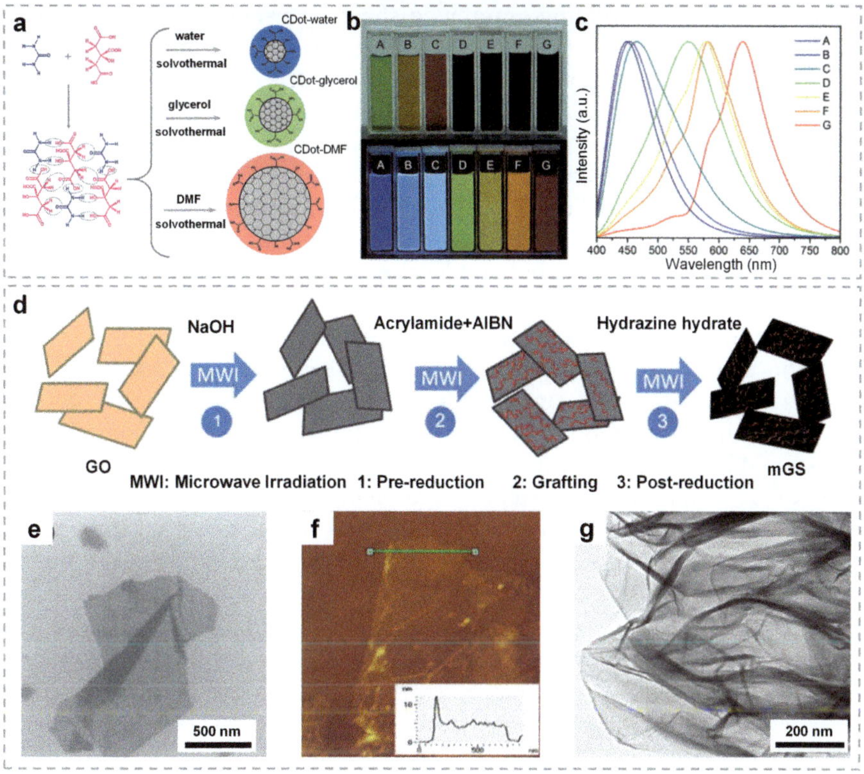

Figure 4.29 (a) Schematic illustration of a possible formation mechanism for CDs-water, CDs-glycerol, and CDs-DMF. (b) Photographs of CD solutions under daylight (top) and UV light (bottom). (c) Photoluminescence spectra at 375 nm excitation. (d) Schematic illustration of preparation procedures for PAM-modified graphene sheets. (e) SEM image of PAM-modified graphene sheets on silicon wafer. (f) Atomic force microscopy image of PAM-modified graphene sheets on freshly cleaved mica. Inset: the height profile along the drawn line. (g) TEM image of few-layered graphene sheets. (a–c) Reproduced from ref. 323 with permission from John Wiley & Sons, Copyright 2017 WILEY-VCH Verlag GmbH & Co. KGaA, Weinheim. (d–f) Reproduced from ref. 330 with permission from the Royal Society of Chemistry. (g) Reproduced from ref. 332 with permission from Elsevier, Copyright 2011.

were incorporated into a silica network, which could effectively prevent aggregation of the CDs.

4.3.4.5.2 Graphene. Long *et al.*[330] reported the microwave-assisted rapid preparation of water-dispersible polyacrylamide (PAM)-modified graphene sheets. As shown in Figure 4.29d, the preparation process included three steps: graphite oxide (GO) was first reduced in alkaline solution; then grafted with PAM chains *via* free-radical polymerisation between acrylamide and azodiisobutyronitrile (AIBN); and finally post-reduced using hydrazine.

The as-prepared PAM-modified graphene sheets are shown in Figure 4.29e and f. Janowska *et al.*[331] reported the microwave-assisted preparation of graphene sheets in an ammonia aqueous solution. The as-prepared graphene sheets mostly consisted of mono-, bi-, or few-layer graphene (less than ten layers). Vadahanambi *et al.*[332] reported the microwave-assisted preparation of graphene sheets using a chemical foaming agent (azodicarbonamide), a green oxidiser (H_2O_2), and a radical initiator (ammonium peroxydisulfate). It was found that the intercalated foaming agent between the graphite oxide layers played a critical role in the rapid and large expansion of the graphene worm along the thickness direction, and in the reduction process of the graphite oxide under microwave irradiation. The as-prepared graphene sheets showed fluffy and expanded accordion-like structures (Figure 4.29g). Voiry *et al.*[341] reported the preparation of high-quality graphene *via* microwave reduction of solution-exfoliated graphene oxide using 1 to 2 second pulses of microwaves. They started by slightly reducing graphene oxide by thermal annealing before exposure to the microwaves, resulting in conductive graphene oxide that could absorb microwaves. They proposed that absorption of microwaves led to rapid heating of graphene oxide, causing desorption of oxygen functional groups and reordering of the graphene basal plane.

4.3.4.6 Metal–Organic Frameworks

Metal–organic frameworks (MOFs) feature a highly porous crystalline structure, which is built by metal sites and organic linkers.[342] One of their most striking features of merit is their structural diversity. Appropriate metal sites and organic linkers can be selected for construction of the as-needed MOFs. MOFs have high specific surface area, which is attractive for adsorption and catalytic applications.[343,344] Among various methods for the synthesis of MOFs, microwave-assisted synthesis is promising for their rapid synthesis. Klinowski *et al.*[345] reviewed the microwave-assisted synthesis of MOFs, and emphasis was given to the preparation of nanosized crystallites of MOFs with promising applications in functional devices. Khan *et al.*[346] provided an overview of the synthesis of MOF materials under microwave and ultrasonic irradiation, which described the degree of acceleration, phase-selective crystallisation, and crystal size reduction in MOFs synthesis. The advantages of microwave and ultrasonic irradiation compared with conventional heating were discussed.

In this section, we will discuss some recent developments in the microwave-assisted synthesis of MOFs, as summarised in Table 4.20.

Choi *et al.*[348] reported the microwave-assisted synthesis of MOF-5 in 1-methyl-2-pyrrolidone (NMP) containing $Zn(NO_3)_2$ and 1,4-benzenedicarboxylate acid (H_2BDC). They compared microwave heating and conventional heating for the synthesis of MOF-5. The experimental results showed that: (1) high-quality MOF-5 crystals could be obtained within 30 min and 24 h for microwave heating and conventional heating, respectively; (2) uniform cubic crystals with an average size of 20–25 μm were obtained by microwave

Table 4.20 Some examples of MOFs synthesised by microwave heating[a]

MOF	Metal	Organic linker	Solvent	Nanostructure	Reference
CoZn-ZIF-8	$Zn(NO_3)_2$, $Co(NO_3)_2$	2-Methylimidazole	Methanol	Polyhedra	347
MOF-5	$Zn(NO_3)_2$	H_2BDC	NMP	Cubes	348
IRMOF1, 2, 3	$Zn(NO_3)_2$	H_2BDC	DEF	Cubes	349
$[Ce_2(pydc)_2(Hpydc)(H_2O)_2]Cl_3\cdot(9+y)H_2O$	$CeCl_3$	2,5-H_2pydc	Water	Crystals	350
HKUST-1	$Cu(NO_3)_2$	H_3BTC	Water, ethanol	Particles	351
Cr/Fe-MOFs (MIL-101, MIL-88B, MIL-53)	$Cr(NO_3)_3$/$Fe(NO_3)_3$	H_2BDC	Water	Octahedra	352
MIL-140A-R	$ZrCl_4$	H_2BDC-R (R = NH_2, NO_2, Br, CL, F)	DMF/acetic acid	Plates	353
MIL-53(Fe)	$FeCl_3$	H_2BDC	DMF	Octahedra, spindles, nanorods	354
Ni-BTC	$Ni(NO_3)_2$	H_3BTC	DMF	Spheres	355
Zn-Doped Ni-MOFs	$NiCl_2$, $Zn(CH_3COO)_2$	H_2BDC	DMF	Hierarchical spheres	356
Cu,Ni-BTC	$Ni(NO_3)_2$, $Cu(NO_3)_2$	H_3BTC	DMF	Spheres	357
MIL-88B/NH_2-MIL-88B	$FeCl_3$	H_2BDC, NH_2-BDC	DMF	Spindles, needles	358
Cu, Co, In MOFs	$Co(NO_3)_2$, $Cu(NO_3)_2$, $In(NO_3)_3$	Bipyen, NDC, H_3BTC, Himdc, H_2dach	Water/ethanol/DMF	Particles	359
ZIF-8	$ZnCl_2$	Methylimidazole	Methanol	Membrane	360

[a]ZIF: zeolitic imidazolate frameworks; H_2BDC: 1,4-benzenedicarboxylate acid; NMP: 1-methyl-2-pyrrolidone; DEF: *N,N*-diethylformamide; 2,5-H_2pydc: 2,5-pyridinedicarboxylic acid; H_3BTC: 1,3,5-benzenetricarboxylic acid; NH_2-BDC: 2-aminoterephthalic acid; Bipyen: *trans*-1,2-bis(4-pyridyl)ethylene; NDC: 2,6-naphthalenedicarboxylic acid; Himdc: 4,5-imidazoledicarboxylate; H_2dach: 4,5-imidazoledicarboxylic acid; IRMOF: iso-reticular metal–organic framework; HKUST: Hong Kong University of Science and Technology; MIL: Materials Institute Lavoisier.

heating, while MOF-5 prepared by conventional heating showed cubic crystals with a size range of 400–500 μm; (3) under an optimum set of synthetic conditions, the Langmuir specific surface area (3008 *vs.* 3200 $m^2 g^{-1}$) and CO_2 adsorption capacity (812 *vs.* 823 mg g^{-1}) were almost the same for the samples prepared by the two heating methods.

Silva *et al.*[350] reported the synthesis of microporous MOFs built using lanthanide (Ce) and 2,5-pyridinedicarboxylic acid, which were called lanthanide–organic frameworks, with prominent channels running parallel to the [001] direction containing disordered charge-balancing chloride anions and water molecules of crystallisation. Ma *et al.*[358] reported the microwave-assisted synthesis of iron-based MIL-88B and NH_2-MIL-88B microcrystals with high dispersibility and uniform size. By carefully controlling the reaction conditions, high-quality MIL-88B crystals with a spindle-like morphology and high yields and high phase purity could be obtained.

MOF membranes are promising as molecular sieve materials for gas separation. Bux *et al.*[360] reported the microwave-assisted solvothermal synthesis of MOF membranes. They selected ZIF-8 (zeolitic imidazolate framework) for study. ZIF-8 has the formula $Zn(mim)_2$ (mim = 2-methylimidazolate) and crystallises with a sodalite-related structure. Due to the narrow size of the six-membered ring pores (~3.4 Å), the ZIF-8 membrane could separate H_2 (kinetic diameter ~2.9 Å) from larger molecules. In addition, ZIF-8 is hydrophobic, whereas microporous zeolites are usually hydrophilic. This feature provided the ZIF-8 membrane with an advantage over other zeolites in the separation of H_2 from steam. They prepared the ZIF-8 membrane on a titania support in methanol containing $ZnCl_2$, 2-methylimidazole, and sodium formate. The as-prepared ZIF-8 membrane exhibited superior properties and could achieve a fine balance between flux and selectivity.

MOFs can be transformed into various inorganic materials, such as carbon–metal composites and metal oxides, by thermal treatment under specific atmosphere, and hollow or core–shell nanostructures can be obtained.[355] For example, Cuo *et al.*[357] reported NiO yolk–shell structures and CuO@NiO multilayer hollow microspheres using Ni-BTC and Cu,Ni-BTC as the template, respectively. (1) Ni-BTC was first prepared in DMF containing $Ni(NO_3)_2$ and H_3BTC under microwave irradiation. Then, Cu,Ni-BTC was obtained after microwave heating of the DMF solution containing $Cu(NO_3)_2$ and the Ni-BTC precursor. Finally, metal oxide structures could be produced through thermal treatment. (2) The formation mechanism was proposed (Figure 4.30a). For the synthesis of NiO yolk–shell structures, Ni-BTC was used as the template, and the yolk–shell structures of NiO were obtained by decomposition of Ni-MOFs. (3) For the synthesis of CuO@NiO multilayer hollow microspheres, Cu,Ni-BTC was used as the template. During the decomposition process by thermal treatment, the Cu,Ni-BTC precursor was totally transformed to CuO and NiO, and three-layer hollow CuO@NiO microspheres were obtained. In another work, Guo *et al.*[354] prepared Fe-based MOFs, MIL-53(Fe), and MOF-derived Fe_2O_3 nanostructures. Various porous Fe_2O_3 nanostructures, including spindles, concave octahedra, solid octahedra, yolk–shell octahedra, and nanorods, were obtained by simply adjusting the microwave irradiation time. As shown in Figure 4.30b, the microwave heating time played an important role in the formation of the MIL-53(Fe) precursor, which also further influenced the morphology of MOF-derived iron oxides. In the structure of MIL-53(Fe), a one-dimensional Fe–O zigzag chain was formed along the *a* axis, which further constructed a 3D network under the linkage of BDC ligands. Therefore, the MIL-53(Fe) was inclined to form spindle or octahedron morphologies, and the rhomb-shaped channels in the

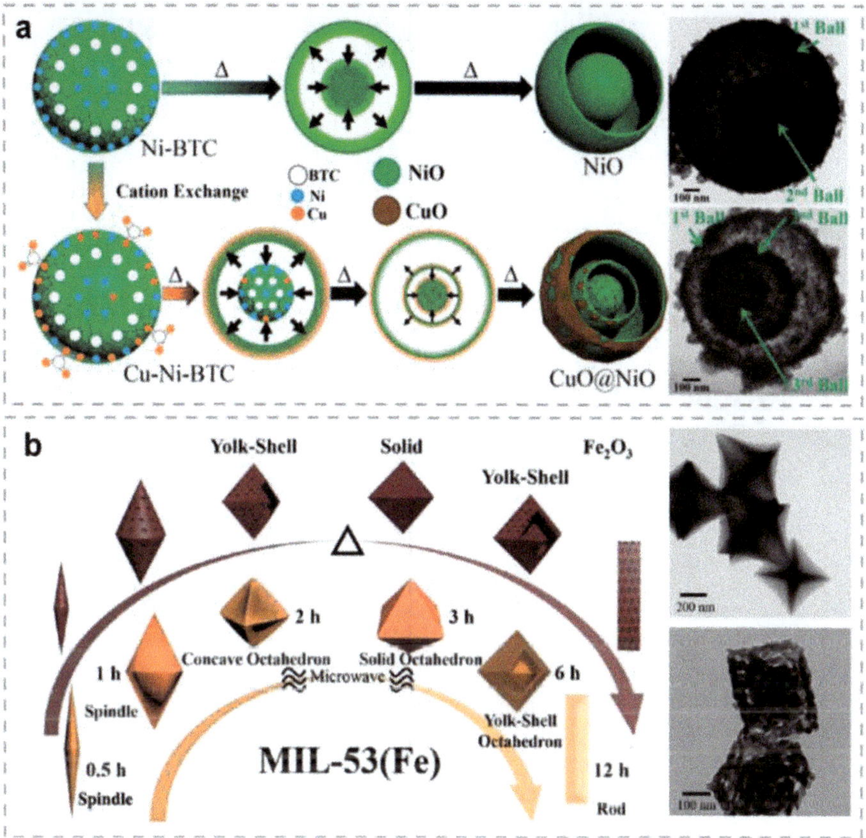

Figure 4.30 (a) Schematic illustration of preparation procedures for MOF-derived NiO yolk–shell structure and CuO@NiO multilayer hollow microspheres and the corresponding TEM images. (b) Schematic illustration of preparation procedures for MIL-53(Fe) and MOF-derived Fe_2O_3 nanostructures, and TEM images showing the as-prepared MIL-53(Fe)-2 (top) and Fe_2O_3-2 (bottom). (a) Reproduced from ref. 357 with permission from Ameican Chemical Society, Copyright 2015. (b) Reproduced from ref. 354 with permission from American Chemical Society, Copyright 2017.

skeleton of MIL-53(Fe) led to the porosity of the derived Fe_2O_3 structures. TEM images (Figure 4.30b) showed the MIL-53(Fe)-2 sample with a concave octahedron morphology and a size around 600 nm, and the Fe_2O_3-2 sample with a particle-assembled octahedron-in-octahedron yolk–shell structure.

4.3.4.7 Other Inorganic Nanostructures

The microwave-assisted preparation of other inorganic nanostructures has also been investigated, and some examples are listed in Table 4.21.

Table 4.21 Some examples of other inorganic nanostructures prepared by microwave heating[a]

Material	Precursor	Solvent	Additive	Nanostructure	Reference
CuC_2O_4	$CuCl_2$, $K_2C_2O_4$	Water	SDB	Square nanostructures	361
$BaCO_3$	$BaCl_2$, $NaHCO_3$	Water	SDS	Nanosheet-assembled structures	362
$Ca_6Si_6O_{17}(OH)_2$	$Ca(NO_3)_3$, Na_2SiO_3	Water		Nanobelts	363
$Bi_6O_6(OH)_3(NO_3)_3$	$Bi(NO_3)_2$, NaOH	Water		Nanosheets	364
$CaSb_2O_5(OH)_2$	$CaCl_2$, $K_2H_2Sb_2O_7$	Water		Particles	365
FeTi coordination complex	$Ti(OC_4H_9)_4$, $FeSO_4$	EG		Yolk–shell microspheres	366
$SrCO_3$, $BaCO_3$	$Sr(NO_3)_2$, $Ba(NO_3)_2$, Na_2CO_3	EG		Particles	367
Li_2FeSiO_4, Li_2MnSiO_4	Tetraethyl orthosilicate, LiOH, $Fe(CH_3COO)_2$, $Mn(CH_3COO)_2$	TEG		Particles	368
$NH_4Fe_2(PO_4)_2OH{\cdot}2H_2O$	$Fe(NO_3)_3$, $(NH_4)_2HPO_4$	Water, [BMIM][BF_4]		Double-shelled hollow microspheres	369
$AlF_x(OH)_{3-x}{\cdot}nH_2O$	$AlCl_3$, $CO(NH_2)_2$	Water, [BMIM][BF_4]		Octahedra	370

[a]SDB: sodium dodecylbenzene sulfonate; TEG: tetraethylene glycol.

CuC_2O_4. Qiu *et al.*[361] reported the microwave-assisted synthesis of CuC_2O_4 square nanostructures in aqueous solution containing $CuCl_2$, $K_2C_2O_4$, and sodium dodecylbenzene sulfonate (SDB). The as-prepared CuC_2O_4 square nanostructures consisted of a large number of assembled primary nanoparticles.

$BaCO_3$. Zhang *et al.*[362] reported the microwave-assisted synthesis of $BaCO_3$ nanosheet-assembled structures starting from an aqueous solution containing $BaCl_2$, $NaHCO_3$, and SDS. The experiments indicated that SDS played an important role in the formation of the $BaCO_3$ nanosheet-assembled structures.

$Bi_6O_6(OH)_3(NO_3)_3{\cdot}1.5H_2O$. Xie *et al.*[364] reported the microwave-assisted hydrothermal synthesis of $Bi_6O_6(OH)_3(NO_3)_3{\cdot}1.5H_2O$ nanosheets in aqueous solution containing $Bi(NO_3)_3$ and NaOH at 160 °C for 1 h. The thickness of the nanosheets was 40–70 nm. Compared with the sample synthesised *via* the conventional hydrothermal method, the sample synthesised by microwave heating had the advantage of shorter reaction time, sheet-like morphology, and larger specific surface area, which resulted in higher photocatalytic activity for the degradation of dyes.

$CaSb_2O_5(OH)_2$. Sun *et al.*[365] reported the microwave-assisted synthesis of $CaSb_2O_5(OH)_2$ nanocrystals in aqueous solution containing $CaCl_2$ and $K_2H_2Sb_2O_7$. The photocatalytic activity was evaluated by the decomposition of rhodamine B, methyl orange, and methylene blue in aqueous solution under UV irradiation. The experimental results showed that $CaSb_2O_5(OH)_2$ nanocrystals had a much higher photocatalytic activity than P25.

FeTi Coordination Complex. Shi *et al.*[366] reported the microwave-assisted synthesis of FeTi yolk–shell microspheres as the catalyst for removal of arsenic and for chemical degradation of methylene blue in polluted water. The synthesis was carried out in EG containing $Ti(OC_4H_9)_4$ and $FeSO_4$ by microwave heating at 180 °C for 30 min. Solid microspheres were first formed, and then morphology evolution occurred from solid microspheres to yolk–shell structures and to hollow microspheres.

4.3.5 Nanocomposites

The term nanocomposites refers to multiphase materials in which at least one phase is a nanostructured material. The nanocomposites are typically endowed with properties that outperform their individual components. The microwave-assisted method is especially attractive for preparing various kinds of nanocomposites in terms of simplicity, speed, high efficiency, low cost, and energy saving.

4.3.5.1 Inorganic/Inorganic Nanocomposites

The microwave-assisted synthesis of inorganic/inorganic nanocomposites has been widely reported in the literature, and some examples are listed in Table 4.22.

Table 4.22 Some examples of inorganic/inorganic nanocomposites synthesised by microwave heating[a]

Material	Precursor	Solvent	Additive	Nanostructure	Reference
Carbon-based nanocomposites					
Ag/C	$Ag(NH_3)_2^+$, porous carbon spheres	Water	PVP	Particles, spheres	371
Se/C	H_2SeO_3, starch	Water		Core/shell	372
Pt/C	H_2PtCl_6, carbon	EG		Particles	373
Pt–Cu/C	H_2PtCl_6, $CuCl_2$, carbon	Water, EG	NaOH	Particles	374
SnO_2/C	$SnCl_4$, graphite	Water	Urea	Particles	375
SnO_2/C	$SnCl_2$, carbon	EG		Particles	376
MnO/C	$Mn(CH_3COO)_2$, sodium citrate	EG		Particles	377
Pt/CNTs	H_2PtCl_6, CNTs	EG		Particles/nanotubes	378
GO/CNTs	CNTs, $KMnO_4$	H_2SO_4/H_3PO_4		Nanotubes, nanosheets	379
ZnS/CNTs	CNTs, $Zn(CH_3COO)_2$, thioacetamide	Water		Nanotubes, particles	380
Bi_2Te_3/CNTs	$Bi(NO_3)_3$, Na_2TeO_3, CNTs	EG		Particles/nanotubes	381
Pd/graphene	$Pd(NO_3)_2$, GO, HNO_3, hydrazine hydrate	Water		Particles, nanosheets	382
Ru, Rh/graphene	Metal carbonyl, graphene	[BMIM][BF_4]		Particles/nanosheets	383
TiO_2/graphene	Graphene, $TiOSO_4$	*tert*-Butyl alcohol		Particles, nanosheets	384
Cu_2O/RGO	$Cu(NO_3)_2$, GO	Water, ethanol	Formic acid	Microspheres, nanosheets	385
SnO_2/RGO	K_2SnO_3, GO, urea	Water, EG		Particles, nanosheets	386
CdS/RGO	$Cd(NO_3)_2$, thiourea, GO	Water		Particles, nanosheets	387
MoS_2/RGO	GO, $(NH_4)_2MoS_4$	DMF		Nanosheets	388
CoS_2/RGO	$Co(NO_3)_2$, $Na_2S_2O_3$, GO	Water, EG		Particles, nanosheets	389
$Ni(OH)_2$/graphene	$Ni(NO_3)_2$, graphene, urea	Water		Flowers, nanosheets	390
UIO-66/graphene	Graphene, 2-amino-terephthalic acid, $ZrCl_4$	DMF		Particles, nanosheets	391
Ag/$ZnFe_2O_4$/RGO	$FeCl_3$, $Zn(CH_3COO)_2$, $AgNO_3$, GO, hydrazine hydrate	Water		Particles, nanosheets	392

Metal oxide-based nanocomposites					
Au/ZnO	$Zn(NO_3)_2$, $HAuCl_4$	Water	HMTA, tri-sodium citrate	Hollow doughnut-like nanostructures	393
Ag/ZnO	$Zn(NO_3)_2$, $AgNO_3$	Water	Citric acid	Particles	394
Ag/ZnO	$Zn(NO_3)_2$, $AgNO_3$	Water, EG	CTAB	Particles, flowers	395
Ag/TiO_2	$AgNO_3$, TiO_2	Water		Particles	396
Pt/TiO_2	$Ti(SO_4)_2$, H_2PtCl_6, urea	Water	NH_4HF_2,	Hollow chains	397
C quantum dots/TiO_2	Thiourea, citric acid, TiO_2	Water		Particles	398
C/MnO	$Mn(CH_3COO)_2$	EG	PVP	Nanodiscs	399
Bi/Bi_2O_3	$BiCl_3$	EG	Glucose	Spheres	400
Ag/SiO_2	$AgNO_3$, subsequent coating of SiO_2 shell	EG	PVP	Core/shell nanoparticles	401
Pd/SnO_2	$SnCl_2$, $PdCl_2$, $Na_3C_6H_5O_7$	Water, ethylenediamine	NaOH	Hierarchical structures	402
Pd/TiO_2, ZrO_2, ZnO, SiO_2	K_2PdCl_6, TiO_2, ZrO_2, ZnO, SiO_2, hydrazine hydrate	Water, ethanol		Particles	403
Cu-Doped iron oxide	$FeCl_3$, $FeCl_2$, $CuCl_2$, $NH_3 \cdot H_2O$	Water		Particles	404
Al_2O_3/ZrO_2	$AlCl_3$, $ZrOCl_2$	Water		Particles	405
TiO_2/ZrO_2	N-Butanol zirconium, butyl titanate	Ethanol	CMCS	Particles	406
$NiCo_2O_4$/NiO	$Ni(NO_3)_2$, $Co(NO_3)_2$	Water	Oxalic acid	Bundle-like aggregates	407
CdO/NiO	$Cd(CH_3COO)_2$, $Ni(CH_3COO)_2$	Water	NaOH	Rod-like aggregates	408
CdO/NiO/ZnO	$Cd(CH_3COO)_2$, $Ni(CH_3COO)_2$, $Zn(CH_3COO)_2$	Water		Sheet-like structures	409
ZnO/CuO, Fe_3O_4/CuO	$Zn(CH_3COO)_2$ or $FeSO_4$, NaOH	Water		Particles, nanowire arrays	410
α-Fe_2O_3/$LaFeO_3$	$La(NO_3)_3$, $Fe(NO_3)_3$	Water	Citric acid	Porous and hollow spheres	411
Metal chalcogenide-based nanocomposites					
ZnO/ZnS	$Zn(NO_3)_2$, thiourea	Water	Dodecyl-amine	Hollow core/shell nanostructures	412
ZnO/ZnSe	$Zn(NO_3)_2$, Na_2SeO_3	Water	HMTA/L-(+)-ascorbic acid/hydrazine hydrate	Porous nanostructures	413

(continued)

Table 4.22 (*Continued*)

Material	Precursor	Solvent	Additive	Nanostructure	Reference
CdSe/ZnS	$CdCl_2$, Se powder, $ZnCl_2$, MPA	Water	$NaBH_4$/$NH_3{\cdot}H_2O$	Core/shell quantum dots	414
CdTe/CdS	CdTe, $CdCl_2$, Na_2S, MPA	Water		Core/shell nanostructures	415
Mn:ZnSe/ZnS	$Zn(NO_3)_2$, $MnCl_2$, Se powder, MPA, Na_2S, $Zn(CH_3COO)_2$	Water	Sodium oleate, $NaBH_4$	Core/shell nanostructures	416
Sb-PbTe/Ag_2Te	$SbCl_3$, $AgNO_3$, $Pb(CH_3COO)_2$, Na_2TeO_3	EG	$NaBH_4$	Core/shell nanocubes	417
CoFe/Co_8FeS_8/CNT	Co–Fe-Prussian blue analogues, thioacetamide	EG		Nanoparticles, nanotubes	418
C_3N_4-based nanocomposites					
Ag_2SO_4/C_3N_4	C_3N_4, $AgNO_3$, Na_2SO_4	Water		Particles, sheets	419
Fe_2O_3/C_3N_4	C_3N_4, Fe_2O_3	Water, ethanol	SDS, NaOH	Aggregates	420
ZnO/C_3N_4	C_3N_4, $Zn(NO_3)_2$	Water	NaOH	Nanosheets	421
$Pt/ZnO/C_3N_4$	C_3N_4, $Zn(NO_3)_2$, H_2PtCl_6, NaOH, $NaBH_4$			Particles, nanorod arrays, nanosheets	422
Bi_2WO_6/RGO/C_3N_4	C_3N_4, GO, $Bi(NO_3)_3$, Na_2WO_4	Water	CTAB	Nanosheets	423
Other inorganic/inorganic nanocomposites					
$SiO_2/Cd(OH)_2$	$KCdCl_3$ nanowires, ethyl silicate	Ethanol	NaOH	Nanotubes	424
MoS_2/PDA/Ag	MoS_2-PDA, $AgNO_3$	Water		Particles, nanosheets	425
BiF_3/BiOBr	KBr, NaF, $Bi(NO_3)_2$	Water	Acetic acid	Flowers	426

[a]RGO: reduced graphene oxide; MPA: 3-mercaptopropionic acid; CMCS: carboxymethyl chitosan; PDA: polydopamine.

4.3.5.1.1 Carbon-based Nanocomposites. *Ag/C*. Tang *et al.*[371] reported the microwave-assisted preparation of Ag/C nanocomposite in aqueous solution containing $Ag(NH_3)_2^+$ ions, nanoporous carbon spheres, and PVP within a short reaction time. By controlling the Ag^+ doping, the composite spheres showed not only a tuneable plasmon resonance shift, but also a high catalytic activity towards the reduction of 4-nitrophenol by sodium borohydride.

Pt–Cu/C. Song *et al.*[374] reported the rapid one-step synthesis of carbon-supported platinum–copper nanoparticles *via* microwave heating of a solution containing H_2PtCl_6, $CuCl_2$, NaOH, carbon, EG, and water. Pt_3Cu nanoparticles with an average diameter of 2.31 nm were uniformly dispersed on the carbon support. The as-prepared Pt–Cu/C nanocomposite had a specific surface area of 483.57 $m^2\ g^{-1}$ and a pore diameter of 12.41 nm. The Pt–Cu/C nanocomposite exhibited a higher performance for the ethanol oxidation reaction and the methanol oxidation reaction compared with Pt/C. For more examples of Pt/carbon, please refer to Section 4.3.1.4.

SnO_2/C. The SnO_2-graphite nanocomposite was prepared by urea-mediated homogeneous hydrolysis of $SnCl_4$ under microwave heating.[375] The as-prepared SnO_2 nanoparticles had a narrow size distribution (1–3 nm, mean: 2.1 nm). Compared with conventional heating, the microwave-assisted synthesis produced smaller SnO_2 nanoparticles that were homogenously dispersed on graphite, resulting in improved electrochemical performance as a lithium storage compound.

Inorganics/graphene. Graphene and its derivatives are an excellent matrix for loading nanostructured materials to form high-performance nanocomposites. They have ultrathin thickness with abundant accessible active sites and unique electrical and thermal properties, enabling wide applications in catalysis, sensors, and energy storage. Many nanostructured materials have been incorporated with graphene derivatives to prepare nanocomposites with the help of microwave irradiation (Figure 4.31a): for example, metals, including Pd,[382] Ru,[383] Rh,[383] Au,[427] Ag;[392] metal oxides, including TiO_2,[384,427] Cu_2O,[385] CuO,[428] Fe_2O_3,[429,430] MnO_2,[431] SnO_2,[386,432] Sn_3O_4,[433] $ZnFe_2O_4$,[392] Bi_2WO_6;[423,434] metal chalcogenides, including CdS,[387,435] ZnS,[436] MoS_2,[388,437,438] CoS_2;[389] and other nanostructured materials, including C_3N_4,[423] $Ni(OH)_2$,[390] Ni,Co-LDH,[439] MOFs.[391] Some typical examples are reviewed below. See also Chapter 5 for a comprehensive survey on graphene/inorganic based nanocomposites.

Pd/graphene. Siamaki *et al.*[382] reported the microwave-assisted preparation of Pd/graphene nanocomposites in aqueous solution containing $Pd(NO_3)_2$, graphene oxide, HNO_3, and hydrazine hydrate. Highly active Pd nanoparticles (7–9 nm) were well dispersed on the surface of the graphene sheets. The as-prepared Pd/graphene nanocomposite exhibited high catalytic activity for the carbon–carbon cross-coupling reactions (Suzuki and Heck) under ambient conditions.

TiO_2/graphene. Le *et al.*[384] reported the microwave-assisted solvothermal synthesis of mesoporous anatase TiO_2 nanostructures anchored on graphene. In the nanocomposite synthesis, graphene was dispersed in *tert*-butyl

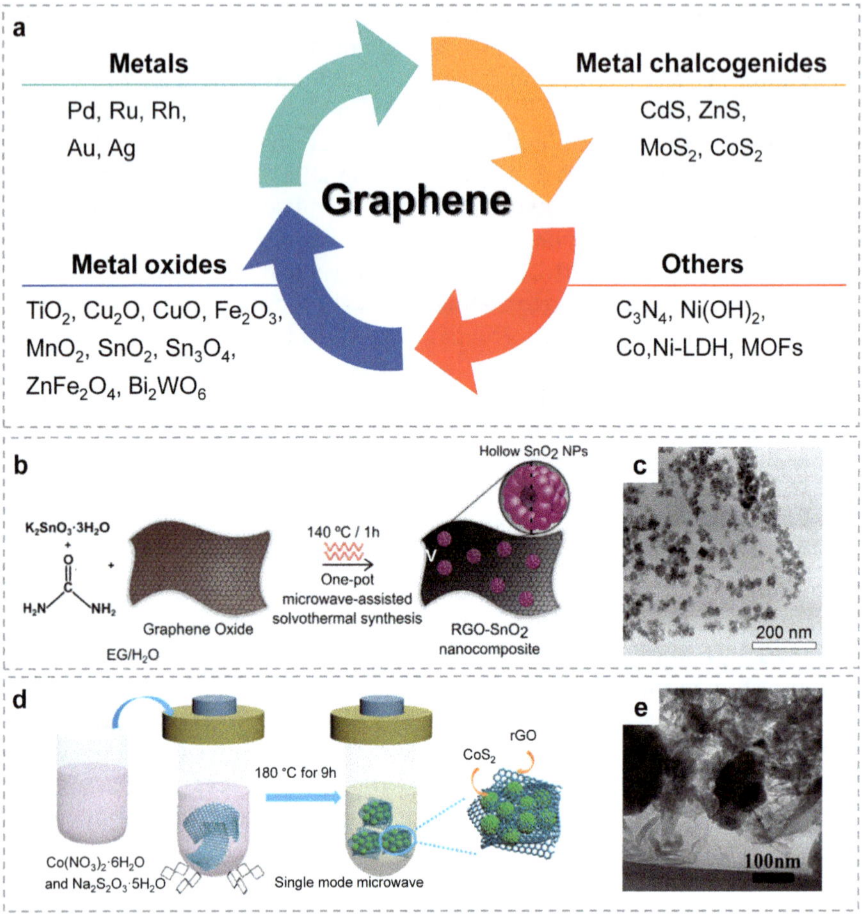

Figure 4.31 (a) Schematic illustration of inorganic/graphene nanocomposites. (b) Schematic illustration of the synthetic procedure for SnO_2/RGO nanocomposite. (c) TEM image of SnO_2/RGO nanocomposite. (d) Schematic illustration of the synthetic procedure for CoS_2/RGO nanocomposite. (e) TEM image of CoS_2/RGO nanocomposite. (b and c) Reproduced from ref. 386 with permission from Elsevier, Copyright 2017. (d and e) Reproduced from ref. 389 with permission from American Chemical Society, Copyright 2017.

alcohol under sonication, and $TiOSO_4$ was added dropwise. The as-prepared TiO_2/graphene nanocomposite exhibited pseudocapacitive charge storage behaviour with fast kinetics, high reversibility, and high stability.

SnO_2/RGO. Zito *et al.*[386] reported a one-step microwave-assisted solvothermal method for the synthesis of SnO_2/reduced graphene oxide (RGO) nanocomposite. RGO was used as the matrix to support SnO_2 nanoparticles. As shown in Figure 4.31b, the SnO_2/RGO nanocomposite was synthesised by the microwave-assisted solvothermal method in water/EG solution containing

K_2SnO_3, urea, and GO. SnO_2 nanoparticles were homogeneously dispersed on the surface of the RGO nanosheets (Figure 4.31c). They also demonstrated the high sensing performance of the SnO_2/RGO nanocomposite for ethanol.

CoS_2/RGO. Zhang *et al.*[389] reported the microwave-assisted solvothermal synthesis of CoS_2/RGO nanocomposite in water/EG solution containing $Co(NO_3)_2$, $Na_2S_2O_3$, and GO (Figure 4.31d). A TEM image is shown in Figure 4.31e. They also demonstrated the potential application of the CoS_2/RGO nanocomposite in microwave absorption. Compared with other RGO-based materials, the CoS_2/RGO nanocomposite showed superior microwave absorption ability. The as-prepared CoS_2/RGO nanocomposite has potential to be used as a microwave absorber.

Ag/$ZnFe_2O_4$/RGO. Mady *et al.*[392] reported the microwave-assisted synthesis of the Ag/$ZnFe_2O_4$/RGO nanocomposite for efficient removal of organic dyes under UV and visible light irradiation. The microwave-assisted synthesis was carried out in an aqueous solution containing $FeCl_3$, $Zn(CH_3COO)_2$, $AgNO_3$, GO, and $N_2H_4{\cdot}H_2O$ at 100 °C under atmospheric pressure for 10 min. The Ag/$ZnFe_2O_4$/RGO nanocomposite with 15.2 wt% RGO exhibited excellent light adsorption properties and high photocatalytic activity for the degradation of methylene blue, rhodamine B, and methyl orange.

4.3.5.1.2 Metal Oxide based Nanocomposites. *Au/ZnO*. Geng *et al.*[393] reported the microwave-assisted preparation of the Au/ZnO nanocomposite in an aqueous solution containing $Zn(NO_3)_2$, $HAuCl_4$, HMTA, and trisodium citrate. Hollow doughnut-like Au/ZnO composite nanostructures were obtained. In addition, various nanostructures, such as Au nanoparticles/ZnO nanorods, Au nanoparticle/ZnO nanodiscs, and Au nanoparticles/ZnO nanospheres were prepared by adjusting the experimental conditions.

Pd/SnO_2. Wang *et al.*[402] reported the microwave-assisted solvothermal synthesis of Pd/SnO_2 nanostructures in water/ethylenediamine solution containing $PdCl_2$ and $SnCl_2$, $Na_3C_6H_5O_7$, and NaOH maintained at 180 °C for 60 min. Pd nanoparticles were grown *in situ* on SnO_2 nanoflakes to form a hierarchical structure. A gas sensor based on the as-prepared Pd/SnO_2 nanocomposite was fabricated and tested for response to carbon monoxide gas. The as-prepared gas sensor showed a fast response performance for detection of carbon monoxide gas. In addition, microwave-assisted preparations of metal/metal oxide nanocomposites, including Ag/ZnO,[394,395] Pt/TiO_2,[397] and Pd/MO_x (M = Ti, Zr, Zn, Si),[403] were also reported for various applications.

C/TiO_2. Li *et al.*[398] synthesised carbon quantum dots/TiO_2 nanocomposites by microwave heating. The carbon quantum dots were synthesised under microwave irradiation in an aqueous solution containing citric acid and thiourea. The obtained carbon quantum dots were mixed with TiO_2 particles in aqueous suspension at 80 °C for 12 h under stirring. In addition, the as-prepared nanocomposite exhibited evident enhancement of the photocatalytic activity for CO_2 photoreduction under simulated solar light irradiation.

ZnO/CuO and Fe_3O_4/CuO. Malwal *et al.*[410] reported a two-step method for the synthesis of metal oxides (ZnO and Fe_3O_4) deposited on CuO nanowire

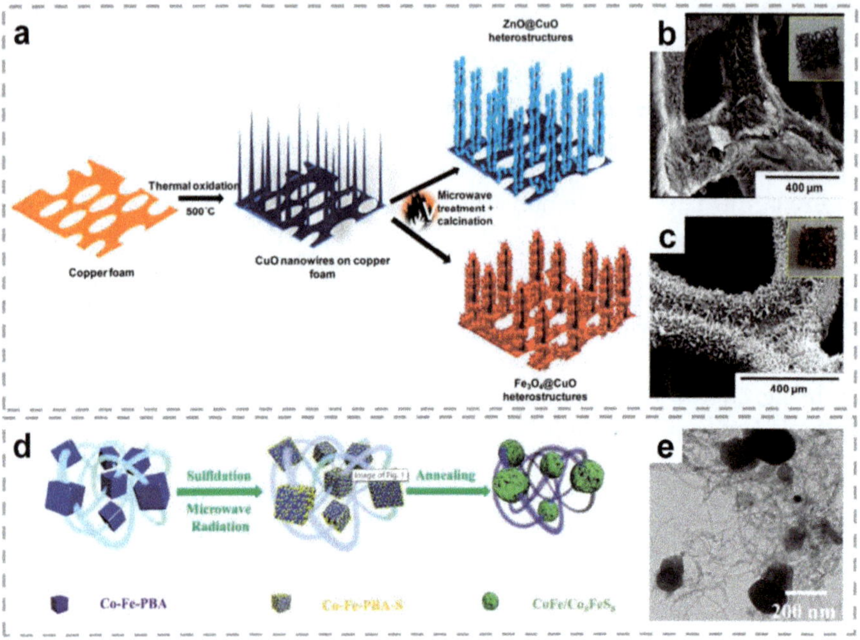

Figure 4.32 (a) Schematic illustration of preparation procedure for ZnO@CuO and Fe_3O_4@CuO nanocomposites. (b) SEM image of the ZnO/CuO nanocomposite and corresponding photograph of the sample. (c) SEM image of Fe_3O_4/CuO nanocomposite and corresponding photograph of the sample. (d) Schematic illustration of preparation procedure for CoFe/Co_8FeS_8/CNT nanocomposite. (e) TEM image of CoFe/Co_8FeS_8/CNT nanocomposite. (a–c) Reproduced from ref. 410 with permission from the Royal Society of Chemistry. (d and e) Reproduced from ref. 418 with permission from Elsevier, Copyright 2019.

arrays over 3D copper foam by thermal oxidation, followed by microwave-assisted deposition. As shown in Figure 4.32a, CuO nanowire arrays were grown on the copper foam by thermal oxidation. For the preparation of ZnO/CuO and Fe_3O_4/CuO nanocomposites, CuO nanowires were decorated with the metal oxide (ZnO or Fe_3O_4) by microwave heating using $FeSO_4$ or $Zn(CH_3COO)_2$ and NaOH in distilled water. Then, the sample was heated at 500 °C for 1 h in air (Figure 4.32b and c). The as-prepared nanocomposite exhibited improved photocatalytic activity for dye degradation compared with the individual components under solar irradiation.

α-Fe_2O_3/$LaFeO_3$. Zhang *et al.*[411] reported the microwave-assisted synthesis of porous and hollow α-Fe_2O_3/$LaFeO_3$ nanocomposite for acetone gas sensing and photocatalytic degradation of methylene blue. The synthesis was carried out in an aqueous solution containing $La(NO_3)_2$, $Fe(NO_3)_3$, and citric acid, followed by thermal treatment at 800 °C for 2 h. With increasing microwave heating time, the morphology of the nanocomposite changed from smooth solid microspheres, to hollow microspheres, to nanoparticle aggregates, and to net-like nanostructures. The as-prepared nanocomposite was investigated as a sensor for acetone gas: the optimised sample showed excellent performance, including a fast response, high selectivity, good reversibility, and good recoverability. In addition, the as-prepared nanocomposite was an efficient catalyst for the degradation of methylene blue.

4.3.5.1.3 Metal Chalcogenide-based Nanocomposites. *ZnO/ZnS*. Chen *et al.*[412] reported the microwave-assisted synthesis of the hollow ZnO/ZnS core–shell nanocomposite in an aqueous solution containing $Zn(NO_3)_2$, thiourea, and dodecylamine. ZnO truncated hexagonal pyramids first formed on the precursor flakes and then evolved into ZnO hexagonal twin crystals, which subsequently grew and dissolved internally. The hollowing process was first controlled by the Kirkendall effect and then underwent an Ostwald ripening process. In addition, the ZnO/ZnS core–shell nanocomposite exhibited enhanced photocatalytic activity and good photocatalytic stability compared with single ZnO or ZnS under UV light irradiation.

CdTe/CdS. He *et al.*[415] reported the microwave-assisted synthesis of CdTe/CdS core–shell nanocrystals in an aqueous solution containing CdTe nanocrystals, $CdCl_2$, Na_2S, and MPA. The photoluminescence quantum yield could be enhanced by the epitaxial growth of the CdS shell. A short time (5 min) was required to form an optimum thickness of the CdS shell in the microwave-assisted synthesis.

CoFe/Co_8FeS_8/CNTs. Wang *et al.*[418] reported CoFe/Co_8FeS_8 nanoparticles embedded in a CNT network, which were synthesised *via* three steps with the help of microwave-assisted sulfidation. As shown in Figure 4.32d, the Co–Fe-Prussian blue analogue (PBA)/CNT composites were first prepared by a co-precipitation reaction of $Co(NO_3)_2$, trisodium citrate, potassium ferrocyanide, and CNTs in deionised water. The obtained composites were dispersed in EG containing thioacetamide, followed by microwave heating. Finally, thermal treatment was applied in argon to obtain the CoFe/Co_8FeS_8/CNT nanocomposite (Figure 4.32e). Compared with CoFe/CNT or Co_8FeS_8/CNT, the as-prepared CoFe/Co_8FeS_8/CNT nanocomposite exhibited a better performance for the oxygen evolution reaction and higher stability.

4.3.5.2 Inorganic/Organic Nanocomposites

Generally, the preparation of inorganic/organic nanocomposites is performed by incorporating inorganic particles into organic polymers. Using this method the dispersion of inorganic particles in the polymer matrix is usually poor. One-step preparation of high-quality inorganic/organic nanocomposites is desirable and microwave heating is an ideal technique to solve this problem. In recent years, the microwave-assisted synthesis of inorganic/organic nanocomposites has been investigated, and various kinds of inorganic/organic nanocomposites have been prepared. Some examples of microwave-assisted synthesis of inorganic/organic nanocomposites are listed in Table 4.23.

Table 4.23 Some examples of inorganic/organic nanocomposites synthesised by microwave heating[a]

Material	Precursor	Solvent	Additive	Nanostructure	Reference
Ag, Pt, Cu/PAM	$AgNO_3$, $PtCl_6$, or $CuSO_4$, acrylamide	EG		Nanoparticles	46
Ag_2S, Cu_2S, HgS/PAM	$AgNO_3$, $Cu(NO_3)_2$, or $Hg(NO_3)_2$, S powder, acrylamide	EG		Nanoparticles	440
Calcium phosphate/PAM	$CaCl_2$, $(NH_4)_2HPO_4$, $(NH_4)_2S_2O_8$, acrylamide	Water	Ammonia	Nanostructures	441
Pt, Cu, In, Pt–In, Ag–Pt, Pt–Fe, Cu–Pd, Pd–Fe/PVA	Na_2PtCl_6, $CuCl_2$, $InCl_3$, $AgNO_3$, $Fe(NO_3)_3$, $PdCl_2$, PVA	Water		Nanospheres, nanodendrities, nanocubes	442
Pd, Cu, Pd–Cu/MIL-101	$Pd(NO_3)_2$, $Cu(NO_3)_2$, hydrazine hydrate, MIL-101	Water		Particles	443
Cu, Ag, In, Fe/CMC	$CuCl_2$, $AgNO_3$, $InCl_3$, $FeCl_3$, CMC	Water		Particles	444
Ag/polycarbazole	$AgNO_3$, carbazole	EG	PVP/Cu^{2+}	Particles	445
V_2O_5/polyaniline	V_2O_5, aniline	Water		Nanosheets	446
SiO_2/polyaniline	SiO_2, aniline, ammonium persulfate	Water	HCl	Particles	447
Fe_3O_4/PF	Fe_3O_4, phenol, formaldehyde	Water		Core/shell spheres	448
Fe_3O_4/polyacid	$FeCl_3$, CH_3COONa, polyacrylic acid, EG			Nanostructures	449
Fe_3O_4/polypyrrole	$FeCl_3$, pyrrole	EG	P123, CH_3COONa, NaDBS	Core/shell nanostructures	450
CdS/bacterial cellulose	$CdCl_2$, bacterial cellulose, thiourea	Water, ethanol		Particles, fibres	451

[a]CMC: carboxymethyl cellulose; PF: phenol formaldehyde; NaDBS: sodium dodecylbenzenesulfonate-6; PAM: polyacrylamide; MIL: Materials Institute Lavoisier; P123: poly(ethylene glycol)-*block*-poly(propylene glycol)-*block*-poly(ethylene glycol).

4.3.5.2.1 One-step Formation of Inorganic/Organic Nanocomposites. *Metal/PAM.* Good dispersion of inorganic nanoparticles in the organic matrix can significantly enhance the optical, electrical, and mechanical properties of the as-prepared nanocomposites. Generally, nanocomposites are prepared from the dispersion of inorganic nanoparticles in polymer solution by solvent evaporation or co-precipitation. However, aggregation of the inorganic nanoparticles is inevitable during the preparation process. In order to solve this problem, Zhu's research group developed a one-step microwave-assisted strategy for simultaneous formation of inorganic nanoparticles and polymerisation of the organic monomer in solution.[46] For example, they reported the microwave-assisted rapid synthesis of metal (Ag, Pt, Cu) nanoparticle/PAM nanocomposites in EG containing metal salts and acrylamide monomer. The microwave-assisted synthesis was rapid and took only 15 min. Ethylene glycol acted as both the reducing reagent and solvent, thus no additional reductant was needed. Another advantage was that the initiator for acrylamide (AM) polymerisation and the surfactant for stabilisation of the metal nanoparticles were not necessary. Metal nanoparticles with a narrow size distribution were homogeneously dispersed in the PAM matrix (Figure 4.33a).

Metal Sulfide/PAM. Zhu's research group extended the one-step microwave-assisted strategy to the simultaneous formation of metal sulfide nanoparticles and polymerisation of the organic monomer in solution, and they prepared various metal sulfide (Ag_2S, Cu_2S, HgS)/PAM nanocomposites.[440] Figure 4.33b shows a TEM image of the as-prepared Ag_2S/PAM nanocomposite. The spherical Ag_2S nanoparticles were homogeneously dispersed in the polyacrylamide matrix. The selected-area electron diffraction pattern indicates the single-crystalline structure of the Ag_2S nanoparticles. The Ag_2S

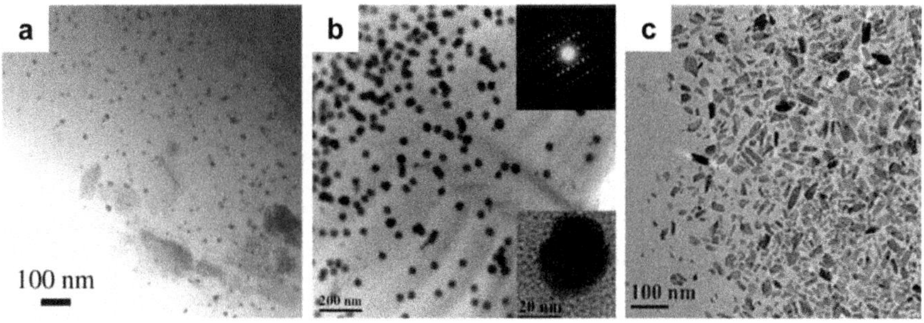

Figure 4.33 (a) TEM image of Ag/polyacrylamide nanocomposite. (b) TEM image of Ag_2S/polyacrylamide nanocomposite. Upper inset: selected-area electron diffraction; lower inset: magnified TEM image of an individual nanoparticle. (c) TEM image of HAP/PAM nanocomposite. (a) Reproduced from ref. 46 with permission from American Chemical Society, Copyright 2006. (b) Reproduced from ref. 440 with permission from American Chemical Society, Copyright 2007. (c) Reproduced from ref. 441 with permission from Elsevier, Copyright 2009.

nanoparticles had a narrow size distribution in the range 36–54 nm and an average diameter of ~43.8 nm.

Calcium Phosphate/PAM. The one-step microwave-assisted synthesis of calcium phosphate/PAM nanocomposite was also reported in an aqueous solution containing $CaCl_2$, $(NH_4)_2HPO_4$, $(NH_4)_2S_2O_8$, ammonia, and acrylamide.[441] It was found that: (1) when the reaction was carried out in basic medium, the calcium phosphate phase consisted of HAP nanorods, and HAP nanorods were homogeneously dispersed in the PAM matrix (Figure 4.33c); (2) when the reaction was carried out in weak acidic medium, amorphous calcium phosphate nanoparticles were formed and dispersed in the PAM matrix.

4.3.5.2.2 Other Nanocomposites. Nadagouda and Varma[442] reported the microwave-assisted preparation of metallic and bimetallic crosslinked PVA nanocomposites. Nanocomposites of PVA crosslinked metallic systems, such as Pt, Cu, and In, and bimetallic systems, such as Pt–In, Ag–Pt, Pt–Fe, Cu–Pd, Pt–Pd, and Pd–Fe, with various morphologies, including nanospheres, nanodendrites, and nanocubes, were prepared by reacting the respective metal salt with PVA under microwave irradiation.

You *et al.*[448] reported the microwave-assisted preparation of magnetic Fe_3O_4/phenol formaldehyde (PF) core–shell microspheres by *in situ* polycondensation of phenol and formaldehyde with magnetic Fe_3O_4 nanoparticles in an aqueous solution without any surfactant. It was found that the PF shell was evenly coated on Fe_3O_4 nanoparticles within 10 min under microwave irradiation, and the PF shell thickness could be tuned in the range of 10–200 nm by adjusting the synthetic parameters, such as the amounts of phenol and formaldehyde, the reaction temperature, and the microwave heating time. The as-prepared nanocomposites exhibited high dispersibility in water, ethanol, acetone, and DMF. Karim *et al.*[447] reported the preparation of nanocomposites consisting of conducting polyaniline and SiO_2. The synthesis was carried out under microwave irradiation in HCl solution containing SiO_2 powder, aniline monomer, and ammonium persulfate. The as-prepared SiO_2/polyaniline nanocomposite was investigated as an electrochemical sensor for 1,4-dioxane detection.

Qian *et al.*[451] reported the microwave-assisted solvothermal *in situ* synthesis of CdS nanoparticles on bacterial cellulose (BC) matrix for the photocatalytic degradation of methylene blue. The BC/CdS nanocomposite was prepared through a designed 'anchoring-reacting-forming' route. CdS nanoparticles were attached to BC fibres *via* the coordination effect. The as-prepared BC/CdS nanocomposite exhibited high photocatalytic activity under visible light irradiation.

El-Shall *et al.*[443] incorporated metallic and bimetallic nanocrystals into highly porous metal–organic framework MIL-101 under microwave irradiation. Pd, Cu, and Pd–Cu nanoparticles (2–3 nm) were incorporated within the pores and larger 4–6 nm particles were supported on the surface of the MIL-101 crystals. The as-prepared Pd/MIL-101 nanocomposite exhibited high catalytic activity towards CO oxidation.

4.4 Conclusions and Prospects

It this chapter, we started with a concise background of microwave chemistry, followed by discussion of microwave heating mechanisms and microwave effects. The roles of the microwave synthetic parameters, including the microwave frequency, fixed or variable frequency, microwave power, heating type, reaction vessel, solvent type, temperature, heating time, and reaction system, were discussed. In addition, developments in the microwave heating apparatus, such as the multimode cavity, single-mode cavity, reactor, and temperature measurement, have been reviewed. In particular, the comparison between the microwave heating and conventional heating has been emphasised considering the following aspects: energy consumption, conversion ratio, crystalline phase, crystallinity, size distribution, and morphology.

In the past three decades, microwave heating has been widely applied in the preparation of inorganic nanostructured materials in the liquid phase. Herein, we provide an overview on various kinds of inorganic nanostructures and their nanocomposites prepared by microwave-assisted wet-chemistry methods.

From the literature survey, it can be rationally concluded that microwave heating is a promising technology with many unique advantages. These remarkable advantages include rapid volumetric heating, higher chemical reaction rate and selectivity, shorter reaction time, and higher product yield compared with conventional heating methods. In many cases, microwave heating can enhance the chemical reaction rate by several orders of magnitude compared with conventional heating methods. In addition, microwave-assisted synthesis can also produce inorganic nanostructured materials with novel structures and properties.[452] Therefore, microwave heating is a novel heat source and can enable rapid production of inorganic nanostructured materials in the liquid phase, leading to cost effectiveness, energy saving, and high efficiency. The microwave heating technology has emerged as a novel heat source for rapid chemical reactions and material preparation within minutes, instead of the hours, or even days, usually needed by conventional heating methods. It is expected that the use of the microwave heating technology will become more popular in many research laboratories and even be scaled up for industrial applications.[12]

Although significant advances have been achieved in the microwave-assisted rapid preparation of inorganic nanostructured materials in the liquid phase, some challenges still exist and need to be solved. For example, one major problem is that many published papers do not provide detailed experimental parameters for microwave heating, and the information on the microwave reactors and reaction containers is rarely reported. In some cases, the microwave-assisted preparation was performed in household microwave ovens with poor reproducibility and in other cases the microwave heating temperatures were not measured accurately or were not reported. Thus, these experimental reports are difficult to reproduce and compare with other research studies. Therefore, the use of microwave reactors specifically designed for chemical reactions and materials synthesis, with precise control over various experimental parameters and good repeatability, is strongly recommended. In addition, it is strongly suggested that the experimental description in publications related to microwave-assisted synthesis should be specific and detailed. Another challenge is that the detailed mechanisms for the chemical reaction, nucleation, and growth of inorganic nanostructures in the liquid phase under microwave irradiation are still not fully understood, and some important issues, such as non-thermal effects or specific microwave effects under microwave irradiation, have not been clarified. More carefully designed comparative studies of microwave heating and conventional heating are needed for further understanding of microwave heating mechanisms.[12]

Some future research trends and directions can be proposed, as follows:[12] (1) in-depth studies of the detailed mechanisms of the chemical reaction, nucleation, and growth of inorganic nanostructures in the liquid phase under microwave irradiation; (2) carefully designed comparative studies between microwave heating and conventional heating, to clarify some important issues, such as non-thermal effects or specific microwave effects under microwave irradiation, which have been controversial and are under intense debate; (3) theoretical and computer simulations of microwave effects and microwave heating mechanisms; (4) studies on the microwave-assisted precise control over the chemical composition, structure, size, morphology, and self-assembly of inorganic nanostructures; (5) development of environmentally friendly, low cost, and high-efficiency microwave-assisted preparation methods; (6) development of microwave-assisted large-scale production of inorganic nanostructured materials for industrial applications, such as continuous-flow microwave synthesis; (7) studies on the relationship between structure and properties and applications of microwave-prepared nanostructured materials;[12] (8) new developments of microwave chemistry. It is worth noting that new applications of microwave irradiation in the preparation and processing of inorganic nanostructured materials have been reported, such as the rapid (4 min) activation of MOFs by microwaves without structural damage,[453] and rapid reduction of GO into pristine graphene using 1 to 2 second pulses of microwaves and removal of the oxygen-containing groups of GO.[341] It is expected that more potential uses of microwave irradiation will be explored for high-efficiency rapid preparation and processing of inorganic nanostructures, both in laboratories and in industry.

References

1. S. Diodati, P. Dolcet, M. Casarin and S. Gross, *Chem. Rev.*, 2015, **115**, 11449.
2. M. B. Gawande, S. N. Shelke, R. Zboril and R. S. Varma, *Acc. Chem. Res.*, 2014, **47**, 1338.
3. H. J. Kitchen, S. R. Vallance, J. L. Kennedy, N. Tapia-Ruiz, L. Carassiti, A. Harrison, A. G. Whittaker, T. D. Drysdale, S. W. Kingman and D. H. Gregory, *Chem. Rev.*, 2014, **114**, 1170.

4. S. Komarneni and R. Roy, *Mater. Lett.*, 1985, **3**, 165.
5. R. Gedye, F. Smith, K. Westaway, H. Ali, L. Baldisera, L. Laberge and J. Rousell, *Tetrahedron Lett.*, 1986, **27**, 279.
6. C. O. Kappe, *Angew. Chem., Int. Ed.*, 2004, **43**, 6250.
7. D. Dallinger and C. O. Kappe, *Chem. Rev.*, 2007, **107**, 2563.
8. M. Nüchter, B. Ondruschka, W. Bonrath and A. Gum, *Green Chem.*, 2004, **6**, 128.
9. D. M. P. Mingos and D. R. Baghurst, *Chem. Soc. Rev.*, 1991, **20**, 1.
10. S. A. Galema, *Chem. Soc. Rev.*, 1997, **26**, 233.
11. M. Baghbanzadeh, L. Carbone, P. D. Cozzoli and C. O. Kappe, *Angew. Chem., Int. Ed.*, 2011, **50**, 11312.
12. Y.-J. Zhu and F. Chen, *Chem. Rev.*, 2014, **114**, 6462.
13. X. Zhang, D. O. Hayward and D. M. P. Mingos, *Catal. Lett.*, 2003, **88**, 33.
14. A. G. Whittaker and D. M. P. Mingos, *J. Chem. Soc., Dalton Trans.*, 1992, 2751.
15. J.-S. Schanche, *Mol. Diversity*, 2003, **7**, 293.
16. C. O. Kappe, *Chem. Soc. Rev.*, 2008, **37**, 1127.
17. A. de la Hoz, Á. Díaz-Ortiz and A. Moreno, *Chem. Soc. Rev.*, 2005, **34**, 164.
18. I. Bilecka and M. Niederberger, *Nanoscale*, 2010, **2**, 1358.
19. M. Godinho, C. Ribeiro, E. Longo and E. R. Leite, *Cryst. Growth Des.*, 2008, **8**, 384.
20. J. Jacob, L. H. L. Chia and F. Y. C. Boey, *J. Mater. Sci.*, 1995, **30**, 5321.
21. D. Obermayer, B. Gutmann and C. O. Kappe, *Angew. Chem., Int. Ed.*, 2009, **48**, 8321.
22. J. M. Kremsner and C. O. Kappe, *J. Org. Chem.*, 2006, **71**, 4651.
23. W. C. Conner and G. A. Tompsett, *J. Phys. Chem. B*, 2008, **112**, 2110.
24. J. Robinson, S. Kingman, D. Irvine, P. Licence, A. Smith, G. Dimitrakis, D. Obermayer and C. O. Kappe, *Phys. Chem. Chem. Phys.*, 2010, **12**, 4750.
25. N. E. Leadbeater and H. M. Torenius, *J. Org. Chem.*, 2002, **67**, 3145.
26. S. Horikoshi, F. Sakai, M. Kajitani, M. Abe and N. Serpone, *Chem. Phys. Lett.*, 2009, **470**, 304.
27. S. Horikoshi, H. Abe, T. Sumi, K. Torigoe, H. Sakai, N. Serpone and M. Abe, *Nanoscale*, 2011, **3**, 1697.
28. K. A. Malinger, Y.-S. Ding, S. Sithambaram, L. Espinal, S. Gomez and S. L. Suib, *J. Catal.*, 2006, **239**, 290.
29. H. Jiang, K.-s. Moon, Z. Zhang, S. Pothukuchi and C. P. Wong, *J. Nanopart. Res.*, 2006, **8**, 117.
30. E. K. Nyutu, C.-H. Chen, P. K. Dutta and S. L. Suib, *J. Phys. Chem. C*, 2008, **112**, 9659.
31. S. K. Seol, D. Kim, S. Jung and Y. Hwu, *Mater. Chem. Phys.*, 2011, **131**, 331.
32. Y. Yang, Y. Hu, X. Xiong and Y. Qin, *RSC Adv.*, 2013, **3**, 8431.
33. P. K. Shen, S. Yin, Z. Li and C. Chen, *Electrochim. Acta*, 2010, **55**, 7969.
34. L. Bonaccorsi and E. Proverbio, *J. Cryst. Growth*, 2003, **247**, 555.
35. W. C. Conner, G. Tompsett, K.-H. Lee and K. S. Yngvesson, *J. Phys. Chem. B*, 2004, **108**, 13913.

36. H. Katsuki, S. Furuta and S. Komarneni, *Mater. Lett.*, 2012, **83**, 8.
37. S. Komarneni and H. Katsuki, *Ceram. Int.*, 2010, **36**, 1165.
38. B. Panzarella, G. A. Tompsett, K. S. Yngvesson and W. C. Conner, *J. Phys. Chem. B*, 2007, **111**, 12657.
39. J. D. Ferguson, *Mol. Diversity*, 2003, 7, 281.
40. M. Nishioka, M. Miyakawa, Y. Daino, H. Kataoka, H. Koda, K. Sato and T. M. Suzuki, *Ind. Eng. Chem. Res.*, 2013, **52**, 4683.
41. S. Hayden, M. Damm and C. O. Kappe, *Macromol. Chem. Phys.*, 2013, **214**, 423.
42. S. Horikoshi, H. Abe, K. Torigoe, M. Abe and N. Serpone, *Nanoscale*, 2010, **2**, 1441.
43. H. Katsuki and S. Komarneni, *J. Ceram. Soc. Jpn.*, 2011, **119**, 525.
44. X. Zhu, Q. Zhang, Y. Li and H. Wang, *J. Mater. Chem.*, 2010, **20**, 1766.
45. J. M. Campelo, T. D. Conesa, M. J. Gracia, M. J. Jurado, R. Luque, J. M. Marinas and A. A. Romero, *Green Chem.*, 2008, **10**, 853.
46. J.-F. Zhu and Y.-J. Zhu, *J. Phys. Chem. B*, 2006, **110**, 8593.
47. I. Bilecka, P. Elser and M. Niederberger, *ACS Nano*, 2009, **3**, 467.
48. H. Katsuki and S. Komarneni, *J. Am. Ceram. Soc.*, 2001, **84**, 2313.
49. E. Haque, N. A. Khan, C. M. Kim and S. H. Jhung, *Cryst. Growth Des.*, 2011, **11**, 4413.
50. S. H. Jhung, T. Jin, Y. K. Hwang and J.-S. Chang, *Chem. – Eur. J.*, 2007, **13**, 4410.
51. N. Dahal, S. García, J. Zhou and S. M. Humphrey, *ACS Nano*, 2012, **6**, 9433.
52. D. A. Lewis, J. D. Summers, T. C. Ward and J. E. McGrath, *J. Polym. Sci., Part A: Polym. Chem.*, 1992, **30**, 1647.
53. K. H. Brosnan, G. L. Messing and D. K. Agrawal, *J. Am. Ceram. Soc.*, 2003, **86**, 1307.
54. X. Fang, R. Hutcheon and D. A. Scola, *J. Polym. Sci., Part A: Polym. Chem.*, 2000, **38**, 2526.
55. E. Haque, C. M. Kim and S. H. Jhung, *CrystEngComm*, 2011, **13**, 4060.
56. B. Gutmann, D. Obermayer, B. Reichart, B. Prekodravac, M. Irfan, J. M. Kremsner and C. O. Kappe, *Chem. – Eur. J.*, 2010, **16**, 12182.
57. A. Pein, M. Baghbanzadeh, T. Rath, W. Haas, E. Maier, H. Amenitsch, F. Hofer, C. O. Kappe and G. Trimmel, *Inorg. Chem.*, 2011, **50**, 193.
58. G. V. Hartland, *Chem. Rev.*, 2011, **111**, 3858.
59. C. Zhu, D. Du, A. Eychmüller and Y. Lin, *Chem. Rev.*, 2015, **115**, 8896.
60. M. N. Nadagouda, T. F. Speth and R. S. Varma, *Acc. Chem. Res.*, 2011, **44**, 469.
61. V. Polshettiwar, M. N. Nadagouda and R. S. Varma, *Aust. J. Chem.*, 2009, **62**, 16.
62. L. Shang, L. Yang, F. Stockmar, R. Popescu, V. Trouillet, M. Bruns, D. Gerthsen and G. U. Nienhaus, *Nanoscale*, 2012, **4**, 4155.
63. A. Pal, S. Shah and S. Devi, *Colloids Surf., A*, 2007, **302**, 51.
64. M. A. Uppal, A. Kafizas, M. B. Ewing and I. P. Parkin, *New J. Chem.*, 2010, **34**, 2906.

65. N. N. Mallikarjuna and R. S. Varma, *Cryst. Growth Des.*, 2007, **7**, 686.
66. S. K. Mehta and S. Gupta, *J. Appl. Electrochem.*, 2011, **41**, 1407.
67. B. Hu, S.-B. Wang, K. Wang, M. Zhang and S.-H. Yu, *J. Phys. Chem. C*, 2008, **112**, 11169.
68. B. Baruwati, V. Polshettiwar and R. S. Varma, *Green Chem.*, 2009, **11**, 926.
69. S. Kundu, L. Peng and H. Liang, *Inorg. Chem.*, 2008, **47**, 6344.
70. B. Baruwati and R. S. Varma, *ChemSusChem*, 2009, **2**, 1041.
71. J. Kou and R. S. Varma, *RSC Adv.*, 2012, **2**, 10283.
72. H. Zhang, Y. Yin, Y. Hu, C. Li, P. Wu, S. Wei and C. Cai, *J. Phys. Chem. C*, 2010, **114**, 11861.
73. N. M. Zain, A. G. F. Stapley and G. Shama, *Carbohydr. Polym.*, 2014, **112**, 195.
74. J. Chen, J. Wang, X. Zhang and Y. Jin, *Mater. Chem. Phys.*, 2008, **108**, 421.
75. R. Bandi, M. Alle, C. W. Park, S. Y. Han, G. J. Kwon, J. C. Kim and S. H. Lee, *Carbohydr. Polym.*, 2020, **240**, 116356.
76. G. A. Kahrilas, L. M. Wally, S. J. Fredrick, M. Hiskey, A. L. Prieto and J. E. Owens, *ACS Sustainable Chem. Eng.*, 2014, **2**, 367.
77. Y.-Q. Liu, M. Zhang, F.-X. Wang and G.-B. Pan, *RSC Adv.*, 2012, **2**, 11235.
78. M. Pradhan, S. Sarkar, A. K. Sinha, M. Basu and T. Pal, *J. Phys. Chem. C*, 2010, **114**, 16129.
79. M. E. El-Naggar, T. I. Shaheen, M. M. G. Fouda and A. A. Hebeish, *Carbohydr. Polym.*, 2016, **136**, 1128.
80. K. Saha, S. S. Agasti, C. Kim, X. Li and V. M. Rotello, *Chem. Rev.*, 2012, **112**, 2739.
81. S. K. Ghosh and T. Pal, *Chem. Rev.*, 2007, **107**, 4797.
82. M.-C. Daniel and D. Astruc, *Chem. Rev.*, 2004, **104**, 293.
83. E. C. Dreaden, A. M. Alkilany, X. Huang, C. J. Murphy and M. A. El Sayed, *Chem. Soc. Rev.*, 2012, **41**, 2740.
84. A. Corma and H. Garcia, *Chem. Soc. Rev.*, 2008, **37**, 2096.
85. S. Kundu, K. Wang and H. Liang, *J. Phys. Chem. C*, 2009, **113**, 5157.
86. S. Joseph and B. Mathew, *J. Mol. Liq.*, 2015, **204**, 184.
87. N. Jayaprakash, J. J. Vijaya, K. Kaviyarasu, K. Kombaiah, L. J. Kennedy, R. J. Ramalingam, M. A. Munusamy and H. A. Al-Lohedan, *J. Photochem. Photobiol., B*, 2017, **169**, 178.
88. M. Valodkar, S. Modi, A. Pal and S. Thakore, *Mater. Res. Bull.*, 2011, **46**, 384.
89. F. Fievet, J. P. Lagier, B. Blin, B. Beaudoin and M. Figlarz, *Solid State Ionics*, 1989, **32–33**, 198.
90. M. Zawadzki, *J. Alloys Compd.*, 2007, **439**, 312.
91. Y. Yu, Y. Zhao, T. Huang and H. Liu, *Mater. Res. Bull.*, 2010, **45**, 159.
92. M. Nishioka, M. Miyakawa, Y. Daino, H. Kataoka, H. Koda, K. Sato and T. M. Suzuki, *Chem. Lett.*, 2011, **40**, 1327.
93. A. M. Raspolli Galletti, C. Antonetti, I. Longo, G. Capannelli and A. M. Venezia, *Appl. Catal., A*, 2008, **350**, 46.
94. M. Tsuji, N. Miyamae, M. Hashimoto, M. Nishio, S. Hikino, N. Ishigami and I. Tanaka, *Colloids Surf., A*, 2007, **302**, 587.

95. T. Zhao, J.-B. Fan, J. Cui, J.-H. Liu, X.-B. Xu and M.-Q. Zhu, *Chem. Phys. Lett.*, 2011, **501**, 414.
96. H. Kawasaki, Y. Kosaka, Y. Myoujin, T. Narushima, T. Yonezawa and R. Arakawa, *Chem. Commun.*, 2011, **47**, 7740.
97. Y. Jung, Y.-H. Son and J.-K. Lee, *RSC Adv.*, 2012, **2**, 5877.
98. W.-T. Cheng and H. W. Cheng, *AIChE J.*, 2009, **55**, 1383.
99. K. P. Donegan, J. F. Godsell, J. M. Tobin, J. P. O'Byrne, D. J. Otway, M. A. Morris, S. Roy and J. D. Holmes, *CrystEngComm*, 2011, **13**, 2023.
100. B. Zhou, T. Ren and J.-J. Zhu, *Int. J. Mod. Phys. B*, 2005, **19**, 2829.
101. J. Jia, J. C. Yu, Y.-X. J. Wang and K. M. Chan, *ACS Appl. Mater. Interfaces*, 2010, **2**, 2579.
102. Y.-J. Zhu and X.-L. Hu, *Mater. Lett.*, 2004, **58**, 1517.
103. H. Zhu, C. Zhang and Y. Yin, *Nanotechnology*, 2005, **16**, 3079.
104. Y. J. Zhu and X. L. Hu, *Chem. Lett.*, 2003, **32**, 732.
105. D. Marquardt, Z. Xie, A. Taubert, R. Thomann and C. Janiak, *Dalton Trans.*, 2011, **40**, 8290.
106. Z. Li, Z. Liu, J. Zhang, B. Han, J. Du, Y. Gao and T. Jiang, *J. Phys. Chem. B*, 2005, **109**, 14445.
107. E. Redel, M. Walter, R. Thomann, C. Vollmer, L. Hussein, H. Scherer, M. Krüger and C. Janiak, *Chem. – Eur. J.*, 2009, **15**, 10047.
108. C. Vollmer, E. Redel, K. Abu-Shandi, R. Thomann, H. Manyar, C. Hardacre and C. Janiak, *Chem. – Eur. J.*, 2010, **16**, 3849.
109. L. Ren, L. Meng, Q. Lu, Z. Fei and P. J. Dyson, *J. Colloid Interface Sci.*, 2008, **323**, 260.
110. Y.-J. Zhu, W.-W. Wang, R.-J. Qi and X.-L. Hu, *Angew. Chem., Int. Ed.*, 2004, **43**, 1410.
111. S. Tang, S. Vongehr, H. Ren and X. Meng, *CrystEngComm*, 2012, **14**, 7209.
112. S. Guo, D. Wen, Y. Zhai, S. Dong and E. Wang, *ACS Nano*, 2010, **4**, 3959.
113. X. Li, W.-X. Chen, J. Zhao, W. Xing and Z.-D. Xu, *Carbon*, 2005, **43**, 2168.
114. W. Tu and H. Liu, *J. Mater. Chem.*, 2000, **10**, 2207.
115. J. W. Doolittle and P. K. Dutta, *Langmuir*, 2006, **22**, 4825.
116. J. H. Chen, R. Cao, J. Zhang and G. Z. Wang, *Mater. Sci. Technol.*, 2005, **21**, 507.
117. X. Tong, Y. Zhao, T. Huang, H. Liu and K. Y. Liew, *Appl. Surf. Sci.*, 2009, **255**, 9463.
118. M. B. Mohamed, K. M. AbouZeid, V. Abdelsayed, A. A. Aljarash and M. S. El-Shall, *ACS Nano*, 2010, **4**, 2766.
119. F. Gao, Q. Lu and S. Komarneni, *Chem. Mater.*, 2005, **17**, 856.
120. Z. Liu, X. Y. Ling, X. Su and J. Y. Lee, *J. Phys. Chem. B*, 2004, **108**, 8234.
121. Q. Xu, Y. Zhao, J. Z. Xu and J.-J. Zhu, *Sens. Actuators, B*, 2006, **114**, 379.
122. W. X. Chen, J. Y. Lee and Z. Liu, *Chem. Commun.*, 2002, 2588.
123. J. Zhao, P. Wang, W. Chen, R. Liu, X. Li and Q. Nie, *J. Power Sources*, 2006, **160**, 563.
124. Z. Liu, L. Hong, M. P. Tham, T. H. Lim and H. Jiang, *J. Power Sources*, 2006, **161**, 831.

125. S. Sharma, A. Ganguly, P. Papakonstantinou, X. Miao, M. Li, J. L. Hutchison, M. Delichatsios and S. Ukleja, *J. Phys. Chem. C*, 2010, **114**, 19459.
126. Y. Boyjoo, M. Wang, V. K. Pareek, J. Liu and M. Jaroniec, *Chem. Soc. Rev.*, 2016, **45**, 6013.
127. A. Mirzaei and G. Neri, *Sens. Actuators, B*, 2016, **237**, 749.
128. S. Faraji and F. N. Ani, *J. Power Sources*, 2014, **263**, 338.
129. M. Parthibavarman, K. Vallalperuman, S. Sathishkumar, M. Durairaj and K. Thavamani, *J. Mater. Sci.: Mater. Electron.*, 2014, **25**, 730.
130. Y. Q. Zhu, H. Z. Guo, H. Z. Zhai and C. B. Cao, *ACS Appl. Mater. Interfaces*, 2015, **7**, 2745.
131. S. Cho, S. Kim, N.-H. Kim, U.-J. Lee, S.-H. Jung, E. Oh and K.-H. Lee, *J. Phys. Chem. C*, 2008, **112**, 17760.
132. X. Hu, J. C. Yu, J. Gong, Q. Li and G. Li, *Adv. Mater.*, 2007, **19**, 2324.
133. C. Blanco-Andujar, D. Ortega, P. Southern, Q. A. Pankhurst and N. T. K. Thanh, *Nanoscale*, 2015, **7**, 1768.
134. S. K. Meher and G. R. Rao, *J. Phys. Chem. C*, 2011, **115**, 25543.
135. T. T. Truong, Y. Liu, Y. Ren, L. Trahey and Y. Sun, *ACS Nano*, 2012, **6**, 8067.
136. S. Islam, M. H. Alfaruqi, V. Mathew, J. Song, S. Kim, S. Kim, J. Jo, J. P. Baboo, D. T. Pham, D. Y. Putro, Y.-K. Sun and J. Kim, *J. Mater. Chem. A*, 2017, **5**, 23299.
137. D. Zhang, G. Li, F. Wang and J. C. Yu, *CrystEngComm*, 2010, **12**, 1759.
138. Y. Yang, G. Z. Wang, Q. Deng, D. H. L. Ng and H. J. Zhao, *ACS Appl. Mater. Interfaces*, 2014, **6**, 3008.
139. A. Phuruangrat, D. J. Ham, S. J. Hong, S. Thongtem and J. S. Lee, *J. Mater. Chem.*, 2010, **20**, 1683.
140. Z. H. Liang and Y. J. Zhu, *Chem. Lett.*, 2005, **34**, 214.
141. C. Yang, F. Xiao, J. D. Wang and X. T. Su, *Sens. Actuators, B*, 2015, **207**, 177.
142. K. Karthik, M. M. Naik, M. Shashank, M. Vinuth and V. Revathi, *J. Cluster Sci.*, 2019, **30**, 311.
143. Y. Lei, J. Li, Y. Y. Wang, L. Gu, Y. F. Chang, H. Y. Yuan and D. Xiao, *ACS Appl. Mater. Interfaces*, 2014, **6**, 1773.
144. R. A. Doong and C. Y. Liao, *J. Hazard. Mater.*, 2017, **322**, 254.
145. Y. Su, B. Zhu, K. Guan, S. Gao, L. Lv, C. Du, L. Peng, L. Hou and X. Wang, *J. Phys. Chem. C*, 2012, **116**, 18508.
146. X.-F. Cao, L. Zhang, X.-T. Chen and Z.-L. Xue, *CrystEngComm*, 2011, **13**, 306.
147. S. Komarneni, M. C. D'Arrigo, C. Leonelli, G. C. Pellacani and H. Katsuki, *J. Am. Ceram. Soc.*, 1998, **81**, 3041.
148. M. Zawadzki, *Solid State Sci.*, 2006, **8**, 14.
149. S. H. Jhung, J.-H. Lee, J. W. Yoon, Y. K. Hwang, J.-S. Hwang, S.-E. Park and J.-S. Chang, *Mater. Lett.*, 2004, **58**, 3161.
150. Z. Zhu, J. Du, J. Li, Y. Zhang and D. Liu, *Ceram. Int.*, 2012, **38**, 4827.
151. C. Xia, J. Guo, Y. J. Lei, H. F. Liang, C. Zhao and H. N. Alshareef, *Adv. Mater.*, 2018, **30**, 1705580.

152. H. D. Liu, J. Wang, X. F. Zhang, D. Zhou, X. Qi, B. Qiu, J. H. Fang, R. Kloepsch, G. Schumacher, Z. P. Liu and J. Li, *ACS Appl. Mater. Interfaces*, 2016, **8**, 4661.
153. S. Khalid, C. B. Cao, L. Wang and Y. Q. Zhu, *Sci. Rep.*, 2016, **6**, 22699.
154. X. Hu, J. Gong, L. Zhang and J. C. Yu, *Adv. Mater.*, 2008, **20**, 4845.
155. S.-W. Cao and Y.-J. Zhu, *J. Phys. Chem. C*, 2008, **112**, 6253.
156. Z. Ai, K. Deng, Q. Wan, L. Zhang and S. Lee, *J. Phys. Chem. C*, 2010, **114**, 6237.
157. Y. Lu, L. Zhang, J. Li, Y.-D. Su, Y. Liu, Y.-J. Xu, L. Dong, H.-L. Gao, J. Lin, N. Man, P.-F. Wei, W.-P. Xu, S.-H. Yu and L.-P. Wen, *Adv. Funct. Mater.*, 2013, **23**, 1534.
158. W. Staszak, M. Zawadzki and J. Okal, *J. Alloys Compd.*, 2010, **492**, 500.
159. E. Solano, L. Perez-Mirabet, F. Martinez-Julian, R. Guzmán, J. Arbiol, T. Puig, X. Obradors, R. Yañez, A. Pomar, S. Ricart and J. Ros, *J. Nanopart. Res.*, 2012, **14**, 1034.
160. L. Wu, J. Bi, Z. Li, X. Wang and X. Fu, *Catal. Today*, 2008, **131**, 15.
161. A. Phuruangrat, T. Thongtem and S. Thongtem, *J. Alloys Compd.*, 2009, **481**, 568.
162. J.-Q. Zheng, Y.-J. Zhu, J.-S. Xu, B.-Q. Lu, C. Qi, F. Chen and J. Wu, *Mater. Lett.*, 2013, **100**, 62.
163. W.-W. Wang and Y.-J. Zhu, *Inorg. Chem. Commun.*, 2004, **7**, 1003.
164. F. T. Li, X. J. Wang, Y. Zhao, J. X. Liu, Y. J. Hao, R. H. Liu and D. S. Zhao, *Appl. Catal., B*, 2014, **144**, 442.
165. Y. Chen, W. Z. Li, J. Y. Wang, Y. L. Gan, L. Liu and M. T. Ju, *Appl. Catal., B*, 2016, **191**, 94.
166. W.-W. Wang, Y.-J. Zhu, G.-F. Cheng and Y.-H. Huang, *Mater. Lett.*, 2006, **60**, 609.
167. L.-X. Yang, Y.-J. Zhu, W.-W. Wang, H. Tong and M.-L. Ruan, *J. Phys. Chem. B*, 2006, **110**, 6609.
168. N. Lee, J. Kwak, J. H. Kwak, S. M. Jung, J. Kim, A. Giri, K. Thiyagarajan, Y. T. Kim, S. Jung, J. K. Kim and U. Jeong, *J. Mater. Chem. A*, 2020, **8**, 9654.
169. L. Xiao, H. Shen, R. von Hagen, J. Pan, L. Belkoura and S. Mathur, *Chem. Commun.*, 2010, **46**, 6509.
170. S.-W. Cao, Y.-J. Zhu, G.-F. Cheng and Y.-H. Huang, *J. Hazard. Mater.*, 2009, **171**, 431.
171. J. G. Geng, C. T. Ma, D. Zhang and X. F. Ning, *J. Alloys Compd.*, 2020, 825.
172. P. Zhu, J. Zhang, Z. Wu and Z. Zhang, *Cryst. Growth Des.*, 2008, **8**, 3148.
173. K. V. Sankar, D. Kalpana and R. K. Selvan, *J. Appl. Electrochem.*, 2012, **42**, 463.
174. M. Zawadzki, *J. Alloys Compd.*, 2008, **454**, 347.
175. G. J. Wilson, G. D. Will, R. L. Frost and S. A. Montgomery, *J. Mater. Chem.*, 2002, **12**, 1787.
176. W. Yan, H. Fan and C. Yang, *Mater. Lett.*, 2011, **65**, 1595.
177. T.-W. Sun, Y.-J. Zhu, C. Qi, G.-J. Ding, F. Chen and J. Wu, *J. Colloid Interface Sci.*, 2016, **463**, 107.

178. S. Q. Chen, Y. F. Zhao, B. Sun, Z. M. Ao, X. Q. Xie, Y. Y. Wei and G. X. Wang, *ACS Appl. Mater. Interfaces*, 2015, **7**, 3306.
179. C. Sun, X. Su, F. Xiao, C. Niu and J. Wang, *Sens. Actuators, B*, 2011, **157**, 681.
180. R. Rahal, A. Wankhade, D. Cha, A. Fihri, S. Ould-Chikh, U. Patil and V. Polshettiwar, *RSC Adv.*, 2012, **2**, 7048.
181. Y. Tao, H. Wang, Y. Xia, G. Zhang, H. Wu and G. Tao, *Mater. Chem. Phys.*, 2010, **124**, 541.
182. E. Hammarberg, A. Prodi-Schwab and C. Feldmann, *Thin Solid Films*, 2008, **516**, 7437.
183. A. K. Mondal, D. W. Su, S. Q. Chen, K. Kretschmer, X. Q. Xie, H. J. Ahn and G. X. Wang, *ChemPhysChem*, 2015, **16**, 169.
184. F. Wang, Y. Liu, Y. F. Zhao, Y. Wang, Z. J. Wang, W. H. Zhang and F. Z. Ren, *Appl. Sci.*, 2018, **8**, 22.
185. V. S. Marques, L. S. Cavalcante, J. C. Sczancoski, A. F. P. Alcântara, M. O. Orlandi, E. Moraes, E. Longo, J. A. Varela, M. Siu Li and M. R. M. C. Santos, *Cryst. Growth Des.*, 2010, **10**, 4752.
186. F. Bensebaa, F. Zavaliche, P. L'Ecuyer, R. W. Cochrane and T. Veres, *J. Colloid Interface Sci.*, 2004, **277**, 104.
187. X. Zhu, J. Wang, Z. Zhang, J. Zhu, S. Zhou, Z. Liu and N. Ming, *J. Am. Ceram. Soc.*, 2008, **91**, 2683.
188. J. S. Xu and Y. J. Zhu, *Mater. Lett.*, 2011, **65**, 2793.
189. S. Chandrasekaran, L. Yao, L. Deng, C. Bowen, Y. Zhang, S. Chen, Z. Lin, F. Peng and P. Zhang, *Chem. Soc. Rev.*, 2019, **48**, 4178.
190. X. He, I. N. Demchenko, W. C. Stolte, A. van Buuren and H. Liang, *J. Phys. Chem. C*, 2012, **116**, 22001.
191. M. Shkir, M. T. Khan, I. M. Ashraf, S. AlFaify, A. M. El-Toni, A. Aldalbahi, H. Ghaithan and A. Khan, *Ceram. Int.*, 2019, **45**, 21975.
192. T. Thongtem, C. Pilapong, J. Kavinchan, A. Phuruangrat and S. Thongtem, *J. Alloys Compd.*, 2010, **500**, 195.
193. C.-F. Mu, Q.-Z. Yao, X.-F. Qu, G.-T. Zhou, M.-L. Li and S.-Q. Fu, *Colloids Surf., A*, 2010, **371**, 14.
194. S. K. Mehta, Khushboo and A. Umar, *Talanta*, 2011, **85**, 2411.
195. H. Pang, C. Z. Wei, X. X. Li, G. C. Li, Y. H. Ma, S. J. Li, J. Chen and J. S. Zhang, *Sci. Rep.*, 2014, **4**, 3577.
196. Y. Hu, Y. Liu, H. Qian, Z. Li and J. Chen, *Langmuir*, 2010, **26**, 18570.
197. Y. Su, D. Ao, H. Liu and Y. Wang, *J. Mater. Chem. A*, 2017, **5**, 8680.
198. S. K. Apte, S. N. Garaje, R. D. Bolade, J. D. Ambekar, M. V. Kulkarni, S. D. Naik, S. W. Gosavi, J. O. Baeg and B. B. Kale, *J. Mater. Chem.*, 2010, **20**, 6095.
199. J. Zhu, O. Palchik, S. Chen and A. Gedanken, *J. Phys. Chem. B*, 2000, **104**, 7344.
200. H. Qian, X. Qiu, L. Li and J. Ren, *J. Phys. Chem. B*, 2006, **110**, 9034.
201. F. Bensebaa, C. Durand, A. Aouadou, L. Scoles, X. Du, D. Wang and Y. Le Page, *J. Nanopart. Res.*, 2010, **12**, 1897.
202. L. Li, H. Qian and J. Ren, *Chem. Commun.*, 2005, **36**, 528.

203. J. Duan, L. Song and J. Zhan, *Nano Res.*, 2009, **2**, 61.
204. Y. He, H.-T. Lu, L.-M. Sai, W.-Y. Lai, Q.-L. Fan, L.-H. Wang and W. Huang, *J. Phys. Chem. B*, 2006, **110**, 13352.
205. F. L. Luo, J. Li, H. Y. Yuan and D. Xiao, *Electrochim. Acta*, 2014, **123**, 183.
206. J. F. A. Oliveira, T. M. Milão, V. D. Araújo, M. L. Moreira, E. Longo and M. I. B. Bernardi, *J. Alloys Compd.*, 2011, **509**, 6880.
207. Z. Wang, W. Lou, X. Wang and W. Liu, *J. Nanosci. Nanotechnol.*, 2011, **11**, 2080.
208. T. Tian, L. Huang, L. H. Ai and J. Jiang, *J. Mater. Chem. A*, 2017, **5**, 20985.
209. M.-L. Li, Q.-Z. Yao, G.-T. Zhou, X.-F. Qu, C.-F. Mu and S.-Q. Fu, *CrystEngComm*, 2011, **13**, 5936.
210. X. He, L. Gao, S. Yang and J. Sun, *CrystEngComm*, 2010, **12**, 3413.
211. K. Aup-Ngoen, S. Thongtem and T. Thongtem, *Mater. Lett.*, 2011, **65**, 442.
212. O. Palchik, R. Kerner, A. Gedanken, A. M. Weiss, M. A. Slifkin and V. Palchik, *J. Mater. Chem.*, 2001, **11**, 874.
213. R. J. Mehta, C. Karthik, W. Jiang, B. Singh, Y. Shi, R. W. Siegel, T. Borca-Tasciuc and G. Ramanath, *Nano Lett.*, 2010, **10**, 4417.
214. B. Zhou, Y. Ji, Y.-F. Yang, X.-H. Li and J.-J. Zhu, *Cryst. Growth Des.*, 2008, **8**, 4394.
215. G.-H. Dong, Y.-J. Zhu and L.-D. Chen, *CrystEngComm*, 2011, **13**, 6811.
216. B. Zhou, Y. Zhao, L. Pu and J.-J. Zhu, *Mater. Chem. Phys.*, 2006, **96**, 192.
217. G.-H. Dong, Y.-J. Zhu, G.-F. Cheng and Y.-J. Ruan, *Mater. Lett.*, 2012, **76**, 69.
218. H. Grisaru, O. Palchik, A. Gedanken, V. Palchik, M. A. Slifkin and A. M. Weiss, *Inorg. Chem.*, 2003, **42**, 7148.
219. J. Ya and Z. Ying-Jie, *Chem. Lett.*, 2004, **33**, 1390.
220. Y. Jiang and Y.-J. Zhu, *J. Phys. Chem. B*, 2005, **109**, 4361.
221. Y. Jiang, Y.-J. Zhu and G.-F. Cheng, *Cryst. Growth Des.*, 2006, **6**, 2174.
222. G. Ji, Y. Shi, L. Pan and Y. Zheng, *J. Alloys Compd.*, 2011, **509**, 6015.
223. Y. Hayakawa, Y. Nonoguchi, H.-P. Wu, E. W. G. Diau, T. Nakashima and T. Kawai, *J. Mater. Chem.*, 2011, **21**, 8849.
224. B. You, N. Jiang, M. L. Sheng and Y. J. Sun, *Chem. Commun.*, 2015, **51**, 4252.
225. M. R. Gao, M. K. Y. Chan and Y. G. Sun, *Nat. Commun.*, 2015, **6**, 7493.
226. Y. Jiang, Y.-J. Zhu and Z.-L. Xu, *Mater. Lett.*, 2006, **60**, 2294.
227. S. Rafai, C. Qiao, M. Naveed, Z. T. Wang, W. Younas, S. Khalid and C. B. Cao, *Chem. Eng. J.*, 2019, **362**, 576.
228. P. Nekooi and M. K. Amini, *Electrochim. Acta*, 2010, **55**, 3286.
229. Y. Jiang, Y.-J. Zhu and L.-D. Chen, *Chem. Lett.*, 2007, **36**, 382.
230. M. Hong, T. C. Chasapis, Z. G. Chen, L. Yang, M. G. Kanatzidis, G. J. Snyder and J. Zou, *ACS Nano*, 2016, **10**, 4719.
231. G.-H. Dong, Y.-J. Zhu and L.-D. Chen, *J. Mater. Chem.*, 2010, **20**, 1976.
232. Z. T. Wang, S. Rafai, C. Qiao, J. Jia, Y. Q. Zhu, X. L. Ma and C. B. Cao, *ACS Appl. Mater. Interfaces*, 2019, **11**, 7046.

233. D. W. Ayele, H.-M. Chen, W.-N. Su, C.-J. Pan, L.-Y. Chen, H.-L. Chou, J.-H. Cheng, B.-J. Hwang and J.-F. Lee, *Chem. – Eur. J.*, 2011, **17**, 5737.
234. M. V. Limaye, S. Gokhale, S. A. Acharya and S. K. Kulkarni, *Nanotechnology*, 2008, **19**, 415602.
235. W. Li, D. Li, W. Zhang, Y. Hu, Y. He and X. Fu, *J. Phys. Chem. C*, 2010, **114**, 2154.
236. D. P. Amalnerkar, H.-Y. Lee, Y. K. Hwang, D.-P. Kim and J.-S. Chang, *J. Nanosci. Nanotechnol.*, 2007, **7**, 4412.
237. Y. Wang, X. Ai, D. Miller, P. Rice, T. Topuria, L. Krupp, A. Kellock and Q. Song, *CrystEngComm*, 2012, **14**, 7560.
238. P. K. Panigrahi and A. Pathak, *Sci. Technol. Adv. Mater.*, 2008, **9**, 045008.
239. D. Han, C. Song and X. Li, *Mater. Chem. Phys.*, 2009, **116**, 41.
240. A. F. Zedan, S. Sappal, S. Moussa and M. S. El-Shall, *J. Phys. Chem. C*, 2010, **114**, 19920.
241. Y. Zhong, F. Peng, F. Bao, S. Wang, X. Ji, L. Yang, Y. Su, S.-T. Lee and Y. He, *J. Am. Chem. Soc.*, 2013, **135**, 8350.
242. X. D. Zhang, X. K. Chen, S. Q. Kai, H. Y. Wang, J. J. Yang, F. G. Wu and Z. Chen, *Anal. Chem.*, 2015, **87**, 3360.
243. Y. He, Y. Zhong, F. Peng, X. Wei, Y. Su, Y. Lu, S. Su, W. Gu, L. Liao and S.-T. Lee, *J. Am. Chem. Soc.*, 2011, **133**, 14192.
244. Y. Zhong, F. Peng, X. Wei, Y. Zhou, J. Wang, X. Jiang, Y. Su, S. Su, S.-T. Lee and Y. He, *Angew. Chem., Int. Ed.*, 2012, **51**, 8485.
245. T. M. Atkins, A. Y. Louie and S. M. Kauzlarich, *Nanotechnology*, 2012, **23**, 294006.
246. J.-W. Liu, F. Chen, M. Zhang, H. Qi, C.-L. Zhang and S.-H. Yu, *Langmuir*, 2010, **26**, 11372.
247. Z. Ying-Jie and H. Xian-Luo, *Chem. Lett.*, 2004, **33**, 760.
248. Y.-J. Zhu and X.-L. Hu, *Mater. Lett.*, 2004, **58**, 1234.
249. Q. Lu, F. Gao and S. Komarneni, *J. Mater. Res.*, 2004, **19**, 1649.
250. L. Ma, W.-X. Chen, Y.-F. Zheng, J. Zhao and Z. Xu, *Mater. Lett.*, 2007, **61**, 2765.
251. L. Ma, W.-X. Chen, X.-Y. Xu, L.-M. Xu and X.-M. Ning, *Mater. Lett.*, 2010, **64**, 1559.
252. S. Wang, S. Su, S. Song, R. Deng and H. Zhang, *CrystEngComm*, 2012, **14**, 4266.
253. H. Schäfer, P. Ptacek, B. Voss, H. Eickmeier, J. Nordmann and M. Haase, *Cryst. Growth Des.*, 2010, **10**, 2202.
254. C.-C. Mi, Z.-H. Tian, B.-F. Han, C.-B. Mao and S.-K. Xu, *J. Alloys Compd.*, 2012, **525**, 154.
255. J.-S. Xu and Y.-J. Zhu, *CrystEngComm*, 2012, **14**, 2630.
256. C. Li, P. a. Ma, P. Yang, Z. Xu, G. Li, D. Yang, C. Peng and J. Lin, *CrystEngComm*, 2011, **13**, 1003.
257. C. Chen, L.-D. Sun, Z.-X. Li, L.-L. Li, J. Zhang, Y.-W. Zhang and C.-H. Yan, *Langmuir*, 2010, **26**, 8797.
258. J. Zhao, Y.-J. Zhu, J. Wu and F. Chen, *J. Colloid Interface Sci.*, 2015, **440**, 39.

259. M. Shkir and S. AlFaify, *Sci. Rep.*, 2017, **7**, 16091.
260. L. Zhang, X.-F. Cao, X.-T. Chen and Z.-L. Xue, *J. Colloid Interface Sci.*, 2011, **354**, 630.
261. R. A. He, S. W. Cao, J. G. Yu and Y. C. Yang, *Catal. Today*, 2016, **264**, 221.
262. A. Dash, S. Sarkar, V. Adusumalli and V. Mahalingam, *Langmuir*, 2014, **30**, 1401.
263. H. Liu, Z. Wu, H. Gao, J. Shao, H. Zou, D. Yao, Y. Liu, H. Zhang and B. Yang, *ACS Appl. Mater. Interfaces*, 2017, **9**, 42919.
264. Q. Pan, H. Hu, Y. Zou, M. Chen, L. Wu, D. Yang, X. Yuan, J. Fan, B. Sun and Q. Zhang, *J. Mater. Chem. C*, 2017, **5**, 10947.
265. Y. X. Li, H. Huang, Y. Xiong, S. V. Kershaw and A. L. Rogach, *Angew. Chem., Int. Ed.*, 2018, **57**, 5833.
266. X.-Y. Zhao, Y.-J. Zhu, F. Chen, B.-Q. Lu and J. Wu, *CrystEngComm*, 2013, **15**, 206.
267. F. Chen, P. Huang, Y.-J. Zhu, J. Wu, C.-L. Zhang and D.-X. Cui, *Biomaterials*, 2011, **32**, 9031.
268. C. Qi, Y.-J. Zhu, B.-Q. Lu, X.-Y. Zhao, J. Zhao, F. Chen and J. Wu, *Chem. – Eur. J.*, 2013, **19**, 5332.
269. M. Shkir, M. Kilany and I. S. Yahia, *Ceram. Int.*, 2017, **43**, 14923.
270. C. Qi, Y.-J. Zhu, X.-Y. Zhao, B.-Q. Lu, Q.-L. Tang, J. Zhao and F. Chen, *Chem. – Eur. J.*, 2013, **19**, 981.
271. C. Qi, Y.-J. Zhu, B.-Q. Lu, J. Wu and F. Chen, *RSC Adv.*, 2015, **5**, 14906.
272. C. Qi, Y.-J. Zhu, C.-T. Wu, T.-W. Sun, F. Chen and J. Wu, *J. Colloid Interface Sci.*, 2016, **462**, 297.
273. C. Qi, Y.-J. Zhu, B.-Q. Lu, G.-J. Ding, T.-W. Sun, F. Chen and J. Wu, *J. Mater. Chem. B*, 2014, **2**, 8576.
274. G. Li, Y. Ding, Y. Zhang, Z. Lu, H. Sun and R. Chen, *J. Colloid Interface Sci.*, 2011, **363**, 497.
275. G. Yang, H. Ji, H. Liu, K. Huo, J. Fu and P. K. Chu, *J. Nanosci. Nanotechnol.*, 2010, **10**, 980.
276. C. R. Patra, G. Alexandra, S. Patra, D. S. Jacob, A. Gedanken, A. Landau and Y. Gofer, *New J. Chem.*, 2005, **29**, 733.
277. D. H. Kim, J. W. Kang, I. O. Jung, J. S. Im, E. J. Kim, S. J. Song, J. S. Lee and J. Kim, *J. Nanosci. Nanotechnol.*, 2008, **8**, 5376.
278. G. Bühler and C. Feldmann, *Angew. Chem., Int. Ed.*, 2006, **45**, 4864.
279. C. Qi, S. Musetti, L.-H. Fu, Y.-J. Zhu and L. Huang, *Chem. Soc. Rev.*, 2019, **48**, 2698.
280. K.-W. Wang, Y.-J. Zhu, F. Chen, G.-F. Cheng and Y.-H. Huang, *Mater. Lett.*, 2011, **65**, 2361.
281. C. Qi, Y.-J. Zhu, B.-Q. Lu, X.-Y. Zhao, J. Zhao and F. Chen, *J. Mater. Chem.*, 2012, **22**, 22642.
282. X. Chen, B. Yang, C. Qi, T.-W. Sun, F. Chen, J. Wu, X.-P. Feng and Y.-J. Zhu, *Dalton Trans.*, 2016, **45**, 1648.
283. H.-P. Yu, Y.-J. Zhu and B.-Q. Lu, *Ceram. Int.*, 2018, **44**, 12352.
284. Y.-D. Yu, Y.-J. Zhu, C. Qi and J. Wu, *Ceram. Int.*, 2017, **43**, 6511.

285. G.-J. Ding, Y.-J. Zhu, C. Qi, B.-Q. Lu, J. Wu and F. Chen, *J. Colloid Interface Sci.*, 2015, **443**, 72.
286. C. Qi, Y.-J. Zhu and F. Chen, *ACS Appl. Mater. Interfaces*, 2014, **6**, 4310.
287. G.-J. Ding, Y.-J. Zhu, C. Qi, B.-Q. Lu, F. Chen and J. Wu, *J. Mater. Chem. B*, 2015, **3**, 1823.
288. C. Qi, Y.-J. Zhu, G.-F. Cheng, Y.-J. Ruan, G.-J. Ding, T.-W. Sun, F. Chen and J. Wu, *Mater. Lett.*, 2015, **140**, 79.
289. J. Zhao, Y.-J. Zhu, J.-Q. Zheng, F. Chen and J. Wu, *Microporous Mesoporous Mater.*, 2013, **180**, 79.
290. C. Qi, Q.-L. Tang, Y.-J. Zhu, X.-Y. Zhao and F. Chen, *Mater. Lett.*, 2012, **85**, 71.
291. F. Chen, P. Huang, C. Qi, B.-Q. Lu, X.-Y. Zhao, C. Li, J. Wu, D.-X. Cui and Y.-J. Zhu, *J. Mater. Chem. B*, 2014, **2**, 7132.
292. C. Qi, Y.-J. Zhu, T.-W. Sun, J. Wu and F. Chen, *Chem. – Asian J.*, 2015, **10**, 2503.
293. C. Qi, Y.-J. Zhu and F. Chen, *Chem. – Asian J.*, 2013, **8**, 88.
294. W. Yu, T.-W. Sun, C. Qi, Z. Ding, H. Zhao, F. Chen, D. Chen, Y.-J. Zhu, Z. Shi and Y. He, *ACS Appl. Mater. Interfaces*, 2017, **9**, 3306.
295. C. Qi, F. Chen, J. Wu, Y.-J. Zhu, C.-N. Hao and J.-L. Duan, *RSC Adv.*, 2016, **6**, 33393.
296. C. Qi, Y.-J. Zhu, F. Chen and J. Wu, *J. Mater. Chem. B*, 2015, **3**, 7775.
297. W. Yu, T.-W. Sun, Z. Ding, C. Qi, H. Zhao, F. Chen, Z. Shi, Y.-J. Zhu, D. Chen and Y. He, *J. Mater. Chem. B*, 2017, **5**, 1039.
298. J. Zhao, Y.-J. Zhu, G.-F. Cheng, Y.-J. Ruan, T.-W. Sun, F. Chen, J. Wu, X.-Y. Zhao and G.-J. Ding, *Mater. Lett.*, 2014, **124**, 208.
299. X.-Y. Zhao, Y.-J. Zhu, C. Qi, F. Chen, B.-Q. Lu, J. Zhao and J. Wu, *Chem. – Asian J.*, 2013, **8**, 1313.
300. X.-Y. Zhao, Y.-J. Zhu, F. Chen, B.-Q. Lu, C. Qi, J. Zhao and J. Wu, *CrystEngComm*, 2013, **15**, 7926.
301. G.-J. Ding, Y.-J. Zhu, C. Qi, T.-W. Sun, J. Wu and F. Chen, *RSC Adv.*, 2015, **5**, 40154.
302. E.-P. Ng, L. Delmotte and S. Mintova, *Green Chem.*, 2008, **10**, 1043.
303. A. Umar, F. Al-Hazmi, G. N. Dar, S. A. Zaidi, R. M. Al-Tuwirqi, F. Alnowaiserb, A. A. Al-Ghamdi and S. W. Hwang, *Sens. Actuators, B*, 2012, **166–167**, 97.
304. S. Ran, Y. Zhu, H. Huang, B. Liang, J. Xu, B. Liu, J. Zhang, Z. Xie, Z. Wang, J. Ye, D. Chen and G. Shen, *CrystEngComm*, 2012, **14**, 3063.
305. X. Tian, C. Cheng, L. Qian, B. Zheng, H. Yuan, S. Xie, D. Xiao and M. M. F. Choi, *J. Mater. Chem.*, 2012, **22**, 8029.
306. Y. Q. Zhu, C. B. Cao, S. Tao, W. S. Chu, Z. Y. Wu and Y. D. Li, *Sci. Rep.*, 2014, **4**, 5787.
307. X. Liu, R. Ma, Y. Bando and T. Sasaki, *Angew. Chem., Int. Ed.*, 2010, **49**, 8253.
308. R. Souleymen, Z. T. Wang, C. Qiao, M. Naveed and C. B. Cao, *J. Mater. Chem. A*, 2018, **6**, 7592.
309. J. Pang, Y. Luan, Q. Wang, J. Du, X. Cai and Z. Li, *Colloids Surf., A*, 2010, **360**, 6.

310. R. K. Chava, H. Y. Cho, J. M. Yoon and Y. T. Yu, *J. Alloys Compd.*, 2019, **772**, 834.
311. A. Phuruangrat, S. Thongtem and T. Thongtem, *Ceram. Int.*, 2012, **38**, 4075.
312. C. Qiao, Y. Zhang, Y. Q. Zhu, C. B. Cao, X. H. Bao and J. Q. Xu, *J. Mater. Chem. A*, 2015, **3**, 6878.
313. W. C. Wang, N. Zhang, Z. Y. Shi, Z. R. Ye, Q. Y. Gao, M. J. Zhi and Z. L. Hong, *Chem. Eng. J.*, 2018, **338**, 55.
314. Y. Wang, D. Yan, S. El Hankari, Y. Zou and S. Wang, *Adv. Sci.*, 2018, **5**, 1800064.
315. M. R. Benzigar, S. N. Talapaneni, S. Joseph, K. Ramadass, G. Singh, J. Scaranto, U. Ravon, K. Al-Bahily and A. Vinu, *Chem. Soc. Rev.*, 2018, **47**, 2680.
316. A. M. Schwenke, S. Hoeppener and U. S. Schubert, *Adv. Mater.*, 2015, **27**, 4113.
317. X. Wang, K. Qu, B. Xu, J. Ren and X. Qu, *J. Mater. Chem.*, 2011, **21**, 2445.
318. Q. Wang, H. Zheng, Y. Long, L. Zhang, M. Gao and W. Bai, *Carbon*, 2011, **49**, 3134.
319. A. Jaiswal, S. S. Ghosh and A. Chattopadhyay, *Chem. Commun.*, 2012, **48**, 407.
320. S. Qu, X. Wang, Q. Lu, X. Liu and L. Wang, *Angew. Chem., Int. Ed.*, 2012, **51**, 12215.
321. R. Tabaraki and N. Sadeghinejad, *Ecotoxicol. Environ. Saf.*, 2018, **153**, 101.
322. Y. Liu, N. Xiao, N. Gong, H. Wang, X. Shi, W. Gu and L. Ye, *Carbon*, 2014, **68**, 258.
323. Z. Tian, X. T. Zhang, D. Li, D. Zhou, P. T. Jing, D. Z. Shen, S. N. Qu, R. Zboril and A. L. Rogach, *Adv. Opt. Mater.*, 2017, **5**, 1700416.
324. H. Zhu, X. Wang, Y. Li, Z. Wang, F. Yang and X. Yang, *Chem. Commun.*, 2009, 5118.
325. S. Chandra, P. Das, S. Bag, D. Laha and P. Pramanik, *Nanoscale*, 2011, **3**, 1533.
326. X. C. Sun, J. K. He, Y. T. Meng, L. C. Zhang, S. C. Zhang, X. Y. Ma, S. Dey, J. Zhao and Y. Lei, *J. Mater. Chem. A*, 2016, **4**, 4161.
327. R. Cui, C. Liu, J. Shen, D. Gao, J.-J. Zhu and H.-Y. Chen, *Adv. Funct. Mater.*, 2008, **18**, 2197.
328. T. Q. Chen, L. K. Pan, T. Lu, C. L. Fu, D. H. C. Chua and Z. Sun, *J. Mater. Chem. A*, 2014, **2**, 1263.
329. M. Zhang, S. Liu, X. M. Yin, Z. F. Du, Q. Y. Hao, D. N. Lei, Q. H. Li and T. H. Wang, *Chem. – Asian J.*, 2011, **6**, 1151.
330. J. Long, M. Fang and G. Chen, *J. Mater. Chem.*, 2011, **21**, 10421.
331. I. Janowska, K. Chizari, O. Ersen, S. Zafeiratos, D. Soubane, V. D. Costa, V. Speisser, C. Boeglin, M. Houllé, D. Bégin, D. Plee, M.-J. Ledoux and C. Pham-Huu, *Nano Res.*, 2010, **3**, 126.
332. S. Vadahanambi, J.-H. Jung and I.-K. Oh, *Carbon*, 2011, **49**, 4449.
333. L. B. Li, B. Yu and T. Y. You, *Biosens. Bioelectron.*, 2015, **74**, 263.

334. M. H. Xu, G. L. He, Z. H. Li, F. J. He, F. Gao, Y. J. Su, L. Y. Zhang, Z. Yang and Y. F. Zhang, *Nanoscale*, 2014, **6**, 10307.
335. Y. Q. Zhang, X. Y. Liu, Y. Fan, X. Y. Guo, L. Zhou, Y. Lv and J. Lin, *Nanoscale*, 2016, **8**, 15281.
336. T. Edison, R. Atchudan, M. G. Sethuraman, J. J. Shim and Y. R. Lee, *J. Photochem. Photobiol., B*, 2016, **161**, 154.
337. Q. Wang, X. Liu, L. Zhang and Y. Lv, *Analyst*, 2012, **137**, 5392.
338. F. Arcudi, L. Dordevic and M. Prato, *Angew. Chem., Int. Ed.*, 2016, **55**, 2107.
339. J. Jiang, Y. He, S. Li and H. Cui, *Chem. Commun.*, 2012, **48**, 9634.
340. X. C. Sun, C. Bruckner and Y. Lei, *Nanoscale*, 2015, **7**, 17278.
341. D. Voiry, J. Yang, J. Kupferberg, R. Fullon, C. Lee, H. Y. Jeong, H. S. Shin and M. Chhowalla, *Science*, 2016, **353**, 1413.
342. Q. Wang and D. Astruc, *Chem. Rev.*, 2020, **120**, 1438.
343. Q. Yang, Q. Xu and H.-L. Jiang, *Chem. Soc. Rev.*, 2017, **46**, 4774.
344. C. A. Trickett, A. Helal, B. A. Al-Maythalony, Z. H. Yamani, K. E. Cordova and O. M. Yaghi, *Nat. Rev. Mater.*, 2017, **2**, 17045.
345. J. Klinowski, F. A. Almeida Paz, P. Silva and J. Rocha, *Dalton Trans.*, 2011, **40**, 321.
346. N. A. Khan and S. H. Jhung, *Coord. Chem. Rev.*, 2015, **285**, 11.
347. F. Hillman, J. M. Zimmerman, S.-M. Paek, M. R. A. Hamid, W. T. Lim and H.-K. Jeong, *J. Mater. Chem. A*, 2017, **5**, 6090.
348. J.-S. Choi, W.-J. Son, J. Kim and W.-S. Ahn, *Microporous Mesoporous Mater.*, 2008, **116**, 727.
349. Z. Ni and R. I. Masel, *J. Am. Chem. Soc.*, 2006, **128**, 12394.
350. P. Silva, A. A. Valente, J. Rocha and F. A. Almeida Paz, *Cryst. Growth Des.*, 2010, **10**, 2025.
351. Y.-K. Seo, G. Hundal, I. T. Jang, Y. K. Hwang, C.-H. Jun and J.-S. Chang, *Microporous Mesoporous Mater.*, 2009, **119**, 331.
352. L. Pukdeejorhor, K. Adpakpang, P. Ponchai, S. Wannapaiboon, S. Itti-sanronnachai, M. Ogawa, S. Horike and S. Bureekaew, *Cryst. Growth Des.*, 2019, **19**, 5581.
353. W. Liang, R. Babarao and D. M. D'Alessandro, *Inorg. Chem.*, 2013, **52**, 12878.
354. W. Guo, W. Sun, L.-P. Lv, S. Kong and Y. Wang, *ACS Nano*, 2017, **11**, 4198.
355. S. Kong, R. Dai, H. Li, W. Sun and Y. Wang, *ACS Sustainable Chem. Eng.*, 2015, **3**, 1830.
356. Y. X. Chen, D. Ni, X. W. Yang, C. C. Liu, J. L. Yin and K. F. Cai, *Electrochim. Acta*, 2018, **278**, 114.
357. W. Guo, W. Sun and Y. Wang, *ACS Nano*, 2015, **9**, 11462.
358. M. Ma, A. Bétard, I. Weber, N. S. Al-Hokbany, R. A. Fischer and N. Metzler-Nolte, *Cryst. Growth Des.*, 2013, **13**, 2286.
359. P. Sarawade, H. Tan and V. Polshettiwar, *ACS Sustainable Chem. Eng.*, 2013, **1**, 66.
360. H. Bux, F. Liang, Y. Li, J. Cravillon, M. Wiebcke and J. Caro, *J. Am. Chem. Soc.*, 2009, **131**, 16000.

361. M. Qiu, L. Zhu, T. Zhang, H. Li, Y. Sun and K. Liu, *Mater. Res. Bull.*, 2012, **47**, 2437.
362. H. Zhang, J. Hong, Y. Ni and Y. Zhou, *CrystEngComm*, 2008, **10**, 1031.
363. J. Wu, Y.-J. Zhu, G.-F. Cheng and Y.-H. Huang, *Mater. Res. Bull.*, 2010, **45**, 509.
364. L. Xie, J. Wang, Y. Hu, Z. Zheng, S. Weng, P. Liu, X. Shi and D. Wang, *Mater. Chem. Phys.*, 2012, **136**, 309.
365. M. Sun, D. Li, Y. Chen, W. Chen, W. Li, Y. He and X. Fu, *J. Phys. Chem. C*, 2009, **113**, 13825.
366. X. F. Shi, C. Wang, Y. Y. Ma, H. Liu, S. D. Wu, Q. Shao, Z. F. He, L. Guo, T. Ding and Z. H. Guo, *Powder Technol.*, 2019, **356**, 726.
367. N. Tipcompor, T. Thongtem, A. Phuruangrat and S. Thongtem, *Mater. Lett.*, 2012, **87**, 153.
368. T. Muraliganth, K. R. Stroukoff and A. Manthiram, *Chem. Mater.*, 2010, **22**, 5754.
369. S.-W. Cao, Y.-J. Zhu and J.-B. Cui, *J. Solid State Chem.*, 2010, **183**, 1704.
370. J.-S. Xu, Y.-J. Zhu and F. Chen, *Mater. Lett.*, 2013, **94**, 104.
371. S. Tang, S. Vongehr and X. Meng, *J. Phys. Chem. C*, 2010, **114**, 977.
372. J. C. Yu, X. Hu, Q. Li, Z. Zheng and Y. Xu, *Chem. – Eur. J.*, 2006, **12**, 548.
373. E. Lebègue, S. Baranton and C. Coutanceau, *J. Power Sources*, 2011, **196**, 920.
374. P. P. Song, Y. J. Lei, X. B. Hu, C. Wang, J. L. Wang and Y. P. Tang, *J. Colloid Interface Sci.*, 2020, **574**, 421.
375. Y. Wang and J. Y. Lee, *J. Power Sources*, 2005, **144**, 220.
376. F. M. Courtel, E. A. Baranova, Y. Abu-Lebdeh and I. J. Davidson, *J. Power Sources*, 2010, **195**, 2355.
377. W. Luo, X. Hu, Y. Sun and Y. Huang, *ACS Appl. Mater. Interfaces*, 2013, **5**, 1997.
378. M. Sakthivel, A. Schlange, U. Kunz and T. Turek, *J. Power Sources*, 2010, **195**, 7083.
379. C.-L. Sun, C.-T. Chang, H.-H. Lee, J. Zhou, J. Wang, T.-K. Sham and W.-F. Pong, *ACS Nano*, 2011, **5**, 7788.
380. H. Wu, Q. Wang, Y. Yao, C. Qian, X. Zhang and X. Wei, *J. Phys. Chem. C*, 2008, **112**, 16779.
381. L. Zhou, X. Zhang, X. Zhao, C. Sun and Q. Niu, *J. Alloys Compd.*, 2010, **502**, 329.
382. A. R. Siamaki, A. E. R. S. Khder, V. Abdelsayed, M. S. El-Shall and B. F. Gupton, *J. Catal.*, 2011, **279**, 1.
383. D. Marquardt, C. Vollmer, R. Thomann, P. Steurer, R. Mülhaupt, E. Redel and C. Janiak, *Carbon*, 2011, **49**, 1326.
384. Z. Y. Le, F. Liu, P. Nie, X. R. Li, X. Y. Liu, Z. F. Bian, G. Chen, H. B. Wu and Y. F. Lu, *ACS Nano*, 2017, **11**, 2952.
385. X. Q. An, K. F. Li and J. W. Tang, *ChemSusChem*, 2014, **7**, 1086.
386. C. A. Zito, T. M. Perfecto and D. P. Volanti, *Sens. Actuators, B*, 2017, **244**, 466.

387. X. Liu, L. Pan, T. Lv, G. Zhu, Z. Sun and C. Sun, *Chem. Commun.*, 2011, **47**, 11984.
388. Y. G. Sun, F. Alimohammadi, D. T. Zhang and G. S. Guo, *Nano Lett.*, 2017, **17**, 1963.
389. C. Zhang, B. Wang, J. Xiang, C. Su, C. Mu, F. Wen and Z. Liu, *ACS Appl. Mater. Interfaces*, 2017, **9**, 28868.
390. J. Yan, Z. Fan, W. Sun, G. Ning, T. Wei, Q. Zhang, R. Zhang, L. Zhi and F. Wei, *Adv. Funct. Mater.*, 2012, **22**, 2632.
391. X. J. Wang, X. L. Zhao, D. Q. Zhang, G. S. Li and H. X. Li, *Appl. Catal., B*, 2018, **228**, 47.
392. A. H. Mady, M. L. Baynosa, D. Tuma and J.-J. Shim, *Appl. Catal., B*, 2017, **203**, 416.
393. J. Geng, G.-H. Song, X.-D. Jia, F.-F. Cheng and J.-J. Zhu, *J. Phys. Chem. C*, 2012, **116**, 4517.
394. I. Matai, A. Sachdev, P. Dubey, S. U. Kumar, B. Bhushan and P. Gopinath, *Colloids Surf., B*, 2014, **115**, 359.
395. A. Phuruangrat, S. Siri, P. Wadbua, S. Thongtem and T. Thongtem, *J. Phys. Chem. Solids*, 2019, **126**, 170.
396. X. Li, L. Wang and X. Lu, *J. Hazard. Mater.*, 2010, **177**, 639.
397. L. Qi, B. Cheng, J. Yu and W. Ho, *J. Hazard. Mater.*, 2016, **301**, 522.
398. M. L. Li, M. Wang, L. F. Zhu, Y. M. Li, Z. Yan, Z. Q. Shen and X. B. Cao, *Appl. Catal., B*, 2018, **231**, 269.
399. Y. Sun, X. Hu, W. Luo and Y. Huang, *J. Mater. Chem.*, 2012, **22**, 19190.
400. X. Liu, H. Cao and J. Yin, *Nano Res.*, 2011, **4**, 470.
401. M. Nishioka, M. Miyakawa, H. Kataoka, H. Koda, K. Sato and T. M. Suzuki, *Chem. Lett.*, 2011, **40**, 1204.
402. Q. J. Wang, C. Wang, H. B. Sun, P. Sun, Y. Z. Wang, J. Lin and G. Y. Lu, *Sens. Actuators, B*, 2016, **222**, 257.
403. M. K. Das, J. A. Bobb, A. A. Ibrahim, A. Lin, K. M. AbouZeid and M. S. El-Shall, *ACS Appl. Mater. Interfaces*, 2020, **12**, 23844.
404. R. M. Wong, D. A. Gilbert, K. Liu and A. Y. Louie, *ACS Nano*, 2012, **6**, 3461.
405. F. Prete, A. Rizzuti, L. Esposito, A. Tucci and C. Leonelli, *J. Am. Ceram. Soc.*, 2011, **94**, 3587.
406. J. Y. Tian, Q. Shao, J. K. Zhao, D. Pan, M. Y. Dong, C. X. Z. Jia, T. Ding, T. T. Wu and Z. H. Guo, *J. Colloid Interface Sci.*, 2019, **541**, 18.
407. A. Shanmugavani and R. K. Selvan, *Electrochim. Acta*, 2016, **189**, 283.
408. K. Karthik, S. Dhanuskodi, C. Gobinath, S. Prabukumar and S. Sivaramakrishnan, *J. Phys. Chem. Solids*, 2018, **112**, 106.
409. K. Karthik, S. Dhanuskodi, C. Gobinath, S. Prabukumar and S. Sivaramakrishnan, *J. Mater. Sci.: Mater. Electron.*, 2018, **29**, 5459.
410. D. Malwal and P. Gopinath, *Catal. Sci. Technol.*, 2016, **6**, 4458.
411. D. M. Zhang, M. P. Chen, H. Y. Zou, Y. M. Zhang, J. C. Hu, H. P. Wang, B. Y. Zi, J. Zhang, Z. Q. Zhu, L. L. Duan and Q. J. Liu, *Nanotechnology*, 2020, **31**, 215601.
412. W. Chen, H. Ruan, Y. Hu, D. Li, Z. Chen, J. Xian, J. Chen, X. Fu, Y. Shao and Y. Zheng, *CrystEngComm*, 2012, **14**, 6295.

413. S. Cho, J.-W. Jang, J. S. Lee and K.-H. Lee, *Nanoscale*, 2012, **4**, 2066.
414. W. Schumacher, A. Nagy, W. J. Waldman and P. K. Dutta, *J. Phys. Chem. C*, 2009, **113**, 12132.
415. Y. He, H.-T. Lu, L.-M. Sai, W.-Y. Lai, Q.-L. Fan, L.-H. Wang and W. Huang, *J. Phys. Chem. B*, 2006, **110**, 13370.
416. D. Zhu, X. Jiang, C. Zhao, X. Sun, J. Zhang and J.-J. Zhu, *Chem. Commun.*, 2010, **46**, 5226.
417. G.-H. Dong and Y.-J. Zhu, *Chem. Eng. J.*, 2012, **193–194**, 227.
418. B. Wang, Y. Hu, B. Yu, X. J. Zhang, D. X. Yang and Y. F. Chen, *J. Power Sources*, 2019, **433**, 126688.
419. A. Akhundi and A. Habibi-Yangjeh, *J. Colloid Interface Sci.*, 2016, **482**, 165.
420. M. A. Karimi, M. Iliyat, M. Atashkadi, M. Ranjbar and A. Habibi-Yangjeh, *J. Chin. Chem. Soc.*, 2020, **67**, 2032.
421. H. Tian, H. Fan, J. Ma, L. Ma and G. Dong, *Electrochim. Acta*, 2017, **247**, 787.
422. H. L. Tian, H. Q. Fan, J. W. Ma, Z. Y. Liu, L. T. Ma, S. H. Lei, J. W. Fang and C. B. Long, *J. Hazard. Mater.*, 2018, **341**, 102.
423. S. H. Liu, W. T. Tang and P. H. Chou, *Chem. Eng. J.*, 2020, 387.
424. X.-L. Liu, Y.-J. Zhu, G.-F. Cheng and Y.-H. Huang, *Mater. Lett.*, 2011, **65**, 343.
425. G. J. Zeng, L. Huang, Q. Huang, M. Y. Liu, D. Z. Xu, H. Y. Huang, Z. Y. Yang, F. J. Deng, X. Y. Zhang and Y. Wei, *Appl. Surf. Sci.*, 2018, **459**, 588.
426. S. J. Zhang, X. X. Chen and L. M. Song, *J. Hazard. Mater.*, 2019, **367**, 304.
427. Y. Wang, J. G. Yu, W. Xiao and Q. Li, *J. Mater. Chem. A*, 2014, **2**, 3847.
428. Z. Y. Wang, Y. Xiao, X. B. Cui, P. F. Cheng, B. Wang, Y. Gao, X. W. Li, T. L. Yang, T. Zhang and G. Y. Lu, *ACS Appl. Mater. Interfaces*, 2014, **6**, 3888.
429. X. Zhu, Y. Zhu, S. Murali, M. D. Stoller and R. S. Ruoff, *ACS Nano*, 2011, **5**, 3333.
430. X. J. Liu, T. Q. Chen, H. P. Chu, L. Y. Niu, Z. Sun, L. K. Pan and C. Q. Sun, *Electrochim. Acta*, 2015, **166**, 12.
431. A. G. El-Deen, N. A. M. Barakat and H. Y. Kim, *Desalination*, 2014, **344**, 289.
432. L. L. Liu, M. Z. An, P. X. Yang and J. Q. Zhang, *Sci. Rep.*, 2015, **5**, 9055.
433. X. Yu, Z. H. Zhao, D. H. Sun, N. Ren, J. H. Yu, R. Q. Yang and H. Liu, *Appl. Catal., B*, 2018, **227**, 470.
434. U. Rajaji, M. Govindasamy, S. M. Chen, T. W. Chen, X. H. Liu and S. Chinnapaiyan, *Composites, Part B*, 2018, **152**, 220.
435. I. Tzanidis, F. Bairamis, L. Sygellou, K. S. Andrikopoulos, A. Avgeropoulos, I. Konstantinou and D. Tasis, *Chem. – Eur. J.*, 2020, **26**, 6643.
436. H. Hu, X. Wang, F. Liu, J. Wang and C. Xu, *Synth. Met.*, 2011, **161**, 404.
437. W. Qin, T. Q. Chen, L. K. Pan, L. Y. Niu, B. W. Hu, D. S. Li, J. L. Li and Z. Sun, *Electrochim. Acta*, 2015, **153**, 55.
438. A. Kagkoura, I. Tzanidis, V. Dracopoulos, N. Tagmatarchis and D. Tasis, *Chem. Commun.*, 2019, **55**, 2078.

439. W. C. Wang, N. Zhang, Z. R. Ye, Z. L. Hong and M. J. Zhi, *Inorg. Chem. Front.*, 2019, **6**, 407.
440. J.-F. Zhu, Y.-J. Zhu, M.-G. Ma, L.-X. Yang and L. Gao, *J. Phys. Chem. C*, 2007, **111**, 3920.
441. Q.-L. Tang, K.-W. Wang, Y.-J. Zhu and F. Chen, *Mater. Lett.*, 2009, **63**, 1332.
442. M. N. Nadagouda and R. S. Varma, *Macromol. Rapid Commun.*, 2007, **28**, 465.
443. M. S. El-Shall, V. Abdelsayed, A. E. R. S. Khder, H. M. A. Hassan, H. M. El-Kaderi and T. E. Reich, *J. Mater. Chem.*, 2009, **19**, 7625.
444. M. N. Nadagouda and R. S. Varma, *Biomacromolecules*, 2007, **8**, 2762.
445. A. Zahoor, T. Qiu, J. Zhang and X. Li, *J. Mater. Sci.*, 2009, **44**, 6054.
446. M. A. Jagtap, M. V. Kulkarni, S. K. Apte, S. D. Naik and B. B. Kale, *Mater. Sci. Eng., B*, 2010, **168**, 199.
447. M. R. Karim, M. M. Alam, M. O. Aijaz, A. M. Asiri, M. A. Dar and M. M. Rahman, *Talanta*, 2019, **193**, 64.
448. L.-J. You, S. Xu, W.-F. Ma, D. Li, Y.-T. Zhang, J. Guo, J. J. Hu and C.-C. Wang, *Langmuir*, 2012, **28**, 10565.
449. S. Liu, F. Lu, X. Jia, F. Cheng, L.-P. Jiang and J.-J. Zhu, *CrystEngComm*, 2011, **13**, 2425.
450. H. Zhang, X. Zhong, J.-J. Xu and H.-Y. Chen, *Langmuir*, 2008, **24**, 13748.
451. X. Qian, Y. J. Xu, X. P. Yue, C. Wang, M. Liu, C. Duan, Y. Xu, C. Zhu and L. Dai, *Cellulose*, 2020, **27**, 5939.
452. L. Branzi, M. Back, P. Cortelletti, N. Pinna, A. Benedetti and A. Speghini, *Dalton Trans.*, 2020, **49**, 15195.
453. E. J. Lee, J. Bae, K. M. Choi and N. C. Jeong, *ACS Appl. Mater. Interfaces*, 2019, **11**, 35155.

CHAPTER 5

In Situ Synthesis of Multifunctional Hybrid Nanocomposites Based on Graphene Derivatives and Inorganic Nanoparticles for Advanced Applications

C. INGROSSO[a*], N. DEPALO[a], E FANIZZA[a,b], A. PANNIELLO[a], R. COMPARELLI[a], A. AGOSTIANO[a,b], M. STRICCOLI[a] AND M. L. CURRI[a,b*]

[a]CNR-IPCF Sez. Bari, c/o Department of Chemistry, Università degli Studi di Bari, via Orabona 4, I-70126 Bari, Italy; [b]Department of Chemistry, Università degli Studi di Bari, via Orabona 4, I-70126 Bari, Italy
*E-mail: c.ingrosso@ba.ipcf.cnr.it, marialucia.curri@uniba.it

5.1 Introduction

Graphene, a polycrystalline single-atom thick framework of sp^2-hybridised carbon atoms, arranged in a two-dimensional (2D) honeycomb lattice structure, is currently one of the most extensively investigated materials, due to its unique structural and electronic properties.

Inorganic Materials Series No. 14
Unconventional Green Synthesis of Inorganic Nanomaterials
Edited by Silvia Gross

Published by the Royal Society of Chemistry, www.rsc.org

The various extraordinary characteristics of graphene, such as large specific surface area,[1] high optical transparency in the visible spectral range,[2] ambipolar field effect,[3] high mechanical strength and stiffness, high unparalleled thermal conductivity,[4] remarkable electronic properties, and, specifically, high carrier mobility and capacity[5] and biocompatibility, account for the enormous technological interest in this material in a variety of fields. The potential applications of such an exceptional material range from (bio) sensors, to biomedicine, to (photo)catalysis, to energy storage devices (*i.e.*, supercapacitors and batteries), to fuel and solar cells, to transparent conductive films and flexible electronics.[6–9]

However, despite the remarkable properties exhibited by graphene, intrinsic limitations are also present, such as the lack of a bandgap, which hinders its applicability as a semiconductor in nanoelectronics, and its transparency in the visible light range (97.7% transmittance), which restricts its integration as a light harvester in solar light energy conversion devices. In addition, the absence of functional groups precludes the occurrence of specific molecular interactions, which are relevant for recognition properties, *e.g.*, for selective sensing.

These limitations motivate the intense efforts of the materials science community in developing doping procedures based on the use of heteroatoms,[10] to perturb the extended aromatic π-network of graphene, in modifying its electronic state distribution, and thus possibly resulting in a bandgap,[11] and in implementing chemical procedures for grafting functional groups onto its basal plane, so as to anchor another functional compound that is able to convey additional properties.

The large specific surface area and the high susceptibility of the graphene basal plane to chemical modifications allow decoration with inorganic nanostructures, polymers and biomolecules (proteins, peptides, aptamers, *etc.*) to impart missing functionalities to graphene or for improving its pristine properties.[12]

Graphene oxide (GO) and reduced graphene oxide (RGO) are graphene derivatives that show several of the outstanding properties of pristine graphene,[7–9,12,13] and, in addition, can be solution processed, making them suitable for the implementation of solution-based chemical procedures for *in situ* decoration with inorganic nanoparticles (NPs) to obtain hybrid nanocomposites. The latter ingeniously combine, in a synergistic manner, the functionalities of graphene with the unique size- and shape-dependent properties of the NPs (*i.e.*, surface plasmon resonance, excellent surface red/ox electrocatalytic activity, UV–visible–near infrared (NIR) light harvesting, (photo)catalytic and magnetic properties), thus boosting the pristine chemical–physical properties of both components and/or bestowing unprecedented, original functionalities, and so finally opening up the prospect of promising advanced applications.

The great potential of such a class of nanocomposites has recently prompted scientists to develop new *ex situ* and *in situ* hybridisation approaches with NPs, exploiting either the intrinsic chemical reactivity of the graphene basal plane, implementing non-covalent modification routes, or grafting, by non-covalent procedures, chemical groups or functional linkers, which can be exploited as anchoring sites of pre-synthesised NPs (*ex situ* routes), or as coordinating sites for their heteronucleation and growth (*in situ* routes).[12]

This chapter aims to provide a comprehensive overview of selected recent unconventional *in situ* chemical and physical synthesis methodologies for obtaining hybrid nanocomposites based on graphene derivatives and inorganic NPs. A discussion of their preparative routes, with a description of the morphology- and structure-related properties, which can be exploited for applications, is also provided.

5.2 Synthesis and Properties of Graphene and Graphene Derivatives

Since the first isolation of graphene single sheets and the assessment of their outstanding electronic properties in 2004,[4] a lot of effort has been devoted to the development and optimisation of the synthesis of graphene in the form of single and few layers.

Polycrystalline graphene sheets can be obtained by mechanical cleavage, liquid phase exfoliation of graphite, chemical vapour deposition (CVD) and epitaxial growth.[12]

In general, the different procedures lead to graphene derivatives with variable quality in terms of structural defects and optoelectronic properties, and also as a function of the substrate on which they are grown. Accordingly, the preparative approach is decisive in defining the optoelectronic and mechanical properties of the final material, and, consequently, its suitability for technological application.[13]

Graphene can be produced from graphite as mono or few layers by micromechanical exfoliation, which results from the mechanical cleavage of graphite by applying a force that is higher than the van der Waals force responsible for sheet stacking.[14] Specifically, the method relies on peeling off a piece of graphite by using an adhesive tape, or rubbing lithographically patterned pillars with 'tipless' atomic force microscopy (AFM) cantilevers.[12,14]

Such an approach provides single-layer graphene samples characterised by high purity, low density of defects and lateral sheet dimensions up to mm. However, these are produced in low yield and low throughput and with poor reproducibility. Therefore, these samples are appropriate only for fundamental studies and are impractical for large-scale applications.[13]

Epitaxial growth of graphene represents an attractive and alternative bottom-up route to mechanical cleavage, relying on the growth of graphene layers by sublimation of single-crystalline SiC substrates at high temperatures, followed by ordering of the carbide layers at these temperatures into a graphene structure with sp^2 bonding that is aligned with the SiC substrate, which triggers reconfiguration of the remaining carbon atoms into epitaxially aligned graphene.[15] However, epitaxial growth is not appropriate for large-scale production of graphene, as the graphene domains do not possess uniform thickness over a large area, unless expensive semiconducting substrates that are atomically flat and contamination-free are used. In addition, the tight bonding of graphene with the supporting substrate may affect its electronic properties due to doping effects.[16,17]

For a steady supply of graphene over large areas and in large quantities, which is suited for mass applications, the preferred approaches are CVD onto catalyst substrates[18,19] and liquid phase exfoliation from graphite.[20,21]

The growth of graphene by CVD from gaseous hydrocarbon precursors onto catalyst Cu foils at high temperatures is currently the preferred route for the industrial production of large-area and uniformly thick polycrystalline graphene sheets, providing coverage of *ca.* 95% of monolayer graphene, with high structural quality and high carrier mobility, and which is suited for fabricating transparent coatings for nanoelectronics.[22] However, such a method still requires tight process control, thus resulting in costs that are much higher than for liquid phase exfoliation.[23]

Thanks to the wide availability and great abundance of graphite in nature and its high exfoliation capability, liquid phase exfoliation of graphite in different solvents[20,21] and intercalative or thermal expansion of graphite[24] are currently among the most cost-effective routes for mass production of graphene nanoplatelets (GNPs) with randomly distributed shapes and lateral dimensions of the order of hundreds of nm up to tens of mm.

One type of very versatile GNP, for diverse applications, is represented by GO, which is achieved by oxidation of graphite that is then exfoliated or expanded by using a combination of strong acids and oxidants, such as potassium chlorate and fuming nitric acid,[25,26] or potassium permanganate and sulfuric acid.[27] The GO nanosheets achieved are then further exfoliated by ultrasonication to form single-, few- and multilayered GO, which can be dispersed in hydrophilic solvents.

The basal plane of GO is formed by randomly distributed aromatic domains characterised by sp^2 C atoms, aliphatic regions of sp^3 C atoms containing hydroxyl, epoxy and carbonyl groups lying above and below the graphene basal plane, and carboxyl functional groups, usually at the plane edges.

Although these types of GO sheets are enriched in oxygen functionalities,[28] which limit electron transfer, they still exhibit interesting electronic,[29] thermal, electrochemical and mechanical properties, and also feature optical transparency, biocompatibility and hydrophilicity. To restore the graphene lattice, GO can be reduced to RGO *via* thermal or chemical treatments. Such a process, called graphitisation or chemical conversion, can only partially remove the oxygen-containing moieties from the GO basal planes and edges, thus partially restoring the graphene lattice by re-hybridisation of the sp^3

C atoms to sp^2.[30] For this reason, the electrical and thermal conductivity of RGO approach those of graphite.[31]

The efficiency of chemical conversion of GO to RGO depends on the experimental conditions applied for the reduction, as well as the strength of the reducing agent and treatment duration and temperature. As assessed by density functional theory (DFT) studies, the chemical conversion leaves oxygen functionalities,[32] mainly epoxy and hydroxyl, as well as damage and defects in the carbon network,[31] thus affecting the RGO properties.

The most common reducing agents used to remove oxygen groups from the GO basal plane are hydrazine, hydroquinone, ascorbic acid and sodium borohydride.[26,30] Oxygen functional groups in the form of water, carbon dioxide and carbon monoxide can also be removed by thermal reduction on exposing GO to microwaves[30] or treating GO in a quartz tube at 1000–1100 °C in the absence of air.[33]

It is worth noting that, while the electrical conductivity and the specific capacitance of RGO are significantly higher than that of GO, the latter presents higher transparency and dispersibility in solvents. In addition, despite the lower structural quality of RGO with respect to graphene, chemical conversion of GO to RGO is considered the most suitable method for scalable production of small graphene sheets, due to its cost effectiveness and temperature and pressure requirements being milder than those needed for CVD and epitaxial growth.[23]

The oxygen functionalities in GO make it dispersible in water and in alcohols, *N,N*-dimethylformamide (DMF), *N*-methyl-2-pyrrolidone (NMP) and tetrahydrofuran (THF), thus opening the way to the implementation of inorganic and organic compound functionalisation routes. In addition, the high surface chemical reactivity of RGO is available for binding other compounds by supramolecular and covalent interactions.

In general, the possibility of handling such compounds as macromolecules, and hence their processability, enable their easy fabrication in freestanding films or coatings on surfaces,[34,35] also assisting with their ability to merge with other species to develop original multifunctional materials.[36]

5.3 Covalent and Non-covalent Functionalisation of Graphene and Related Derivatives

Covalent and non-covalent routes for functionalisation of graphene and its derivatives are reported here. Covalent approaches rely on grafting chemical species onto the graphene basal plane by binding free radicals or dienophiles, and converting sp^2 C atoms into sp^3. Non-covalent approaches exploit supramolecular interactions, without affecting the extended π-conjugation of the basal plane and hence its electric conductivity and mechanical strength.[12]

5.3.1 Covalent Functionalisation Routes

Covalent functionalisation encompasses both: (1) formation of covalent bonds between free radicals or dienophiles and the –C=C– bonds of graphene; and (2) formation of covalent bonds between organic molecules and oxygen groups of GO.

Reactive free radicals produced by diazonium molecules, for instance, can graft onto graphene, forming a covalent bond and generating n- or p-type doping effects,[37] resulting in some cases in modification of the distribution of the electronic state of graphene and therefore in generation of a bandgap.[38]

Although the chemical doping of graphene with diazonium salts typically leads to partial interruption of the delocalised aromatic system, which decreases the planar conductivity,[39] such an effect can be controlled by tuning the reaction conditions and monitoring the process by Raman spectroscopy, which provides an indication of the degree of the covalent functionalisation from the ratio between the intensities of the D and G peaks (I_D/I_G ratio), which is typical of graphitic materials, at *ca.* 1340 cm^{-1} and 1590 cm^{-1}, respectively. The D peak, originates from the breathing modes of the sp^2 C atoms in hexagonal rings and is a second-order Raman mode, which involves the inelastic and elastic scattering of an electron with phonons and defects, respectively.[40] After functionalisation with diazonium salts, the intensity ratio between the D and G peaks increases proportionally.[39]

Dienophiles are compounds that react with the sp^2 C atoms of graphene, and include commonly used azomethine ylides, which react with the carbon network of graphene by 1,3-dipolar cycloaddition with the evolution of simultaneous bond cleavage and formation.[39] This reaction is highly flexible as it offers a wide choice of available aldehydes or α-amino-substituted acid precursors, and yields a variety of reactive functional groups grafted onto graphene.

Some example are offered by the large number of visible light-responsive nanocomposites based on the phthalocyanine (Pc) and porphyrin surface-modified graphene compounds prepared by Bottari *et al.*[41] and Ragoussi *et al.*,[42] which are effective for applications in optoelectronics, solar energy conversion and biomedicine.

The oxygen moieties of GO are exploited for the synthesis of nanocomposites with diverse peripherally substituted chromophores (*i.e.*, phorphyrins, phthalocyanine, polythiophenes) and polymers featuring hanging hydroxyl and amine groups (*e.g.*, poly(ethylene glycol), polylysine, polyallylamine and poly(vinyl alcohol)), by amidation, esterification and atom transfer radical polymerisation (ATRP) reactions, which are particularly suited for improving GO dispersion in solvents and for preparing multifunctional nanocomposites with promise for photovoltaics, (bio)sensors and in biomedicine.[12]

Functionalisation of GO with amine polyethylene glycol, subsequent esterification with folic acid and, finally, π–π stacking of the anticancer drug camptothecin resulted in a drug delivery system characterised by pH-dependent drug release with high anticancer activity.[43]

GO was refluxed at 70 °C in a DMF solution of $SOCl_2$ and treated at 120 °C with octadecylamine (ODA) to achieve GO with a high density of –COOH groups that were available for binding the $-NH_2$ moieties of ODA by amidation.[44] $NaBH_4$ and the $HAuCl_4$ precursor were added to synthesise Au NPs of 3–30 nm in size and to reduce GO to RGO.[44] In this process, the $HAuCl_4$:GO-ODA w/w controls the morphology and size dispersion of the Au NPs, as well as the coating density, ultimately resulting in modulation of the surface plasmon resonance absorption band of the graphite (G)/Au NP composite.[45]

ATRP can be effective for functionalising GO nanoplatelets with polymers that grow directly at the surface of GO *via* living radical polymerisation, after covalent attachment of the initiator monomer (styrene, methyl methacrylate or butyl acrylate). The obtained nanocomposites show significant enhancement of thermal and mechanical properties compared to the bare polymers.[12]

5.3.2 Non-covalent Functionalisation

The non-covalent or supramolecular functionalisation of graphene derivatives relies on the establishment of π–H bonds and van der Waals forces, electrostatic and π–π interactions between the graphene basal plane and the organic molecules, polymers[12,14] or inorganic nanostructures, without affecting the aromatic electronic network.

The different surface chemistry of pristine graphene, GO and RGO translates into differences in their interaction with aromatic species.[14] Pristine graphene monolayer has a rich extended aromatic system with an almost planar geometry that undergoes strong π–π interactions with small aromatic molecules[14] (*i.e.*, pyridine and fluorinated benzene derivatives, quinoline, pyrene, anthracene, triphenylene, coronene, ionic liquids and polymer ionic liquids or porphyrin derivatives).

RGO has an aromatic character that is lower than pristine graphene, whereas GO can be described as a network of sp^3 C atoms featuring hydroxyl, carboxyl and epoxy moieties, which are responsible for a deviation from planarity and are able to interact with isolated and small aromatic domains of sp^2 C atoms.[14] Wang *et al.* demonstrated that the higher hydrophobic character of RGO makes it able to bind small aromatic hydrocarbons (*i.e.*, naphthalene, phenanthrene and pyrene)[46] much more effectively than GO, which is characterised instead by a polar surface that limits π–π interactions with such species.

Pyrene is an aromatic compound that can undergo strong π–π stacking and van der Waals interactions with the graphene basal plane, allowing concomitantly the establishment of an effective electron coupling. The π–π interactions of such a class of aromatic molecules are influenced by the presence of electron-donating or electron-withdrawing groups, as well as by their steric hindrance.[14]

The electron donor pyrene-1-sulfonic acid and the electron acceptor 3,4,9,10-perylenetetracarboxylic diimidebisbenzene sulfonic acid interact strongly with RGO by π–π interactions, inducing a remarkable doping effect[47] and making the resulting functionalised RGO sheets highly dispersible in water, as they are stabilised by repulsive interactions between the dangling sulfonic groups.[47]

Pyrene-based molecules are also able to exfoliate graphite and/or stabilise the functionalised GO nanosheets in solvents, acting as 'molecular wedges', which penetrate deeper between the graphene layers upon sonication, without using strong acids, reducing agents or high-temperature annealing, which are typically used in other exfoliation routes.

Using 1-pyrenecarboxylic acid (PCA) graphite was effectively exfoliated into single-, few- and multilayered flakes, which retained the sheet structural quality, and concomitantly grafted polar functional groups onto the graphene surface *via* non-covalent π–π stacking.[48]

The functionalisation of graphene with pyrene and perylene derivatives can tailor its surface chemical reactivity, and thus enable template growth or deposition of other materials. Water-soluble negatively charged thioglycolic acid/2-mercaptoethanesulfonate co-stabilised CdTe quantum dots anchored by electrostatic interactions onto positively charged trimethyl-(2-oxo-2-pyrene-1-yl-ethyl)-ammonium bromide decorated graphene nanosheets, showed effective charge transfer upon photoexcitation.[49]

On the other hand, the surface functionalisation of nanostructures with functional pyrene molecules allows their chemical immobilisation on graphene. As an example, a solution-based procedure was implemented for decorating large-area monolayer CVD graphene films with size-tuneable UV–vis–NIR light-absorbing colloidal PbS nanocrystals (NCs).[50] The immobilisation, obtained by soaking the graphene films in a solution of pyrene-1-butyric acid (PBA) surface-coated PbS NCs, led to π–π stacking interactions between PBA and graphene, resulting in stable anchoring of the NCs. The sheet resistance of such hybrid structures decreased by 3% with respect to the bare graphene after illumination with a white light-emitting diode, due to hole transfers from the photoexcited PbS NCs, interconnected to form a multilayer film, to graphene, which was channelled by pyrene.[50] The same approach was used for fabricating an optically transparent and UV light-active anode based on a CVD graphene surface decorated with PBA-coated TiO_2 NCs and characterised by high (photo)conductivity and charge mobility.[51]

Pristine graphene and RGO are also characterised by their hydrophobic character, which can be exploited for binding hydrophobic or partially hydrophobic organic molecules, such as surfactants, ionic liquids or macromolecules, with the aim of increasing the dispersion of graphene nanosheets in solvents, adding chemical functionalities to the basal plane or improving the incorporation in polymers.[12,14]

On the other hand, GO, which instead presents a more hydrophilic character, can interact electrostatically, or *via* hydrogen bonds, with other molecules or macromolecules. The negative carboxylate groups of GO, for instance, were exploited for binding electrostatically positive complexes of Au NPs, first modified by using Nafion™ (Au@Nafion) and then by poly(diallyldimethylammonium chloride) (PDDA), and then linked to L-012 luminol (Au@Nafion@PDDA@L012) to achieve a self-electrochemiluminescent nanocomposite of GO and Au@L012 having a self-luminescence *ca.* 740-fold higher than that of the bare Au NPs coated with L012. The GO/Au@L012 nanocomposite was used as a label for the carcinoembryonic antibody (CEA) of MCF-7 cells to detect attomole concentrations of CEA antigen from 72 cells.[52]

5.4 Chemical Routes for Nanocomposites Based on Inorganic Nanoparticle-decorated Graphene Derivatives

Hybrid nanocomposites based on inorganic NPs and graphene derivatives can be prepared following both *ex situ* and *in situ* approaches. The former rely on anchoring a molecular linker onto the graphene basal plane that features functional groups (–NH_2, –COOH, –SH) capable of binding to the surface of pre-synthesised NPs or ligand molecules coordinated to the NP surface, by using covalent or non-covalent routes.[50,51] The latter approaches exploit the functional groups of the molecular linkers anchored onto the graphene basal plane as coordinating sites for the heteronucleation and growth of the NPs starting from metal precursors.[53]

The *in situ* bottom-up and top-down routes, which rely on both chemical and physical methods, allow hybrid nanocomposites formed by inorganic NP-decorated graphene derivatives to be achieved. The multifunctional components are combined by means of different unconventional approaches, yielding nanocomposites organised in hierarchical nanostructures. Use of sacrificial organic and inorganic templating agents provide novel paths to nanostructures with original morphologies, which are then transferred to the final nanocomposite, which is characterised by unprecedented mechanical strength, chemical stability, high density of surface-active sites, high energy storage capability, high electron transfer properties, high electrical conductivity,[54] (photo)catalytic activity[55] and electrocatalytic activity.[56] All of these characteristics hold great promise for a wide range of applications, including batteries,[57] (bio)sensors,[58] supercapacitors, fuel cells and catalytic systems.[55]

5.4.1 Bottom-up Chemical Approaches

5.4.1.1 Organic Polymer-assisted Routes

A facile and environmentally friendly approach for synthesising both noble metal and metal oxide NPs grafted onto graphene derivatives relies on the use of a polymer precursor that polymerises and acts as both reducing and coordinating agent for the formation of the NPs. This route does not require high temperatures, high pressures, harsh chemicals or even organic surfactants for the reduction of the NP precursor, and the organic matrix improves dispersion of the resulting hybrid nanocomposite in solvents, also favouring graphene–NP electron coupling.

Conducting polymers have attracted attention for the synthesis of this kind of ternary nanocomposite, not only for their chemical moieties, which behave as coordinating sites to effectively control the NP morphology, but also for their high electrocatalytic activity when deposited onto electrodes, and provide good environmental and chemical stability, relatively high electrical conductivity and high processability in combination with other materials.[59]

A nanocomposite based on NiO NPs and polypyrrole (PPy)-coated RGO (NiO@ PPy/RGO) was synthesised by an *in situ* oxidative polymerisation technique,[60] with the aim of increasing the low, inherent electrical conductivity and low Seebeck coefficient of PPy, in order to prepare an efficient thermoelectric material capable of producing an appreciable thermo-electromotive force from a suitable low temperature difference.[54]

For the synthesis of this nanocomposite, the oxidising agent ammonium persulfate was added dropwise at room temperature to a mixture of RGO, NiO NPs and hydrochloric acid and pyrrole (Py) monomer, to initiate the oxidative polymerisation. The NiO NPs were distributed homogeneously in the polymer matrix, being coordinated by the amine groups of the PPy, with almost uniform size distribution and an average size of 160 nm. Meanwhile, PPy anchored the RGO sheets by π–π stacking interactions.

X-ray diffraction (XRD) investigation showed that the π–π interactions between the conjugated chains of PPy and RGO led to the formation of more ordered polymer regions with better crystallinity.

Raman measurements displayed two strong characteristic peaks at 1325 cm^{-1} and 1589 cm^{-1}, corresponding to the D and G bands of RGO, respectively, and two peaks of PPy at 1330 cm^{-1} (D band) and 1593 cm^{-1} (G band), corresponding to the ring and the –C=C– backbone stretching vibrations of neat PPy, respectively (see Figure 5.1A).[54]

In the PPy-coated RGO complex, the signals for PPy were shifted to 1323 cm^{-1} and 1586 cm^{-1}, with an intensity higher than that of neat PPy, due to π–π interactions between PPy and RGO. On the other hand, in the NiO@PPy/RGO nanocomposite these signals were further shifted to 1332 cm^{-1} and 1570 cm^{-1}, respectively, and were more intense, probably due to the occurrence of interactions between NiO NPs and PPy. In addition, the I_D/I_G intensity ratio decreased to 0.921 and 0.909 for the PPy-coated RGO and NiO@PPy/RGO nanocomposite, respectively, due to reduction in the π–π conjugated defects in the polymer backbone chain and to the more ordered molecular arrangement of PPy, compared to the neat polymer, both of which were responsible for an improvement in the carrier mobility along the polymer backbone chain (see Figure 5.1A).[54]

The UV–vis spectrum of PPy presented absorption bands at 267, 356 and 500 nm, attributed to the π–π* transition of the aromatic ring, the π–π* transition of the polymer chain and transitions from the valence band to the uppermost bipolaron band and antibipolarons of the oxidised form of PPy,

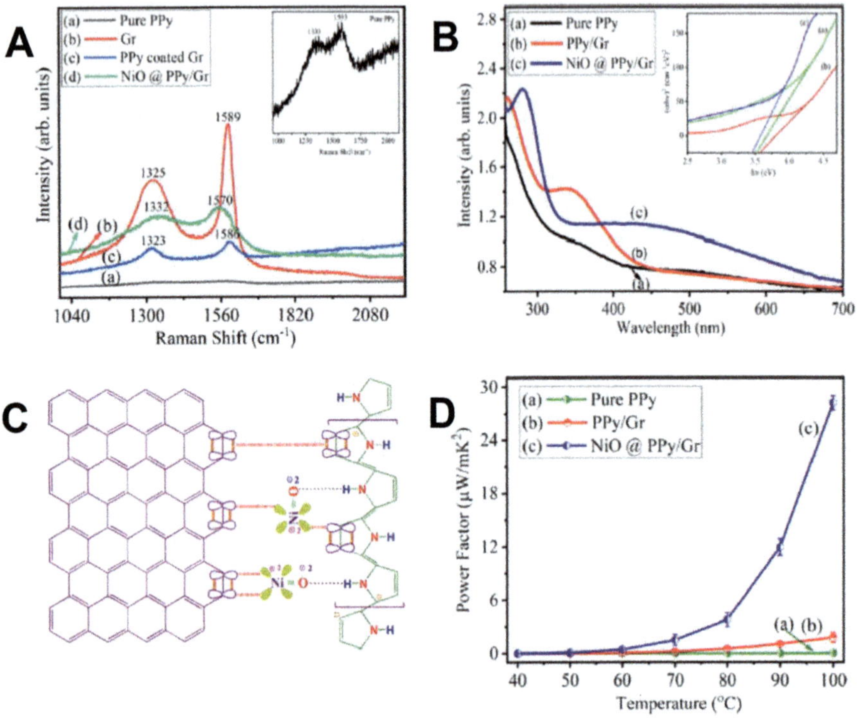

Figure 5.1 (A) Raman spectra of neat PPy, RGO, PPy-coated RGO and NiO@PPy/RGO composite. (B) UV–vis spectra of neat PPy, PPy-coated RGO and NiO@PPy/RGO nanocomposite. (C) Sketch of the PPy, RGO and NiO NP interactions in NiO@PPy/RGO. (D) Temperature dependence of the power factor for pure PPy, PPy-coated RGO and NiO@PPy/RGO nanocomposite. Reproduced from ref. 54 with permission from American Chemical Society, Copyright 2020.

respectively (see Figure 5.1B). In the absorption spectra of PPy-coated RGO the blue shift of the characteristic π–π* transition peaks from 267 nm to 260 nm and from 356 nm to 344 nm, confirmed the strong PPy-RGO interaction (see Figure 5.1B).[54]

The incorporation of NiO NPs in PPy-coated RGO led to a shift of the ring π–π* transition at 281 nm due to the extension of the conjugation length of PPy and formation of an effective network through electronic transitions from the NiO NPs to RGO. In addition, the peaks at 356 nm and 500 nm were replaced by a broad signal at 425 nm (see Figure 5.1B).[54]

The strong interactions between NiO NPs, PPy and RGO, which can be attributed to the bond between the oxygen atoms of NiO and the H atoms of PPy (see Figure 5.1C), behaved as a charge transport channel between PPy and RGO, leading to the formation of defect levels in the bandgap of PPy, and finally resulting in a shift of the energy bandgap of the nanocomposite towards lower energies (see inset of Figure 5.1B).[54]

As a result, the electrical conductivity and Seebeck coefficient of the nanocomposite were enhanced, as assessed by the maximum value of the power factor, which was 28.22 μW mK^{-2} at a temperature difference of 100 °C (see Figure 5.1D), *ca.* 855 and 15.7 times higher than that of PPy and PPy-coated RGO, respectively.[54]

Oxidative polymerisation of Py was exploited by Balakumar *et al.*[55] for the preparation of a nanocomposite formed by Ag NPs and PPy decorating GO sheets, as a selective catalyst for the oxidation of hydroquinone (HQ) to benzoquinone (BQ) in the presence of hydrogen peroxide (H_2O_2). Such a catalyst has been designed as a ternary nanocomposite because the combination of graphene, metal NPs and conducting polymer endows the resulting material with a stability higher than that of the neat polymer–graphene complex and the metal NP–polymer nanocomposite.[55]

Py polymerises to PPy in the presence of the oxidant $AgNO_3$, which is simultaneously reduced *in situ* to Ag NPs in the presence of GO, behaving as a coordinating agent with its oxygen functionalities.[55]

Spectroscopic and morphological investigation of the nanocomposite demonstrated the effective growth of Ag NPs, of 10 nm in diameter, uniformly distributed onto the surface of PPy, as well as the occurrence of strong π–π interactions between GO and the Ag NPs-PPy nanocomposite. Ag NPs/PPy/GO was found to be able to catalyse the oxidation of HQ to BQ, with H_2O_2 in aqueous solution, more efficiently than neat GO, PPy-GO and PPy-Ag NPs.[55]

Another ternary nanocomposite based on Au NPs coordinated to PPy anchored onto nitrogen-doped (N-doped) RGO (Au NPs/PPy/N-RGO) was prepared starting from $HAuCl_4$ by *in situ* oxidative polymerisation of Py, which was previously covalently anchored by amidation to the carboxyl groups of the N-RGO sheets, for the highly sensitive and selective electrochemical detection of catechol (CC).[61]

Nitrogen-doped RGO was covalently functionalised by Py and, after dropwise addition of an aqueous $HAuCl_4$ solution, the heteronucleation of Au NPs onto the –NH groups of PPy occurred concomitantly with its oxidative polymerisation.[61]

Glassy carbon electrodes (GCEs) modified with the hybrid nanocomposite showed high electrocatalytic activity towards CC oxidation, which was higher than for pristine N-doped RGO and N-doped RGO/PPy. This was attributed to the occurrence of PPy-mediated electron transfers from the Au NPs to N-RGO, to the high conductivity and to the high specific surface area of the nanocomposite.[61]

The CC sensor exhibited a wide, linear sensing range, from 0.1–0.9 μM, with a limit of detection (LOD) and quantification (LOQ) of 0.0016 μM and 0.0053 μM, respectively (S/N = 3 and 10). The proposed approach was reliable and was successfully applied for the selective determination of CC in real samples, with good recovery and stability.[61]

A ternary nanocomposite formed by Au NPs, RGO and polydopamine (PDA) was prepared, exploiting the self-polymerisation properties of dopamine (DA), as well as the capability of PDA to reduce metal salts to metal NPs, and GO to RGO, by binding its basal plane by π–π interactions and stabilising the final nanocomposite in solution.[62]

Such a ternary nanocomposite was obtained by non-covalently functionalising GO sheets with DA, which self-polymerises to PDA in alkaline conditions at room temperature. Then, $HAuCl_4$ was added, and the $[AuCl_4]^-$ ions, coordinated by the DA amino groups, were reduced *in situ* by the DA catechol groups, still at room temperature and without harsh reducing reagents and organic surfactants, to Au NPs, which were then stabilised by the same catechol moieties.[62] The transmission electron microscopy (TEM) investigation showed that the size and morphology of the Au NPs synthesised onto the RGO sheets depended on the PDA-RGO : $HAuCl_4$ w/w%, thus confirming the role of the PDA amino groups in behaving as coordinating agents.

The obtained PDA/RGO/Au NP nanocomposite was tested as a catalyst for the reduction of 4-nitrophenol (4-NP) to 4-aminophenol in the presence of $NaBH_4$. The results showed that the catalytic efficiency depended on the size and coating density of the Au NPs on RGO, and hence on the PDA-RGO : $HAuCl_4$ w/w%.[62]

Oxidative self-polymerisation of DA non-covalently anchored onto the GO sheets was also used for the *in situ* synthesis of Ag and Pt NPs, as well as for magnetic Fe_3O_4 NPs and semiconducting TiO_2 NPs.[63]

Ag and Pt NPs were synthesised *in situ* at room temperature onto the GO surface coated with DA, by using $AgNO_3$ and chloroplatinic acid ($H_2PtCl_6{\cdot}2H_2O$) as precursors, respectively to achieve PDA/RGO/Ag NPs and PDA/RGO/Pt NPs nanocomposites.[63] Fe_3O_4 NPs were obtained by adding an aqueous precursor solution of Fe^{2+} and Fe^{3+} to the DA-GO complex and performing its controlled oxidation under ambient conditions at the low temperature of 90 °C.

The PDA/RGO/Fe_3O_4 NPs nanocomposite showed good dispersibility in water, as well as a good response to an external magnet. Its magnetic hysteresis curve, recorded at 300 K, showed a superparamagnetic state with a saturation magnetisation of 19.4 (emu g).[63]

PDA/RGO/TiO_2 NPs nanocomposites were prepared by hydrolysis of a Ti alkoxide onto the PDA/RGO complex, after dropwise addition into an ethanol solution of DA-GO.[63]

The oxidative polymerisation of aniline to polyaniline (PANI), anchored non-covalently onto GO sheets, and occurring at 30 °C in the presence of the oxidant $CuNO_3$, which was then reduced to Cu NPs, was exploited for the preparation of the ternary nanocomposite Cu NPs/PANI/GO for the electrochemically sensitive and selective determination of the neurotoxic and carcinogenic hydrazine.[58] Cu NPs, 20 nm in size, heteronucleated and grew coordinated to the aniline amine groups with simultaneous polymerisation of PANI.[58] UV–vis absorption spectra demonstrated the formation of the Cu NPs and the occurrence of strong interactions between PANI and Cu NPs. Infrared spectroscopy showed that PANI was adsorbed onto the surface of GO *via* hydrogen bonding and π–π stacking, and the XRD assessed the decrease in the GO aggregation with the occurrence of *in situ* synthesis of the PANI-Cu NPs nanocomposite.[58] Amperometry measurements of the PANI/GO/Cu NPs-modified GCEs showed a linear relationship between the catalytic current and hydrazine concentration, with a LOD, LOQ and sensitivity of 0.0045, 0.015 μM (S/N = 3 and 10) and 359.93 μA mM^{-1} cm^{-2}, respectively, making the prepared electrodes among the most successful reported.[58] Finally, they exhibited stability, reproducibility and an outstanding recovery towards hydrazine in diverse real-water samples.[58]

A nanocomposite formed by a coordination polymer (COP) achieved by polymerisation of 1,2,4,5-benzenetetramine (BTA) onto RGO, in the presence of palladium(II) chloride ($PdCl_2$), was synthesised at room temperature after *in situ* reduction of Pd^{2+} to Pd NPs by $NaBH_4$, as an electrocatalyst for the hydrogen evolution reaction (HER).[56]

COPs have recently attracted interest for hydrogen storage and evolution,[64] since they possess high specific surface area and metal centres, which favour electrocatalytic HERs, as well as high stability at diverse pH. However, most of the catalysts derived from COPs display high overpotential and slow kinetics for water splitting,[65] and hence their combination with graphene can improve the stability and activity of the metal centre catalyst.

The nanocomposite [Pd(BTA)-RGO] was prepared by adding NH_4OH, RGO and BTA tetrahydrochloride to an aqueous solution of $PdCl_2$ at room temperature. BTA polymerised, assembling by π π interactions onto RGO, and bound Pd^{2+} ions in linear chains formed by Pd^{2+}–NH- bridging motifs, which were reduced to Pd after addition of $NaBH_4$, to allow the formation of $[Pd(BTA)\text{-}RGO]_{red}$.[56]

XRD and field-emission scanning electron microscopy (FESEM) analyses showed that the reduction by $NaBH_4$ led to the formation of a crystalline, 2D sheet-like $[Pd(BTA)\text{-}RGO]_{red}$ layered network, with Pd NPs embedded into the COP.[56] Moreover, XRD showed an increase in the interplanar distance of the face-centred cubic (fcc) unit cell of Pd NPs *via* lattice expansion, which was attributed to the absorption of hydrogen during the borohydride reduction.[66] The TEM images of $[Pd(BTA)\text{-}RGO]_{red}$ showed the Pd NPs deposited onto the sheet-like structures, with a size of *ca.* 3–5 nm (see Figure 5.2a), whereas the electron diffraction from a selected area (SAED) of $[Pd(BTA)\text{-}RGO]_{red}$ presented diffraction rings with *d*-spacings of 0.2330 nm, 0.2082 nm, 0.1481 nm and 0.1236 nm, which could be ascribed to the *d*(111), *d*(200), *d*(220) and *d*(311) planes, respectively (see inset of Figure 5.2a).

Such evidence, which was also in agreement with the results of XRD, Fourier-transform infrared (FTIR) and X-ray photoelectron spectroscopy (XPS) analyses, confirmed the occurrence of lattice expansion due to hydrogen absorption, evidencing the establishment of Pd–H interactions between Pd NPs and the –NH– groups of the $[Pd(BTA)\text{-}RGO]_{red}$ composite.[56]

The high-resolution TEM (HRTEM) image presented the lattice fringes of the *d*(111) and *d*(200) planes of the Pd NPs having an interplanar distance of 0.233 nm and 0.208 nm, respectively (see Figure 5.2b). In addition, TEM-energy dispersive X-ray spectroscopy (EDS) maps of the region reported in the high-angle annular dark-field (HAADF) image (see Figure 5.2c) showed

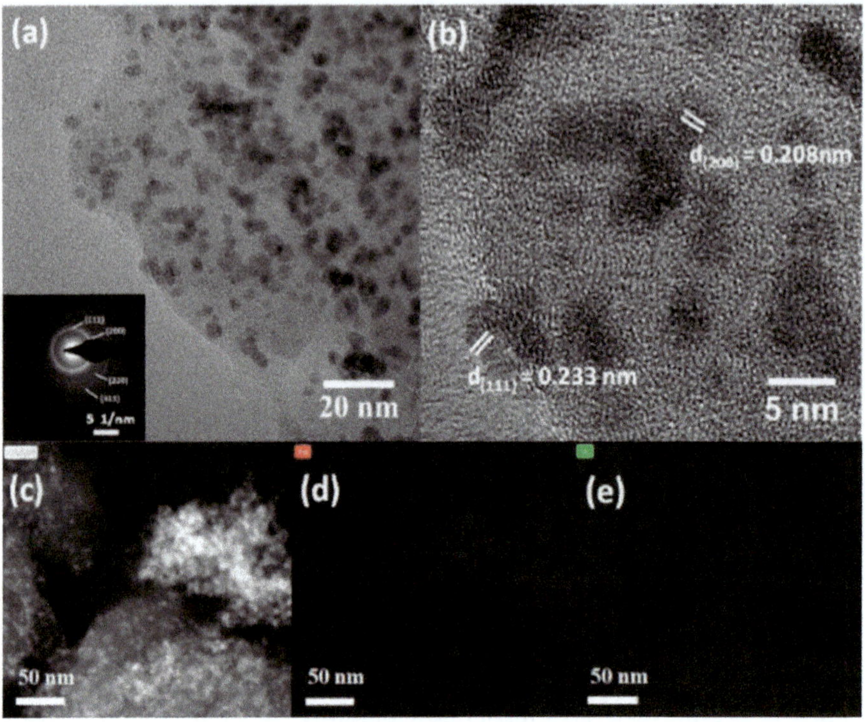

Figure 5.2 HRTEM, HAADF and EDS images of $[Pd(BTA)\text{-}RGO]_{red}$: (a) low-magnification TEM image with SAED image in the inset; (b) high-magnification HRTEM image; (c) HAADF image; (d) EDS map of Pd; (e) EDS map of N. Reproduced from ref. 56 with permission from American Chemical Society, Copyright 2019.

bright red spots on the sheets, supporting the formation of Pd NPs (see Figure 5.2d), and green spots corresponding to N of BTA (see Figure 5.2e) uniformly distributed onto the sheets.[56]

Linear sweep voltammetry measurements demonstrated that $[Pd(BTA)\text{-}RGO]_{red}$-modified GCEs presented an overpotential that was three-fold lower than Pd(BTA), $Pd(BTA)_{red}$ and Pd(BTA)-RGO complexes, with a current density of −10 mA cm^{-2} (η_{10}), which confirmed the crucial role of RGO in determining the η_{10} value.[56] Analysis of the Tafel plot showed that both $[Pd(BTA)\text{-}RGO]_{red}$ and Pd(BTA)-RGO nanocomposites had Tafel slope values significantly lower than for the neat COP, indicating that RGO accelerated the kinetics of the HER.[56]

The double layer capacitance (C_{dl}) of $[Pd(BTA)\text{-}RGO]_{red}$ was higher (0.097 F cm^{-2}) than that of Pd(BTA)-RGO (0.015 F cm^{-2}). C_{dl} was directly proportional to the electrochemical surface area (ECSA), which was higher for $[Pd(BTA)\text{-}RGO]_{red}$ because of the presence of Pd NPs participating actively in the HER,[66] whereas Pd(BTA)-RGO contained only Pd^{2+} ions.[56]

5.4.1.2 Co-precipitation-assisted Routes

Nanocomposites formed by N-doped graphene (NG) and Fe_3O_4 NPs were synthesised by an edge-functionalised ball milling approach, followed by a co-precipitation method as a catalyst for the colorimetric reaction between hydrogen peroxide (H_2O_2) and 2,2′-azino-bis(3-ethylbenzo-thiazoline-6-sulfonic acid) diammonium salt (ABTS).[67]

An intrinsic mimicking activity of peroxidase enzymes[68] was observed in Fe_3O_4 NPs and in graphene oxide-Fe_3O_4 NPs, which was effectively applied for the detection of H_2O_2 and glucose. N-doping of graphene was demonstrated to be an efficient strategy to further improve the peroxidase-like activity of graphene-based peroxidase mimetics[69] because of the presence of electron-acceptor nitrogen atoms.[70]

N-doped graphene was prepared by milling a mixture of graphite and chitosan at 500 rpm min^{-1} for 12 h in a planetary ball mill. In this process, graphite exfoliated and chitosan bound graphene edges, introducing N atoms onto the basal plane.[67] Then, NG was added to $FeCl_3 \cdot 6H_2O$ and $FeCl_2 \cdot 4H_2O$ and kept at 80 °C with overnight stirring under N_2. NaOH was subsequently added dropwise and the solution was further kept at 80 °C under stirring to allow the Fe_3O_4 NPs to precipitate onto the NG sheets.[67]

AFM investigation showed that the NG had an average thickness of *ca.* 1.8 nm, which increased up to 5.8 nm after functionalisation with chitosan. After co-precipitation of the Fe_3O_4 NPs, the NG sheets presented an average thickness of 15.7 nm, indicating successful immobilisation of the Fe_3O_4 NPs.[67]

The nanocomposite formed a homogeneous dispersion in water and could be magnetically actuated towards an external magnetic field.[67] Both Fe_3O_4 NPs and the related nanocomposite with NG demonstrated coercivity/remnant magnetisation and superparamagnetic behaviour. The saturated magnetisation of Fe_3O_4 NPs and Fe_3O_4 NPs/NG was *ca.* 62 and 40 emu g^{-1}, respectively.[67]

The resulting Fe_3O_4 NPs/NG nanocomposite presented outstanding intrinsic peroxidase-like activity, as assessed by the enhanced oxidation of ABTS by H_2O_2. Colour change, as well as an increase in the UV absorption signal intensity at 417 nm, due to oxidation of ABTS after addition of the nanocomposite, were observed.[67]

The optical biosensor showed a good performance in detecting H_2O_2, showing linearity up to 10×10^{-3} M H_2O_2 and a LOD of 17.1×10^{-6} M. The catalytic activity depended on pH and temperature, with optimal conditions at pH 4.0 and 40 °C, respectively. Finally, the nanocomposite demonstrated a glucose concentration-dependent linear optical signal (up to 18.0×10^{-3} M) with a LOD of 57.9×10^{-6} M and a sensitivity higher than that required for the determination of glucose in blood for the clinical diagnosis of diabetes.[67]

A ternary nanocomposite based on RGO, Fe_2O_3 and SnO_2 NPs was synthesised as the anode for lithium-ion batteries (LIBs) by *in situ* precipitation of Fe_2O_3 NPs onto GO, which was reduced by $SnCl_2$ to RGO, resulting in the formation of SnO_2 NPs, which were then added to the nanocomposite.[57]

For the preparation of the nanocomposite, $FeCl_3 \cdot 6H_2O$ and urea were added to an alkaline solution of GO and, after sonication, the suspension was transferred to a Teflon-lined autoclave, sealed and maintained at 120 °C for 4 h.

During the hydrothermal process, the oxygen functional groups of the GO sheets electrostatically interacted with and coordinated Fe^{3+} ions, which then precipitated onto the GO sheets due to the presence of OH^- groups, with concomitant decomposition of urea. Then, after cooling down the mixture, the pH of the suspension was adjusted to 7.0 and $SnCl_2 \cdot H_2O$ was added to reduce GO at a temperature of 30 °C for 1 h. Finally, the product was dried at 80 °C for 8 h, followed by annealing at 400 °C for 1 h under Ar, to enhance the degree of crystallisation of Fe_2O_3 and SnO_2 in the ternary nanocomposite.[57]

The reduction mechanism of GO by Sn^{2+} depended on the temperature,[55] but this has not been fully clarified; upon addition of Sn^{2+}, the carboxyl groups on GO were expected to chelate Sn^{2+} ions, leading to formation of interlinked GO complexes, which could, however, be easily separated by sonication.

The ternary anode showed a high stable capacity during discharge–charge cycling tests in LIBs. The anchoring of Fe_2O_3 and SnO_2 NPs onto RGO hindered the stacking of graphene sheets and, in addition, SnO_2 NPs avoided the agglomeration of Fe_2O_3 NPs during Li^+ uptake/release. As a result, the novel nanocomposite exhibited initial discharge and charge capacities of 1179 and 746 mA h g^{-1}, respectively, at a current density of 400 mA g^{-1}, and a nearly 100% discharge–charge efficiency, which was maintained after 100 cycles with a specific capacity of 700 mA h g^{-1}.[57]

In a further paper, a nanocomposite formed by FeS NPs wrapped with RGO (FeS@RGO) was prepared as the anode for LIBs by co-precipitation followed by annealing.[71]

The nanocomposite was obtained by dissolving $Fe(NO_3)_3 \cdot 9H_2O$ in deionised water and adding it to a GO suspension under vigorous stirring. The mixture was sonicated and dropwise reacted with an aqueous solution of $Na_2S \cdot 9H_2O$ reducing agent. The obtained precipitate containing Fe_2S NPs and S was collected by vacuum filtration and washed with water. It was finally transferred into a quartz tube furnace and annealed at 400 °C under Ar to reduce and exfoliate GO, as well as to increase the crystallinity of the FeS NPs.[71]

The addition of the GO suspension to the $Fe(NO_3)_3$ solution under ultrasonication promoted the electrostatic attraction of Fe^{3+} ions to the functional groups (*e.g.*, –OH, –COOH) of the GO surface. Then, the addition of Na_2S led to precipitation of the FeS NPs, whereas stirring and ultrasonication enabled the NPs to wrap with the GO sheets. In the subsequent annealing step in Ar, FeS NPs crystallised, whereas GO was partially converted to RGO.[71] XRD investigation demonstrated the formation of FeS NPs in the nanocomposite without decomposition to Fe in the annealing process, and TEM measurements showed that the RGO sheets wrapped multiple NPs, like peas in a pod.[71]

The FeS NPs prepared without GO had an average diameter of *ca.* 150 nm, whereas in the nanocomposite their size decreased, suggesting that the inclusion of GO limited aggregation of newly formed FeS NPs during precipitation, as well as limiting coalescence during annealing. High-resolution

TEM measurements showed that the NPs were surrounded by wrinkled RGO sheets.[71] The nanocomposite showed a very stable cyclic performance with a high capacity of 978 mA h g^{-1}, which remained constant even after 40 cycles and was threefold higher than that of a commercial graphite anode (372 mA h g^{-1}) and comparable to those reported in the literature.[72]

Nanocomposites formed by metal NPs, namely Pt, Pd and Ru, respectively supported onto graphene nanosheets and covered by mesoporous silica ($mSiO_2$) layers, were prepared by a facile and scalable wet-chemical approach.[73] The metal hydroxide (MOH) was loaded onto the surface of GO nanosheets by a general deposition–precipitation method,[74] and NH_4OH was used to catalyse the hydrolysis of tetraethyl orthosilicate (TEOS) directly to form a mesoporous $mSiO_2$ shell covering the surface of the MOH-GO composite, in which cetyltrimethylammonium bromide (CTAB) behaved as a pore-generating agent.[75] Finally, the as-prepared nanosheets of MOH-GO@$mSiO_2$ were annealed under H_2 to convert GO into RGO and MOH into metal NPs, thus producing M-RGO@$mSiO_2$ composite nanostructures.

The Pt-RGO@$mSiO_2$ nanocomposite showed a nanosheet-like structure, with lateral dimensions ranging from hundreds of nm to a few μm, with a thickness of approximately 28–30 nm. TEM investigation indicated that mesopores with an average size of *ca.* 3 nm formed uniformly in the silica layers, whereas Pt NPs were well encapsulated between the RGO and $mSiO_2$ layers, without aggregation. The removal of the outer $mSiO_2$ layer by etching with an aqueous solution of NaOH revealed the morphology of the RGO and Pt NPs, which was characterised by uniform Pt NPs with an average size of 1.65 nm that were well dispersed onto the RGO supporting nanosheets.[73]

The ultrafine Pt NPs of the Pt-RGO@$mSiO_2$ composite did not undergo sintering during annealing in H_2 at 240 °C, thanks to the protection afforded by the $mSiO_2$ layers, as well as during calcination at high temperatures (300–700 °C). Similarly, the graphene nanosheets were shown to act as a space barrier, thus hindering Pt NP sintering at high temperatures. The retention of the ultrasmall size of the Pt NPs allowed significant improvement in the catalytic performance for various reactions, thanks to the large specific surface area of the catalyst and the high density of active sites at the edges and corners of the Pt NPs[76] anchored onto graphene.[73] In addition, the $mSiO_2$ layers avoided π–π stacking of graphene, so that the Pt NPs in the Pt-RGO@$mSiO_2$ composite exposed more catalytic active sites, with a consequent higher catalytic conversion efficiency. Pt-RGO@$mSiO_2$ showed a high catalytic activity both in gas and in liquid phase, exhibiting excellent stability in high-temperature catalysis processes with excellent recycling and reusability in gas- and solution-phase reactions.[73]

5.4.1.3 Colloidal Chemistry Routes

Among the colloidal chemistry routes that are typically used for the preparation of inorganic NPs, the microemulsion approach has been extensively applied. Nanocomposites based on Ag/AgCl and Ag/AgBr NPs, respectively, wrapped by gauze-like GO nanosheets were synthesised by oil/water microemulsion at room temperature, to be used eventually as visible light-driven plasmonic photocatalysts.[77] The synthesis of Ag/AgBr/GO was carried out by adding dropwise a chloroform solution of CTAB to an aqueous dispersion of GO and $AgNO_3$, while for Ag/AgCl/GO cetyltrimethylammonium chloride (CTAC) was added.[77] TEM investigation showed that the Ag/AgX/GO nanocomposites (X = Br, Cl) consisted of Ag/AgX NPs with an average diameter of *ca.* 200 nm, wrapped with gauze-like GO nanosheets (see Figure 5.3A–D). Such a structure is formed thanks to coordination of the C=O groups of GO to the NP surface, as assessed by FTIR spectroscopy.[77]

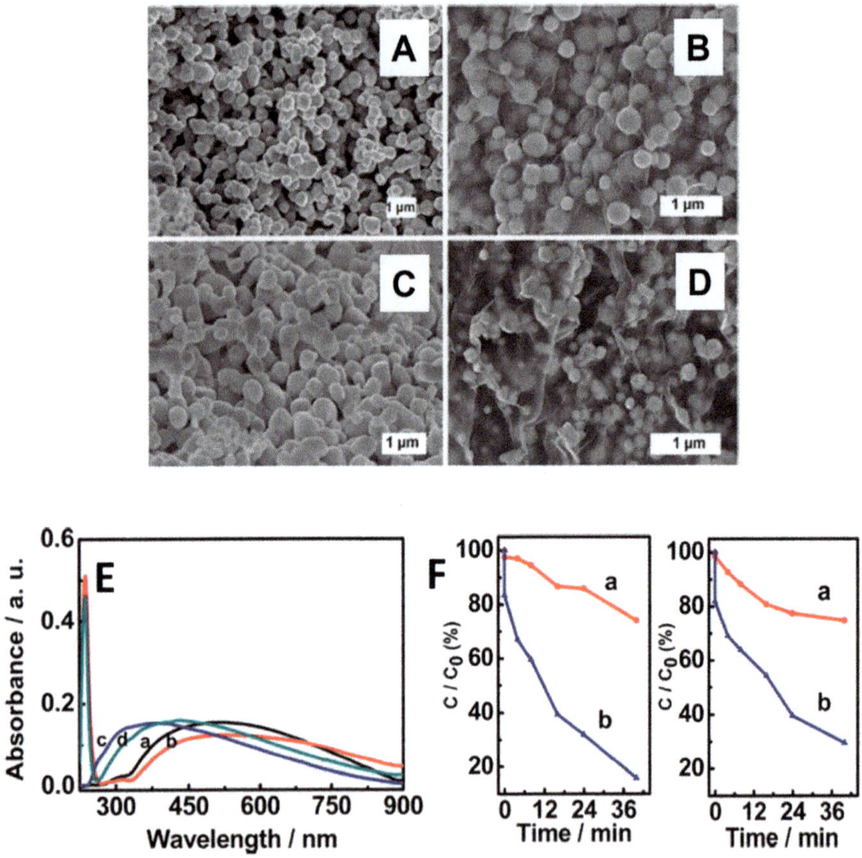

Figure 5.3 SEM images of the synthesised Ag/AgBr (A), Ag/AgBr/GO (B), Ag/AgCl (C) and Ag/AgCl/GO (D) nanostructures. (E) UV–vis absorption spectra of Ag/AgBr (a), Ag/AgBr/GO (b), Ag/AgCl (c) and Ag/AgCl/GO (d) nanostructures. (F) Photocatalytic activities of Ag/AgBr (a) and Ag/AgBr/GO (b) (on the left) and of Ag/AgCl (a) and Ag/AgCl/GO (b) (on the right) in the photodegradation of methyl orange under visible light irradiation. Reproduced from ref. 77 with permission from American Chemical Society, Copyright 2011.

In the absence of GO, the nanospheres had a diameter of *ca.* 500 nm, and hence the smaller Ag/AgX NPs observed in the nanocomposite could be due to the GO oxygen functionalities behaving as coordinating sites in the oil/water microemulsion.[77,78]

The Ag/AgX-based nanostructures showed a broad absorption signal, spanning from the UV to the visible spectral region, which was ascribed to the plasmon resonance absorption of the metallic Ag,[79] as well as the distinct absorption of GO at *ca.* 238 nm[80] (see Figure 5.3E).

The Ag/AgCl/GO and Ag/AgBr/GO hybrid nanocomposites were used as stable plasmonic photocatalysts for the photodegradation of methyl orange (MO) under visible light irradiation, and displayed photocatalytic activities (see Figure 5.3F) 2.8 and 3.5 times higher than the corresponding Ag/AgCl or Ag/AgBr compounds, respectively.[77]

This enhanced performance was ascribed to: (1) π–π non-covalent interactions of MO with GO, resulting in an increase of the adsorptive capability of the Ag/AgX/GO photocatalysts to MO; (2) the smaller size of the Ag/AgX NPs in Ag/AgX/GO, resulting in photocatalysts with a higher surface area; and (3) a more favourable charge separation/transfer at the hybrid junction, enabled by the direct coordination of the GO oxygen moieties to the NP surface, which suppresses electron–hole pair recombination in Ag/AgX/GO, with GO and Ag/AgX behaving as electron acceptor and donor, respectively.[77]

A coral-like nanocomposite based on RGO sheets wrapping NiCo alloy NPs ($Co_{0.5}Ni_{0.5}$/RGO) was synthesised as a catalyst for the oxygen evolution reaction (OER) by simultaneous reduction of hydrophobic complexes of Ni^{2+} and Co^{2+} with bis(2-ethylhexyl) phosphate (HDEHP) in a water/oil emulsion containing hydrophilic GO, followed by thermal annealing.[81]

Co- or Ni-based catalysts have been widely studied because of their high OER activity, and hierarchical nanocomposites of such nanoalloys with graphene have demonstrated enhanced catalytic activity.[82,83]

The nanocomposite was prepared by mixing an organic solution of $Ni(DEHP)_2$ and $Co(DEHP)_2$ with an aqueous solution of GO and after addition of $NaBH_4$, the Ni^{2+} and Co^{2+} ions electrostatically attracted and coordinated to the carboxylate groups of GO, thus reducing to alloy NPs alongside conversion of GO to RGO.[81]

The Co atoms in the NiCo alloy NPs were coordinated by the carboxylate sites of the graphene basal plane and, after annealing at 600 °C under Ar, NiCo alloy NPs with sizes less than 100 nm were effectively encapsulated by an RGO shell to form a $Co_{0.5}Ni_{0.5}$/RGO coral-like structure.[81] Microscopy investigation demonstrated that the RGO sheets decreased the tendency of the NiCo NPs to aggregate, thus also preventing their oxidation.[81] A small amount of Ni modified the electronic structure of the Co species involved in the OER process, synergistically improving the OER activity of the NiCo alloy NPs.[81] Finally, the RGO shell facilitated charge transfer for intermediate species and provided a higher catalyst/electrolyte interphase area, with more active sites available for surface processes since the NP aggregation was prevented. Indeed, $Co_{0.5}Ni_{0.5}$/RGO showed a better OER performance compared to $Co_{0.5}Ni_{0.5}$ NPs, with good electrochemical durability.[81]

The overpotential was lower (288 mV) than that of the standard RuO_2 catalyst (299 mV) at a current density of 10 mA cm^{-2}, and lower than that of the bare alloy NPs, with a dependence on the RGO content, finally indicating that the conductivity of the graphene shell improved the efficiency of the OER process.[81]

Effective electron coupling between the nanocomposite components, with concomitant reliable control of the NP morphology, can be achieved when suitable linker molecules and effective coordinating agents are used.[53]

A novel and facile *in situ* colloidal chemical approach was recently reported for the synthesis of a new hybrid nanocomposite formed by PCA surface-functionalised RGO flakes decorated with a high density of 3,4-dimethylbenzenethiol (DMBT)-capped Au NPs and characterised by a monodisperse size distribution.[53]

Commercial RGO was exfoliated in single-, few- and multilayered sheets and functionalised with PCA, following a procedure reported in the literature with minor modifications,[48] with stirring and sonication in an ice bath of a mixture of PCA and RGO powder in NMP.[53]

The PCA-RGO/Au NPs hybrid nanocomposite was prepared by modifying the two-phase method of Brust *et al.*,[84] which is typically used for the synthesis of thiol-coated Au NPs. PCA-modified RGO sheets were dispersed in a solution of tetraoctylammonium bromide (TOAB) and toluene, and then $HAuCl_4 \cdot 3H_2O$ was added. After transfer of the Au precursor from water to toluene, assisted by TOAB, water was removed from the reaction flask and DMBT was added to allow a controlled reduction of Au^{3+} to Au^{+}. Finally, a solution of $NaBH_4$ in MilliQ water was added and growth of the Au NPs was allowed to proceed overnight.

In such a synthesis, the short aromatic thiol: (1) yielded additional binding sites for the Au NPs to the PCA-RGO surface; (2) granted NPs size monodispersion; (3) potentially enhanced electron conductivity and promoted charge transfers; and (4) enabled dispersion in organic solvents.[53]

The TEM investigation showed that the nanocomposite PCA-RGO/Au NPs was mainly formed by micrometre-sized, large sheet-like structures ascribed to the PCA-RGO flakes, homogeneously and densely coated by Au NPs of 2.8 ± 0.6 nm mean diameter. The Au NPs heteronucleated and grew onto the flakes from Au^{3+} species stabilised by the –COOH groups of PCA, with uniform morphology and narrow size distribution (see Figure 5.4A).[53]

Control experiments demonstrated the crucial role of the carboxyl functionalities of PCA in coordinating and hence anchoring the Au NPs onto the surface of the flakes. The Au NPs self-assembled onto the RGO sheets in porous multilayer stacks formed by π–π interactions with the DMBT ligand molecules, with the Au NPs behaving also as nano-spacers between the PCA-RGO sheets (see Figure 5.4A).[85]

SEM images of the PCA-RGO flakes, collected before and after decoration with the Au NPs, showed that the surface morphology of the PCA-RGO flakes

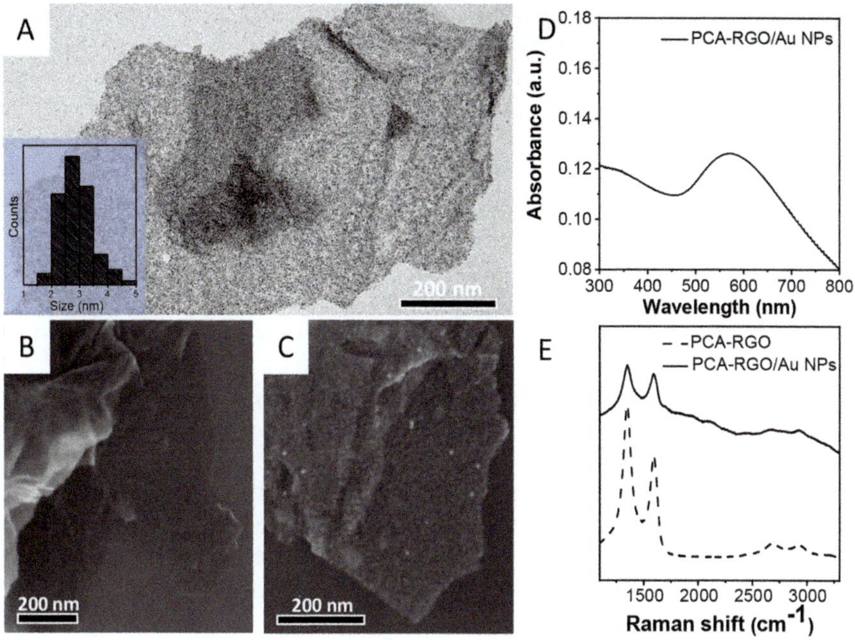

Figure 5.4 (A) TEM micrograph of PCA-RGO/Au NPs flakes with the corresponding Au NPs size distribution (inset). SEM images of (B) bare PCA-RGO (×160.22) and (C) PCA-RGO/Au NPs flakes (×174.12), respectively. (D) UV–vis absorption spectrum of 0.15 mg mL^{-1} PCA-RGO/Au NPs flakes, 7 $\times 10^{-3}$ M in Au and 2 mg mL^{-1} in PCA-RGO in toluene. (E) Raman spectra of PCA-RGO and PCA-RGO/Au NPs flakes. Reproduced from ref. 85 with permission from the Royal Society of Chemistry.

was characterised by low contrast and sheet-like structures (see Figure 5.4B), while for the PCA-RGO/Au NPs samples, a grain-type and bright contrast network of NPs was observed on the sheets, which was ascribed to the highly electron dense Au NPs (see Figure 5.4C).

The UV–vis absorption spectrum of the hybrid nanocomposite presented the typical localised surface plasmon resonance (LSPR) absorption signal of Au NPs at 564 nm,[84] which was shifted towards the lower energy side of the spectrum with respect to the expected position of the LSPR band of comparable size and shape Au NPs (see Figure 5.4D).[85]

The Raman spectra of both the PCA-RGO complex and the PCA-RGO/Au NPs hybrid nanocomposite presented the characteristic G and D peaks of RGO (see Figure 5.4E). The latter displayed a broad background signal due to the PCA photoluminescence and a lower intensity ratio between the D and G peaks (see Figure 5.4E), which is representative of a less defective material. This can be explained by considering that the nanocomposite had domains of conjugated sp^2 C atoms, which were larger than those in the PCA-RGO complex. Such a feature was ascribed to the reduction of RGO defects induced by addition of $NaBH_4$ that occurred at room temperature because of the catalysis by immobilised Au NPs.[85]

The cyclic voltammograms (CVs) of PCA-RGO and PCA-RGO/Au NPs dispersed in toluene and drop cast on screen-printed carbon electrodes (SPCE) presented a decrease in ΔE_p with respect to bare SPCEs, which was accounted for by the higher conductivity and the enhancement of the electron transfer capability at the electrode/electrolyte interface upon SPCE surface modification. In addition, the peak current of the PCA-RGO/Au NPs hybrid-modified SPCEs increased with respect to that of bare and PCA-RGO-modified SPCEs, which was reasonable due to the increase in the electroactive surface area, and attested to the higher electrochemical reactivity of the hybrid-modified SPCEs.[85]

The electroactive area of the PCA-RGO/Au NPs-modified SPCEs was twofold higher than that of neat SPCEs and the apparent heterogeneous electron transfer rate constant of the hybrid nanocomposite-modified SPCEs was threefold higher than that of bare SPCEs, indicating that the hybrid nanostructure modification led to increased conductivity and faster heterogeneous electron transfer kinetics at the electrode surface,[85] which both accelerate the electron transfers at the electrode/electrolyte interface. Indeed, in the nanocomposite, the Au NPs were anchored onto the RGO sheets by means of the aromatic PCA linker and the DMBT capping ligand, ensuring electron coupling due to particle–RGO and interparticle π–π interactions.[85]

The manufactured colloidal hybrid-modified SPCEs were a well-suited platform for developing alkaline phosphatase (AP)-, horse radish peroxidase- and glucose oxidase (GOD)-based electrochemical biosensors for enzyme amplified bioassays. In particular, the platform was highly reliable and sensitive for the AP-mediated signal amplification assay for the miRNA-221 biomarker detected in spiked human blood serum samples, with a LOD of 0.7 pM (7 amol in 10 mL) and an average percentage standard deviation (%RSD) of 13% in the range 1–5000 pM.[85]

The PCA-RGO/Au NPs SPCEs were also applied for the detection of vitamin C in infant food and formula as a sensing tool for quality control laboratories, as an alternative to routine enzymatic-based spectrophotometric methods for determining vitamin C in complex food matrices.[86] They showed a good linear response over the 50–500 μM concentration range, with a LOD of 17 μM and a LOQ of 57 μM with a %RSD of 8% ($n = 5$), which was lower than for other graphene/Au NPs-modified electrodes reported in the literature, and lower than those of routine enzymatic-based spectrophotometric methods.

5.4.1.4 Annealing in a Controlled Atmosphere

One scalable and facile *in situ* synthesis approach for hybrid nanocomposites relies on the thermal treatment, in a controlled atmosphere, of mixtures of inorganic precursors and GO, with the addition of sacrificial templating organic molecules, silica nanostructures or metal–organic frameworks (MOFs). This route was used for obtaining nanocomposites with a hierarchical porous morphology, which endows the final material with a high specific surface area that can be effectively exploited in batteries, electromagnetic wave (EMW) absorbing devices and in the electrocatalytic oxygen reduction reaction (ORR) and the OER in fuel cells.

These materials have a high capacity for energy storage, but have low conductivity, and undergo large volume expansion during charge–discharge processes, which produces an unstable solid–electrolyte interphase (SEI) layer, and which can even lead to pulverisation of the electrode surface and NP aggregation, resulting in fast capacity fading, low rate capability and unsatisfactory electrode cycle life.[87,88]

To solve these issues, the inorganic nanostructures were combined with carbonaceous materials such as graphene and/or carbon layers generated after thermal treatment in a controlled atmosphere, by degradation of organic molecules.

Specifically, GO was used because the oxygen moieties can coordinate with the metal atoms of the inorganic NPs, thus avoiding aggregation and allowing effective electron coupling between the nanocomposite components. Upon thermal treatment in a controlled atmosphere, GO is converted to RGO, making the nanocomposite highly conductive. In addition, thanks to their mechanical flexibility, RGO sheets act as a backbone to maintain the structural integrity of the electrode, and under suitable conditions allow wrapping of the NPs, thus accommodating their volume expansion, providing an extra protective layer to prevent direct contact with the electrolyte, avoiding the risk of formation of unstable SEI layers and contributing to an increase in the Coulombic efficiency.

A conductive carbon layer generated after calcination of sacrificial organic molecules or MOFs, possibly doped with N and S heteroatoms, has also been used in such nanocomposites, in combination with RGO for improving the conductivity. In addition, this presents voids or nanocages, which are effective for accommodating volume changes of the active material at the electrode surface and avoiding direct contact with the electrolyte.

Various MOF-derived materials have been used as anodes for high-performance batteries.[89]

When MOFs are used as sacrificial templates, the corresponding nanocomposites obtained after thermal treatment perfectly inherit their original morphology, exhibiting sandwich, yolk–hollow, core–shell, coral-like, crispy rice-like, pod-like, polyhedral, mulberry-like and pomegranate-like structures characterised by pores, voids and nanocages.

An antimony/amorphous carbon/RGO (Sb/C/RGO) nanocomposite was synthesised by an *in situ* route with a yolk–hollow structure as the anode for LIBs.[90] Metallic Sb has received attention for this application due to its high theoretical capacity of *ca.* 660 mA h g^{-1}, low Li^{+} insertion potential (~0.5–0.8 V *vs.* Li/Li^{+})[91] and remarkable Li storage at diverse high current densities. Sb was combined with graphene in an unusual morphology that resembled watermelon flesh, and which stemmed from a typical yolk–hollow structure with numerous seeds distributed in the continuous flesh.

The nanocomposite was prepared by mixing GO, antimony chloride ($SbCl_3$), citrate and cyanamide (CA) in water. Citrate formed a chelate compound with Sb^{3+} ions, inhibiting their hydrolysis, whereas the positively charged CA interacted with the functional groups of GO and citrate by means of covalent bonds, hydrogen bonding and electrostatic interactions. The calcination in Ar at 800 °C caused GO graphitisation and N-doping, thermal decomposition of citrate and CA, and release of Sb^{3+} ions. These species interacted electrostatically with the negatively charged oxygen groups of GO, which then acted as heteronucleation and growth sites for the *in situ* reduction of Sb NPs promoted by the surrounding N-doped carbon layers, generated upon oxidation of citrate and CA, and arranged in nanocages.

The Sb NPs formed strong Sb–O–C bonds with the N-doped RGO, and the resultant Sb/C/RGO nanocomposite was composed of a conductive N-doped amorphous carbon matrix and RGO, as well as voids in the carbon nanocages that were evenly distributed around the Sb NPs. The size of the voids and the density of the Sb–O–C bonds were tuned by varying the content of CA, because the positively charged CA bound part of the oxygen functional groups of GO which were also involved in the coordination of Sb^{3+}, and so, at a given Sb content, increasing the CA concentration resulted in a higher density of Sb NPs, and fewer Sb–O–C bonds in the Sb/C/RGO nanocomposite.

TEM investigation and elemental mapping images indicated that the nanocomposite presented blocky morphologies, with Sb NPs homogeneously embedded in the carbon matrix, while the watermelon flesh-like structure was observed only for certain CA concentrations and also affected the void size.[90]

In the absence of CA, the morphology of the Sb/C/RGO-0 nanocomposite was very different. When the CA content was 200 mg, the Sb NPs had a mean size of *ca.* 20 nm and were sparsely embedded in the carbon matrix. When the CA content increased up to 400 mg, the Sb NPs had a size of *ca.* 80 nm and were in the carbon nanocages with a mean diameter of *ca.* 100 nm.

When the Sb NPs were alloyed with Li^+ to form Li_3Sb, a volume expansion of *ca.* 150% occurred in the discharge process.[91] The calculated size of the Li_3Sb NPs was *ca.* 92 nm, which was very close to that of the nanocages of *ca.* 100 nm, which therefore could accommodate the Li_3Sb NPs. When the CA concentration increased up to 600 mg, Sb NPs with a size of *ca.* 120 nm were in the carbon nanocages with an average diameter of 200 nm, and the calculated size of the Li_3Sb NPs was *ca.* 136 nm, which was still far smaller than the size of the nanocage.[90]

Another original nanocomposite formed by MoO_2 NPs assembled in crispy rice-like mesoporous microrods, derived from a MOF and wrapped with RGO (MoO_2/RGO) was fabricated as a sulfur host for long-life Li–S batteries, which are currently considered to have high potential as a solution for next-generation electrochemical energy storage systems, thanks to their high theoretical energy storage density.[92]

The MoO_2/RGO crispy rice-like mesoporous microrods were prepared by mixing Mo-based MOF (Mo-MOF) with GO, followed by annealing in Ar at 600 °C. In this process, Mo^{2+} centres were coordinated by the oxygen groups of GO and, upon thermal treatment, they were converted into MoO_2 NPs with concomitant formation of RGO.[92] Microscopy investigation demonstrated that the MoO_2/RGO nanocomposite retained the microrod morphology of the Mo-MOF structure, and the MoO_2 NPs assembled in crispy rice-like microrods coated by RGO.[92]

The mesoporous morphology of the microrods enabled efficient impregnation of sulfur, blocked diffusion of soluble lithium polysulfides (LiPSs) by physical confinement, limited dissolution and consequent shuttling of LiPSs and enhanced the transport of electrons and Li^+ diffusion. Moreover, fast electron transfer that occurred from RGO to LiPSs *via* MoO_2 promoted redox conversion of LiPSs to thiosulfates (polythionates).

The MoO_2/RGO cathode presented a discharge capacity of 1145 mA h g^{-1} at 0.5C, which remained at 1027 mA h g^{-1} after 500 cycles, corresponding to an average capacity decay rate of 0.02% per cycle. Specifically, the cathode had an ultralong cycling life with a low capacity decay of only 0.016% over 3200 cycles at 3 °C.[92]

In a further study, a novel nanocomposite formed by a stable 3D conductive network based on iron phosphide (FeP) NPs embedded in a polyhedral carbon network arranged onto RGO sheets (FeP@C/RGO) was prepared as the anode for LIBs.[93] Metal phosphides have recently become regarded as promising anode materials because of their high theoretical capacity, small electrode polarisation and low charge–discharge potential.[94] The lithium superconductor Li_3P, which is produced during the charge and discharge of metal phosphides, has an ionic conductivity that is higher than that of lithium conductors produced in other LIB anode materials (*i.e.*, Li_2O, Li_2S),[95] whereas FeP has broad application prospects due to its abundant storage and large theoretical capacity.[96]

The FeP@C/RGO nanocomposite was prepared using the novel MIL-101(Fe) MOF as a sacrificial template, which was anchored onto the oxygen moieties of the GO basal plane after solvothermal treatment at 120 °C in DMF, and then converted to a well-defined porous carbon polyhedral skeleton, which resembled the MOF morphology after annealing at 450 °C in Ar. Then, FeP NPs were achieved after a phosphiding treatment with $NaH_2PO_2{\cdot}H_2O$ at 350 °C.

The FeP NPs were found to be uniformly encapsulated in the carbon skeleton and effective in confining the volume changes (*ca.* 202%). The resulting FeP@C nanocomposite anchored to the RGO sheets to construct a highly conductive 3D network.[93]

The nanocomposite showed a discharge capacity of 414.7 mA h g^{-1} at a current density of 8 A g^{-1}, reaching a capacity of 949.7 mA h g^{-1} after 100 cycles at 0.1 A g^{-1}. Even when cycled at a current density of 1 A g^{-1}, it had a capacity of 737.7 mA h g^{-1} after 450 cycles.[93]

Hierarchical Si/SCF@RGO nanocomposites formed by Si NPs dispersed in a 3D carbon network of S-doped carbon fibres (SCF) wrapped by RGO sheets (Si/SCF@RGO) were synthesised as anodes for LIBs,[97] with Si being attractive for its high theoretical capacity and low working potential.[98]

Such nanocomposites were achieved by an *in situ* soft-templated chemical oxidative polymerisation starting from a solution of Si powder dispersed in water with sodium dodecyl sulfate (SDS), and adding the oxidant $FeCl_3{\cdot}6H_2O$ and then heating up to 80 °C. 3,4-Ethylenedioxythiophene (EDOT) monomer was added under moderate stirring and oxidative polymerisation to PEDOT was allowed to proceed at 80 °C. SDS formed micelles with a hollow rod-like morphology upon addition of $FeCl_3{\cdot}6H_2O$, the EDOT molecules anchored the inner hollow SDS micelles and rapidly polymerised following the SDS template morphology, giving rise to nanofibres (NFs).

The SDS concentration was crucial for determining the size of the hollow regions and of the NFs. SDS also improved the dispersion of NF-PEDOT in aqueous solution and provided S-doping of the carbon network upon thermal treatment. S-doped carbon is expected to be more effective than N-doped carbon for the insertion/extraction of Li^+ because of the larger covalent radius of the S atom, which can increase the interlayer distance between the carbon atoms.

To synthesise Si/SCF@RGO, the as-prepared Si/NF-PEDOT composite was added to GO, which was reduced to RGO and wrapped around the Si/NF-PEDOT composite after addition of sodium ascorbate under vigorous stirring at 70 °C. The intermediate was dried by cryodesiccation and finally annealed at 800 °C under Ar to decompose PEDOT and achieve S-doped carbon fibres in the Si/SCF@RGO.[97]

The hierarchical structure of the Si/SCF@RGO nanocomposite was formed by a consecutive network of one-dimensional (1D) SCF and 2D flexible RGO sheets, with high specific surface area, high chemical stability and high conductivity, as well as good electrochemical performance.

When integrated in LIBs, the Si/SCF@RGO nanocomposite showed a reversible capacity of 1441.0 mA h g^{-1} at 400 mA g^{-1}, which was retained after 100 cycles. This enhanced cycling performance was attributed to the synergistic effect of the randomly stacked SCF conductive network and the wrinkled, flexible RGO nanosheets.[97]

Graphene-wrapped porous Fe_3O_4/N-doped carbon frameworks (Fe_3O_4/NC@RGO) were synthesised by using a sacrificial MOF template by electrostatic-induced assembly for fabricating anodes to integrate into LIBs.[99] Fe_3O_4 is regarded as a promising candidate for LIBs due to its high theoretical capacity (924 mA h g^{-1}),[100] but it has to be combined with carbon-based materials such as graphene in porous/hollow structures because it suffers from dissolution and agglomeration, limiting electrochemical activity.[101]

This hybrid nanocomposite was prepared by mixing amino-functionalised Fe-MOFs (NH_2-MIL-101(Fe)) with GO in ethanol, followed by freeze-drying at 80 °C. The positive surface charges of the amino Fe-MOFs anchored onto the negatively charged GO sheets through electrostatic attraction, which provided the driving force for the subsequent assembly, facilitating the formation of a hydrogel structure due to covalent bonds between the MOF amine groups and the GO carboxyl moieties. Finally, annealing at 500 °C under Ar thermally reduced GO and led to the formation of an N-doped carbon framework around the NPs wrapped by RGO.

The obtained Fe_3O_4/NC@RGO showed excellent electrochemical performance, delivering a reversible capacity of 764 mA h g^{-1} after 100 cycles at a current density of 0.2 A g^{-1}, showing a rate capability of 8.0 A g^{-1} with a reversible capacity of 370 mA h g^{-1} and displaying an enhanced long-term cycling stability of 441 mA h g^{-1} after 800 cycles at 2.0 A g^{-1}.

Sb NPs embedded in hollow carbon/RGO (Sb/RGO-C) microspheres were prepared by spray pyrolysis and subsequent annealing as an anode material for sodium-ion batteries (SIBs).[102] Sb has reversible alloying reactions with Na[103] and spray pyrolysis was chosen as an effective technique to fabricate the spherical microspheres[104] with carbonaceous materials.[105]

An aqueous solution prepared by dissolving antimony oxide, tartaric acid, and GO nanosheets was used for the preparation of the Sb/Sb_2O_3/RGO-C microspheres by spray pyrolysis at 500 °C and at an N_2 flow rate of 5 L min^{-1}. In this step, thermal reduction of the carbon species converted some Sb_2O_3 to Sb NCs and the decomposition of antimony tartrate chelates, at 500 °C, yielded hollow microspheres with numerous Sb NCs embedded in shells containing wide voids.

The growth of the heteronucleated Sb NCs was found to be inhibited by the surrounding RGO matrix. During spray pyrolysis, hollow microspheres formed due to the emission of large amounts of gas from the decomposition of organic materials, and GO was thermally reduced to RGO. Sb_2O_3 was subsequently completely reduced at 350 °C in H_2/Ar, resulting in Sb/RGO-C microspheres (see Figure 5.5). Figure 5.5a showed well-dispersed microspheres with Sb NCs confined in each empty void (arrows) (see Figure 5.5b and c).

The high-magnification TEM image of Sb/RGO-C clearly revealed that the Sb NCs were located at the graphitic carbon wall (Figure 5.5d), and the HRTEM image showed lattice fringe spacing of 0.31 nm and 0.37 nm, corresponding to the (012) and (003) planes of Sb, respectively (Figure 5.5e). The SAED pattern and elemental mapping are shown in Figure 5.5f and g. The Sb/RGO-C composite microspheres presented remarkable electrochemical properties, with a discharge capacity at the 2nd cycle of 433 mA h g^{-1} at a current density of 0.5 A g^{-1}, and capacity retention after 500 cycles, calculated from the 5th cycle, of 80%. The microspheres showed excellent rate capability and a discharge capacity as high as 323 mA h g^{-1} at a current density of 3.0 A g^{-1}.

Composite microspheres formed by RGO sheets decorated by $Li_4Ti_5O_{12}$ NPs and coated with a uniform carbon layer were synthesised *in situ* by using $Ti(OC_2H_5)_4$ as precursor and ethylene glycol (EG) as both stabiliser and carbon source, by a polyol-mediated spray-dry synthesis, followed by annealing.[106]

$Ti(OC_2H_5)_4$ was mixed with EG and anhydrous ethanol and an aqueous LiOH solution at 25 °C. EG was used as a stabiliser to form glycolates with the metal alkoxide precursor, thus limiting their hydrolysis. This solution was then homogeneously mixed with a GO suspension, and the final mixture was sprayed and dried to form C/Li–Ti–O/GO (C/LTO/GO) microspheres at 200 °C. The 2–3 μm microspheres, containing *ca.* 20 nm LTO NPs coated with a 3 nm thick carbon layer were then decorated on RGO sheets behaving as a

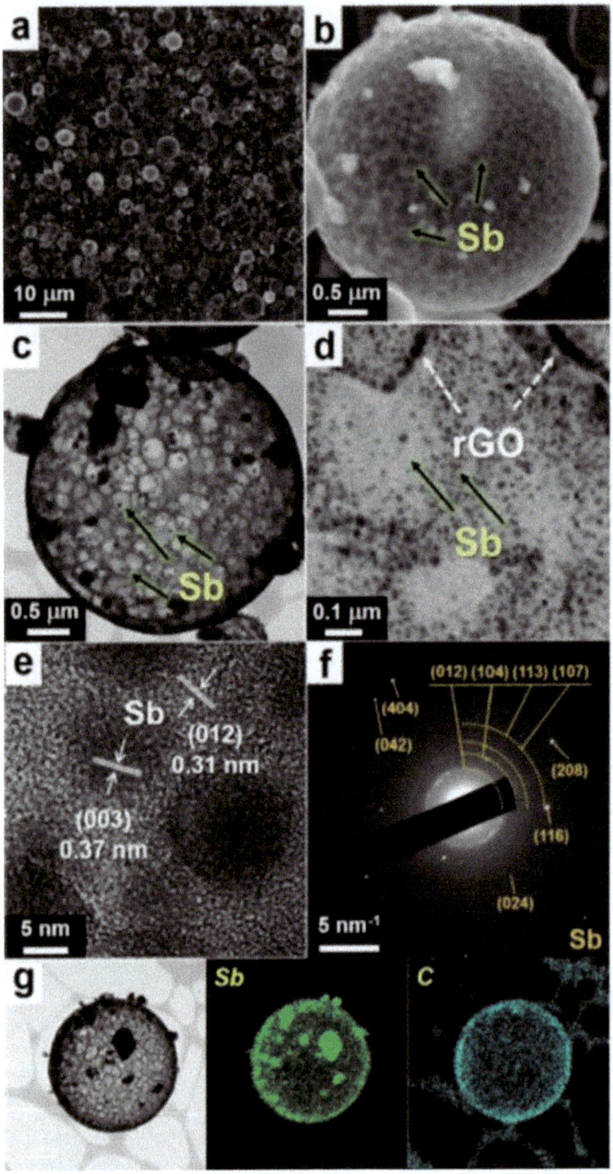

Figure 5.5 (a and b) SEM, (c and d) TEM and (e) HRTEM images and (f) SAED pattern and (g) elemental mapping of Sb/RGO-C nanocomposites. Reproduced from ref. 102 with permission from Elsevier, Copyright 2019.

conductive skeleton, and providing fast conduction pathways. In addition, C/LTO/RGO presented micropores of 5–50 nm in between the RGO sheets and the C/LTO NPs. These microspheres showed excellent performance as anodes in LIBs (99.7 mA h g^{-1} at 80C *vs.* 175 mA h g^{-1} at 0.1C) and SIBs (82 mA h g^{-1} at 50C *vs.* 219 mA h g^{-1} at 0.1C). Most importantly, C/LTO/RGO exhibited excellent cycling performance for both LIBs and SIBs at high and room temperature.[106]

Mulberry-like 3D hollow composite microspheres, based on RGO modified with CoO NPs (CoO/MGra), were synthesised by spray-drying followed by thermal treatment as cathodes for lithium-sulfur batteries (LSBs).[107] Although these batteries are extremely appealing for manufacturing portable devices, due to their energy storage capability, high theoretical specific capacitance and low cost, their large-scale commercial applications are hindered by the insulating properties of sulfur at room temperature, which limit fast electron transfer, by the tendency of its intermediates (Li_2S_x; $4 < x \leq 8$) to dissolve in the electrolyte solution (shuttle effect) and by the significant changes in volume during the charge and discharge of the sulfur products, which result in electrode collapse. Graphene is a good sulfur host for its capability to store LiPSs by van der Waals forces and physical adsorption, thus evading the shuttle effect. Similarly, metal oxides (Fe_3O_4, Fe_2O_3, Co_3O_4, CoO, NiO, MnO_2) are proper mediators for high energy density LSBs capable of anchoring LiPSs by –O–S– bonding, and avoiding the shuttle effect.[108]

CoO/MGra composite microspheres were generated by spray-drying at 120 °C with a feed rate of 5 mL min^{-1}, and mixing poly(methyl methacrylate) (PMMA), $Co(CH_3COO)_2$ and GO. The obtained product was placed in a tube furnace and heated to 500 °C in Ar with a heating rate of 5 °C min^{-1}.

The subsequent thermal treatment allowed: (1) PMMA removal, which served as a pore former leading to the mulberry-like hollow structure with concomitant bending and folding of GO; (2) $Co(CH_3COO)_2$ decomposition into CoO NPs; and (3) GO reduction.

The obtained 2–5 μm-sized CoO/MGra microspheres presented high crystallinity, with a mulberry-like 3D hollow RGO framework, with CoO NPs of size 4.8 nm that were uniformly distributed onto RGO, and cavities having a diameter of 400 nm with a high Brunauer–Emmett–Teller (BET) specific surface area of 108.03 m^2 g^{-1}, indicating the existence of a high density of active sites that were useful for storing LiPSs. Indeed, the CoO/MGra microspheres showed excellent cycle stability (1010 mA h g^{-1} after 150 cycles at 0.2C) and outstanding performance (745 mA h g^{-1} at 3C).[107]

A composite of Si NPs and RGO (Si/RGO), having a micrometre-sized porous pomegranate morphology, was successfully prepared as the anode for LIBs by spray-drying and subsequent magnesiothermic reduction and calcination.[109]

PDDA-coated monodisperse SiO_2 NPs, with an average size of approximately 150 nm, and few-layered GO sheets were used as precursors for the Si NPs and RGO, respectively. PDDA coated the SiO_2 NPs surface with a positive surface charge, electrostatically interacting with the negatively charged GO sheets.[109] An aqueous suspension of SiO_2 NPs and GO was spray dried at 180 °C, and the SiO_2 NPs immobilised onto the GO supporting matrix (SiO_2/GO) were further mixed with Mg powder to allow formation of smaller Si NPs and RGO. Then, a 3D porous RGO framework embedding the Si NPs was generated upon annealing at 700 °C under Ar. As a result, a micrometre-sized pomegranate-like Si/RGO composite, with a BET surface area of 384.2 $m^2 g^{-1}$ and pore size of 4 nm, was obtained. In this structure, the 3D porous RGO protecting framework provided high Li^+ storage, excellent electronic and ionic conductivity and immobilised the Si NPs. The voids left by the SiO_2 NPs, which transformed into smaller Si NPs, accommodated volume changes occurring during Li^+ storage. As an anode material for LIBs, the Si/RGO-S composite exhibited outstanding long-term cycle performance and rate capability. Under a current density of 500 mA g^{-1}, the Si/RGO composite showed a reversible capacity of 1200 mA h g^{-1} for 300 consecutive cycles.[109]

A 3D heterostructure based on RGO sheets supporting NiFe alloy NPs encapsulated in arrays of 1D N-doped CNTs (NCNTs), here indicated as NiFe/N-GCTs, was fabricated as a novel EMW absorbing material,[110] for limiting the impact of EMWs on the normal operation of communication equipment and for human health.

Carbon nanostructures are considered promising EMW absorbing materials owing to their robust mechanical strength, low density and large specific surface area, but their weak dielectric losses and non-magnetic losses have prompted combining them with magnetic loss materials in 1D and 2D nanocomposites, which are advantageous for improving EMW absorption.[111]

The nanocomposite was fabricated by mixing GO with melamine, $Fe(NO_3)_3 \cdot 9H_2O$ and $Ni(NO_3)_2 \cdot 6H_2O$, and then the solution was freeze-dried and annealed at 500 °C and 900 °C in Ar. SEM investigation highlighted that the NCNT array length was in the range of 3–10 μm (see Figure 5.6a) and the NCNTs were in tight contact with 1 nm thick RGO sheets, forming a 3D interconnected network that was effective in avoiding RGO sheet restacking.

TEM investigation showed that the NiFe alloy NPs, with a diameter of 8–40 nm, were encapsulated in the NCNTs (see Figure 5.6b).[110] TEM images revealed that the NCNTs exhibited bamboo-like shapes and their diameters were in the range of 15–45 nm (see Figure 5.6c), the wall thickness was *ca.* 0.5–5 nm (see Figure 5.6d) and the NCNTs bamboo joints were *ca.* 2 nm in thickness, with a periodic spacing of *ca.* 13 nm (see Figure 5.6d).[110] The NiFe alloy NPs were encapsulated within the NCNTs (see Figure 5.6e) with an interplanar spacing between the NPs of *ca.* 0.20 nm, ascribed to the *d*-spacing of the (111) plane of the NiFe NPs. The N-doping into carbon-based materials was favourable for enhancement of the oxidation resistance of metal NPs, increasing the stability of the overall 3D heterostructure. The resulting NiFe alloy NPs were surrounded by 3–5 layers of RGO having an interplanar spacing of *ca.* 0.34 nm, corresponding to the (002) plane of the graphitic carbon, with various structural defects (blue circles in Figure 5.6e and f).[110]

The 3D NiFe/N-GCT presented the highest dielectric loss, mainly resulting from its largest polarisation relaxation values. In such a material, multiple relaxations occurring upon EMW irradiation were accounted for by: (1) the nanosized NPs, which improved the specific surface area of 3D NiFe/N-GCT, leading to many atoms with a lot of dangling bonds on the outermost surfaces, which caused polarisation relaxation;[112] (2) the numerous interfaces between the metal NPs, NCNTs and RGO sheets, which were regions in which

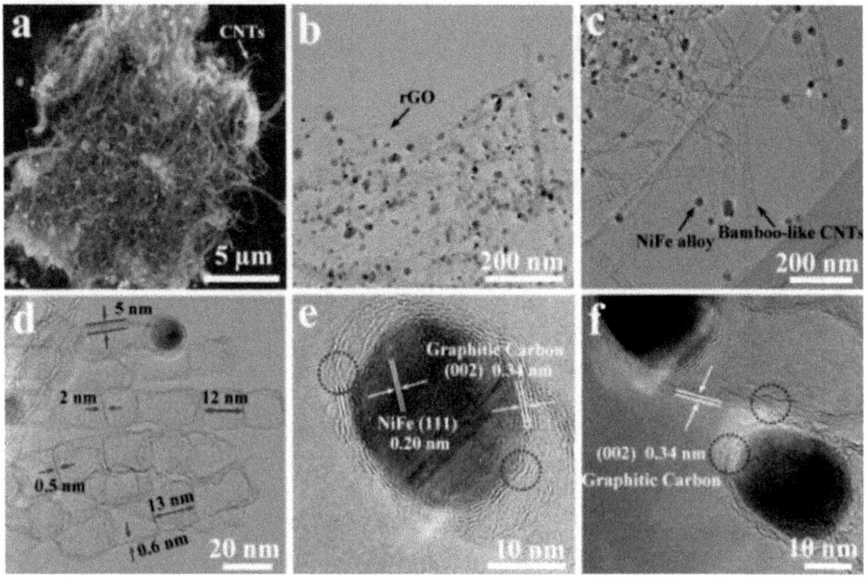

Figure 5.6 (a) SEM and (b) TEM images of 3D NiFe/N-GCT. (c) TEM image of the NiFe alloy NPs embedded in bamboo-like NCNTs. (d) TEM image of NCNTs. (e and f) HRTEM images of the NiFe alloy NPs embedded in graphitic carbon. Reproduced from ref. 110 with permission from American Chemical Society, Copyright 2019.

free carriers accumulated under an external electromagnetic field, leading to strong interface polarisation relaxations;[113] (3) the presence of polar groups (C N, C–O and N dopants) and structural defects that acted as polarised centres, enhancing polarisation relaxations upon EMW irradiation; (4) the unique 3D interconnected network structure that caused multiple scattering towards EMW, contributing to dielectric losses; and (5) the presence of voids enclosed by the walls of NCNTs and the bamboo joints in an individual NCNT, which increased dielectric loss.[114]

The 3D NiFe/N-GCT displayed excellent minimal reflection loss (−40.3 dB) and outstanding efficient absorption bandwidth (4.5 GHz), showing EMW properties comparable to and higher than those of other reported absorbers. The same approach was extended to the fabrication of similar nanocomposite architectures integrating CoNi NPs.[115]

A nanocomposite based on Ni_3Fe alloy NPs and N-doped porous graphene (NPG) was prepared by a phase transformation method, for rechargeable solid-state Zn–air batteries (ZABs).[116] ZABs have attractive prospects for flexible wearable electronics, because of their higher theoretical energy density (1086 Wh kg^{-1}) and the excellent durability of Zn in alkaline solution.[117] At present, however, due to the slow kinetics of the ORR and OER processes, rechargeable ZABs need to be improved by integration of new bifunctional electrocatalysts.[118]

NiFe-based compounds have been extensively studied as the most effective OER electrocatalysts, which are capable of achieving a better OER catalytic performance than neat Ni or Fe species,[119] since alloying can modify their surface electronic structures, thus improving catalytic activity and reducing kinetic energy barriers,[120] and further promoting conversion from hydroxide to oxygen.[121]

Graphene has attracted attention for promoting the ORR because of its large specific surface area, high conductivity, corrosion resistance and easy surface functionalisation.[122] Moreover, owing to the high electronegativity of N, N-doping could adjust the spin and charge distribution of adjacent C atoms,[123] thus facilitating adsorption of oxygen-containing species (OH^- or O_2) on the surface of the electrocatalyst, and thus enhancing the reaction effectiveness.[124]

For this purpose, a novel NPG with high surface area (784.9 $m^2\ g^{-1}$) and rich porosity was prepared and used as support to increase growth sites of metal compounds and to accelerate diffusion of oxygen and of the electrolyte. Negatively charged GO was mixed with the positively charged hard template NH_2-MCM41, formed from an amino-modified silica sphere, to allow electrostatic interactions, driving GO wrapping onto the surface of NH_2-MCM41. N-doping was performed by adding urea, which was decomposed upon thermal treatment in a Teflon-lined stainless steel autoclave at 180 °C. After freeze-drying, the product was treated first at 420 °C in Ar, and subsequently at 800 °C, and then etched with HF to remove the hard NH_2-MCM41 template, leaving the N-doped graphene skeleton.

Then, Ni_3Fe/NPG was prepared by the phase transformation method, relying on ultrasonically dispersing NPG with $Fe(NO_3)_3{\cdot}9H_2O$, $Ni(NO_3)_2{\cdot}6H_2O$ and urea in H_2O, to allow Fe^{3+} and Ni^{2+} ions to absorb onto the oxygen functionalities of the NPG, followed by thermal treatment at 120 °C, and finally at 750 °C in NH_3/Ar, to obtain Ni_3Fe/NPG. In the annealing process, NH_3 acted as reductant, promoting Ni_3Fe alloy NPs formation, and also acted as an N source for doping graphene during thermal treatment. In the obtained structure, NPG not only possessed good electrical conductivity and rich porosity, but also enabled anchoring of Ni_3Fe NPs at N sites, thus mitigating NP aggregation and detachment issues. In addition, the N-doped structure was expected to create additional active sites to promote catalytic activity.

SEM and TEM investigation of the nanocomposite showed that the NPG basal plane presented a 3D flake-like skeleton and a wrinkled bulk surface, which was beneficial to the further homogeneous deposition of Ni_3Fe NPs (see Figure 5.7a–f). The Ni_3Fe/NPG maintained the overall architecture of pristine NPG (see Figure 5.7b), indicating a negligible effect of phase transformation on the morphology of the nanocomposite upon addition of NH_3.[116]

TEM investigation confirmed these findings, along with the successful anchoring of the Ni_3Fe NPs at the NPG surface (see Figures 5.7d and e).

The HRTEM image in Figure 5.7f showed lattice spaces of 0.34 nm and 0.205 nm, which corresponded to the (002) plane of the NPG and the (111) plane of cubic Ni_3Fe, respectively. According to the elemental mapping

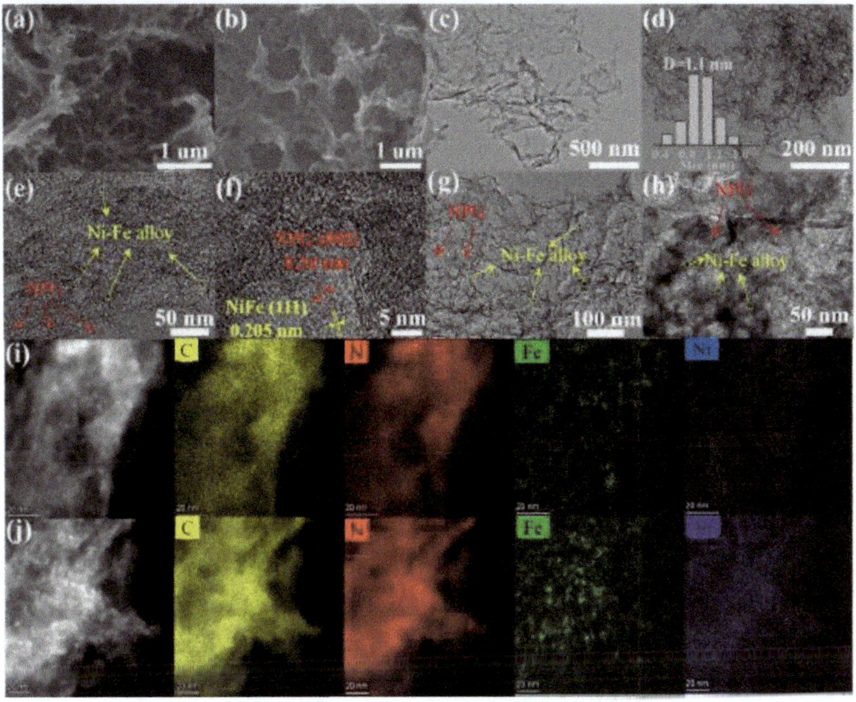

Figure 5.7 SEM images of (a) NPG and (b) Ni_3Fe/NPG. TEM image of (c) NPG and (d–e) Ni_3Fe/NPG and (f) HRTEM image of Ni_3Fe/NPG. TEM images of Ni_3Fe/NPG after the OER test (g and h). High-angle annular dark-field scanning transmission electron microscopy (HAADF-STEM) and corresponding elemental mapping of Ni_3Fe/NPG-1 before (i) and after (j) the OER test. Reproduced from ref. 116 with permission from American Chemical Society, Copyright 2020.

images of Ni_3Fe/NPG, shown in Figure 5.7i and j, the Ni_3Fe NPs were coupled with the layer of NPG, presenting uniform overlap of Ni and Fe elements. The C and N elements were also evenly distributed in the whole Ni_3Fe/NPG composite, indicating successful doping with nitrogen.[116]

The decoration of the NPG with the NiFe NPs was demonstrated to substantially improve the OER activity, and control of the ratio of Ni_3Fe NPs to NPG further enhanced the OER and ORR performances. The excellent OER stability mainly originated from the fact that the integration of highly active Ni_3Fe NPs in the ultrathin NPG prevented their dissolution.[116] TEM, HAADF-STEM and the corresponding elemental mapping images of the catalyst collected after the OER test (see Figure 5.7j) demonstrated that the Ni, Fe, C and N elements were homogeneously distributed on the selected mapping area, further indicating that the alloy NPs remained stable throughout the long OER cycling process and did not dissolve from the NPG surface into the electrolyte solution. The Ni_3Fe/NPG catalyst showed an ORR electrocatalytic activity higher than that of commercial Pt/C, with an onset potential of 0.874 V and a limited current density of 6.04 mA cm^{-2}, which was better than that of neat Ni_3Fe NPs and NPG, and comparable to commercial 10 wt% and 20 wt% Pt/C.

The fabricated liquid ZAB with optimised bifunctional catalyst showed a high voltage of 1.50 V, maximum power density of 143 mW cm^{-2}, outstanding cycling and mechanical stability and superior charge–discharge performance. The Ni_3Fe/NPG-based all-solid-state ZAB also presented a superb open-circuit voltage of 1.45 V, high peak power density of 50.0 mW cm^{-2} and cyclic durability for 150 cycles.[116]

MoS_2 has been considered a promising alternative to Pt as an HER catalyst owing to its abundance, low cost and good stability.[125] In order to make its catalytic activity comparable to that of Pt, the most common strategy has relied on enhancing the MoS_2 active edge sites,[126] both at the edge of the plane and at the basal plane, making them accessible for catalytic processes. Therefore, the number of stacked layers and the layer dimensions have been reduced, increasing the interlayer spacing and adjusting the sheet alignment. In addition, the improvement in the conductivity of MoS_2 by incorporation of graphene has been considered to be effective in enhancing the HER performance.[127] A nanocomposite based on few-layered MoS_2 sheets and RGO (MoS_2-RGO) was prepared by spray-drying a solution of $(NH_4)_6Mo_7O_{24}\cdot 4H_2O$, thiourea, ammonium carbonate and GO in deionised water, followed by calcination at 800 °C under H_2/N_2.[128]

XRD analysis of the nanocomposite showed the formation of pure phase and small MoS_2 crystallites arranged onto RGO in few-stacked layers, with a layer thickness and an interlayer spacing that decreased and increased, respectively, when passing from neat MoS_2 to the nanocomposite, and with increase in the RGO content.

Raman and XPS spectroscopy of the MoS_2-RGO demonstrated the presence of sulfur vacancies, a higher extent of incorporation of oxygen into the nanocomposite than in neat MoS_2, as well as the presence of both 2H and 1T phases of MoS_2, with the 1T phase being the major phase.[128]

TEM images of the nanocomposite showed MoS_2 sheets well dispersed onto RGO. In addition, magnified images showed 4–8 stacked MoS_2 layers with small sizes (5–15 nm) on RGO, indicating the exceptionally high degree of edge defects in the MoS_2-RGO samples. HRTEM images evidenced that the interlayer distance was quite a bit higher than the interlayer spacing of bulk MoS_2 (0.61 nm), thus facilitating accessibility of the inner active sites and defects of the basal planes of the MoS_2 crystal structure.[128] During spray-drying, the molybdate ions were reduced by thiourea, which acted also as a sulfur source, and the oxygen functional groups of GO behaved as heteronucleation sites, which bound the MoS_2 clusters during growth.[128]

Thiourea stabilised the MoS_2 surface binding of molybdenum, limited growth and helped the formation of nanosized MoS_2 domains along the GO basal planes. Ammonium carbonate determined the formation of thin sheet structures, enhanced interlayer spacing and favoured formation of the 1T phase, which has been assessed to be more catalytically active than the 2H phase for the HER.[129] NH_3 intercalated in between the MoS_2 layers, thus enhancing the interlayer distance, but during calcination under H_2/N_2 it escaped from the interlayer space and MoS_2 tended to restack. The layer stabilising activity of GO, which was converted to RGO during annealing, avoided migration of MoS_2 for restacking, leading to a high interlayer spacing in the MoS_2–RGO nanocomposite.[128]

Due to the strain stabilisation by RGO, the MoS_2-RGO composite presented enhanced sulfur vacancies compared to neat MoS_2. During thermal treatment, the oxygen functional groups of GO were decomposed and, at high temperatures, replacement of the surface sulfur by oxygen atoms was thermodynamically favourable, leading to a high degree of surface-incorporated oxygen atoms in the MoS_2-RGO sample, effectively enhancing the HER activity. The MoS_2-RGO nanocomposite exhibited excellent electrocatalytic HER activity and showed a current density of 10 mA cm^{-2} at 168 mV, a Tafel slope as small as 62 mV dec^{-1} and activity retention higher than 98% after 1000 cycles.[128]

5.5 Physical Routes to Nanocomposites Based on Inorganic Nanoparticle-decorated Graphene Derivatives

5.5.1 Microwave-assisted Irradiation

A novel nanocomposite based on carved multilayer graphene with nested Fe_3O_4 NPs (MUG-Fe_3O_4) was synthesised by ionic liquid microwave-assisted exfoliation, starting from a solution of iron(II) acetate in 1-ethyl-3-methylimidazolium acetate (EMIMAc) and graphite, as an anode for LIBs.[130] The nanocomposite was prepared starting from microplatelets of expanded graphite, non-covalently functionalised by the negatively charged EMIMAc, which electrostatically bound Fe^{2+} ions during ultrasonication. Under microwave (MW) irradiation, thoroughly described in Chapter 4, Fe_3O_4 NPs, 5 to 100 nm in size, heteronucleated and grew onto EMIMAc, and meanwhile the graphite microplatelets generated graphene multilayers.

Scanning transmission electron microscopy (STEM) investigation showed some of the carved paths and grooves traced by the Fe_3O_4 NPs along the edge planes of the MUG. The bright-field TEM micrograph showed 5 and 30 nm sized dark contrast Fe_3O_4 NPs nested in the carved multilayer graphene matrix, and the bright contrast grooves generated by the NPs (red arrows in Figure 5.8a).[130] In the image, differently carved paths, 10 nm to 40 nm wide, on different jagged edge planes of the MUG structure were seen (see Figure 5.8b). In Figure 5.8c, the SAED pattern showed *d*-values fitting the XRD reflections of the Fe_3O_4 structure with lattice spacings of 0.295 nm, while the peaks at *ca.* 0.207 nm and *ca.* 0.125 nm corresponded to Fe_3O_4 (400) and (622), as well as the graphite (100) and (110) reflections. The HRTEM image distinctly showed the Fe_3O_4 NPs covered by thin graphitic carbon layers (marked with red arrows) (see Figure 5.8d).

Such a morphology was accounted for by the NPs, which could carve different edge planes of MUG and caused fragmentation of the carbonaceous

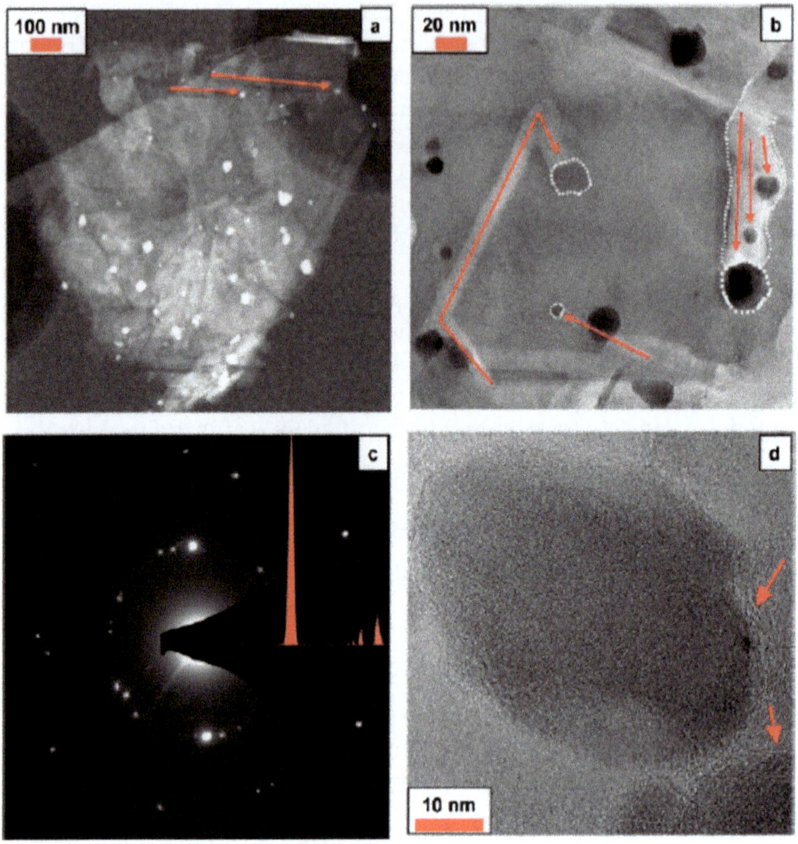

Figure 5.8 (a) STEM micrograph of the MUG-Fe_3O_4 composite. The carved paths created by the Fe_3O_4 NPs are clearly distinguishable (marked with red arrows). (b) Bright-field TEM micrograph of the composite where Fe_3O_4 NPs are highlighted with white dashed lines and the carving paths are indicated with red arrows. (c) Corresponding SAED pattern for the image in (b). (d) HRTEM micrograph of a Fe_3O_4 NP covered by a thin graphitic carbon layer originated by the carving of the multilayer graphene (marked with red arrows). Reproduced from ref. 130, https://doi.org/10.1038/srep23585, under the terms of the CC BY 4.0 license https://creativecommons.org/licenses/by/4.0/.

matrix, especially of the flakes with nm lateral size, thus resulting in wrapping by thin graphitic carbon layers,[130] which endowed the nanocomposite with Li^+ storage properties, enhanced with respect to those of bare graphene multilayers and bare Fe_3O_4 NPs, synthesised through the same approach.

Differential capacity analysis and electrochemical impedance spectroscopy measurements demonstrated that the peculiar morphology of the nanocomposite, designed to contain a minimal amount of Fe_3O_4 (*i.e.*, <3 wt%) for limiting volume changes during reaction with Li^+, provided high energy efficiency, relatively low lithiation/delithiation voltages and superior power performance.[130]

A key role was played by the carved structure of the multilayer graphene and by the confinement, upon cycling, of the Fe_3O_4 NPs. When the delithiation voltage was limited to 1 V *vs.* Li/Li^+, the carved paths traced by the Fe_3O_4 NPs, and the unconverted metallic iron formed *in situ* upon the first lithiation, resulted in a highly enhanced rate capability and, especially at high specific currents (*i.e.*, 5 A g^{-1}), remarkable cycling stability (99% of specific capacity retention after 180 cycles), low average delithiation voltage (0.244 V) and a substantially increased volumetric capacity with respect to commercial graphite (58.8 Ah L^{-1} *vs.* 9.6 Ah L^{-1}).[130]

Microwave irradiation was also used to prepare a hybrid nanocomposite formed by void-induced graphene-wrapped NiO NPs (VGWN) to integrate in supercapacitors.[131]

GO was partially exfoliated by microwave irradiation at a relatively low temperature and it showed a crumpled, curly and wavy morphology,[131] with edges with an open porous structure having several voids likely contributing to wrap the NiO NPs.

Wrapping the microwave-exfoliated GO (MEGO) around the NiO NPs was achieved homogeneously by mixing aqueous suspensions of MEGO and $Ni(NO_3)_2$ using ultrasonication, in order to provide electrostatic interactions of Ni^{2+} ions with the oxygen functional groups of MEGO. Then, after addition of ammonia and H_2O_2, void-assisted MEGO wrapped NiO NPs were synthesised under microwave irradiation.[131]

In the VGWN hybrids, the NiO NPs had an average size of 30–60 nm, were homogeneously distributed inside the voids and no detectable NP aggregation was observed. Owing to their relatively larger lateral size, the mechanical flexibility of MEGO and the occurrence of electrostatic interactions between oppositely charged NiO NPs and GO, a network of NiO NPs wrapped and tightly interconnected by the MEGO sheets formed, providing ion transfer and conducting channels for electrons.[131]

The voids embedded in the hybrid interconnected network increased the active surface area, and hence VGWN behaved as a promising electrode for applications in energy storage and conversion systems, owing to its enhanced host capability, structural stability and electrical conductivity. Specifically, the VGWN hybrid material showed a BET surface area of 398 m^2 g^{-1} with pore dimensions in the range 5–15 nm. It displayed an enhanced specific capacitive performance of 549 F g^{-1} at a scan rate of 10 mV s^{-1}, and over 88% of the initial capacitance was retained after repeating cyclic voltammetry tests for 2500 cycles.[131]

5.5.2 Sonochemical Routes

For a theoretical introduction, see Section 6.5. Cu NPs were synthesised onto nanometre-sized RGO sheets with a monodisperse size and controlled morphology by using an eco-friendly and surfactant-free, *in situ*, sonochemical method. Such a nanocomposite was a robust catalyst of the hydrogenation reaction of dimethyl oxalate (DMO), to produce methyl glycolate (MG), EG and ethanol (EtOH).[132]

GO was sonicated with $Cu(NO_3)_2{\cdot}6H_2O$ in an aqueous solution with a frequency of 20 kHz in pulsed mode, and the pH of the reaction was adjusted by adding ammonium hydroxide (NH_4OH). Then, the prepared CuO NPs/RGO nanocomposite was reduced by H_2 at 280 °C into the active Cu NPs/RGO catalyst for DMO hydrogenation.

In this process, the reactive $H^{\bullet}$ and $OH^{\bullet}$ free radicals, which were produced during sonication from water, avoided the use of any reducing agent or surfactant in the synthesis. Such radicals enhanced decomposition of $Cu(NO_3)_2{\cdot}6H_2O$ into Cu atom clusters, which subsequently were oxidised *in situ* to $[Cu(NH_3)_4]^{2+}$ in the presence of NH_4OH. The oxygen groups of the GO basal plane acted as anchoring sites of the CuO NPs, *via* electrostatic binding, physisorption or charge-transfer interactions, leading to *in situ* synthesis of the CuO NPs onto the RGO sheets (see Figure 5.9a and b).

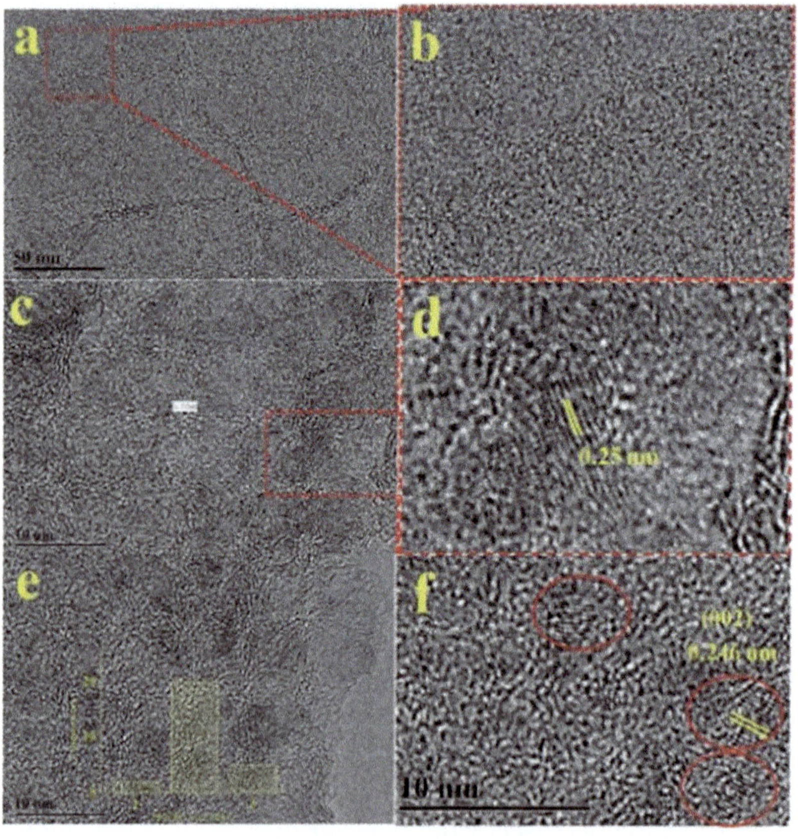

Figure 5.9 TEM images of (a and b) RGO sheets, (c and d) 10 wt% CuO/RGO catalyst and (e and f) 25 wt% CuO/RGO catalyst synthesised by US in the presence of ammonia. Reproduced from ref. 132 with permission from the Royal Society of Chemistry.

The direct attachment of the CuO NPs onto GO in a highly uniform manner was provided by the abrupt liquid jet streams (*ca.* 400 km h^{-1}) and by shock waves, which were produced from the bubbles collapsing during ultrasonication (see Chapter 6).[133] The bubbling gases adsorbed onto the surface of the nuclei reduced their interfacial free energy with the solution, and rapid nucleation occurred leading to grain growth into small NPs.[133]

When both ultrasound (US) and NH_4OH were used in the reaction with a Cu loading of 10 wt% and 25 wt%, CuO NPs with quite a small average size of 2 and 3 nm were achieved (see Figure 5.9c–f).

HRTEM images showed clearly the lattice fringes having an interplanar distance of 0.25 nm and 0.246 nm in the samples with a CuO NP loading of 10 and 25 wt%, respectively, reflecting the highly crystalline structure of the NPs (see Figure 5.9d and f). The homogeneous dispersion of the CuO NPs onto the RGO sheets was confirmed through SEM-EDS mapping analysis and N_2 adsorption/desorption isotherms of the CuO/RGO catalyst, demonstrating the macroporous material structure.[134] With increase in the Cu loading up to 45 wt%, dramatic evolution of the morphology of the CuO NPs onto the RGO sheets into a leaf-like shape was observed.[132] XRD analysis showed that NH_4OH was responsible for the complete reduction of GO to RGO and suggested the co-existence of Cu^0 and Cu^+ species in the catalyst upon reduction with hydrogen. The prepared CuO/RGO NPs catalyst was used in the DMO hydrogenation after being reduced by H_2 at 280 °C into the active Cu/RGO catalyst.

MG is the first product of the DMO hydrogenation which gives, by further hydrogenation steps, EG and EtOH, the complete hydrogenation being thermodynamically favoured.[135] Cu^0 active sites led to the complete hydrogenation of DMO to EG and EtOH. Conversely, the presence of Cu^+ in the catalyst stabilised the intermediate MG product.[136] The results demonstrated that the Cu/RGO catalyst, prepared by using 20 kHz of US in the presence of ammonia with 25 wt% of Cu content, displayed the highest MG selectivity (98.8%) at a relatively low reaction temperature (210 °C), while the Cu/RGO catalyst synthesised with 45 wt% Cu content showed the highest EtOH selectivity (94%) at 240 °C.[132]

A simple sonochemical route was used to synthesise a 3D Pd nanosponge-shaped network, wrapped by graphene dots (Pd@G NSs), as a catalyst for the HER.[137]

Such a nanocomposite was achieved by sonicating K_2PdCl_4 and carbon dots in a NaBr aqueous solution, adding L-ascorbic acid and heating up to 30 °C.[137]

DFT calculations suggested that Pd ions penetrated the interlayer space of the carbon dots and formed Pd clusters, which could not however initiate carbon dot exfoliation. Hence, Pd clusters and carbon dots were randomly aggregated. Conversely, the exfoliation of carbon dots was more energetically favoured after coordination of the Br^- ions to preformed Pd_3 clusters inside the carbon dots layers. Thus, Br^- ions played an essential role in accelerating the exfoliation of carbon dots supplying graphene layers, and the synergistic interplay between K_2PdCl_4 and Br^- ion concentrations determined the carbon dots exfoliation into graphene sheets.[137]

Multilayered graphene sheets were produced in various thicknesses and the Pd precursor nucleated and grew into the nanosponge shape, interacting with the π system of the exfoliated graphene sheets. In the absence of Br^- ions only aggregated Pd NPs were observed, randomly mixed with carbon dots to make raspberry-shaped nanocomposites (Pd-rNPs).[137]

The XRD pattern showed the high crystallinity of the Pd NS in Pd@G-NS (see Figure 5.10a).

The TEM images of Pd@G-NSs showed formation of the 3D network of Pd NSs (see Figure 5.10b and c) and the HRTEM images (see inset of Figure 5.10c) confirmed the crystallinity of the Pd NSs, showing the (111) planes of Pd with a *d*-spacing of *ca.* 0.226 nm.[137]

Raman investigation showed that the relative intensity of the D band (disorder) with respect to the G band (crystalline) of exfoliated carbon dots was 0.35, which was lower than that of the as-synthesised carbon dots (I_D/I_G = 0.54) (see Figure 5.10d).

The XPS spectrum of the Pd 3d core level exhibited peaks at binding energies of *ca.* 335.54 eV and 340.80 eV, corresponding to Pd $3d_{5/2}$ and Pd $3d_{3/2}$ of metallic Pd (see Figure 5.10e).

UV–vis absorption spectra of Pd precursor solutions, before and after addition of carbon dots and Br^-, showed the formation of $[PdBr^4]^{2-}$ or

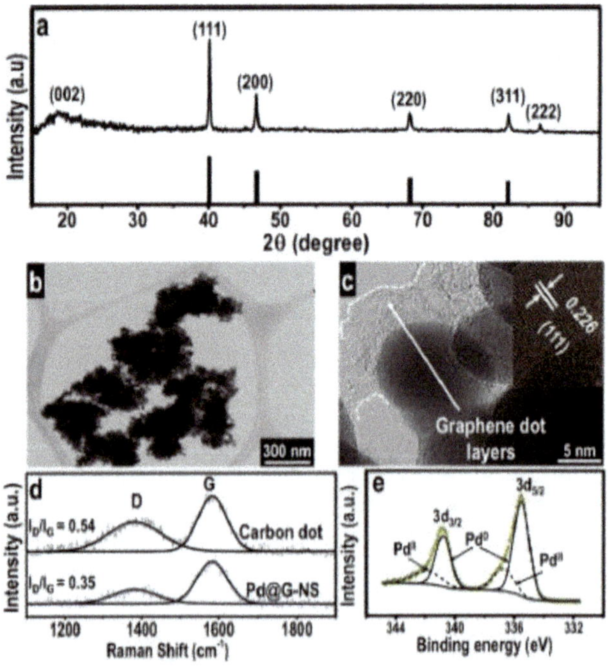

Figure 5.10 (a) XRD pattern, (b) TEM image and (c) HRTEM image of Pd@G-NSs. (d) Raman spectra of Pd@G-NSs and carbon dots. (e) High-resolution XPS spectrum of Pd 3d of Pd@G-NSs. Reproduced from ref. 137 with permission from American Chemical Society, Copyright 2020.

$[PdCl_{4-n}Br_n]^{2-}$ complexes in the reaction solution already containing $[PdCl_4]^{2-}$ and $PdCl_2$,[138] which slowed down the Pd precursor reduction.[139] As a consequence, it was concluded that the Pd precursors and Br^- ions could penetrate and exfoliate carbon dots to form graphene dot layers that armoured Pd NSs, leading to the generation of Pd@G-NSs nanostructures.[136]

The Pd@G-NSs catalyst had a higher HER performance with respect to Pd NPs and Pd-rNPs, exceeding that of the commercial Pt/C catalyst (cPt/C), with a surface porosity resistance (R_p) and charge-transfer resistance (R_{ct}) approximately 50% and 25% smaller than that of cPt/C, respectively.[136]

The overpotential (η) of the Pd@G-NSs catalyst at 10 mA cm^{-2} and the Tafel slope were lower than those of cPt/C, namely 32 mV dec^{-1} and 33 mV dec^{-1}, against 40 mV dec^{-1} and 35 mV dec^{-1}, respectively, confirming the higher HER kinetics of Pd@G-NSs, which were attributed to the synergistic effect of the accelerated electron transfers and of the higher electrochemically active surface area of the Pd@G-NS.[136]

The HER performance of Pd@G-NSs was preserved after 3000 cycles of CV and maintained better than that of Pd-rNP, confirming that the superior long-term durability of Pd@G-NSs was due to the thin layers of graphene. The current density retention of Pd NPs, Pd-rNPs and cPt/C was *ca.* 43, 75 and 84% in 0.5 M H_2SO_4 after 12 h, respectively. In contrast, the current density of Pd@G-NSs surprisingly remained at 96% for up to 15 h, thus showing a stability much higher than that of cPt/C in acidic solution. This result could be attributed to the protective action of the graphene layers, which avoided Pd NS dissolution, agglomeration and poisoning in acidic media under long-term operation.[140]

5.5.3 Chemical Vapour Deposition-based Routes

A series of nanocomposites was synthesised encapsulating NPs of 3d transition metals (TMs), namely Fe, Co, Ni and their alloys, into a single-layer graphene shell by CVD,[141] to prevent corrosion of 3d TMs from harsh environments, and simultaneously promote catalytic reactions (ORR, HER and the triiodide reduction reaction in dye-sensitised solar cells) on the carbon surface due to electron transfers from encapsulated 3d TMs.[142]

The channels of the SBA-15 template, with a pore size distribution of 7–9 nm, were filled by the metal precursors by using an impregnation method from methanol solution. Then, the samples were heated up to 700 °C under H_2 to allow reduction of the metal precursor to metal NPs, followed by CVD growth of graphene from acetonitrile onto the NPs at high temperature.

Due to the confinement of the SBA-15 channels, the metal precursors generated uniform NPs onto the SBA-15 walls, and single graphene layers grew between the metal NPs surface and the pore walls of the SBA-15, whereas multiwalled CNTs formed onto the residual metal NPs outside the SBA-15 channels. The final single-layer graphene encapsulating the metal NPs onto the interlaced CNTs was obtained after etching SBA-15 and bare metal NPs by using a HF solution.[141]

A series of single graphene layers encapsulating 3d TMs (M@NCs) was prepared by applying this method, including Fe@NC, Co@NC, Ni@NC, FeCo@NC, FeNi@NC and CoNi@NC.[141]

For the FeNi@NC sample, HRTEM, XRD and XPS investigation demonstrated that, after the CVD process and before treatment with HF, the FeNi@NC/SBA-15 clearly presented numerous small NPs inside the ordered mesoporous channel of SBA-15, corresponding to FeNi NPs encapsulated by graphene.

After removal of SBA-15, uniform metal NPs with an average diameter of 6–10 nm distributed onto the interlaced CNTs could be observed.[141]

Further HRTEM analysis indicated that the metal NPs were completely encapsulated by a 3.4 Å thick graphene shell, that is, a single-layer graphene. The FeNi NPs exhibited a *d*-spacing of 2.1 Å, in good agreement with the (111) plane of the FeNi alloy, and the crystal structure investigated by XRD showed characteristic peaks at 51.81° and 76.41°, corresponding to the (200) and (220) planes of the FeNi alloy, respectively. In addition, the XRD pattern assessed the presence of multiwalled CNTs. XPS suggested that both Fe and Ni in FeNi@NC maintained their metallic state, in agreement with the XRD analysis and HRTEM results.[141]

DFT calculations indicated that the high OER performance originated from the presence of the highly conductive single-layer graphene undergoing electron transfers from the encapsulated metals. The NPs tuned the electronic structure of graphene, thereby triggering the OER activity on the inert graphene surface.[141]

Electrochemical measurements indicated that the trend in the overpotential at a current density of 10 mA cm^{-2} was FeNi@NC < CoNi@NC < FeCo@NC < Ni@NC < Fe@NC < Co@NC, with FeNi@NC being the most durable and the best performing catalyst, and superior to commercial IrO_2.[141]

Following the same approach, a nanocomposite formed by N-doped graphene encapsulating CoNi nanoalloy metal catalyst was prepared as anode for the highly efficient electrocatalytic production of H_2 and S from H_2S.[143] The electrocatalytic decomposition of H_2S deriving from industrial by-products is a promising technology for the removal of this environmental pollutant and for energy conversion by H_2 production.[144]

The CoNi@NGs catalyst was prepared by a template-assisted CVD process at 600 °C, which relied on inducing precipitation of the CoNi hydroxide onto a SiO_2 template to obtain the precursor $CoNi(OH)_x$-SiO_2, which was then reduced to CoNi NPs under a reducing atmosphere and after etching the SiO_2 template.

The TEM images in Figure 5.11a–c showed that the precursor $CoNi(OH)_x$ layers were homogeneously distributed onto the SiO_2 template and then underwent a morphology change after CVD. Numerous small NPs, together with the formation of CNTs, appeared on the SiO_2 template (see Figure 5.11b) and, after removal of the SiO_2 template, the CoNi alloy NPs uniformly distributed onto the interlaced CNTs (see Figure 5.11c). The HRTEM image confirmed that the metal NPs, with an average diameter of 4–6 nm, were

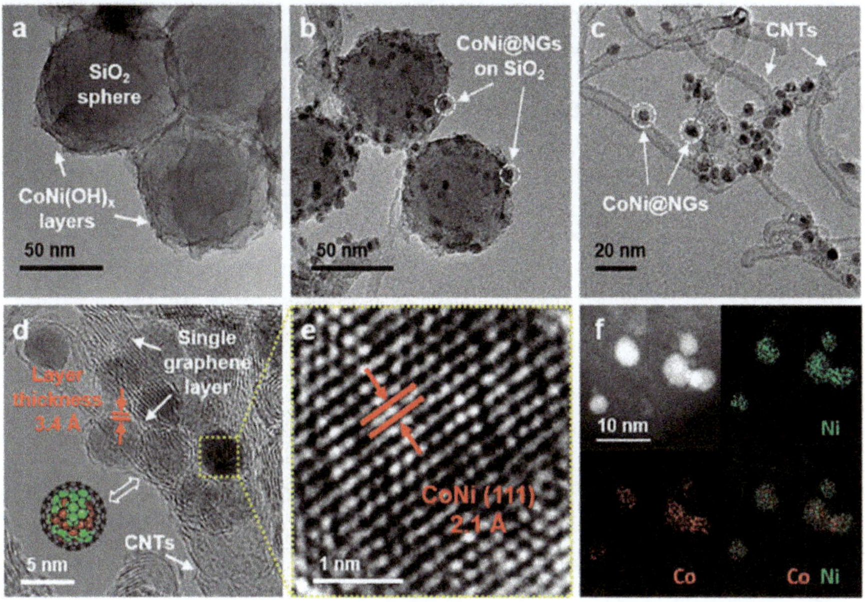

Figure 5.11 (a and b) TEM images of CoNi(OH)*x*-SiO_2 and CoNi@NGs-SiO_2. (c–e) HRTEM images of CoNi@NGs. (d) The red marks show the graphene layer with a layer thickness of 3.4 Å. The inset schematic illustration shows the model structure of CoNi@NGs. (e) Enlargement of (d) showing the (111) crystal plane of the CoNi alloy. (f) EDS maps of Co and Ni elements in CoNi@NGs. Reproduced from ref. 143 with permission from the Royal Society of Chemistry.

completely encapsulated by a single-layer graphene shell with a layer thickness of 3.4 Å (see Figure 5.11d). According to the statistical analysis of the HRTEM images, almost 96% of the graphene layers on the CoNi NPs were single layers.

The lattice fringes of the internal metal NPs displayed an interplanar spacing of 2.1 Å (see Figure 5.11e), which matched with the (111) plane of the CoNi alloy. The elemental mapping images of the metal NPs, obtained by EDS, indicated that the elemental distributions of Co and Ni encapsulated in the graphene shells were homogeneous (see Figure 5.11f). This catalyst showed an ultralow onset potential of 0.25 V for the anodic sulfur oxygen reaction (SOR), which was lower than that required for the OER. It exhibited almost twice the current density of the cPt/C catalyst at a 0.5 V potential. The optimised catalyst showed excellent durability without degradation, even after 500 h at 0.317 V (*vs.* reversible hydrogen electrode (RHE)) with a current density of *ca.* 30 mA cm^{-2}, and approximately 98% Faradaic efficiency towards H_2 production. HRTEM investigation showed that the CoNi NPs encapsulated in the graphene shells retained their structure and morphology after 500 h, confirming the robustness of the catalyst in harsh environments. DFT calculations demonstrated the high activity of the CoNi@NG catalyst for the SOR, originating from the modulation of the electronic structure of the graphene surface by encapsulation of a metal alloy and doping with N, which facilitated the adsorption of S and formation of polysulfide intermediates onto the graphene surface. Furthermore, a demonstration showed the high reliability of the CoNi@NG electrocatalyst for removal of H_2S impurities from industrial syngas to produce H_2, and it showed high stability, which was maintained for more than 1200 h at a current density of 20 mA cm^{-2} at *ca.* 0.5 V (*vs.* RHE).

A core–shell nanocomposite was synthesised incapsulating ZnS NPs in graphene-like multilayers through CVD, as anode material for LIBs.[145] ZnS NPs of different sizes were prepared by mixing different amounts of SDS in a $Zn(CH_3COO)_2$ solution with Na_2S added under stirring, and after purification and freeze-drying.[145] The as-obtained ZnS NPs were transferred to a quartz tube and heated up to 760 °C under Ar/H_2. Pyridine was added as the carbon source, which first evaporated and dehydrogenised at 760 °C and then polymerised, forming graphene layers onto the surface of the ZnS NPs, which behaved concomitantly as graphitisation catalyst and template. Microscopy investigation demonstrated the formation of uniformly sized and monodisperse ZnS NPs, with a diameter of *ca.* 30 nm, coated by graphene-like layers,[146] which formed carbon-based nano-cells enveloping each ZnS NP. In addition, the presence of tiny voids or cracks distributed in the graphene-like layers, which behaved as channels for ions, was observed. Such an unconventional morphology led to high reversible capacities of 1134 mA h g^{-1} after 100 cycles at 0.5 A g^{-1}, 890 mA h g^{-1} after 200 cycles at 1.0 A g^{-1} and an excellent rate capability of 701 mA h g^{-1} at 8.0 A g^{-1}.[145]

An *in situ* confined growth route was developed for fabricating Pt NP@graphene nanobox composites to integrate into fuel cells.[147] Although nanosized Pt-based catalysts with a high surface area are very effective for fuel cells,[148] they were combined with graphene as a support to improve electrical conductivity and to prevent aggregation of the Pt NPs, which are responsible for a decrease in the catalytic activity.[149]

The Pt NP@graphene nanoboxes were synthesised using a Pt-loaded nano-A zeolite as template, which was prepared starting from Zeolite A NCs having a uniform and monodisperse cubic morphology, with a size of *ca.* 120 nm, and achieved by hydrothermal treatment of aluminium isopropoxide, tetramethylammonium, tetraethylorthosilicate and NaOH at 80 °C, and then calcination at 550 °C in air.[147]

Pt ions were exchanged into the Zeolite A NCs by soaking in $Pt(NH_3)_4Cl_2$ aqueous solution at room temperature. The graphene nanoboxes grew onto the external crystal surfaces of the nano-A zeolite by CVD, at 800 °C or 900 °C, in the presence of Ar, using methane as the carbon precursor.[147] In such a process, CH_4 gas molecules were adsorbed onto the crystal surface of the Zeolite A NC substrates and the acidity on the outer surface led to the formation of carbon clusters, thus generating few-layered graphene in a unique 3D nanostructure.[147]

Meanwhile, during the high-temperature CVD process, Pt ions adsorbed in the inner micropores of the Zeolite A NCs diffused and aggregated together, and were then reduced to NPs in methane atmosphere, which were confined into the internal surfaces of the nanoboxes, resulting in sizes of 2–3 nm with a content of *ca.* 28 wt% when the synthesis was performed at 800 °C, and of 4–5 nm with a content of *ca.* 52 wt% when performed at 900 °C.[147]

On the other hand, 3–4 graphene layers grew tightly bound onto the external surfaces of the zeolite support. The final nanocomposite was washed with a 20 wt% HF solution, followed by refluxing at 60 °C in 2 M HCl to completely remove the zeolite substrates, and the monodisperse Pt NPs were finally encapsulated in the graphene boxes.[147]

Microscopy and XRD investigation demonstrated that the Pt NPs were homogeneously dispersed in the graphitic carbon hollow box and spatially separated from each other, indicating that such a confined growth method, integrating a high load of NPs available for contact with the reactants, effectively prevented agglomeration of the NPs, and the hollow box morphology provided large volumes for accommodating oxygen, when utilised as ORR catalysts.[147]

After etching off the zeolite substrate, the resulting Pt NPs were effectively arranged and protected by the graphene hollow box, and the nanoboxes showed excellent electrocatalytic activity for the ORR, including a high current density (175 mA mg^{-1} Pt), electron transfer number of *ca.* 4 and durability (above 1000 cycles).[147]

Such a synthesis method was extended to the preparation of Au-NPs@graphene nanobox composites with 20 wt% of Au.[147]

A nanocomposite based on 3D N-doped graphene (3DNG)-supported NiMo NPs, encapsulated by N-doped graphene layers (NGLs), was synthesised as a HER catalyst with long-term stability and high activity in an acidic environment.[150]

NiMo alloys exhibited a near-zero onset potential and remarkable HER activity,[151] but the corresponding catalysts were rarely utilised in acidic media due to their dissolution.[152] Therefore, since long HER lifetimes are required in acid electrolytes for proton exchange membrane-type water splitting,[153] and HER in acidic conditions is much faster than that in basic conditions,[154] the stability of non-noble catalysts in acid media is a crucial requirement for HER applications.

The encapsulation of non-noble metal NPs in NGLs was a particularly promising strategy to address this goal, but encapsulation by using thick NGLs provided high chemical stability with a reduced catalytic activity,[155] thus, the right number of layers for balancing both HER efficiency and stability was required.

3DNG-supported NiMo NPs, encapsulated by NGLs (ENNiMoNPs/3DNG), were prepared starting from reduction of $NiMoO_4$ nanofibres at 950 °C in H_2/Ar to achieve a NiMo alloy with a 3D porous structure, followed by growth of NGLs onto the as-obtained 3D porous NiMo at 700 °C in H_2/Ar and pyridine atmosphere by CVD, where pyridine was used as the carbon and nitrogen source. The samples were then etched by HNO_3 to achieve the 3D porous N-doped graphene (see Figure 5.12A).

The NiMo NPs supported onto, and encapsulated in, the 3DNG were prepared by soaking the 3DNG in a $[Ni(NH_3)_6]MoO_4$ solution. During drying

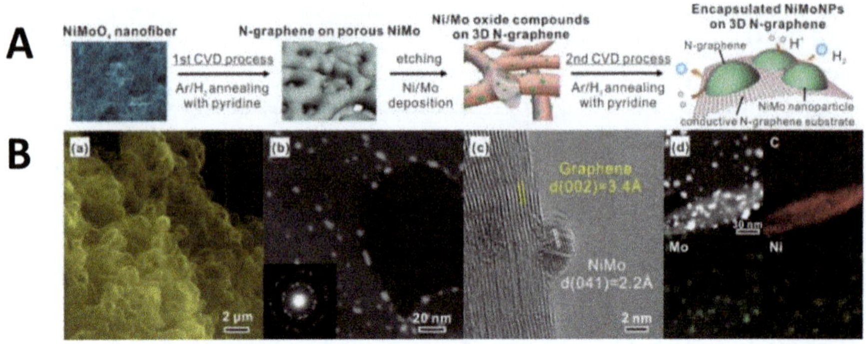

Figure 5.12 (A) Sketch of the preparation procedure for ENNiMoNPs/3DNG. (B) (a) Cross-sectional SEM image of ENNiMoNPs/3DNG. (b) DF-STEM image of uniformly dispersed NPs on 3DNG and the corresponding SAED pattern (inset). (c) Typical HRTEM image of the NiMo NPs encapsulated by a monolayer graphene sheet on 3DNG. (d) DF-STEM image and corresponding EDS elemental mappings of ENNiMoNPs/3DNG. Reproduced from ref. 150 with permission from American Chemical Society, Copyright 2018.

of the resulting product under vacuum, NH_3 was removed, and the Ni/Mo oxide compounds were deposited onto the 3DNG (see Figure 5.12A). Then, the 3DNG decorated with the compounds was again treated at 950 °C in H_2/Ar and at 700 °C under H_2/Ar and pyridine, to convert the Ni/Mo oxide compounds into NiMo NPs and sequentially grow NGLs onto them (see Figure 5.12A).

Microscopy investigation showed that the ENNiMoNPs/3DNG nanocomposite had a bicontinuous and open porous structure, with a pore radius of 300–500 nm (see Figure 5.12B(a)), containing interconnected graphene sheets; whereas the dark-field scanning TEM (DF-STEM) image (see Figure 5.12B(b)) showed uniformly distributed NPs with an average diameter of 11.3 nm on the 3DNG surface. The SAED patterns presented sharp diffraction spots, confirming the high crystallinity of the multilayer graphene (see inset of Figure 5.12B(b)), and the HRTEM images reported lattice fringes with distances of 2.2, 2.1 and 3.4 Å, corresponding to the (041) facet of NiMo, the (121) facet of $MoNi_4$ and the (002) facet of graphene, respectively. NiMo NPs encapsulated by 2–3 and 3–5 NGLs were also observed (see Figure 5.12B(c)).

EDS confirmed that Ni and Mo atoms were homogeneously distributed in the NPs, with a mismatch between the C and Mo/Ni distributions (see Figure 5.12B(d)).

Electrochemical investigation and DFT calculations showed that for 3–5 N-doped graphene protective layers encapsulating the NiMo NPs, the right balance between the HER performance and catalyst lifetime was achieved, specifically with a stable current density of 100 mA cm^{-2} for 25 h in 0.5 M H_2SO_4, which also attested to the remarkable protection capability of the NPs in acidic electrolytes.

5.6 Top-down Approaches

5.6.1 Mechanochemical Routes

Mechanochemistry has recently emerged as one of the most interesting methodologies for synthesising new nanostructured materials for a wide range of applications.

Mechanical-assisted solvent-free reactions are induced by direct absorption of mechanical energy. Ball milling is an efficient solid-state method for grinding various materials into fine powders and during such a 'non-reactive grinding'[156] mechanical forces can provide various processes requiring or releasing different energies, such as mass transfer, generation and relaxation of mechanical stress associated with disruption of crystalline lattices, bond fracture and release of radicals, in a different way that which occurs in solution, thus creating new reaction pathways that complement conventional chemical routes. Such solvent-free synthesis methods can significantly reduce reaction time compared to conventional solvent-based approaches and also save energy and cost.[157]

A nanocomposite formed by SnO_x ($1 \leq x \leq 2$) NPs and Sn NPs, wrapped by few-layered graphene ((SnO_x-Sn)@FLG), was synthesised by oxygen plasma-assisted milling, a simple and eco-friendly approach for mass production of nanomaterials, for SIBs.[158] In this process, the synergistic effect of the rapid plasma heating and ball mill grinding led to the formation of SnO_x ($1 \leq x \leq 2$) NPs tightly wrapped by the FLG nanosheets, simultaneously exfoliated from graphite, thanks to the occurrence of interactions between Sn atoms and the oxygen functionalities at the basal plane of the oxidised graphene nanosheets (see Figure 5.13A).

Sn and SnO_2 NPs are attractive species for SIBs due to their high theoretical capacity of 847 mA h g^{-1}[159] and 1387 mA h g^{-1},[160] respectively, but their reversible capacity faded rapidly due to the extremely large volume expansions occurring during sodiation, and hence they were combined with carbon-based compounds.

Expanded graphite and Sn powder were mixed with the ball at a mass ratio of 1:50, and the oxygen plasma milling (O_2–P-milling) was performed under pure oxygen atmosphere with a vibration-type ball mill.[161] During the O_2–P-milling, oxygen (~0.08 mol) was supplied to the system every 5 h, and the rapid heating generated by the plasma, along with the impact stress provided by the ball milling, led to the *in situ* synthesis of the SnO_x ($1 \leq x \leq 2$) NPs with concomitant exfoliation of graphite into FLG nanosheets, which wrap the SnO_x and Sn NPs (see Figure 5.13A).[158]

Figure 5.13B shows the XRD patterns of the composites prepared by O_2–P-milling and the composites achieved by conventional planetary ball milling in the presence of O_2 (O_2–C-milling) after 15 h, 20 h and 30 h. When the O_2–P-milling time increases, peaks corresponding to Sn NPs became weaker and those corresponding to SnO_2 NPs became stronger. As shown in the XRD pattern, the O_2–P-milling-15 h and O_2–P-milling-20 h samples were composed

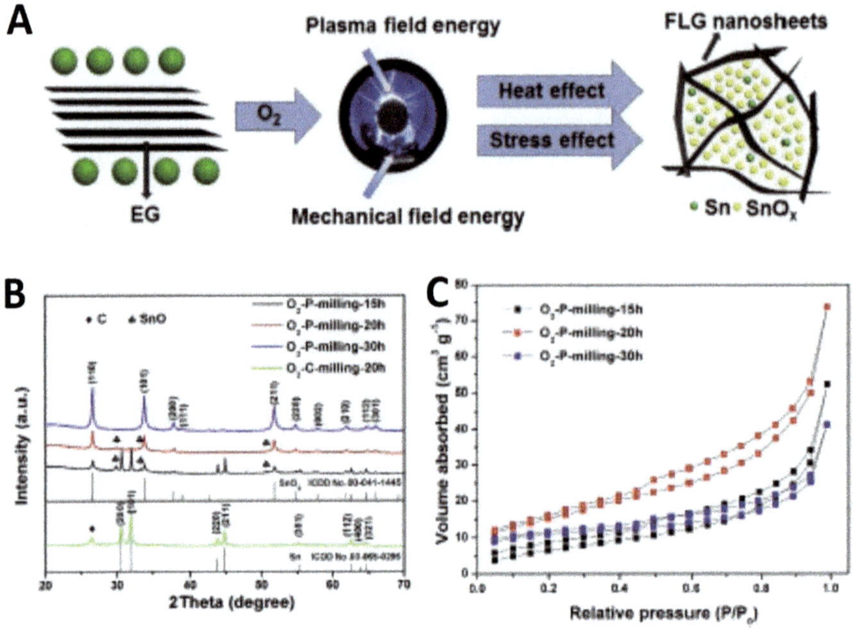

Figure 5.13 (A) Preparation of the (SnOx-Sn)@FLG composite. (B) XRD pattern of the O_2–P milling samples and O_2–C-milling-20 h. (C) N_2 absorption isotherms of the O_2–P-milling samples. Reproduced from ref. 158 with permission from Elsevier, Copyright 2019.

of SnO_2, SnO and Sn phases. When the O_2–P-milling time increased up to 30 h, Sn was fully oxidised to SnO_2. By contrast, Sn was not oxidised at all in the O_2–C-milling-20 h, indicating that the oxygen plasma was much more reactive than the treatment with bare oxygen.[158] The O_2–P-milling time was found to influence the crystalline size of the SnO_2 NPs. Raman investigation demonstrated the beneficial effect of the combination of the rapid heating provided by the plasma and the impact stress of the milling in exfoliating graphite in FLG nanosheets. The N_2 adsorption isotherms of the O_2–P-milling samples showed that the O_2–P-milling-20 h had the largest specific surface area, also due to the smallest crystalline size of the SnO_2 NPs and the effective exfoliation of graphite (see Figure 5.13C).[158] The small size of the SnO_x NCs enabled fast kinetics for Na^+ transfer. On the other hand, the *in situ*-formed FLG, along with the residual Sn NPs, improved the electrical conductivity of the nanocomposite, as well as enhanced the cyclic stability due to mitigation of the SnO_x NP aggregation, thus accommodating volume changes during cycling.[158] The (SnO_x-Sn)@FLG composite showed superior Na^+ storage. When discharged/charged at 100 mA g^{-1}, it showed a high reversible capacity of 448 mA h g^{-1} in the first cycle and a retention of 82.6% of the initial capacity after 250 cycles. Even at a high current density of 1000 mA g^{-1}, the composite exhibited a capacity of 327.5 mA h g^{-1} in the initial cycle and retained 316.5 mA h g^{-1} after 250 cycles.[158]

An onion-like nanocomposite formed by crystalline WS_2 NPs uniformly decorating graphene sheets (WS_2@Gs) was prepared as anode for LIBs, by combining ball milling with sulfidation.[162] WS_2 has received considerable attention as an electrode for LIBs owing to its high theoretical specific capacity (432 mA h g^{-1}) and rapid ion diffusivity, which originate from the 2D layered structure.[163] It has, however, been combined with graphene to improve its conductivity and to mitigate the large volume expansion during conversion with Li^+ ions.[163] WO_3 NPs were synthesised by the electrical explosion of a wire (EEW process)[164] at a charging voltage of 320 V, and at a feeding distance of 2 cm in distilled water (see Figure 5.14A). The obtained WO_3 slurry was freeze-dried and the resulting powder collected.

The obtained WO_3 NPs were mixed with graphene dispersed in anhydrous ethanol and were ball milled with ZrO_2 balls, resulting in a homogeneous mixture of WO_3 NPs, of 15 nm in diameter, and graphene sheets (WO_3@Gs).[162]

Then, the WO_3@Gs suspension was dried at 80 °C in an oven at atmospheric pressure, and subsequently WO_3@Gs and S powder were loaded into two individual boats and placed upstream of a quartz tube, which was heated up to 600 °C under Ar/H_2 and the WS_2@Gs were collected. Additionally, carbon-coated WS_2@Gs (C@WS_2@Gs) were prepared *via* the same process, except that glucose was added during ball milling.

FESEM investigation demonstrated that the neat WO_3 NPs, prepared *via* the EEW process, had an irregular bulky structure due to agglomeration (see Figure 5.14a–c).[164] The SEM image acquired after sulfidation of the neat WO_3 NPs showed not only aggregated NPs, but also bulk NPs with a diameter of *ca.* 150 nm (see Figure 5.14b), demonstrating that each NP sintered and subsequently grew during sulfidation.[162]

By contrast, in the case of the WS_2@Gs nanocomposite, the SEM images showed WS_2 NPs uniformly anchored onto the graphene sheets without aggregation or sintering (see Figure 5.14c) with a spherical morphology and a diameter of *ca.* 15 nm (see Figure 5.14d),[162] like that of the initial WO_3 NPs, indicating that the graphene basal plane mitigated particle aggregation during sulfidation.[162]

The HRTEM image (Figure 5.14e) and SAED patterns (Figure 5.14f) showed onion-like coiled atomic layers corresponding to the (002) plane of WS_2 with a *d*-spacing of 0.62 nm. The SAED pattern of WS_2@Gs matched well with the (002), (101) and (110) planes of the WS_2 phase (Figure 5.14f), indicating that sulfidation of the WO_3 NPs anchored onto the graphene sheets proceeded at 600 °C in H_2/Ar/S.

The unique structure of the WS_2@Gs nanocomposites showed excellent Li^+ ion storage, with a high reversible capacity of 587.1 mA h g^{-1} at a current density of 200 mA g^{-1}. To enhance the cyclic stability of WS_2@Gs, a carbon coating was employed simply by adding glucose during ball milling. Even at a high current density of 1000 mA g^{-1}, the C@WS_2@Gs modified electrode exhibited a remarkably high reversible capacity of 371.9 mA h g^{-1} and appreciable cycling stability, with a capacity retention of 62% without any drastic capacity fading after 500 cycles.[162]

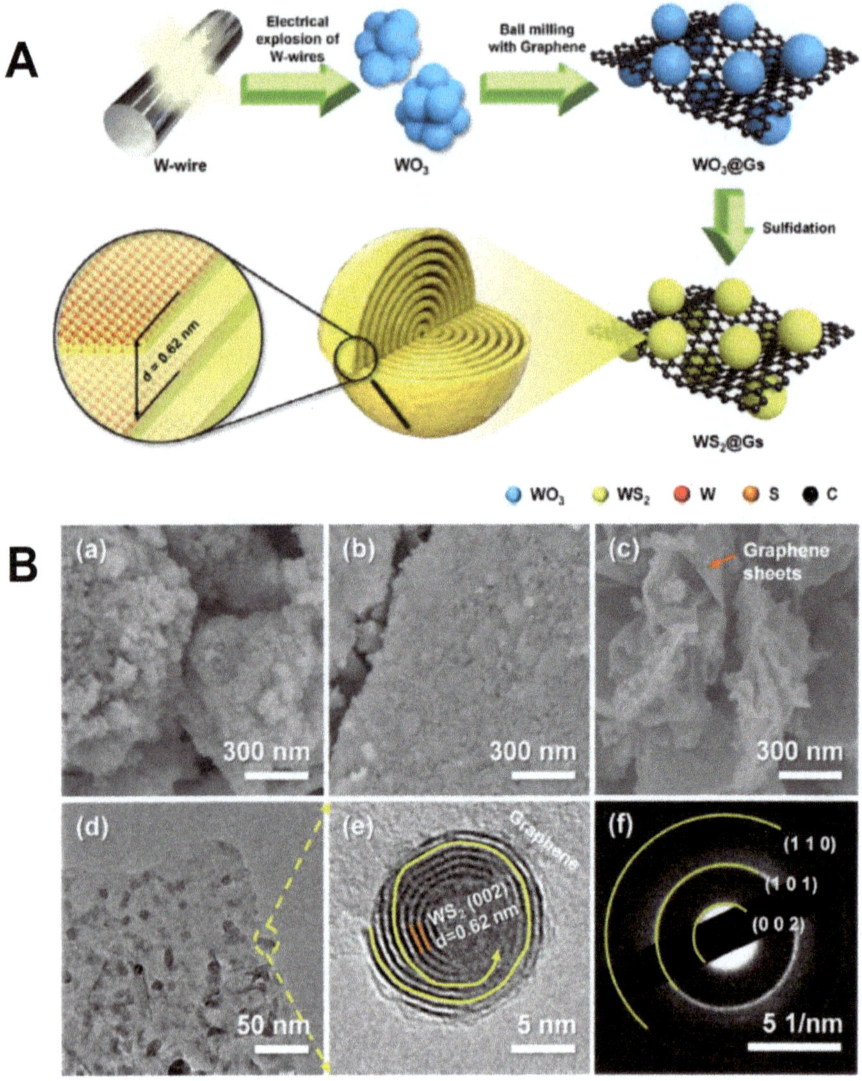

Figure 5.14 (A) Sketch of the synthesis process and crystalline features of WS_2@Gs. (B) SEM images of (a) WO_3, (b) WS_2 and (c) WS_2@Gs. (d) Low-magnification TEM image, (e) HRTEM image and (f) SAED pattern of WS_2@Gs. Reproduced from ref. 162 with permission from Elsevier, Copyright 2013.

FeO(OH)-nanoflake/graphene and nano-Fe_3O_4/graphene nanocomposites were achieved by solid-state milling from commercial Fe powders and GO, followed by heat treatment, as anodes for LIBs and as electrocatalysts for the ORR and OER.[165]

Nanosized iron oxides demonstrated a lithium storage capability that is twofold higher than that of commercial graphite, but their low electrical conductivity required their combination with graphene to further improve the lithium storage capability and electrocatalytic activity.[166]

The nanocomposite was synthesised by adding GO and Fe powders into a stainless steel grinding bowl and grinding them in deionised water under Ar using a planetary mill. The grinding balls were ZrO_2, 5 mm in diameter, with a ball to powder (GO + metal) w/w of 34 : 1 to 40 : 1. After being dried at room temperature, the samples were calcined at 600 °C under Ar to obtain nano-Fe_3O_4/graphene, or calcined at 500 °C under air to obtain nano-Fe_2O_3 (see Figure 5.15A).[165]

Fe powders, irregular in shape and with particle sizes of 10–30 µm (see Figure 5.15B(a)), were milled with the addition of GO under Ar (see Figure 5.15A). The mechanochemical process promoted intimate contact between metallic Fe and GO, as well as repeated fracturing and rewelding of reactant powders, facilitating the activation of the GO reaction sites and, in turn, formation of nanosized iron oxide onto the GO surface. The latter was reduced by metallic Fe and simultaneously exfoliated in RGO sheets, whereas nanosized iron-based oxide (*i.e.*, FeO(OH) nanoflakes) was found to heteronucleate and grow onto the surface of RGO (see Figure 5.15A). Fe_3O_4/RGO and neat Fe_2O_3 NPs were obtained after calcination of FeO(OH)/RGO under Ar and air, respectively. SEM images showed that, after the reactive grinding,[156] the Fe powder transformed into FeO(OH) nanostructured flakes (see Figure 5.15B(b)), with a size much smaller than those obtained in the absence of GO. This suggests that the oxygen-containing functional groups of GO facilitated the formation of a strong bond between the as-formed Fe-based oxide and GO and led to nano-FeO(OH)/RGO composites. In comparison with the results achieved by grinding Fe powders alone, GO also acted as a grinding medium in the ball milling process, reducing the size of Fe powders.[165]

TEM images showed transparent and thin RGO sheets almost coated by flake-like FeO(OH) species (see Figure 5.15B(c)), suggesting that GO was exfoliated during ball milling and that the Fe-based compounds prevented restacking.

EDS mapping showed the homogeneous distribution of the Fe and C elements, confirming that the graphene sheets were coated by flake-like Fe-based compounds (see Figure 5.15B(d)), and the HRTEM image of the composite (Figure 5.15B(e)) clearly showed the RGO layers.[165]

The XRD pattern presented the characteristic peak of GO located at 2θ = 10.4° and of FeO(OH) compounds at 2θ = 44.7, 65.1 and 82.4°, corresponding to the (110), (200) and (211) lattice planes of metallic-phase Fe powder (Figure 5.15B(f)).

After grinding, the peaks of the GO and metallic Fe phases disappeared and the XRD pattern of the composite was assigned to lepidocrocite FeO(OH), attesting to the successful oxidation of GO during grinding (Figure 5.15B(f)). This was also confirmed by high-resolution XPS spectra of the FeO(OH)/RGO composite, which showed the two main peaks centred at 725.6 (Fe $2p_{1/2}$) and 711.9 (Fe $2p_{3/2}$) eV attributed to FeO(OH), as well as the successful reduction of GO to RGO by metallic Fe during ball milling (see Figure 5.15B(g and h)).[165]

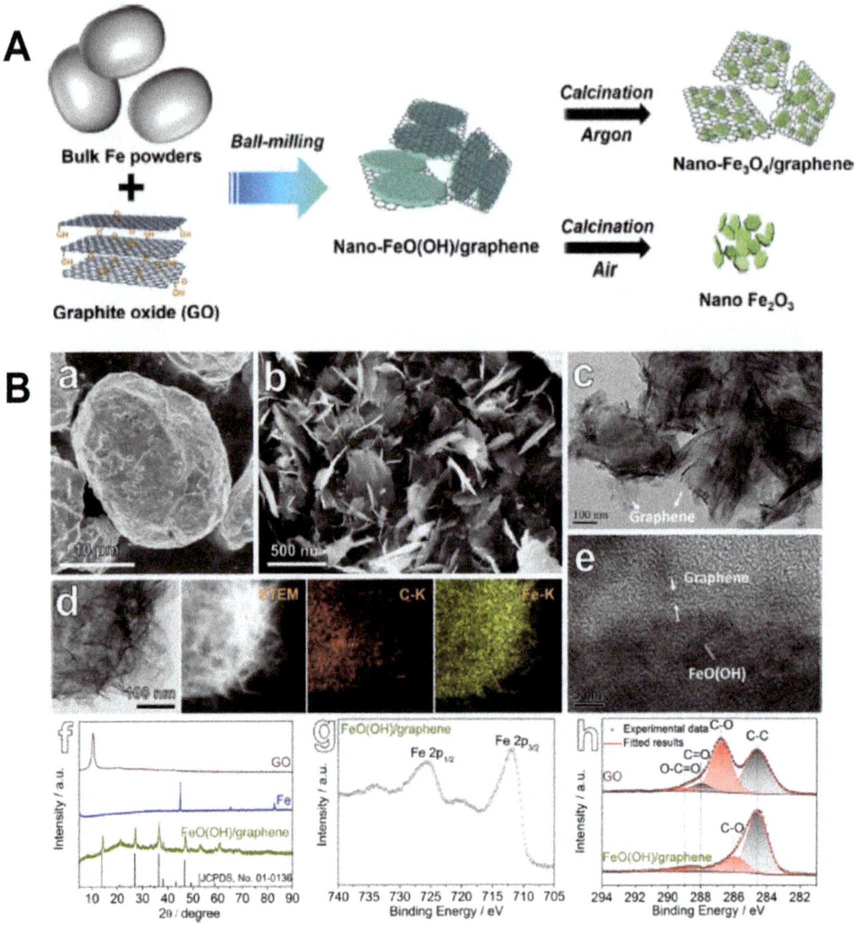

Figure 5.15 (A) Sketch of the preparation of the nanocomposites consisting of iron oxides and graphene by a reactive solid-state milling process. (B) (a) FESEM image of bulk Fe. (b) FESEM image, (c) TEM image and (d) TEM image of FeO(OH)/graphene composite prepared *via* ball milling from Fe and GO, with the corresponding STEM image and EDS elemental distribution of C and Fe. (e) HRTEM image of FeO(OH)/graphene composite. (f) XRD patterns of GO, bulk Fe and FeO(OH)/graphene composite. The XRD intensities of Fe and FeO(OH)/graphene samples are multiplied by a factor of 2 and 10, respectively. (g) XPS spectrum of Fe 2p in FeO(OH)/graphene composite. (h) XPS spectra of C 1 s in GO and FeO(OH)/graphene composite. Reproduced from ref. 165 with permission from American Chemical Society, Copyright 2015.

Notably, the FeO(OH)/RGO nanocomposite showed a specific capacity of 977 mA h g^{-1} after 50 cycles, indicating a reversible conversion reaction of the as-prepared Fe_3O_4/graphene nanocomposite. Specifically, the electrode delivered capacities of 913, 850 and 802 mA h g^{-1} at current densities of 200, 500 and 1000 mA g^{-1}, respectively. At a current density as high as 2000 mA g^{-1}, a specific capacity of 629 mA h g^{-1} was achieved. More importantly, decreasing the current density back to 100 mA g^{-1} after the cycling test, at various current densities of 100–2000 mA g^{-1}, the specific capacity of the Fe_3O_4/graphene electrode recovered to its initial value, confirming the reversible conversion reaction and excellent cycling stability.[165] Fe_3O_4/RGO electrodes showed both a higher ORR current density and a more positive ORR onset potential than those of FeO(OH)/graphene electrodes, suggesting an enhanced ORR catalytic activity after calcination. Conversely, the FeO(OH)/graphene composite displayed a slightly lower onset potential compared to that of the Fe_3O_4/graphene nanocomposite, but the Fe_3O_4/graphene nanocomposite was a more efficient electrocatalyst at higher current densities.[165]

5.6.2 Laser Ablation-assisted Routes

Laser ablation in liquid (LAL), which is thoroughly described in Chapter 6, is an easy-to-use, environmentally friendly and attractive top-down physical method to produce colloidal NPs at laboratory scale.[167] In this approach, a laser beam is typically focused on a solid target immersed in a properly selected liquid medium, which gives rise to NPs with controlled sizes, morphologies and chemical composition (see Section 6.7).[167]

LAL is waste-free, does not require any catalysts and is typically performed in aqueous media, proceeding under ambient conditions and at room temperature.[167]

A nanocomposite based on ZnO NPs wrapped by a GO shell was prepared by ablation of zinc and carbon targets in water and in ambient atmosphere at 22 °C. A nanosecond pulsed Nd:YAG laser (Surelite), with a wavelength of 1064 nm and frequency of 10 Hz, was used for irradiating the targets in deionised water.[168]

The laser beam was focused on the target, which was fixed inside a quartz cuvette filled with 15 mL of deionised water, and the solution was left to stir at 750 rpm. The Zn target was ablated for 30 min with a pulse energy of 50 mJ $pulse^{-1}$, while the C target was ablated for 5 or 10 min by using a pulse energy of 100 mJ $pulse^{-1}$.[168]

Nanosecond-long laser pulses were used to ablate the metallic Zn target immersed in deionised water for 30 min to achieve a ZnO colloid. During Zn plate ablation, the laser beam induced a high-pressure and high-temperature Zn plasma plume, which reacted with vaporised water molecules to form ZnO nuclei.[169] Adiabatic expansion and condensation of the plasma caused formation of cavitation bubbles where the formed nuclei continue to grow, with the growth kinetics related to the dynamics of the cavitation bubble.

The size of the ZnO NPs was found to depend on the experimental parameters,[167] while their morphology depended on the temperature of the liquid medium.[170] Spherical ZnO NPs were observed because no significant temperature rise of the medium was achieved with the nanosecond-pulse laser, and ablation followed the thermal evaporation mechanism of Zn species, which then formed spherical NPs.[168] The GO shell was obtained by replacing the Zn plate with a graphite target, which was ablated in the dispersion of the ZnO colloid for either 5 or 10 min. Similarly, ablation of the graphite plate was found to occur following thermal evaporation and by exfoliation of carbon layers to produce a plasma plume rich in C ions.[169] Simultaneously, plasma-induced water vapour and ionised ZnO NPs deriving from the ZnO colloid interacted with the ablated carbon ions, which assembled on the surface of ZnO in a honeycomb-like structure, leading to the formation of GO-encapsulated ZnO NPs. For longer carbon ablation times, thicker GO layers formed onto the ZnO NPs core.

Both Raman spectra and XRD patterns indicated that formation of the layered GO sheets increased the stress inside the ZnO lattice and reduced the density of defects and disorders in the ZnO crystal lattice. On the other hand, the minimal interfacial disorder between the layered GO and ZnO core resulted in the formation of a continuous GO layer. XPS investigation demonstrated that surface oxygen groups of GO bound Zn atoms of the ZnO NPs surface.[168]

A nanocomposite formed by Pt NPs, *in situ* synthesised by laser ablation onto the basal plane of GO, was prepared and its optical linear, non-linear and thermal properties were investigated.[171] The nanocomposite was prepared by soaking a Pt plate in an aqueous GO suspension, with the Pt plate then being ablated using an Nd : YAG Q-Switch pulsed laser beam. The effect of wavelength, energy and repetition rate of the laser beam, along with ablation time, was investigated. The synthesised NPs were dispersed in the GO suspension under stirring, during the Pt plate ablation. TEM investigation showed the formation of spherical-shaped Pt NPs distributed onto the GO surface. The concentration and volume fraction of the Pt NPs was found to increase with increase in the ablation time, whereas the size decreased from 22 nm to 12 nm. FTIR spectroscopy demonstrated that the Pt NPs were strongly coordinated by the carboxylic and hydroxylic functional groups at the GO basal plane and edges.

5.7 Conclusions and Future Perspectives

Since its first isolation and the demonstration of its superior electrical properties, graphene has been considered by the scientific community to be a 'miracle' material with excellent technology potential, which is further enhanced thanks to its high surface chemical reactivity. The latter has allowed the implementation of a plethora of functionalisation approaches with NPs to achieve multifunctional nanocomposite materials with promise for applications ranging from solar cells, to supercapacitors, to membranes, to photocatalysts, to batteries, to fuel cells and to sensors.

This chapter has provided insights into the most recent bottom-up and top-down chemical and physical *in situ* unconventional approaches for achieving hybrid nanocomposites based on graphene derivatives and inorganic NPs, which exhibit unprecedented properties that cannot be found in pristine components.

The plethora of routes presented here indicates that the preparation of new graphene/NPs composite materials with diverse chemical compositions, morphologies and chemical–physical properties can benefit significantly from ingenious mixing and matching of original and diverse preparative strategies into novel protocols, for the identification of sustainable, up-scalable and cost-effective synthetic routes. This is essential to bridge the gap between laboratory experiments and industrial manufacture of this class of nanocomposites, which is necessary to realise their great technological potential.

Acknowledgements

The authors acknowledge TOPSIS bilateral CNR-BAS Project (n. prot. 0089480/2018), TOOL bilateral MHESR-CNR Project 'Analytical Toolkit for Monitoring Water Pollution', the Ricerca e Innovazione 2014–2020 Italian projects PON BEST-4U (n. prot. ARS01_00519) and ECOTEC (ARS01_00951), the 'Multi-stacked intercalating hybrid PbS quantum dot film/graphene architectures for enhanced photodetectors' project (Prot. MAE0065807, MAE00691402020-06-26) funded by Ministero degli Affari Esteri e della Cooperazione Internazionale and TRANSISTOR project funded by INAIL (BRIC 2019, ID:33).

References

1. M. R. Horn, F. Williams, D. Dubal, J. MacLeod and N. Motta, *ChemSusChem*, 2019, **13**, 1613.
2. R. R. Nair, P. Blake, A. N. Grigorenko, K. S. Novoselov, T. J. Booth, T. Stauber, N. M. R. Peres and A. K. Geim, *Science*, 2008, **320**, 1308.
3. D. S. L. Abergel, V. Apalkov, J. Berashevich, K. Ziegler and T. Chakraborty, *Adv. Phys.*, 2010, **59**, 261.
4. K. S. Novoselov, A. K. Geim, S. V. Morozov, D. Jiang, Y. Zhang, S. V. Dubonos, I. V. Grigorieva and A. A. Firsov, *Science*, 2004, **306**, 666.
5. C. Gomez-Navarro, R. T. Weitz, A. M. Bittner, M. Scolari, A. Mews, M. Burghard and K. Kern, *Nano Lett.*, 2007, **7**, 3499.
6. M. T. Hwang, M. Heiranian, Y. Kim, S. You, J. Leem, A. Taqieddin, V. Faramarzi, Y. Jing, I. Park and A. M. van der Zande, *Nat. Commun.*, 2020, **11**, 1543.
7. G. Darabdhara, M. R. Das, S. P. Singh, A. K. Rengan, S. Szunerits and R. Boukherrou, *Adv. Colloid Interface Sci.*, 2019, **271**, 101991.
8. M. J. Rivero, P. Ribao, B. Gomez-Ruiz, A. Urtiaga and I. Ortiz, *Sep. Purif. Technol.*, 2020, **240**, 116637.
9. B. Fang, D. Chang, Z. Xu and C. Gao, *Adv. Mater.*, 2020, **32**, 1902664.

10. G. V. Bianco, A. Sacchetti, A. Mallardi, M. Grande, A. D'Orazio, P. Capezzuto and G. Bruno, *Carbon*, 2020, **170**, 75.
11. G. Bae, D. S. Song, Y. R. Lim, I. S. Jeon, M. Jang, Y. Yoon, C. Jeon, W. Song, S. Myung, S. S. Lee, C.-Y. Park and K.-S. An, *ACS Appl. Mater. Interfaces*, 2020, **42**, 47802.
12. V. Georgakilas, M. Otyepka, A. B. Bourlinos, V. Chandra, N. Kim, K. C. Kemp, P. Hobza, R. Zboril and K. S. Kim, *Chem. Rev.*, 2012, **112**, 6156.
13. K. S. Novoselov, V. I. Falko, L. Colombo, P. R. Gellert, M. G. Schwab and K. Kim, *Nature*, 2012, **490**, 192.
14. V. Georgakilas, J. N. Tiwari, K. C. Kemp, J. A. Perman, A. B. Bourlinos, K. S. Kim and R. Zboril, *Chem. Rev.*, 2016, **116**, 5464.
15. J. Hass, W. A. de Heer and E. H. Conrad, *J. Phys.: Condens. Matter*, 2008, **20**, 323202.
16. J. Aprojanz, Ph. Rosenzweig, T. T. N. Nguyen, H. Karakachian, K. Küster, U. Starke, M. Lukosius, G. Lippert, A. Sinterhauf, M. Wenderoth, A. A. Zakharov and C. Tegenkamp, *ACS Appl. Mater. Interfaces*, 2020, **12**, 43065.
17. P. W. Sutter, J.-I. Flege and E. A. Sutter, *Nat. Mater.*, 2008, **7**, 406.
18. J. Zhang, L. Sun, K. Jia, X. Liu, T. Cheng, H. Peng, L. Lin and Z. Liu, *ACS Nano*, 2020, **14**, 10796.
19. K. Jia, H. Ci, J. Zhang, Z. Sun, Z. Ma, Y. Zhu, S. Liu, J. Liu, L. Sun, X. Liu, J. Sun, W. Yin, H. Peng, L. Lin and Z. Liu, *Angew. Chem.*, 2020, **59**, 17214.
20. Y. Hernandez, V. Nicolosi, M. Lotya, F. M Blighe, Z. Sun, S. De, I. T. McGovern, B. Holland, M. Byrne, Y. K. Gun'Ko, J. J. Boland, P. Niraj, G. Duesberg, S. Krishnamurthy, R. Goodhue, J. Hutchison, V. Scardaci, A. C. Ferrari and J. N. Coleman, *Nat. Nanotechnol.*, 2008, **9**, 563.
21. M. Matsumoto, Y. Saito, C. Park, T. Fukushima and T. Aida, *Nat. Chem.*, 2015, **7**, 730.
22. X. Li, W. Cai, J. An, S. Kim, J. Nah, D. Yang, R. Piner, A. Velamakanni, I. Jung, E. Tutuc, S. K. Banerjee, L. Colombo and R. S. Ruoff, *Science*, 2009, **324**, 1312.
23. W. Kong, H. Kum, S.-H. Bae, J. Shim, H. Kim, L. Kong, Y. Meng, K. Wang, C. Kim and J. Kim, *Nat. Nanotechnol.*, 2019, **14**, 927.
24. Y. Liu, B. Qu, X. Wu, Y. Tian, K. Wu, B. Yu, R. Du, Q. Fu and F. Chen, *Carbon*, 2019, **153**, 565.
25. C. K. Chua and M. Pumera, *Chem. Soc. Rev.*, 2014, **43**, 291.
26. D. R. Dreyer, S. Park, C. W. Bielawski and R. S. Ruoff, *Chem. Soc. Rev.*, 2010, **39**, 228.
27. W. S. Hummers and R. E. Offeman, *J. Am. Chem. Soc.*, 1958, **80**, 1339.
28. D. Chen, H. Feng and J. Li, *Chem. Rev.*, 2012, **112**, 6027.
29. Y. W. Zhu, S. Murali, W. W. Cai, X. S. Li, J. W. Suk, J. R. Potts and R. S. Ruoff, *Adv. Mater.*, 2010, **22**, 3906.
30. I. K. Moon, J. Lee, R. S. Ruof and H. Lee, *Nat. Commun.*, 2020, **1**, 73.
31. S. F. Pei and H. M. Cheng, *Carbon*, 2012, **50**, 3210.

32. S. Gilje, S. Dubin, A. Badakhshan, J. Farrar, S. A. Danczyk and R. B. Kaner, *Adv. Mater.*, 2010, **22**, 419.
33. T. Kuila, A. K. Mishra, P. Khanra, N. H. Kim and J. H. Lee, *Nanoscale*, 2013, **5**, 52.
34. G. Eda, G. Fanchini and M. Chhowalla, *Nat. Nanotechnol.*, 2008, **3**, 270.
35. D. Li, M. B. Müller, S. Gilje, R. B. Kaner and G. G. Wallace, *Nat. Nanotechnol.*, 2008, **3**, 101.
36. S. Stankovich, D. A. Dikin, G. H. B. Dommett, K. M. Kohlhaas, E. J. Zimney, E. A. Stach, R. D. Piner, S. T. Nguyen and R. S. Ruoff, *Nature*, 2006, **442**, 282.
37. G. Ambrosio, A. Brown, L. Daukiy, G. Drera, G. Di Santo, L. Petaccia, S. De Feyter, L. Sangaletti and S. Pagliara, *Nanoscale*, 2020, **12**, 9032.
38. A. Sinitskii, A. Dimiev, D. A. Corley, A. A. Fursina, D. V. Kosynkin and J. M. Tour, *ACS Nano*, 2010, **4**, 1949.
39. A. Wang, J. Ye, M. G. Humphrey and C. Zhang, *Adv. Mater.*, 2018, **30**, 1705704.
40. A. C. Ferrari and J. Robertson, *Phys. Rev. B: Condens. Matter Mater. Phys.*, 2000, **61**, 14095.
41. G. Bottari, G. la Torre, D. M. Guldi and T. Torres, *Coord. Chem. Rev.*, 2021, **428**, 213605.
42. M. E. Ragoussi, J. Malig, G. Katsukis, B. Butz, E. Spiecker, G. Torre, T. Torres and D. M. Guldi, *Angew. Chem., Int. Ed.*, 2012, **51**, 6421.
43. D. Ananya and R. Vimala, *J. Drug Delivery Sci. Technol.*, 2018, **43**, 333.
44. S. Niyogi, E. Bekyarova, M. E. Itkis, J. L. McWilliams, M. A. Hamon and R. C. Haddon, *J. Am. Chem. Soc.*, 2006, **128**, 7720.
45. R. Muszynski, B. Seger and P. V. Kamat, *J. Phys. Chem. C*, 2008, **112**, 5263.
46. J. Wang, Z. Chen and B. Chen, *Environ. Sci. Technol.*, 2014, **48**, 4817.
47. Q. Su, S. Pang, V. Alijani, C. Li, X. Feng and K. Muellen, *Adv. Mater.*, 2009, **21**, 3191.
48. X. An, T. Simmons, R. Shah, C. Wolfe, K. M. Lewis, M. Washington, S. K. Nayak, S. Talapatra and S. Kar, *Nano Lett.*, 2010, **10**, 4295.
49. G. Katsukis, J. Malig, C. Schulz-Drost, S. Leubner, N. Jux and D. M. Guldi, *ACS Nano*, 2012, **6**, 1915.
50. C. Ingrosso, G. V. Bianco, M. Corricelli, R. Comparelli, D. Altamura, A. Agostiano, M. Striccoli, M. Losurdo, M. L. Curri and G. Bruno, *ACS Appl. Mater. Interfaces*, 2015, **7**, 4151.
51. C. Ingrosso, G. V. Bianco, V. Pifferi, P. Guffanti, F. Petronella, R. Comparelli, A. Agostiano, M. Striccoli, L. Palchetti, L. Falciola, M. L. Curri and G. Bruno, *J. Mater. Chem. A*, 2017, **5**, 9307.
52. C. Chen, Y. Chen, D. Jiang, J.-J. Zhu and H.-Y. Chen, *Anal. Chem.*, 2017, **89**, 2418.
53. C. Ingrosso, M. Corricelli, A. Disha, E. Fanizza, G. V. Bianco, N. Depalo, A. Panniello, A. Agostiano, M. Striccoli and M. L. Curri, *Carbon*, 2019, **152**, 777.
54. A. Debnath, K. Deb, K. L. Bhowmik and B. Saha, *ACS Appl. Energy Mater.*, 2020, **3**, 7772.

55. V. Balakumar and P. Prakash, *J. Catal.*, 2016, **344**, 795.
56. C. Alex, S. A. Bhat, N. S. John and C. V. Yelamaggad, *ACS Appl. Energy Mater.*, 2019, **2**, 8098.
57. G. Xia, N. Li, D. Li, R. Liu, C. Wang, Q. Li, X. Lü, J. S. Spendelow, J. Zhang and G. Wu, *ACS Appl. Mater. Interfaces*, 2013, **5**, 8607.
58. B. Vellaichamy, P. Periakaruppan and S. K. Ponnaiah, *Sens. Actuators, B*, 2017, **245**, 156.
59. H. D. Tran, D. Li and R. B. Kaner, *Adv. Mater.*, 2009, **21**, 1487.
60. W. Sun, S. Meng, R. Zang, L. Chen and Y. Wang, *Polym. Compos.*, 2013, **34**, 997.
61. V. Balakumar, K. P. Sathish and P. Prakash, *Sens. Actuators, B*, 2017, **253**, 392.
62. J. Luo, N. Zhang, R. Liu and X. Liu, *RSC Adv.*, 2014, **4**, 64816.
63. L. Guo, Q. Liu, G. Li, J. Shi, J. Liu, T. Wanga and G. Jiang, *Nanoscale*, 2012, **4**, 5864.
64. B. Zhu, R. Zou and Q. Xu, *Adv. Energy Mater.*, 2018, **8**, 1801193.
65. G. Yilmaz, K. M. Yam, C. Zhang, H. J. Fan and G. W. Ho, *Adv. Mater.*, 2017, **29**, 1606814.
66. M. Yamauchi, H. Kobayashi and H. Kitagawa, *ChemPhysChem*, 2009, **10**, 2566.
67. W. Zhang, C. Chen, D. Yang, G. Dong, S. Jia, B. Zhao, L. Yan, Q. Yao, A. Sunna and Y. Liu, *Adv. Mater. Interfaces*, 2016, **3**, 1600590.
68. F. Yu, Y. Huang, A. J. Cole and V. Yang, *Biomaterials*, 2009, **30**, 4716.
69. Y. Wang, Y. Shao, D. W. Matson, J. Li and Y. Lin, *ACS Nano*, 2010, **4**, 1790.
70. T. Xing, Y. Zheng, L. Li, B. C. C. Cowie, D. Gunzelmann, S. Z. Qiao, S. Huang and Y. Chen, *ACS Nano*, 2014, **8**, 6856.
71. L. Fei, Q. Lin, B. Yuan, G. Chen, P. Xie, Y. Li, Y. Xu, S. Deng, S. Smirnov and H. Luo, *ACS Appl. Mater. Interfaces*, 2013, **5**, 5330.
72. C. Xu, Y. Zeng, X. Rui, N. Xiao, J. Zhu, W. Zhang, J. Chen, W. Liu, H. Tan, H. H. Hng and Q. Yan, *ACS Nano*, 2012, **6**, 4713.
73. L. Shang, T. Bian, B. Zhang, D. Zhang, L.-Z. Wu, C.-H. Tung, Y. Yin and T. Zhang, *Angew. Chem., Int. Ed.*, 2014, **53**, 250.
74. B. Fang, N. K. Chaudhari, M. S. Kim, J. H. Kim and J. S. Yu, *J. Am. Chem. Soc.*, 2009, **131**, 15330.
75. S. B. Yang, X. L. Feng, L. Wang, K. Tang, J. Maier and K. Mullen, *Angew. Chem.*, 2010, **122**, 4905.
76. X. Chen, G. Wu, J. Chen, X. Chen, Z. Xie and X. Wang, *J. Am. Chem. Soc.*, 2011, **133**, 3693.
77. M. Zhu, P. Chen and M. Liu, *ACS Nano*, 2011, **5**, 4529.
78. J. Kim, L. J. Cote, F. Kim, W. Yuan, K. R. Shull and J. Huang, *J. Am. Chem. Soc.*, 2010, **132**, 8180.
79. C. An, S. Peng and Y. Sun, *Adv. Mater.*, 2010, **22**, 2570.
80. D. Li, M. B. Müller, S. Gilje, R. B. Kaner and G. G. Wallace, *Nat. Nanotechnol.*, 2008, **3**, 10.
81. X. Zhang, K. Ding, B. Weng, S. Liu, W. Jin, X. Ji and J. Hu, *J. Colloid Interface Sci.*, 2020, **573**, 96.

82. D. Yin, C. Han, X. Bo, J. Liu and L. Guo, *J. Colloid Interface Sci.*, 2019, **533**, 578.
83. Y. Feng, H. Zhang, L. Fang, Y. Mu and Y. Wang, *ACS Catal.*, 2016, **6**, 4477.
84. M. Brust, M. Walker, D. Bethell, D. J Schiffrin and R. J. Whyman, *J. Chem. Soc., Chem. Commun.*, 1994, 801.
85. C. Ingrosso, M. Corricelli, F. Bettazzi, E. Konstantinidou, G. V. Bianco, N. Depalo, M. Striccoli, A. Agostiano, M. L. Curri and I. Palchetti, *J. Mater. Chem. B*, 2019, **7**, 768.
86. F. Bettazzi, C. Ingrosso, P. S. Sfragano, V. Pifferi, L. Falciola, M. L. Curri and I. Palchetti, *Food Chem.*, 2020, **344**, 128692.
87. Z. Zhang, B. Lu, J. Hao, W. Yang and J. Tang, *Chem. Commun.*, 2014, **50**, 11554.
88. Y. Yao, J. Zheng, Z. Gong, Z. Ding, J. Zhang, W. Yu, D. A. M. Bengono, H. Li, B. Zhang and H. Tong, *J. Alloys Compd.*, 2019, **790**, 288.
89. V. Shrivastav, S. Sundriyal, P. Goel, H. Kaur, S. K. Tuteja, K. Vikrant, K. Kim, U. K. Tiwari and A. Deep, *Coord. Chem. Rev.*, 2019, **393**, 48.
90. Z.-Z. Wang, J. Qu, S.-M. Hao, Y.-J. Zhang, F.-Q. Kong, D. Yang and Z.-Z. Yu, *ChemElectroChem*, 2018, **5**, 2653.
91. J. Liu, L. Yu, C. Wu, Y. Wen, K. Yin, F. K. Chiang, R. Hu, J. Liu, L. Sun, L. Gu, J. Maier, Y. Yu and M. Zhu, *Nano Lett.*, 2017, **17**, 2034.
92. R. Razaq, D. Sun, Y. Xin, Q. Li, T. Huang, Z. Zhang and Y. Huang, *Adv. Mater. Interfaces*, 2019, **6**, 1801636.
93. Y. Huang, R. Yu, G. Mao, W. Yu, Z. Ding, Y. Cao, J. Zheng, D. Chu and H. Tong, *J. Alloys Compd.*, 2020, **841**, 155670.
94. Y. Fu, Q. Wei, G. Zhang and S. Sun, *Adv. Energy Mater.*, 2018, **8**, 1702849.
95. F. Han, C. Zhang, J. Yang, G. Ma, K. He and X. Li, *J. Mater. Chem.*, 2016, **4**, 12781.
96. X. Wang, K. Chen, G. Wang, X. Liu and H. Wang, *ACS Nano*, 2017, **11**, 11602.
97. X. Cheng, R. Na, X. Wang, N. Xia, Z. Shan and J. Tian, *Inorg. Chem. Front.*, 2019, **6**, 1996.
98. X. Zuo, J. Zhu, P. Müller-Buschbaum and Y.-J. Cheng, *Nano Energy*, 2017, **31**, 113.
99. W. Wang, D. Chen, H. Xu, G. Yu, S. Sun, W. Zhang and J. Chen, *Sustainable Energy Fuels*, 2020, **4**, 3519.
100. Y. Huang, Z. Xu, J. Mai, T. Lau, X. Lu, Y. Hsu, Y. Chen, A. C. Lee, Y. Hou, Y. S. Meng and Q. Li, *Nano Energy*, 2017, **41**, 426.
101. L. Yang, G. Guo, H. Sun, X. Shen, J. Hu, A. Dong and D. Yang, *Electrochim. Acta*, 2016, **190**, 797.
102. J.-S. Park and Y. C. Kang, *Chem. Eng. J.*, 2019, **373**, 227.
103. Y. Zhu, X. Han, Y. Xu, Y. Liu, S. Zheng, K. Xu, I. Hu and C. Wang, *ACS Nano*, 2013, **7**, 6378.
104. J.-S. Park, J. S. Cho, J. H. Kim, Y. J. Choi and Y. C. Kang, *J. Alloys Compd.*, 2016, **689**, 554.
105. G. D. Park, J. S. Cho and Y. C. Kang, *Nano Energy*, 2015, **17**, 17.
106. H.-K. Roh, G.-W. Lee, S. Haghighat-Shishavana, K. Y. Chung and K.-B. Kim, *Chem. Eng. J.*, 2020, **385**, 123984.

107. X. Wang, G. Wang, G. Liu, Z. Bakenov and Y. Zhang, *J. Electroanal. Chem.*, 2020, **873**, 114375.
108. X. Liang, C. Y. Kwok, F. Lodi-Marzano, Q. Pang, M. Cuisinier, H. Huang, C. J. Hart, D. Houtarde, K. Kaup, H. Sommer, T. Brezesinski, J. Janek and L. F. Nazar, *Adv. Energy Mater.*, 2016, **6**, 1501636.
109. X. Men, X. Kong, X. Yang, B. Wang, Y. Wang, Y. Liu, L. Yu, H. Li and B. Xu, *Int. J. Hydrogen Energy*, 2020, **45**, 29492.
110. X. Zhang, J. Xu, H. Yuan, S. Zhang, Q. Ouyang, C. Zhu, X. Zhang and Y. Chen, *ACS Appl. Mater. Interfaces*, 2019, **11**, 39100.
111. F. Ye, Q. Song, Z. Zhang, W. Li, S. Zhang, X. Yin, Y. Zhou, H. Tao, Y. Liu, Y. L. Cheng, L. Zhang and H. Li, *Adv. Funct. Mater.*, 2018, **28**, 1707205.
112. X. Jian, W. Tian, J. Li, L. Deng, Z. Zhou, L. Zhang, H. Lu, L. Yin and N. Mahmood, *ACS Appl. Mater. Interfaces*, 2019, **11**, 15869.
113. B. Quan, X. Liang, X. Zhang, G. Xu, G. Ji and Y. Du, *ACS Appl. Mater. Interfaces*, 2018, **10**, 41535.
114. B. Qu, C. Zhu, C. Li, X. Zhang and Y. Chen, *ACS Appl. Mater. Interfaces*, 2016, **8**, 3730.
115. X. Zhang, X. Zhang, H. Yuan, K. Lia, Q. Ouyang, C. Zhu, S. Zhang and Y. Chen, *Chem. Eng. J.*, 2020, **383**, 123208.
116. X. Hao, Z. Jiang, X. Tian, C. Song, S. Chen, X. Hao and Z.-J. Jiang, *ACS Appl. Energy Mater.*, 2020, **3**, 12148.
117. J. Fu, Z. P. Cano, M. G. Park, A. Yu, M. Fowler and Z. Chen, *Adv. Mater.*, 2017, **29**, 1604685.
118. D. Bin, B. Yang, C. Li, Y. Liu, X. Zhang, Y. Wang and Y. Xia, *ACS Appl. Mater. Interfaces*, 2018, **10**, 26178.
119. C. Lai, M. Gong, Y. Zhou, J. Fang, L. Huang, Z. Deng, X. Liu, T. Zhao, R. Lin, K. Wang, K. Jiang, H. Xin and D. Wang, *Appl. Catal., B*, 2020, **274**, 119086.
120. Q. Tang, F. Chen, T. Jin, L. Guo, Q. Wang and H. Liu, *J. Mater. Chem. A*, 2019, **7**, 22996.
121. A. Wong, Q. Liu, S. Griffin, A. Nicholls and J. R. Regalbuto, *Science*, 2017, **358**, 1427.
122. Y. Wang, Q. Cao, C. Guan and C. Cheng, *Small*, 2020, **16**, 2002902.
123. P. Du, K. Hu, J. Lyu, H. Li, X. Lin, G. Xie, X. Liu, Y. Ito and H.-J. Qiu, *Appl. Catal., B*, 2020, **276**, 119172.
124. T. T. Gebremariam, F. Chen, Y. Jin, Q. Wang, J. Wang and J. Wang, *J. Catal. Sci. Technol.*, 2019, **9**, 2532.
125. J. Wang, M. Yan, K. Zhao, X. Liao, P. Wang, X. Pan, W. Yang and L. Mai, *Adv. Mater.*, 2017, **29**, 1604464.
126. G. Zhang, J. Wang, Z. Wu, R. Shi, W. Ouyang, A. Amini, B. N. Chandrashekar, N. Wang and C. Cheng, *ACS Appl. Mater. Interfaces*, 2017, **9**, 763.
127. S. E. Skrabalak and K. S. Suslick, *J. Am. Chem. Soc.*, 2005, **127**, 9990.
128. A. Mondal, A. Paul, D. N. Srivastava and A. B. Panda, *ACS Appl. Nano Mater.*, 2018, **1**, 4622.

129. K. Chang, X. Hai, H. Pang, H. Zhang, L. Shi, G. Liu, H. Liu, G. Zhao, M. Li and J. Ye, *Adv. Mater.*, 2016, **28**, 10033.
130. R. Raccichini, A. Varzi, V. S. K. Chakravadhanula, C. Kübel and S. Passerini, *Sci. Rep.*, 2016, **6**, 23585.
131. R. Kumar, R. K. Singh, R. Savu, P. K. Dubey, P. Kumar and S. A. Moshkalev, *RSC Adv.*, 2016, **6**, 26612.
132. M. Abbas, Z. Chen and J. Chen, *J. Mater. Chem. A*, 2018, **6**, 19133.
133. S. Zhu, J. Guo, J. Dong, Z. Cui, T. Lu, C. Zhu, D. Zhang and J. Ma, *Ultrason. Sonochem.*, 2013, **20**, 872.
134. J. Wu, L. Wang, B. Lv and J. Chen, *ACS Appl. Mater. Interfaces*, 2017, **9**, 14319.
135. J. Zheng, H. Lin, X. Zheng, X. Duan and Y. Yuan, *Catal. Commun.*, 2013, **40**, 129.
136. C. Wen, Y. Cui, X. Chen, B. Zong and W. L. Dai, *Appl. Catal., B*, 2015, **162**, 483.
137. V.-T. Nguyen, H. Ha, N.-A. Nguyen, H. An, H. Y. Kim and H.-S. Choi, *ACS Appl. Mater. Interfaces*, 2020, **12**, 15500.
138. Q. C. Tran, H. An, H. Ha, V. T. Nguyen, N. D. Quang, H. Y. Kim and H. S. Choi, *J. Mater. Chem. A*, 2018, **6**, 8259.
139. W. Gao, P. Tieu, C. Addiego, Y. Ma, J. Wu and X. Pan, *Sci. Adv.*, 2019, **5**, 9590.
140. H. Ataee-Esfahani and S. E. Skrabalak, *Chem. Commun.*, 2016, **52**, 10783.
141. X. Cui, P. Ren, D. Deng, J. Deng and X. Bao, *Energy Environ. Sci.*, 2016, **9**, 123.
142. J. Deng, P. J. Ren, D. H. Deng and X. H. Bao, *Angew. Chem., Int. Ed.*, 2015, **54**, 2100.
143. M. Zhang, J. Guan, Y. Tu, S. Chen, Y. Wang, S. Wang, L. Yu, C. Ma, D. Deng and X. Bao, *Energy Environ. Sci.*, 2020, **13**, 119.
144. G. J. Ma, H. J. Yan, J. Y. Shi, X. Zong, Z. B. Lei and C. Li, *J. Catal.*, 2008, **260**, 134.
145. H. Du, X. Gui, R. Yang, H. Zhang, Z. Lin, B. Liang, W. Chen, H. Zhu and J. Chen, *J. Mater. Sci.*, 2018, **53**, 14619.
146. H. Du, X. Gui, R. Yang, Z. Lin, B. Liang, W. Chen, Y. Zheng, H. Zhu and J. Chen, *Nanoscale*, 2018, **10**, 3877.
147. Y. Lv, Y. Fang, Z. Wu, X. Qian, Y. Song, R. Che, A. M. Asiri, Y. Xia, B. Tu and D. Zhao, *Small*, 2015, **11**, 1003.
148. B. Fang, N. K. Chaudhari, M. S. Kim, J. H. Kim and J. S. Yu, *J. Am. Chem. Soc.*, 2009, **131**, 15330.
149. S. J. Guo and S. H. Sun, *J. Am. Chem. Soc.*, 2012, **134**, 2492.
150. K. Hu, T. Ohto, L. Chen, J. Han, M. Wakisaka, Y. Nagata, J.-I. Fujita and Y. Ito, *ACS Energy Lett.*, 2018, **3**, 1539.
151. J. Zhang, T. Wang, P. Liu, Z. Liao, S. Liu, X. Zhuang, M. Chen, E. Zschech and X. Feng, *Nat. Commun.*, 2017, **8**, 15437.
152. T. Wang, Y. Guo, Z. Zhou, X. Chang, X. J. Zheng and X. Li, *ACS Nano*, 2016, **10**, 10397.

153. M. Carmo, D. Fritz, J. Mergel and D. Stolten, *Int. J. Hydrogen Energy*, 2013, **38**, 4901.
154. W. Sheng, H. Gasteiger and Y. Shao-Horn, *J. Electrochem. Soc.*, 2010, **157**, B1529.
155. J. Deng, P. Ren, D. Deng and X. Bao, *Angew. Chem., Int. Ed.*, 2015, **54**, 2100.
156. V. T. Nguyen, N. A. Nguyen, Y. Ali, Q. C. Tran and H. S. Choi, *Carbon*, 2019, **146**, 116.
157. N. Kosova, E. Devyatkina and D. Osintsev, *J. Mater. Sci.*, 2004, **39**, 5031.
158. M. J. Munoz-Batista, D. Rodriguez-Padron, A. R. Puente-Santiago and R. Luque, *ACS Sustainable Chem. Eng.*, 2018, **6**, 9530.
159. D. Cheng, J. Liu, X. Li, R. Hu, M. Zeng, L. Yang and M. Zhu, *J. Power Sources*, 2017, **350**, 1.
160. J. W. Wang, X. H. Liu, S. X. Mao and J. Y. Huang, *Nano Lett.*, 2012, **12**, 5897.
161. J. Cui, Z. L. Xu, S. Yao, J. Huang, J. Q. Huang, S. Abouali, M. A. Garakani, X. Ning and J. K. Kim, *J. Mater. Chem. A*, 2016, **4**, 10964.
162. H. Liu, R. Z. Hu, W. Sun, M. Q. Zeng, J. W. Liu, L. C. Yang and M. Zhu, *J. Power Sources*, 2013, **242**, 114.
163. I. Kim, S.-W. Park and D.-W. Kim, *Chem. Eng. J.*, 2019, **375**, 122033.
164. M. Chowalla, H. S. Shin, G. Eda, L.-J. Li, K. P. Loh and H. Zang, *Nat. Chem.*, 2013, **5**, 263.
165. S.-W. Park, I. Kim, S.-I. Oh, J.-C. Kim and D.-W. Kim, *J. Catal.*, 2018, **366**, 266.
166. B. Zhao, Y. Zheng, F. Ye, X. Deng, X. Xu, M. Liu and Z. Shao, *ACS Appl. Mater. Interfaces*, 2015, **7**, 14445.
167. Q. Zhou, Z. Zhao, Z. Wang, Y. Dong, X. Wang, Y. Gogotsi and J. Qiu, *Nanoscale*, 2014, **6**, 2286.
168. H. Zeng, X. W. Du, S. C. Singh, S. A. Kulinich, S. Yang, J. He and W. Cai, *Adv. Funct. Mater.*, 2012, **22**, 1333.
169. P. Shankara, M. Q. Hafzan Ishak, J. K. Padarti, N. Mintcheva, S. Iwamori, S. O. Gurbatov, J. H. Lee and S. A. Kulinich, *Appl. Surf. Sci.*, 2020, **531**, 147365.
170. A. De Giacomo, A. De Bonis, M. Dell'Aglio, O. De Pascale, R. Gaudiuso, S. Orlando, A. Santagata, G. S. Senesi, F. Taccogna and R. Teghil, *J. Phys. Chem. C*, 2011, **115**, 5123.
171. Y. Ishikawa, Y. Shimizu, T. Sasaki and N. Koshizaki, *J. Colloid Interface Sci.*, 2006, **300**, 612.

CHAPTER 6

Further Sustainable and/ or Unconventional Low-temperature Wet-chemical Synthetic Methods

F. BARBON[a], F. TAJOLI[a], F. SPOLAORE[a] AND S. GROSS[a,b*]

[a]Department of Chemistry, University of Padova, Via Marzolo 1, 35131, Padova, Italy; [b]Karlsruher Institut für Technologie (KIT), Institut für Technische Chemie und Polymerchemie (ITCP), Engesserstr. 20, 76131 Karlsruhe, Germany
*E-mail: silvia.gross@unipd.it

6.1 General Introduction

This chapter focuses on a series of further sustainable and/or unconventional wet-chemical methods to prepare single metal, alloys, oxides, chalcogenides and other inorganic compounds in the form of nanoparticles (NPs) at low temperature (<200 °C). In particular, the aim of the chapter is to provide the reader with an overview of further methods not specifically addressed by other chapters of the book, such as hydrothermal (Section 6.2), polyol-assisted (Section 6.3), continuous-flow (Section 6.4), sonochemical (Section 6.5) methods, as well as radiochemistry (Section 6.6) and laser ablation in liquid media (Section 6.7). The theoretical background of each method, a description of the synthetic procedure and a discussion of the

Inorganic Materials Series No. 14
Unconventional Green Synthesis of Inorganic Nanomaterials
Edited by Silvia Gross

Published by the Royal Society of Chemistry, www.rsc.org

synthetic parameters involved, and their influence on the features of the final products, are given, while also outlining the pros and cons of the presented synthetic approaches. In addition, a description of the state-of-the-art of the compounds obtainable through each approach is presented.

6.2 Hydrothermal Synthesis

Hydrothermal synthesis is regarded as one of most versatile and green methods to develop inorganic materials with tuneable structural and morphological properties. Historically, the name derives from the geological processes of crystal growth in aqueous environment under high temperature.[1] The most widespread definition is 'a heterogeneous chemical reaction in aqueous media above room temperature (normally above 100 °C) and at a pressure greater than 1 atm in a closed system'.[2,3] Under these conditions, materials that otherwise may not be prepared through a classical wet-chemistry route can be synthesised. Moreover, crystallisation of new phases with high purity can be obtained under mild temperatures and short reaction times, without applying high temperature post-treatments, such as annealing.[4–6]

Many theories have been developed to describe accurately both the thermodynamics and the kinetics of the crystallisation in solution under hydrothermal conditions.[7–9] The classical nucleation theory (proposed by LaMer[10,11] in the mid of the 20th century) is widely acknowledged as the first, rough picture of this process, but it does not describe in particular the nucleation kinetics. Many non-classical models (*e.g.*, the two-step model,[12–15] oriented attachment[16,17] and pre-nucleation cluster[18–20]) were therefore proposed to fill this knowledge gap.[21] From a general point of view, the growth rate depends on a number of factors involving both the conditions and the system under investigation, which will be discussed in more detail in the following sections. It is worth mentioning the possibility of using the hydrothermal process not only to synthesise fine-particle products, but also to grow large single crystals and to leach ores to recover valuable metals.[22] At the industrial scale, this method shows great potential as a valuable synthesis route in a low range of temperature and pressure with limited waste, answering the increasing demand for mild solution processes and low energy-demanding processes. In this regard, greater compatibility with the upscale requirements, such as total avoidance of corrosive Cl^- anions in the feedstock solution, has been optimised.[23] Of particular interest, at the large scale, is the synthesis of zeolites, which has been reviewed extensively elsewhere.[24–27] Nevertheless, many advances are still ongoing, partly because of the intrinsic 'black-box' nature of these systems. Indeed, the possibility of monitoring the process of the reaction directly inside a closed reactor is a current challenge, which has been addressed by employing synchrotron radiation to follow the crystallisation process in a time- or temperature-resolved fashion.[28–31]

An organic solvent may be used (in the so-called *solvothermal* synthesis), as well as other solvents (ionic liquids,[32] molten salts[33]), and many excellent reviews have been dedicated to these methodologies.[33–37] However, the focus of the following sections is on water-based syntheses, considering that water is the most environmentally benign solvent, and in subcritical conditions. Hydrothermal syntheses taking place under supercritical conditions (*i.e.*, above 647 K and 21.8 MPa for water),[38] despite having advantages in terms of short reaction time and nanosizing, require a large amount of energy and very high temperatures and are addressed in more detail in Chapter 4.

6.2.1 Fundamentals of Hydrothermal Technology

Under hydrothermal conditions, water (liquid state) is in equilibrium with the gas phase, even above its boiling temperature, due to the increase of the vapour pressure inside the closed vessel: accordingly, the properties of water under these conditions are remarkably different from the standard ones.[39,40] In particular, as can be seen in Figure 6.1 as a function of temperature, the viscosity, the dielectric constant and the density decrease, while the ionisation constant increases.

The effects of these changes are relevant already at subcritical temperatures (*e.g.*, the dielectric constant changes from 78 at ambient temperature and pressure to *ca.* 40 at 423 K and at 0.1 MPa of pressure).[40] This leads to marked changes in the solubility of the reagents, reaction rates and crystal growth kinetics, resulting in different crystallisation processes occurring during the hydrothermal treatment.[41,42] Generally speaking, crystallisation occurs when the concentration of the solute exceeds its solubility in a solvent (the ratio between these two quantities is called *supersaturation*).[43,44] From a thermodynamic point of view, the dissolution process of a reagent continues until the chemical potential of the solute both in solid form and

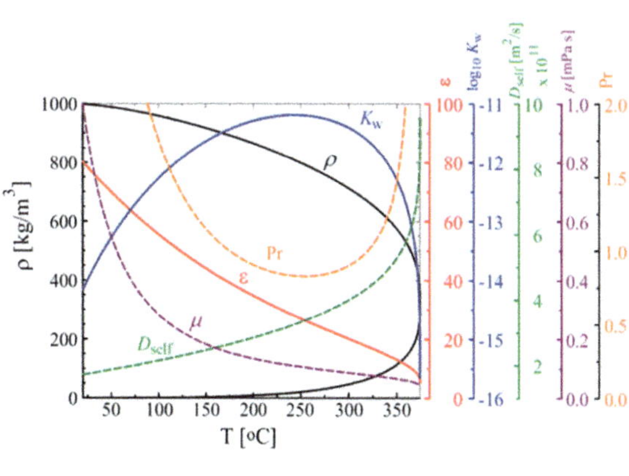

Figure 6.1 Summary of the behaviour of the principal water properties at 24 MPa as a function of temperature inside the hydrothermal reactor. ρ is the density, ε the dielectric constant, K_w the ionisation constant, D_{self} the self-diffusion coefficient, μ the viscosity and Pr the Prandtl number (*i.e.*, the ratio between the kinematic viscosity and the thermal diffusivity). Reproduced from ref. 40 with permission from American Institute of Physics, Copyright 2010.

in solution reaches the same value and therefore the system is at equilibrium. However, it must be noted that these considerations are strictly valid only when the chemical potential of the solute does not depend on the concentration of other components. Indeed, solubility is heavily influenced by the change in the properties of water. Decrease of the dielectric constant is associated with a higher solubility of non-polar species and, *vice versa*, with a lower solubility for polar ones, because electrolytes that would be completely dissociated at lower temperatures start to behave like undissociated molecules.[45] As a consequence, the solvation of ionic species is reduced: experimental studies on $ZnBr_2$[46] and $AgNO_3$[47] evidenced that the number of water dipoles in the first solvation shell decreases from 5 and 4 to about 3, respectively, as a function of temperature up to the supercritical conditions. This is accompanied by a contraction of the solvation shell and by a decrease in the average coordination number of the water molecules (*i.e.*, hydration). Clustering of solute molecules is favoured, alongside the formation of a gradient of the solute concentration in the solution and, in accordance with the 'two-step model' of crystallisation in solution,[14] this is followed by the nucleation of a new phase.[48] This dehydration effect of liquid water above its boiling point can be regarded as the reason for crystallisation of metastable phases that it would not be possible to obtain under standard conditions, due to a high reaction barrier and/or low diffusion rate. It is worth noting that, even if the solubility of ionic species is reduced, the improved mobility of the solubilised species due to the decrease in the viscosity compensates for this effect. Indeed, mass transfer throughout the suspension does not impose any restrictions in the speed of growth, as the activation energies of dissolution and diffusion are both usually lower than 10 kcal mol^{-1}, far less than the energies required for the growth of a new phase.[49] Experimentally, it has been shown that the growth rate reaches a plateau once a high enough concentration is used.[50]

6.2.1.1 Role of Synthetic Parameters

A general overview of the principal synthetic parameters influencing hydrothermal synthesis is provided below. This list should not be considered exhaustive, as many other factors not directly cited here (*e.g.*, time of reaction, concentration of the precursors, different reactor designs, presence of stirring) also play a relevant role. Indeed, the toolkit available to the experimentalist approaching this synthetic method is almost limitless, highlighting the extreme versality and level of adaptation of the technique.

6.2.1.2 Temperature, Pressure and Filling Ratio

Temperature and pressure are two essential parameters that characterise every hydrothermal reaction. As discussed, the properties of the medium, as well as the reaction kinetics and thermodynamics, are strongly correlated to the chosen reaction conditions. From a fundamental point of view, temperature determines whether a reaction (for example, the nucleation of a new phase) is thermodynamically possible.[51] The well-known equation

$$\Delta G^{\circ}_{sol} = \Delta H^{\circ}_{sol} - T\Delta S^{\circ}_{sol} \tag{6.1}$$

where ΔG°_{sol}, ΔH°_{sol} and ΔS°_{sol} are the standard changes in free energy, enthalpy and entropy of dissolution, respectively, establishes the conditions for a successful synthesis. For example, if both the enthalpy and entropy of dissolution are positive, then a high enough temperature is required for a spontaneous reaction to occur (*i.e.*, the change in the free energy of dissolution must be negative). In a similar fashion, the nucleation of a new phase takes place if the free energy of nucleation, expressed as follows according to the Van't Hoff equation, is negative:

$$\Delta G = -RT\log\left(a_i / a_i'\right) \tag{6.2}$$

where a_i is the non-equilibrium activity and a'_i is the equilibrium activity. When the activity coefficients can be considered to be unitary, activities are replaced by concentration and this ratio (a_i/a'_i) is called the *degree of supersaturation*. The temperature also strongly affects the actual pressure inside the autoclave in autogenous system (*i.e.*, sealed vessels without external control over the pressure), where the vapour tension of each component of the solution at a selected temperature determines the final pressure.

Generally, a feedstock solution containing all the components (solvent, precursor, mineraliser and other possible additives) is prepared and then poured into the vessel. Obviously, along with the temperature, the volume of solution poured inside the vessel compared to the vessel internal volume, called the *filling ratio*, determines the autogenous pressure and has to be carefully controlled to avoid safety risks. The filling ratio is also important to determine the reaction medium physical properties. It was demonstrated that if the vessel is filled exactly at 32% of its capacity, the meniscus between the liquid phase and the gas one remains fixed in the starting position, also increasing the temperature. Once the critical point is reached, it will simply vanish. However, if the filling ratio is higher than 32%, the meniscus will move upwards since the liquid phase will expand, gradually filling the vessel. The higher the filling ratio, the lower the temperature required to completely fill the container. In contrast, a value lower than 32% leads to a decrease of the meniscus level, increasing the temperature due to solvent evaporation.[52]

6.2.1.3 *pH and Mineraliser*

While the previous parameters have a common, general effect on all hydrothermal processes, regardless of the actual system studied, the pH and the choice of mineraliser are closely related to the specific reaction performed. *Mineraliser* is a term that comes from geological science, where the aim of hydrothermal treatment is usually to grow a single, large crystal rather than to perform synthesis of a polycrystalline powder. In a typical crystal growth setup, the autoclave is divided into two sections, separated by a perforated disc: the upper one, where a single crystal seed is placed, and a lower one, which is filled with the solution containing the nutrient.[53,54] Taking quartz as an example, the nutrient would be polycrystalline quartz. However, its solubility in water is very low under standard conditions and does not change much even at higher temperature and pressure. Therefore, the time that would be required to grow a large crystal would be unreasonable. Thus, a second compound (the mineraliser) must be introduced to increase the level of dissolution, forming soluble species and speeding up the crystallisation rate. Typically, these are hydroxides of the alkali metals, alkali salts of weak acids or even strong acids. The higher the concentration of the mineraliser, the higher the solubility and therefore the greater the crystallisation speed. A high concentration (over 1 M) is frequently needed to obtain a quantitative yield under mild conditions and short treatment times, but reactions with a lower concentration are possible. The concentration also has an important impact on the final morphology of the product. Mai *et al.*[55] achieved shape and size control of cerium oxide (CeO_2) simply by adjusting the amount of NaOH employed. They obtained cubes and rods with a concentration of 0.01 M, 6 M and 9 M, respectively, as can be seen in Figure 6.2.

Once the nucleation and crystallisation have started, growth kinetics determine the morphology and the degree of crystallisation of the material. Of the many parameters that have a significant impact in these aspects, the pH is of particular relevance in hydrothermal synthesis,[56] compared, for example, with solvothermal synthesis, where it cannot be exactly defined. For many oxide and hydroxide materials, a strongly basic pH is needed to obtain crystalline powders in a very short time. For example, Bi_2WO_6, known for its applications in photocatalysis and optics, can be easily crystallised at pH above 7, while at lower values only weak and broad X-ray diffraction (XRD) reflections are observed.[57] Moreover, the final morphology depends on the speed of crystal growth, and in particular on the relative growth of the single facets. The particle morphology of boehmite (AlO(OH)) can be finely tuned by choosing the

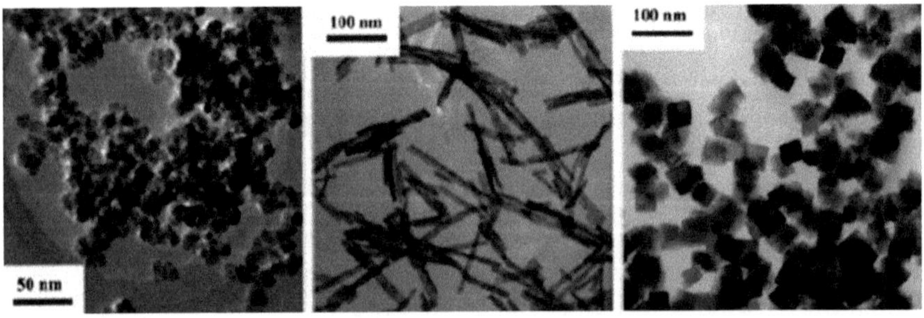

Figure 6.2 TEM images of CeO_2 NPs as obtained using different concentrations of NaOH: 0.01 M (left), 6 M (centre) and 9 M (right). Reproduced from ref. 55 with permission from American Chemical Society, Copyright 2005.

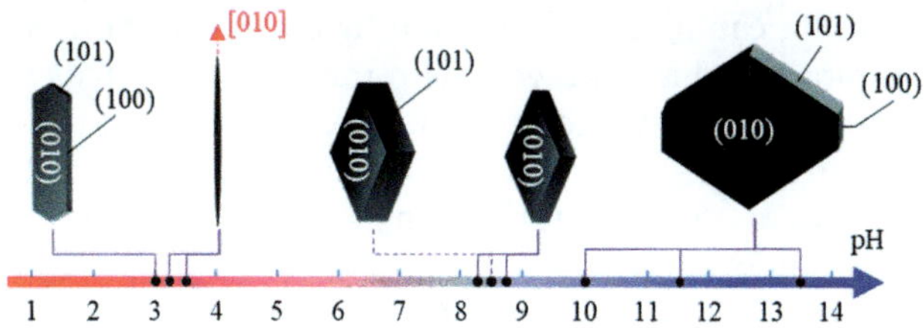

Figure 6.3 Schematic representation of the morphology evolution of boehmite NPs as a function of the reaction mixture pH. Reproduced from ref. 58 with permission from the Royal Society of Chemistry.

appropriate pH, changing from a needle-like one at pH = 3 to rhombic platelets at pH = 8 and finally to hexagonal platelets at pH = 13 (Figure 6.3).[58] In a similar fashion, the morphology of $BiVO_4$, important for its photocatalytic activity, can be easily tuned from random aggregates to rod-like grains and to long rods on changing from pH 1 to pH 4 and 9, respectively.[59]

6.2.1.4 *Additives*

The morphology of the products in all syntheses can be finely tuned with addition of other components, such as surfactants, structure-directing agents (SDA), capping and templating agents and different types of counterions, to selectively control size, shape, phase homogeneity and particle dispersion. Surfactants are amphiphilic molecules, with a polar head and a non-polar tail, and are usually classified into cationic, anionic, amphoteric and non-ionic classes. In particular, ionisable surfactants are characterised by a specific pK_a. Depending on the pH of the solution, they may be completely deprotonated (above the pK_a) or protonated (below the pK_a). When in dissociated form, they may interact with the surfaces of the suspended NPs through hydrogen bonding, electrostatic or van der Waals interactions. On the other hand, NPs may also have a charged surface, as determined by the isoelectric point (IEP)† of the newly forming phase. If the pH is higher than the IEP, the charge of the particles is negative, whereas in the opposite case a positive charge arises. Thus, the experimental conditions that determine the status of the components in suspension must be considered carefully.[60] Surfactants may be selectively adsorbed onto one or more facets, limiting the growth of the crystal in that specific crystallographic direction. Thermodynamically, the rate of facet growth is driven by the necessity to minimise the total surface energy, thus high-energy surfaces grow rapidly, taking up a small fraction of the surface in the final crystal. However, when the mass transport of new material to a specific surface is blocked by an adsorbed species, growth along the free surfaces with a different orientation is favoured and the final percentage of the exposed facets can be kinetically controlled.

†The isoelectric point (IEP) represents the pH value at which the net electric charge of an elementary entity is zero.

A similar situation can also be promoted by the counterion of the reagent, which may interact in different ways with the surfaces.[61] This approach has the advantage of simplicity and cost effectiveness, since one component of the system is eliminated, compared to the use of a surfactant. These considerations are particularly important for nanotechnology, where nanometric powders with a high degree of a selected exposed facet are highly desirable for their specific properties.[62,63] As an example, much effort has been devoted to the hydrothermal synthesis of nanostructured titania (TiO_2, anatase phase), and specifically to the maximisation of high-energy {001} facets, because of the promising properties exhibited by this surface, such as the dissociative adsorption of water molecules.[64] Surfactants, such as oleic acid, were employed, but the best results were obtained when employing a fluoride source for the synthesis, either directly from the precursor or as an additive. F^- anions act as capping agents, selectively adsorbing on the {001} facets and blocking their growth.[65,66] Yang *et al.* studied in detail the stability of these surfaces in the hydrothermal environment. They concluded that, even using fluoride ions, the oriented attachment along the [001] direction is the prevailing growth mechanism, and, over time, it leads to a truncated bipyramidal shape exposing mainly the {101} facet. However, this can be effectively suppressed when a strong acid medium is employed (1.5 M HCl) because the hydrolysis of the Ti–F bonds is prevented and therefore the high-energy surfaces remain exposed even after 88 hours of hydrothermal treatment at 200 °C.[67]

6.2.1.5 Reaction Containers

Hydrothermal reactions are generally performed in sealed containers, called *autoclaves* (Figure 6.4a). The number of available designs is very high, differing in how the pressure and temperature are controlled, the type of material and the dimensions; however, there is not a uniform standard for the classification of these systems. In most cases, the material of choice is stainless steel because it can easily withstand pressures of hundreds of bars and temperatures up to 400–500 °C and possesses a high resistance to different chemical environments.

While the pressure can be finely controlled in the autoclaves, in more simple systems (called *hydrothermal reactors* or *bombs*, Figure 6.4b) the pressure is instead determined by the filling ratio. Usually, this is between 40 and 80%, and in the latter case it corresponds to a pronounced pressure variation, from a few bars to *ca.* 100 bar at 250 °C (Figure 6.4c).[68] Although some particular alloys can be used to increase the resistance to highly corrosive reactants (*e.g.*, by forming a passivating oxide surface), most syntheses are not performed directly in the autoclaves, to reduce the risk of reaction with the material of the container itself and, thus, reducing the purity of the product. Therefore, vessels called *liners* are employed to prevent direct contact between the solution and the containers, creating an internal chamber where the reaction is performed. Of the many different designs and materials

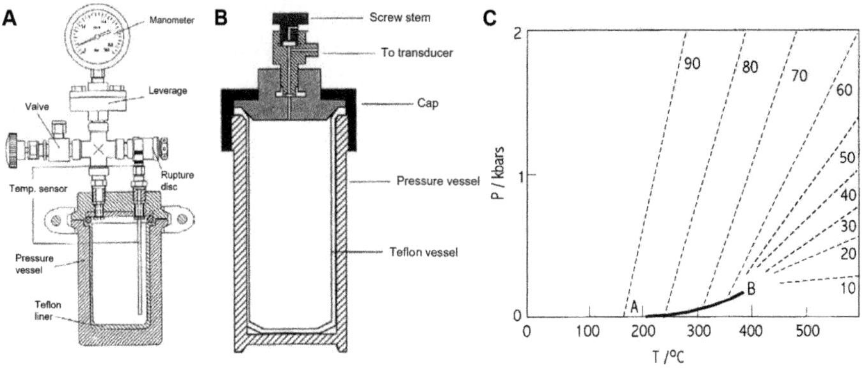

Figure 6.4 Schematic representation of (A) an autoclave with control over the pressure, and (B) a hydrothermal bomb with autogenous pressure. Reproduced from ref. 1 with permission from Elsevier, Copyright 2013. (C) Variation of the autogenous pressure as a function of the temperature. Dotted lines represent the degree of filling, while the solid line represents the water vapour–liquid equilibrium. Adapted from ref. 3 with permission from John Wiley & Sons, Copyright © 1985 by VCH Verlagsgesellschaft mbH, Germany.

that can be employed, Teflon® has been the most widely used for very corrosive environments and with temperature–pressure conditions in the subcritical region (below 250 bar and 250 °C).[69] However, it must be taken into account that using such a setup may lead to severe temperature gradients, both between the oven and the reactor and within the liner itself. Moreover, Teflon® is a porous material, so any contamination from absorbed species must be avoided when the same liner is used for different reactions.

6.2.2 Hydrothermal Synthesis of Inorganic Materials

Inorganic materials prepared through the hydrothermal process have been extensively studied and continue to receive much attention, both in form of nanostructures and as large single crystals.[36,70] The focus in this section is on the synthesis of the most common nano- and micrometre-sized systems obtained under mild, subcritical conditions, with particular emphasis on the most recent advances. The classification is based on the nature of the product, starting from native elements and alloys (Section 6.2.2.1) and metal oxides (Section 6.2.2.2), and then discussing metal sulfides (Section 6.2.2.3) and phosphates (Section 6.2.2.4).

6.2.2.1 Metals and Alloys

Metal NPs possess unique properties compared to the bulk materials that can be exploited in a wide range of applications, from catalysis to sensing and optoelectronics. The crystallisation of metal NPs under hydrothermal condition is possible both in acidic and alkali media starting from their respective metal salts and with a suitable reducing agent.

Gold. Nanostructured Au has long been investigated, and it has been synthesised in different shapes and sizes.[71,72] However, the majority of these syntheses were performed in a reductive environment to promote the formation of metallic Au, starting from Au(III) precursors, for example, through solvothermal or polyol-mediated syntheses, which are less environmentally friendly solvents then water. One of the first contributions in this regard was offered by Huang *et al.*,[73] who were able to control the morphology and the size of gold NPs just by varying the concentration of hexadecyltrimethylammonium bromide (CTAB) in an aqueous system. They obtained triangular nanoplates, nanoprisms and nano-octahedra with an average size of 50–150 nm. In a similar procedure, recently reported by Liu *et al.*,[74] the authors used cetyltrimethylammonium chloride (CTAC) to control the size of triangular NPs from tens to hundreds of nanometres.

Silver. A reduction approach can be used to tune the growth of different morphologies of metallic silver NPs, as demonstrated by Yu *et al.*,[75] where glucose was employed as a reducing agent and different concentrations of CTAB, to control the final morphology (from cubes to rods and triangular prims). In a more recent report, Bari *et al.*[76] obtained unique morphologies of extra-long nanowires (45–65 nm in diameter and 200–500 nm in length) by using glucose as reducing agent and poly(vinylpyrrolidone) (PVP) as stabilising ligand. This morphology is particularly indicated for optoelectronic applications, such as touch panels and wearable electronics, where it showed promising results compared to the commercial counterpart. McNulty *et al.*[77] synthesised $AgVO_3$ decorated with Ag NPs on the surface through a one-step hydrothermal method, as a result of the reduction of the salt precursor, to produce the noble metal NPs, with an average diameter of 5 nm and uniform dispersion on the substrate.

Bimetallic Systems. Combination of two or more metals into alloys can be beneficial not only because it combines the advantages of each element, but also because synergistic effects can arise.[78] For example, PdRh alloys with a precise hollow nanocubic morphology with a very narrow size distribution (14.4 ± 0.6 nm) were synthesised employing PVP and KBr or KI under mild conditions (180 °C for 4 hours).[79,80] In particular, the iodide ion was essential to obtain the hollow nanostructure because of its adsorbate-induced reconstruction effect. When the iodide ion was not introduced, only solid nanocubes were obtained, while increase in the concentration of KBr led to the formation of different morphologies, from nano-icosahedra to nano-truncated octahedra, because of the different adsorption of Br^- ions on the different surfaces. With regard to catalytic applications, a recent example is represented by AuPt nanoclusters.[81] Nanoclusters of *ca.* 2.5 nm in diameter were obtained from a one-pot hydrothermal treatment at 120 °C for 4 hours starting from acidic precursors and employing glutathione as a capping agent, with an optimised transition metal to glutathione molar ratio of 1:3. These nanostructures showed improved catalytic performance for the oxidation of 3,3′,5,5′-tetramethylbenzidine and for the detection of H_2O_2 in sensing applications. Shi *et al.*[82] recently adopted a one-step hydrothermal strategy to synthesise PtPd alloy NPs supported on graphene oxide hydrogels. They found that optimisation of the concentration of the acidic precursors and of the temperature was crucial in stabilising the structure and the dispersion of the NPs on the support. Interestingly, an average diameter of 5 nm with a narrow size distribution was obtained only at 180 °C, while at lower temperatures severe growth (up to 250 nm at 100 °C) was observed.

6.2.2.2 Metal Oxides

Oxides are the largest class of inorganic compounds that has been prepared by hydrothermal processes. As a side note, hydroxides can also be made through this method, and can be easily obtained by avoiding the complete dehydration towards oxides under mild conditions. A number of reviews are available on this topic.[83,84]

ZnO. Zinc oxide, as a typical semiconductor with a wide direct bandgap (3.37 eV), possesses many applications, from optics to photocatalysis.[85–88] Therefore, an efficient, green and reproducible way to synthesise it, with a high control over morphology and size dispersion, is desirable.[89] The hydrothermal growth of ZnO NPs was studied in detail through *in situ* XRD by Bøjesen *et al.*[90] They concluded that a higher synthesis temperature and a prolonged reaction time lead to more isotropic crystallites, whereas an anisotropic shape was more probable with lower temperature and a shorter reaction time. Diodati *et al.*[91] optimised a hydrothermal route at 135 °C for 24 hours to obtain crystalline, urchin-like ZnO grains with a 80 nm section, starting from oxalate precursors. These were tested for H_2S sensing and showed an interesting dosimeter-type behaviour when an operating temperature of 150 °C was employed. The morphology can also be readily tuned. Zhu *et al.*[92] varied the concentration of CTAB and that of the mineraliser (NaOH) in the synthesis, to obtain nanoflowers of *ca.* 2 μm and single nanoplates with a thickness of 30 nm. Nanoflowers exhibited the best ethanol gas-sensing performances, with rapid response and recovery speed.

TiO_2. The extremely wide range of applications of titanium dioxide has made this binary system the most studied oxide in the past few years.[93–95] Of particular interest, as previously mentioned, is the selective growth of specific facets of the crystal, which strongly affects the interaction with molecules/ions, and the hydrothermal method is one of the most versatile approaches to achieve this. Typically, fluoride anions are employed to expose the highly energetic and reactive {001}[96] facet in the anatase phase (depicted in Figure 6.5a). Menzel *et al.*[97] studied the influence of the concentration of Ti source and fluoride source on the final morphology (as summarised in the ternary diagram reported in Figure 6.5b) and identified the right stoichiometry to maximise the plate-like morphology. However, this approach has the drawback of the use of highly toxic and corrosive HF. Recent developments demonstrated that it is possible to obtain nanocubes

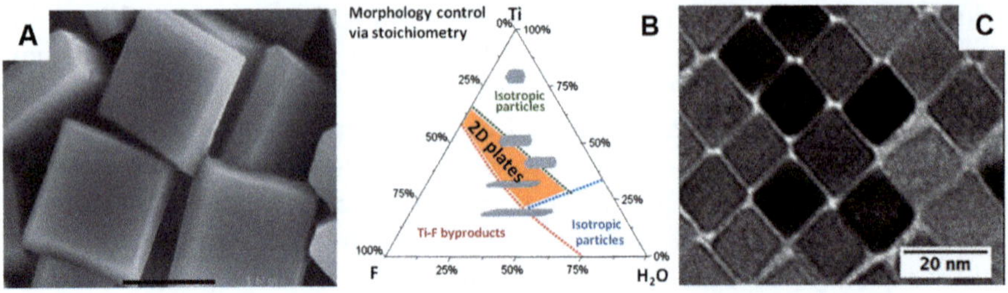

Figure 6.5 (A) SEM image of anatase TiO_2 obtained employing F^- ions as structure-directing agent to increase the percentage of {001} facets. Reproduced from ref. 96 with permission from Springer Nature, Copyright 2008. (B) Ternary diagram illustrating the influence of titanium and fluoride sources and of water on the TiO_2 morphology. Reproduced from ref. 99 with permission from John Wiley & Sons © 2019 WILEY-VCH Verlag GmbH & Co. KGaA, Weinheim. (C) TEM image of anatase TiO_2 employing TMAOH. Reproduced from ref. 97 with permission from American Chemical Society, Copyright 2013.

with almost only this surface exposed by employing suitable organic alkali peptising agents. Lekphet *et al.*[98] and Huesmann *et al.*[99] synthesised nanocubes of *ca.* 20 nm, with extreme control over the size distribution, employing tetramethylammonium hydroxide (TMAOH) starting from titanium alkoxides (Figure 6.5c). Moreover, it was possible to selectively achieve the different phases of titania (anatase, brookite and rutile) by varying the temperature and the pH.[100] Anatase was obtained at 200 °C and at pH higher than 1, while at lower pH brookite was obtained at 175 °C and rutile at 200 °C, respectively.

CeO_2. The ability to remove/uptake oxygen from the lattice, switching the cerium oxidation state between +3 and +4 while maintaining its structural integrity, has made cerium oxide of particular interest for catalytic and optical applications.[101–103] In particular, the amount of oxygen that can be exchanged with a gas phase is called the oxygen storage capacity (OSC).[104] The morphology has a profound influence on the reactivity of the NPs and the role of the hydrothermal conditions on this feature was extensively reviewed by Walton's groups.[34,37] In a study by Li *et al.*[105] nano-octahedra, nanocubes and nanorods with high control over the size distribution were synthesised by changing the temperature (from 200 °C to 180 °C and 100 °C, respectively) and the mineraliser (Na_3PO_4 in the first case and NaOH in the latter cases). The products were tested as catalysts for HCl oxidation and the nanorods showed not only the best stabilities over time (as expected given the exposure of the {110} facet with the lowest surface energy), but also the highest activity. In a recent paper,[106] it was shown how to obtain nano-octahedra with an extremely short reaction time (1 hour at 150 °C in a glass tube) and control over the edge size from 50 to 100 nm upon varying the concentration of the precursor (cerium nitrate). Perhaps more importantly, no base was added during the synthesis, while very high concentrations of this

component are generally used. This allowed tuning of the bandgap, displaying a marked redshift for the bigger NPs.

Fe_2O_3. The stable phase of Fe_2O_3 under 550 °C is hematite (α-Fe_2O_3) and the hydrothermal method has been widely employed for its synthesis,[107,108] owing to the easily tuneable characteristics in terms of morphology and size. A particularly active field of application resides in its use as an anode for Li-ion batteries, due to its high theoretical specific capacity compared to traditional graphite (1007 mA h g^{-1} *vs.* 372 mA h g^{-1}). Wang *et al.*[109] synthesised hollow hematite nanotubes in a graphene–aerogel matrix employing PVP and *N*-methyl pyrrolidone (NMP). When a treatment for 2 hours (at 160 °C) was used, the shape resembled spheroid aggregates on top of the graphene sheet. With longer treatment times, these started to dissolve along the major axis, forming nanotubes of *ca.* 0.5 micrometres when the time was at least 10 hours. Moreover, Tadic *et al.*[110] were able to greatly reduce the size (down to 8 nm, with a narrow size distribution) of spheric NPs by adding oleic acid as a capping agent.

Fe_3O_4. The presence of iron in both +2 and +3 oxidation states makes magnetite very interesting for a variety of applications, such as catalysis,[111] target-drug delivery[112] and magnetic resonance imaging.[113] The particle size and shape directly affect the magnetic properties and therefore the final application. Sun *et al.*[114] synthesised NPs with selective tuning of sizes from 4 to 16 nm by changing the feed concentration of sucrose, which had a bifunctional role. First, it acted as a precursor for the reducing agent, to form magnetite from the Fe(III) precursor, and, second, it formed a coating layer that prevented further growth and agglomeration. However, the reaction was conducted at 250 °C for 24 hours. More recently, Wu *et al.*[115] tuned the shape of carbon-coated magnetite under milder conditions (180 °C for 3 hours) using Fe_3O_4 nanosheets coated with PVP as precursor. The addition of D-glucose induced both the reduction and carbonisation of the precursor and subsequent growth into a nanoring shape through Ostwald ripening.

Ternary Oxides. Of the many ternary metal oxides, perovskites, with chemical composition ABO_3 (where A, B are cations), have been of increasing interest for their dielectric and piezoelectric properties and use as capacitors, transistors and transducers.[36,116] $BaTiO_3$ is a classic example of this class of materials. Inada *et al.*[117] reported modification from the cubic to the more reactive tetragonal structure by adding 10 vol% ethylene glycol to the feedstock solution (treated at 200 °C for 24 h). This phase transformation was accompanied by a change in morphology, from nanocubes of 50–100 nm to nanorods that grew along the [001] direction. In a subsequent study,[118] they proposed that the additive promotes a uniform substitution of the OH^- groups in the O^{2-} sites of the structure instead of only on the outer layer of the crystal. Compared to the cubic structure, the tetragonal one can better accommodate the consequent lattice deformations.

Other perovskite compounds that have attracted particular interest, also for their magnetic properties, are the ferrites, where B is Fe^{3+}.[119] Control over the surface facets is challenging and in the case of $YbFeO_3$, Zhang *et al.*[120]

were able to prove the fundamental role of the mineraliser. The morphology was highly dependent on the adsorbed cation and its concentration: by increasing the amount of KOH, the exposure of {002} facets was maximised, whereas the use of a low and high amount of NaOH was favourable for the formation of cubes and rods, respectively. Ferrites also crystallise with different phases, such as spinel, ilmenite and hexagonal structures.[119] Spinels, with the general formula MFe_2O_4 (M = metal ion), have been the subject of intense research due to their very wide range of applications. A portfolio of different ferrites (M = Ni, Mn, Co and Zn) was synthesised by Diodati *et al.*[121] *via* decomposition of the oxalate precursors. In particular, it was possible to achieve complete crystallisation of nanosized products under very mild conditions (down to 135 °C for as short as 1 hour treatment)[28] with a good control over size distribution. Noteworthy, is the ability to obtain the inverse spinel structure through this procedure. The inversion degree describes the distribution of Fe^{3+} ion over the tetrahedral and octahedral sites and this influences the electrical and magnetic properties. Most of the synthesised nano-ferrites possess only a partial inversion (tetrahedral and octahedral sites both occupied in different percentages) of the spinel structure, while a nearly complete inversion structure was demonstrated recently in $CoFe_2O_4$[122] and $ZnFe_2O_4$.[123] Furthermore, quaternary spinel ferrites (with the presence of two different metal, such as $Zn_xCo_{1-x}Fe_2O_4$ and $Zn_xNi_{1-x}Fe_2O_4$) were also synthesised at 135 °C for 24 hours.[124,125]

6.2.2.3 *Metal Sulfides*

Hydrothermal synthesis of sulfides is second only to that of oxides and silicates as a research topic in this field. A brief overview of the principal binary systems will be given below.

ZnS. As a II–VI group semiconductor, zinc sulfide has long been studied for applications in photocatalysis as well as various other uses, from nonlinear optical devices to light-emitting diodes and infrared windows.[126,127] It mainly occurs as two polymorphs: sphalerite, which is stable at room temperature, and wurtzite, which is stable above 750 °C. In particular, the stability of sphalerite NPs upon oxidation to ZnO under elevated temperatures has been studied,[128] as it deeply affects the optical and catalytic properties. A study by Yang *et al.*[129] showed the possibility of preparing ZnO/ZnS heterostructures by using ZnO nanoflowers as a template and thiourea as a sulfur source, with very short treatment time (2–5 hours) at 100 °C. The structures were composed of flower-like ZnO covered with NPs of ZnS of about 50 nm. The optimum photocatalytic activity was obtained through optimisation of ZnS loading over the ZnO template. Moreover, owing to its relevance in the modern energy transition, ZnS has been proposed as an anode for Li-ion and Na-ion batteries.[130] Through the hydrothermal method, ZnS quantum dots smaller than 10 nm on graphene have been prepared. The presence of graphene was decisive in preventing agglomeration of the NPs, acting as a support substrate and it did not interfere with the nucleation of ZnS.

The reduced dimensions were decisive in improving the electrode–electrolyte contact and in decreasing the length of the ion path during charge and discharge of the battery.

NiS_2. Dong *et al.*[131] reported the morphological control of NiS_2 from nanosphere assemblies of nanoplates to flower-like structures starting from thiourea and $Ni(NO_3)_2$ by addition of increasing concentrations of 1,2-ethylenediamide. Moreover, by addition of sodium dodecyl sulfate (SDS), rods were produced, which self-assembled into urchin-like structures with a longer reaction time. In a more recent report,[132] hollow nanospheres of about 3 μm with a thickness of 200 nm were prepared, adding ethylene glycol and oxalic acid and employing the same Ni source as in the previous example, together with sodium thiosulfate.

MoS_2. Due to its layered structure, MoS_2 has become a topic of growing interest. Wu *et al.*[133] employed a 'sequential two-step hydrothermal growth' to synthesise nanocomposites of MoS_2 nanosphere cores (from the first step), which acted as a templating agent for the hydrothermal growth of CdS NPs of about 50 nm. Proving the ability of the hydrothermal synthesis to access metastable phases, Geng *et al.*[134] were able to produce pure, metallic MoS_2 with a single-layer S–Mo–S structure. Starting from MoO_3, it was possible to maintain the structure during the conversion to MoS_2 by adding urea and optimising the temperature (200 °C)

6.2.2.4 Metal Phosphates

$LiFePO_4$. As one of the most widespread cathode materials for Li-ion batteries, optimisation of the structure–performance relationship of $LiFePO_4$ has been the focus of intense research. As recently reviewed,[135] the precise morphological control allowed by the hydrothermal process can lead to marked improvement in the final properties of the phosphate. A study by Sharikov *et al.*[136] highlighted the inverse relationship between the concentration of the precursor and the particle size. This trend is observed also for other kinds of material, as a consequence of the increased number of nucleation centres. However, in this study it was observed that the decomposition of the iron precursor starts at higher temperature when its concentration is low, forming bigger and better crystallised particles. Moreover, the heating and cooling ramps, which are usually not controlled in a hydrothermal synthesis (especially the cooling one), also have a major influence. Ruiz-Jorge *et al.*[137] found that the heating rate has the highest impact on the crystallinity, while a very fast cooling rate induces morphological changes (from NPs with a hexagonal base to a cubic one).

Lanthanide Phosphates. Compounds based on lanthanides show unique characteristics due to the presence of 4f electrons. $CePO_4$ was synthesised with NaH_2PO_2 as mineraliser, which proved to be crucial to achieve uniform and regular structures.[138] Changing the mineraliser concentration, it was possible to vary the morphology from rods to column sheet-assembled and spindle-like particles, with a length of about 16 μm, 1.5 μm and 3 μm, respectively. Considering $LaPO_4$, the phase selection between the rhabdophane and monazite polymorphs is a challenge. Through hydrothermal synthesis, it was recently demonstrated that, with a temperature over 210 °C, it is possible to select the monazite phase, with nanocrystals of about 7–8 nm in size.[139] The mechanism implies first the nucleation of the rhabdophane phase and then the nucleation of the monazite, with a constant mass transfer from the first to the second. The process was completed (*i.e.*, without rhabdophane phase left) after 7 hours of treatment.

6.3 Polyol Synthesis

Polyols, or polyalcohols, are a very broad class of compounds and encompass short-chain diols, such as ethylene glycol (EG), diethylene glycol (DEG), triethylene glycol (TEG) and increasingly longer molecules up to the polyethylene glycol (PEG), but also molecules such as glycerol (GLY), pentaerythritol (PE) and other polyalcohols also belonging chemically to the polyol group.[140] Polyols are safe to handle, green, biocompatible, biodegradable, sustainable and environmentally compatible solvents, which can be employed to carry out chemical reactions in a broad temperature window, thanks to their high boiling points.[140] They possess higher viscosity (16.1, 30.2 or 49.0 mP s for EG, DEG and TEG, respectively), higher dipole moments (2.28, 2.30 and 5.58 D for EG, DEG and TEG, respectively) and lower dielectric constants compared to water (41.4, 31.8 and 23.7 ε for EG, DEG and TEG, respectively).[141]

These molecules possess more than one hydroxyl substituent, which provides them with both reducing properties as well as coordinating ability.[140] For these reasons, diols or glycols are nowadays the most used polyols employed in a multitude of industrial processes, also thanks to their low cost, with EG being by far the most used.[136] Indeed, EG is the simplest and most studied diol for polyol syntheses: it possesses several unique properties owning two hydroxyl groups at adjacent positions, being colourless, odourless, relatively non-volatile, hygroscopic and possessing low viscosity. It tastes sweet and is completely miscible with many polar solvents (*e.g.*, water, alcohols, glycol ethers and acetone) and slightly soluble in non-polar solvents (*e.g.*, benzene, toluene, dichloroethane and chloroform).[142]

The term *polyol process* or *polyol synthesis* was initially coined in the late 1980s by Fiévet, Lagier and Figlarz to describe a liquid-phase synthesis route where polyols, such as 2-diols and ether glycols, were used both as a solvent and a reducing agent to prepare suspended NPs of Co, Ni, Cu and Pt.[141,143,144] This process was first employed to obtain a variety of metal NPs from their oxides, hydroxides or salts precursors, and recently it has been adopted for the preparation of other metals, as well as for the preparation of intermetallics and alloyed compounds.[145]

As mentioned before, typically employed polyols for the synthesis of inorganic materials are EG, DEG or TEG, which are characterised by high boiling points of about 197 °C, 244 °C and 285 °C, respectively.[140] These high boiling points ensure the preparation of well-crystallised materials, and the low to zero flammability hazard facilitates their use in large volumes in industry.[146] The viscosity of the media, but also their density, dielectric constant, and thermal conductivity, can all be finely tuned by diluting the polyol solvents with other miscible solvents, such as water, providing an additional tool for the synthetic chemist to set the right conditions to modify the nucleation of NPs.[140,147]

It is worth noting that the boiling points of polyol–water mixtures follow a non-linear trend, with mixtures of low polyol content (20–30 wt% of EG for example) boiling at a few degrees above 100 °C, a trend that is suddenly inverted when the proportions are switched.[140,148] The properties of common polyols, such as diols, follow a general trend: the boiling point increases with the number of OH functionalities and with increasing molecular weight, whereas the viscosity of the polyols increases with increasing molecular weight.[140] Despite very high boiling points and thermal stability, polyols can undergo decomposition at high temperatures. It has been suggested that, at high temperatures, the thermal decomposition of the polyol generates a series of by-products that impact the nucleation and growth of the resulting NPs.[142] For EG, for example, the breakdown generates molecules such as glycolaldehyde, glyoxal, glycolic acid, glyoxylic acid and oxalic acid, as presented in Figure 6.6.

Even at high temperatures, the reducing power of polyols is decreased and the thermal decomposition of such organic solvents can interfere with the production of monodisperse or shape-controlled NPs.[140] Generally, a polyol process employed for the synthesis of metal NPs involves a reduction of metal salts or metal oxide precursors by polyol species. Indeed, the redox potential difference between polyols and metal precursor is not always high enough to directly reduce the metal: in this case, temperature can be raised to increase the redox potential gap between the solvent and the metal precursor. The reduction power of polyols can be raised by heating to high temperatures (up to the boiling point of the employed polyol). As a rule of thumb, the standard electrode potentials measured

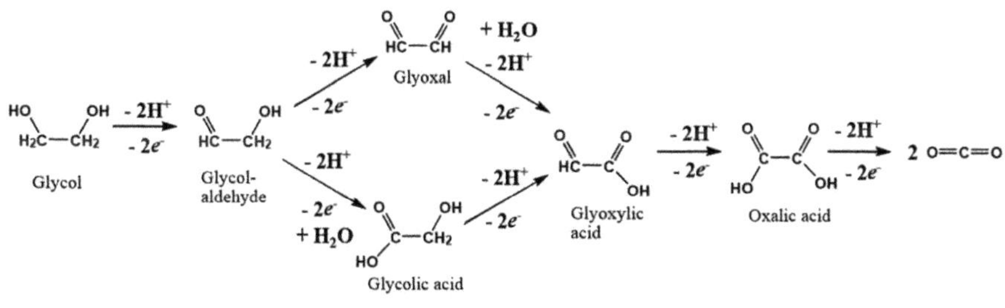

Figure 6.6 Schematic representation of the redox pathway for ethylene glycol. Adapted from ref. 141 with permission from the Royal Society of Chemistry.

in aqueous solutions at 25 °C can be used as a guide for assessing the relative reducibility of the metal precursor used. Larcher and Patrice, for example, studied the reducibility of different metals in different polyols.[149] If the temperature is not high enough, polyols are compatible with most of the common reducing agents that can be added in the solution. Compatible reducing agents that can be added in polyol solvents are the common metal borohydrides, often used in the past in the preparation of metallic and bimetallic core–shell particles.[150] The rate of polyol reduction is also slower compared to that of other reducing agents, for example hydrazine or sodium borohydride, allowing more precise control over the final sizes and morphologies.[151]

The high boiling points of polyols can also be exploited to enhance the dissolution of metal salts. Indeed, hydroxides or acetylacetonates are generally hard to solubilise: one can therefore easily increase the reaction temperature to allow their dissolution. Additionally, similarly to the different metals, each polyol has a different oxidation potential, which defines the temperature at which a specific metal reagent can be reduced by the polyol.[152] Therefore, by varying the reaction temperature, the reaction can be first induced and then also its rate can be finely tuned, resulting in the production of controlled structures and morphologies.[153–155] For example, the formation of single metal particles by reduction in EG starting from Pt and Pd precursors is thermodynamically unfavourable at room temperature, and therefore not spontaneous. On the other hand, Au and Ag salts can be reduced easily to generate single metal NPs in polyol media even at RT. Interestingly, Bonet *et al.* observed that the temperature has a direct impact on the redox reaction occurring between the EG and different precious metal precursors, and the energy required for the reduction of noble metal species can be reached by increasing the temperature. Additionally, the reduction potential of the metal salt remains independent of the temperature for the system under investigation.[156] Moreover, the Bonet group showed that the redox potential difference between EG and the dissolved precious metals was not high enough to initiate the reduction at room temperature. Nevertheless, even at room temperature, small Ag and Au NPs are formed in EG from the respective salts. They proposed that the metal reduction in the polyol process is aided by the overpotential, a kinetic contribution of the measured potential that induces a shift of the measured potential compared to E°, as the measured electrochemical potentials are the sum of both contributions (thermodynamic and kinetic).[156]

Given the intrinsic oxidation potential that is unique to each polyol solvent (therefore defining the temperature at which particle formation occurs, as mentioned above), the selection of the right polyol allows fine control over the nanocrystal formation.[157] The selection of the polyol also impacts the rate of reaction and the nucleation and growth. Indeed, the control over the resulting nanostructures morphologies is achieved as a result of diffusion-controlled particle formation regimes, ensured by the high viscosity of the medium.[141,146,158] This allows the formation of nanostructures with finely tuned size and shape.[140,141,159,160]

Given the water compatibility and solubility of most metal salts, dissolved metal cations can be readily reduced in most high boiling point polyols. The coordinating properties of polyols improve the solubility of metal salts even more, extending the possibility to use all common metal precursors (*e.g.*, chlorides, nitrates, sulfates, hydroxides, acetates or acetylacetonates), by forming metal complexes, at the same time inhibiting coalescence of the generated particles during the nucleation and growth events. The deprotonation of the hydroxyl group by addition of a base or under alkaline conditions facilitates the complexation of the polyol with the metal centre.[161,162] The O–CH_2–CH_2–O moiety, present in most polyols, strongly favours this process, as five-membered metallocycles are highly thermodynamically stable.[146]

Moreover, the chelating properties of the hydroxyl groups also provide additional stabilisation for the suspended NPs, allowing, in certain cases, a reduction in the need for additional stabilising additives (*e.g.*, polymers or capping agents) or even avoiding them completely.[146] For example, quasi-spherical Pd NPs (*ca.* 6 nm) were prepared by Arora *et al.* by reducing $PdCl_2$ in EG. They evidenced in their study the coordination of monoalkoxides at the surface of the isolated Pd NPs by a combination of techniques (*i.e.*, Fourier-transform infrared, thermogravimetric and atomic force microscopy analyses).[163]

To sum up, among the different solvents that can be considered to prepare inorganic materials by wet-chemistry approaches, polyols combine a unique set of advantages (Figure 6.7), determined by the unique chemical features of these compounds: high boiling points, chelating properties, redox properties, high viscosity and low toxicity.[146]

In the following sections, a general description of the state-of-the-art will be provided, focusing mainly on single metals and alloys (Section 6.3.1), metal oxides (Section 6.3.2) and other compounds (Section 6.3.3).

6.3.1 Metals and Alloys

Polyols are widely used to reduce dissolved metal cations in a one-pot reaction, and for this reason metallic NPs were the very first compounds to be synthesised using this approach (*e.g.*, Pb, Co, Ir, Os, Cu, Ag, Au, Pd, Pt and Cd), as highlighted by a large number of reviews.[146,164] When sufficient stabilisation is provided (either by the polyol itself or by additional stabilising agents), the generated suspensions are stable for a long time (*e.g.*, years). As a consequence of the limited reducing power of polyols at lower temperatures, metal salts that can be reduced are those based on noble metals (*e.g.*, Pd, Pt, Rh, *etc.*). For instance, less noble metals, such as Co and Ni, can re-oxidise after the reaction, also thanks to the chelating features of polyols and the presence of many OH groups in the surroundings.[165,166] Other metals, such as In or Sn, can only be obtained at high temperatures, and only the

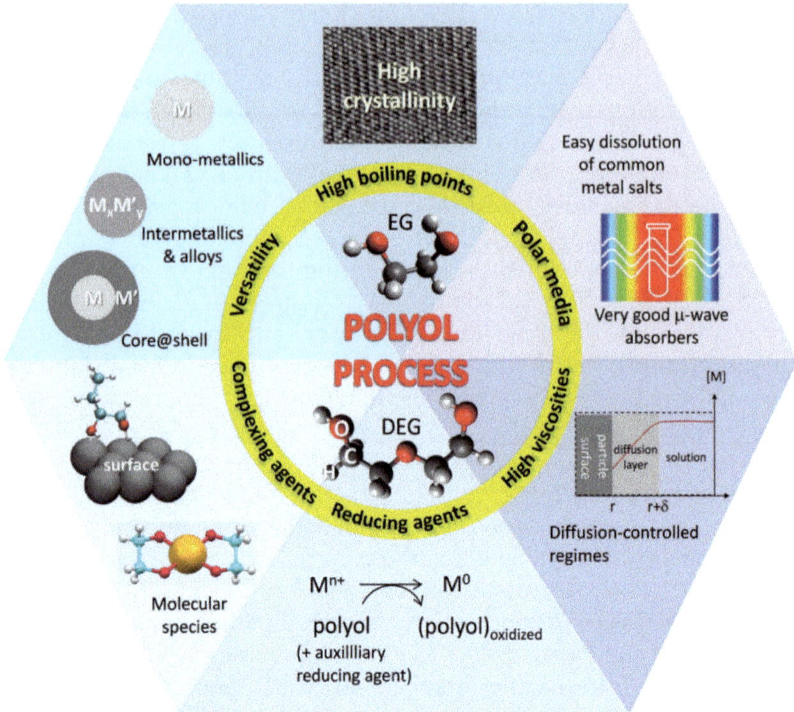

Figure 6.7 Schematic representation of the advantages of the polyol process. Reproduced from ref. 141 with permission from the Royal Society of Chemistry.

presence of additional reducing agents (*e.g.*, borohydrides) prevents their re-oxidation.[167,168] For example, upon injection of concentrated aqueous $NaBH_4$ into a solution of $InCl_3$ and sodium citrate in DEG at 100 °C, In NPs were obtained.[169,170] Alternative polyols to those obtained by petrochemical processes were also explored for the synthesis of metal particles: a 'bio-polyol', such as castor oil, was employed for the preparation of Ni NPs with similar size and shape properties to those obtained from EG or DEG, for example.[171]

Fine control over the final size of metal NPs using the polyol method has been accomplished and a rich variety of nanocrystal shapes are available nowadays. Depending on the type of (anionic) ligands and the oxidation state of the metal cation, the reducibility of precious metal salts in polyols follows the general trend Ru > Pt, Pd, Rh > Au, Ag.[172] For this reason, NPs of a few nm are more easily generated for Ru and Pt, whereas for Ag and Au the size can be easily tuned to dozens of nm. In other words, the high redox gap between the polyol and the metal salt is a first indication of the expected size of the NPs: very small Ru NPs are normally obtained with the polyol process, while it is easier to control the size of Ag and Au NPs.[146] For example, size-controlled Rh NPs were obtained by Chen *et al.* only through a seeded-growth approach (Figure 6.8).[160]

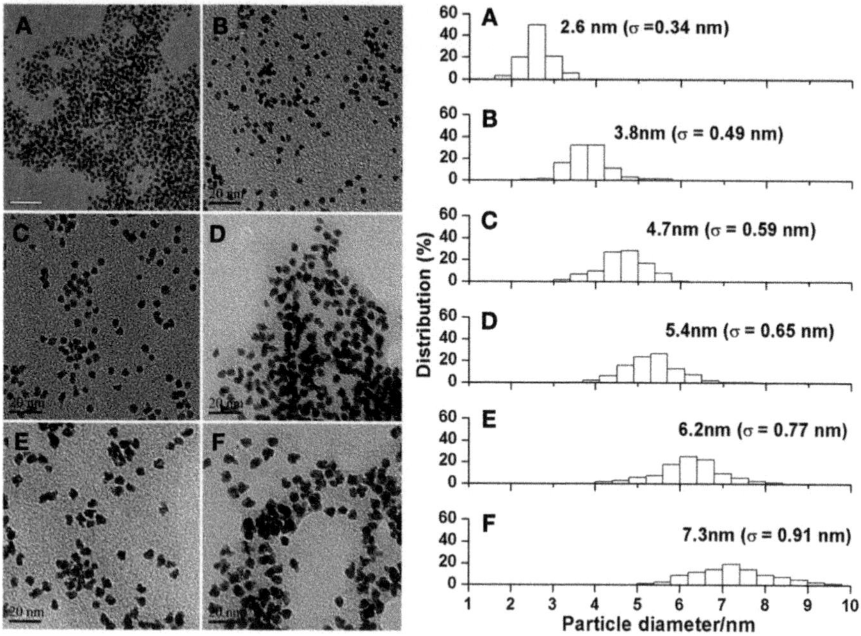

Figure 6.8 TEM micrographs (left) of Ru NPs synthesised by the seeded-growth approach with polyols, and estimated average size distributions (right) with ranges of (A) 2.6 nm, (B) 3.8 nm, (C) 4.7 nm, (D) 5.4 nm (E) 6.2 nm and (F) 7.3 nm. Reproduced from ref. 160 with permission from Elsevier, Copyright 2008.

In general, the shape of the NPs is determined mainly by the growth rates of different nuclei facets.[62] For precious metals like Pd, for example, various shapes have been obtained, such as spherical NPs,[173] tetrahedra,[173] nanocubes with {100} facets,[174–176] nano-octahedra with {111} facets,[177–179] rhombic dodecahedra with {110} facets,[180,181] nanoplates[182,183] and even concave[184,185] Pd nanocubes. The role of many of the reaction factors on the nucleation and growth in determining the final shape was thoroughly investigated by Xia *et al.* in recent years.[62,186–188] Upon addition of additives, different nucleation and growth regimes can be accessed and different shapes for Pd and Pt NPs can therefore be obtained in EG suspensions, as recently demonstrated.[189,190] Silver NPs with a precise shape (*i.e.*, nanocubes) were obtained with a tailored size ranging from 18 to 32 nm by exploiting the different viscosities of EG and DEG.[191]

Other examples involving precious metals are those of Rh^0 NPs with different sizes and shapes obtained by Biacchi and Schaak, by selecting the appropriate polyols and reaction temperatures.[152] A similar size control strategy was also achieved for Ru^0 NPs by Yan *et al.*,[192] as well as by Im *et al.* for Ag^0 NPs.[193] Cubes, rods and cuboctahedra Pd NPs exhibiting a single-crystallite structure were also obtained, mainly due to the fast reduction in EG at relatively high temperatures (close to the boiling point).[140] In general, the coordinating features of most polyols allow them to coordinate the dissolved metal cations in solution, as well as to coordinate on the metal facets of the generated particles in suspension after reduction.[140] For instance, Carroll *et al.* obtained Cu NPs with different morphologies by changing the solvent system from a short-chain polyol (EG) to a long-chain polyol (tetraethylene glycol), demonstrating that the polyols play a major role as an *in situ* capping agent, and that the polyol chain length can directly affect the growth and aggregation steps. A model for the interaction of the polyols with the surface of the metal particles was also proposed.[194] Many of these aspects related to polyol-made metal NPs have been comprehensively reviewed by Fiévet *et al.*[146]

Indeed, to deviate from the thermodynamically stable single nanocrystal, it is necessary to slow down the reduction event by playing with the reaction parameters, and therefore inducing the formation of twin faults by breaking down the original symmetry. This can be done by working at moderate temperature, controlling the precursor concentration and supply rate and introducing ions or capping agents, which can preferentially adsorb over certain crystallographic facets, thereby effectively controlling the growth of the nanostructures.[62]

Alloys and bimetallic particles have also been extensively prepared in polyols. Indeed, by dissolving more than one metal salt in polyol solutions, alloyed and intermetallic NPs can also be prepared, whose stability depends on the overall energy of the M-M^I system. Different metals often display different redox potentials and therefore different kinetics for the nucleation reaction, which makes their preparation challenging. The systems obtained by this approach are typically non-ordered alloys, whereas the formation of pure intermetallic systems is not straightforward and often a second reorganisation of the internal crystal structure upon subsequent thermal treatment is required.[195,196]

A few examples of alloys and intermetallic systems have been produced in polyols, and ternary metal particles have also been synthesised: binary and ternary compounds can indeed be prepared upon reduction of different metal precursors dissolved together, even if they display different redox potentials. Since different redox potentials result into different reduction kinetics, most of the nanosized binary and ternary alloys synthesised by the polyol process are made of precious metals, since they all share similar redox potentials. A few examples have demonstrated that this is not strictly necessary: for example, using a single-step polyol method, Baranova *et al.* synthesised disordered and bi-phase PtSn alloyed NPs with size control from 2.8 to 6.5 nm by adjusting the solution pH. They suggested that the presence of NaOH in the synthesis influenced the NPs size as well as the degree of Pt and Sn alloying, as a result of phase separation and formation of monometallic Pt and amorphous SnO_x phases, therefore highlighting the intrinsic difficulty of coupling metals with different redox potentials.[197] The control over the final size is possible for single metal particles, as well as for bimetallic particles, even when tuning the control over the composition:

recently, Bundli *et al.* reported the preparation of finely monodispersed phase-pure PtRh alloyed NPs with control over the final composition based on the simultaneous reduction of Pt and Rh metal precursors.[172] FePt NPs were also prepared by taking advantage of the polarity and high boiling points of polyols, with a simple and quick microwave procedure.[198] Control over the composition was shown by Kim *et al.* to enhance the performance of Pd-containing materials for catalytic applications: to induce changes in the electronic structure of Pd upon alloying, PdIrY, PdIrLa and PdIrSc alloy catalysts were prepared.[199] By introducing elements that have larger atomic sizes and much lower reduction potentials, such as Y, Sc and La, and alloying them with Pd and Ir in a polyol process, more active catalysts for the oxygen reduction reaction (ORR) were prepared. In particular, they suggested that, by alloying Ir with other noble metals, the ORR activity was enhanced.[199]

Examples of shape-controlled syntheses with alloyed or intermetallic systems have been reported with polyol reactions in the literature, as seen previously for single metals. For example, ordered intermetallic AuCu nanocrystals were synthesised with spherical or ellipsoidal morphology by varying the solvent temperature. Replacing tetraethylene glycol with lower length chain diols, such as EG, DEG, TEG and GLY, yielded highly branched nanowire networks. Whereas the nanowire network morphology remained similar for all of the solvents, the alloy type was tuned from fully disordered alloy to fully ordered intermetallic AuCu, depending on the boiling point of the solvent (the higher the boiling point, the more ordered the alloy obtained). Sra *et al.* proposed that the nanowire networks probably formed through a NP coalescence mechanism. By changing the stoichiometry of Au and Cu in solution, intermetallic $AuCu_3$ nanocrystals and nanowire networks were also obtained using TEG and glycerol.[196] FeCo, FeNi and FePt alloys were also synthesised: elemental iron was obtained by polyol reduction by Joseyphus *et al.* who prepared Fe-rich NPs with diameters of 10–150 nm in boiling EG with only small Pt seeds to support the nucleation.[200]

Bimetallic PtPd alloyed core–shell nanodendrites were prepared with a one-pot polyol method adding a polypeptide (colistin sulfate) as stabiliser and structure directing agent.[201] Different Au–Ag core–shell nanostructures with variable shapes were synthesised by Park *et al.* (Figure 6.9).[202]

Subsequent reduction of Ag salts leads to a controlled deposition over the Au core, turning the shape from spherical to cubic. Moreover, structure-controlled alloys, such as segmented Cu nanorods with a Pd tip, or decahedral Au seeds located at the centre of Pd rods, were also prepared. Other examples of metal-alloyed materials prepared by the polyol method are summarised in greater detail in the recent reviews from the groups of Dong, Huo and Fiévet.[140,203,204]

Following similar procedures, even ternary precious metal-containing compounds have been synthesised: for example, $AuCuSn_2$, $AuNiSn_2$, AuAgCu, Pt–Pd–Ru or PtPdCu.[205–208] FeNiCo ternary alloy NPs with variable

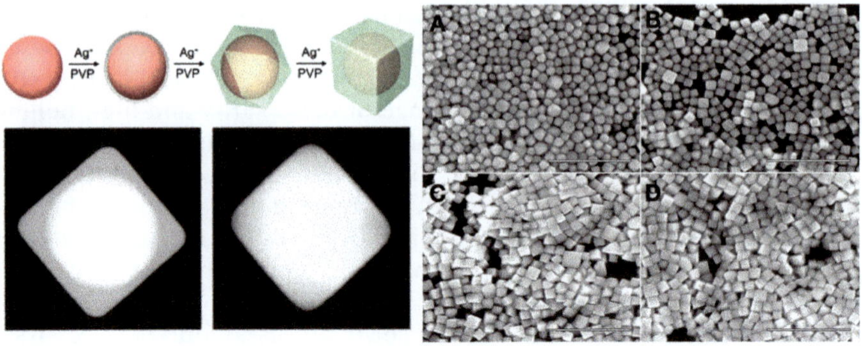

Figure 6.9 Proposed growth mechanism (left) for the seeded growth core–shell Au–Ag particles by Park *et al.* SEM images (right) of the samples quenched at (A) 10, (B) 15, (C) 20 and (D) 25 min after the addition of the silver precursor. The bars represent 500 nm. Adapted from ref. 202 with permission from American Chemical Society, Copyright 2011.

compositions were prepared with the polyol method with good control over the average NP size in a size range from 60 to 230 nm.[209] Other examples of trimetallic compounds are the $CuInTe_2$ and $CuInSe_2$ alloys with body-centred tetragonal structures, which were prepared under microwave radiation using TEG as both solvent and reducing agent.[210]

With regards to bimetallic core–shell structures, two other types of procedures can be applied: a two-step reduction synthesis involving a first reduction of single metal NPs, followed by the addition of a second metal, which becomes reduced at the surface of the previously synthesised particles, or a one-step route based on the subsequent reduction ruled by the redox differences between the two metals.[186] Indeed, to prepare the core–shell structures, a second metal that grows at the surface of the seeds has to be added, carefully controlling the injection rate of the shell metal precursor into the seed medium. The supersaturation concentration of shell metal species should be sufficiently low to suppress self-nucleation and ensure their deposition on the seeds.[146] Wang *et al.* were able to finely control the reaction parameters to deposit a shell of Pt on Pd facets (Figure 6.10a–c).[211] The procedure involved a first step of seed generation by the polyol process, and a second step where Pt was added and reduced on top of the existing NPs, avoiding the nucleation of Pt-only NPs.

As displayed in the scanning transmission electron microscopy (STEM) image in Figure 6.10a and b, the Pd–Pt core–shell material prepared was characterised by a uniform coating of Pt atoms deposited on top of the vertices of the icosahedra. The Pt/Pd ratio was tuned to thicken or shrink the outer shell, therefore generating catalysts of different activity.[211] The same group also prepared core–shell nanocubes of PdPt (Figure 6.10d and e), with a uniform, ultrathin shell of Pt on Pd NPs, showing enhancements in specific activity and durability towards the ORR.[212]

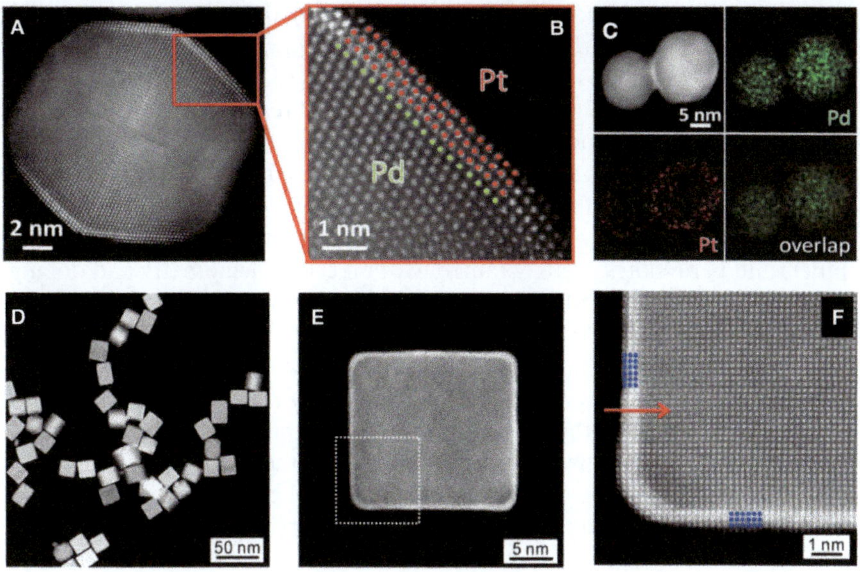

Figure 6.10 (A) High-angle annular dark-field (HAADF)-STEM image of from a PdPt core–shell NPs. (B) A magnified edge of the previous particle, revealing the outer shell of Pt deposited on the Pd core. (C) HAADF-STEM image of two PdPt icosahedra and the corresponding EDX spectroscopy mapping of Pd and Pt, confirming a core–shell structure, Adapted from ref. 211, https://doi.org/10.1038/ncomms8594, under the terms of the CC BY 4.0 license https://creativecommons.org/licenses/by/4.0/. (D) HAADF-STEM image of nanocubes of Pd–Pt core–shell NPs and relative magnification (E) and (F), where the Pt layer is visible (brighter colour of the shell). Adapted from ref. 212 with permission from Springer Nature, Copyright 2014.

6.3.2 Metal Oxides

Given the intrinsic properties of polyols, such as water compatibility and chelating features, these solvents are also ideal for the synthesis of metal oxide NPs, starting from simple metal salt precursors, such as halides, nitrates and acetates.[140] For the formation of oxides, the presence of water is favourable and the polyol/water concentration can be varied by using hydrated precursors or by adding aliquots of water. Of course, the amount of water has an impact on the overall properties of the solution, modifying its boiling temperature, the viscosity and the dielectric constant.

Despite polyol synthesis being initially focused on metal NPs, ZnO and CoO NPs had already been reported by the Fiévet group in 1994.[165,213,214] In more recent years, the range of metal oxides that have been synthesised has grown significantly, with various compositions, shapes and particle sizes achieved.[215] The possible applications are very broad, including catalysis (*e.g.*, with MnO_2, Mn_3O_4, Fe_2O_3, Cu_2O, $NiCo_2O_4$), photocatalysis (*e.g.*, with Cu_2O, ZnO, $BiVO_4$), magnetic materials (*e.g.*, with Fe_2O_3, Fe_3O_4, Gd_2O_3, $MgFe_2O_4$,

$CoFe_2O_4$, $ZnFe_2O_4$), sensors (*e.g.*, with Fe_2O_3, CuO), dye-sensitised solar cells (*e.g.*, with ZnO, TiO_2) or high-power batteries (*e.g.*, with V_2O_5, MnO, Mn_2O_3, Fe_2O_3, SnO_2, $CoMn_2O_4$, $CuFe_2O_4$, $LiFePO_4$, $LiMnPO_4$).[140]

Reaction conditions (*e.g.*, amount of water in the reaction medium) strongly affect the resulting size and shape of the NPs: for example, anatase TiO_2 was prepared in 1,4-butanediol, 1,5-pentanediol or 1,6-hexanediol under microwave irradiation, obtaining crystallite sizes smaller than 10 nm by changing the amount of water.[216] As well as binary and ternary oxides, multi-oxide phases, such as spinels and perovskites, were obtained with precise stoichiometry and doping: for example, fluorescent NPs (such as $LaPO_4$:Ce,Tb, CaF_2:Ce,Tb, $In_2(MoO_4)_3$:Eu), colour pigments (*e.g.*, $CoAl_2O_4$, Cr_2O_3, ZnO:Co, $(Ti_{0.85}Ni_{0.05}Nb_{0.10})O_2$, α-$Fe_2O_3$, $Cu(Fe,Cr)O_4$, TiO_2), transparent oxide conductors (*e.g.*, In_2O_3:Sn, SnO_2:F, ZnO:Al, ZnO:In) and catalysts (*e.g.*, $CuMoO_4$,CeO_2, Mn_3O_4,V_2O_5).[140,217] Fiévet *et al.* recently reviewed in detail the synthesis of spinel phases and their relative magnetic properties.[146] Ungelenk *et al.* used a polyol-mediated synthesis to prepare a range of transition metal tungstates (Figure 6.11) with small crystallite sizes (<20 nm), high surface area (up to 200 $m^2\ g^{-1}$) and phase purity (>85%).[218]

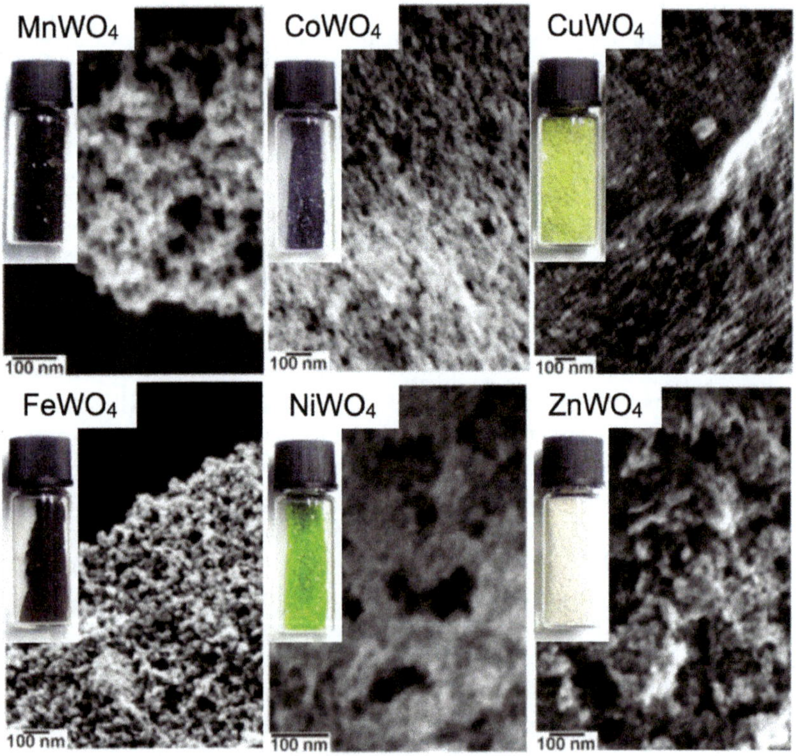

Figure 6.11 SEM images and photographs of samples of the as-prepared transition metal tungstate NPs MWO_4 (M = Mn, Fe, Co, Ni, Cu, Zn). Adapted from ref. 218 with permission from Elsevier, Copyright 2014.

For the preparation of battery cathode materials, layered $LiNi_{0.4}Mn_{0.4}Co_{0.2}O_2$, spinel $LiNi_{0.5}Mn_{1.5}O_4$ and olivine $LiCoPO_4$ were synthesised using the polyol method by Chung *et al.*[219] The morphology control achieved with this synthetic approach impacted positively the lithium-ion transport due to short diffusion lengths.

The main limitation of the polyol synthesis of metal oxides is the boundary that exists between elements that form elemental metals and those that can form metal oxides, which is associated with the redox potential of the metal under investigation. The boundary is not clear cut, as for selected systems, for example, it is possible to obtain elemental metals or metal oxides by adjusting the reaction temperature.[140] Notable examples of temperature-dependent particle compositions are copper ($Cu^{II}O/Cu^{I}_2O/Cu^0$), cobalt ($Co^{II}O/Co^0$) or bismuth ($Bi^{III}_2O_3/Bi^0$) oxides.[208,220–222] Generally, metal oxides are obtained at lower temperatures (*e.g.*, 100–180 °C) whereas elemental metals are obtained at higher temperatures (*e.g.*, 180–250 °C).[140]

6.3.3 Other Compounds

Tellurides and fluorides can also be synthesised by the polyol process.[223] Interesting alloys that are not prepared from precious metals, for example, are the binary metal tellurides, such as PbTe, Ni_2Te_3 and Cu_7Te_5, which have also been synthesised in the form of NPs only under microwave radiation (and not with conventional heating techniques) in EG.[224] Other inorganic compounds can be prepared through the polyol procedure.[223] For example, metal carbides (*e.g.*, Co_3C, Co_2C, Ni_3C) were recently prepared *via* polyol synthesis as highly interesting materials, in particular for their high-coercivity magnetic properties.[225] Metal sulfides are another relevant class of compounds that have recently been under the spotlight for their appealing properties.[223] Among the first metal sulfides to be reported through a polyol process were NiS, CuS, PbS, Ag_2S, In_2S_3, Fe_3S_4, prepared through direct heat-up protocols or solvothermal heat-up procedures through reactions of thiourea and the corresponding metal salts.[140,168,226] Feldmann and Metzmacher also prepared metal sulfite NPs with different metals (Zn, Cd, Hg) using the polyol method, starting with metal acetates and thiourea as precursors, and heating at 180 °C in DEG to obtain crystalline particles of about 30–250 nm in size.[227] Wang *et al.* prepared NiS NPs *via* the polyol route, starting from $NiCl_2$ and thiourea as precursors in EG solutions and heating under reflux for 2 hours.[228] Shen *et al.* prepared SnS and SnS_2 using similar procedures.[229]

CuS NPs were obtained by Castillón-Barraza *et al.* from copper nitrate and sodium sulfide in EG. The reaction took place within seconds and agglomerated particles with crystallites of mean size 10 nm were obtained, with the possibility to slightly tune their size and morphology depending on the reaction temperature.[226] PbS NPs were prepared taking advantage of the microwave heat-up procedure in EG: PbS colloids with sizes <10 nm were homogeneously prepared by Paliwal *et al.*[230]

6.4 Continuous-flow Approach

Originally developed for analytical purposes and later used for organic synthesis and adopted by the chemical industry, the past two decades have seen continuous-flow approaches gaining growing interest in inorganic synthesis. Nowadays, remarkable technological advances have been reached in the field, establishing this method as one of the most acknowledged for preparing high-quality nanocrystals.[231–235]

With respect to a simple stirred vessel (*batch approaches*), flow methods provide several advantages, such as facile automation, generally higher safety and better reproducibility of the reaction conditions, enabling higher reliability for the process, which come at the cost of a more elaborate setup. Continuous-flow reactions are typically carried out within tubular reactors fabricated from glass, polymer or metal materials, where precursors are continuously pumped and eventually mixed in an *ad hoc* engineered mixer, and the products are continuously withdrawn. As described in more in detail in the following, the reaction setup can be conveniently modified in many different configurations, also adding mass flow controllers, check valves to prevent back flow, heating/cooling control, back pressure regulators and even online detectors.[236–238] The relative ease of assembly of homemade flow setups and their versatility also provide the opportunity for transforming elaborate multistep syntheses into a single input–output process.[239,240]

As a consequence of their continuous nature, the concepts of reactant stoichiometry and reaction time in flow approaches differ markedly from conventional batch methods. Indeed, whereas in batch approaches the *molar ratio* between the precursors is determined by the volumetric ratio of their starting solutions, given their concentration, in flow approaches the relative amount of precursor at the mixing point is determined by the flow rate. The *reaction time* under batch conditions can simply be considered as the time interval between the mixing of reactants and the end of the reaction. On the other hand, in flow approaches, the reaction time reflects the residence time in the reactor, determined by the ratio between the reactor volume (*i.e.*, related to the tubing diameter and length) and the total flow rate (*i.e.*, the sum of the individual flow rates of solutions fed into the reactor). Therefore, direct and precise correlation between time and space is established in a continuous-flow approach, enabling 'access' to different reaction times by simply considering different points in space of the reactor after the mixing point. This peculiarity is particularly interesting for: (1) designing multistep reactions, since a subsequent step can be implemented in flow after a specific length of the reactor (*i.e.*, after a specific reaction time); and (2) performing *in situ* time-resolved studies, accessing also extremely short reaction times, by including an online detector (*e.g.*, infrared (IR),[241,242] Raman,[241,243] surface-enhanced Raman spectroscopy (SERS),[244] ultraviolet,[245] photoluminescence,[246–248] X-ray absorbtion (XAS)[249] and nuclear magnetic resonanace (NMR)[250] spectroscopies) in the setup.[237] This last opportunity disclosed by continuous-flow approaches can be exploited for easy and rapid (at the μm length scale and ms time scale) screening of reaction conditions (*e.g.*, flow rate, temperature, reaction time, reactant concentration and molar ratio, and so on) and a quasi-simultaneous evaluation of their effect on the reaction outcome, thus enabling facile and fast optimisation of the synthesis.

Another distinguishing feature of continuous-flow methods concerns the *amount of product* obtained, which is determined by the length of time at the entire flow regime (given the flow rate and the reactor volume, in addition to precursor concentration and reaction yield), while in batch processes the highest amount of product that can be obtained is predetermined by the amount of reactant (concentration and volume) and by the reaction yield. Consequently, the discontinuous nature of conventional batch approaches limits production and scaling up of the reaction, since syntheses need to be cyclically repeated to accumulate product, or the vessel dimensions need to be increased at the expense of homogeneity of the reaction mixture and thus of synthetic control. Conversely, continuous-flow approaches allow the production of high amounts of the target material even with low-volume reactors, by simple accumulation (reactant feeds running for long times) or by numbering up, *i.e.*, by using parallel reactor systems (scaling out) rather than by increasing the dimensions of the channel (scaling up). Any increase in reactor size may lead, *inter alia*, to loss of the optimised reaction parameters and ultimately to the loss of advantages that make the synthetic approach unique. Therefore, the facile scale up of the reaction, and thus its possible implementation in industry, represents one of the major advantages of flow methods. Moreover, the reduced volume of flow reactors enables more efficient engineering controls and heat transfer control between inside and outside the reactor, reducing safety risks by avoiding accumulation of potentially hazardous intermediates and by confining highly reactive materials.[231,238]

Because of these intrinsic advantages, in recent times attention has been drawn to miniaturisation of continuous-flow reactors towards *meso-* or *minifluidic* (from 500 μm to a few mm) and *microfluidic* (10–500 μm) setups.[251] In addition, several (and also more exotic) continuous-flow systems have been developed, working at ambient conditions or at high temperature and/or pressure, and even employing supercritical fluids, such as in continuous hydrothermal flow synthesis (CHFS). More details may be found in Chapter 4 and in specific reviews.[252–254] A particular focus will be given here to microfluidic technology, which has attracted wide attention and achieved much progress for inorganic synthesis in recent years, thanks to its intrinsic advantages related to the unique characteristics of fluid flow on the microscale.

As mentioned, microfluidic setups generally handle small amounts (10^{-9} to 10^{-18} L)[255] of fluids using channels with diameters ranging from tens to hundreds of μm. A general and flexible definition of a microfluidic reactor can be made based on the Reynolds number (Re)‡ of the fluids flowing in the devices, which is generally below 250 for microfluidic devices, characteristic of a laminar flow regime.[231] Thereby, reactors with dimensions in the range 500–1000 μm (overlapping the definition of mesofluidic reactors) can be termed as microfluidic devices if operating at low flow rate. At low Reynolds numbers, inertial forces are negligible and viscous forces dominate: a regular parallel flow of the fluids within the channels is ensured, without any turbulence or back-mixing within the reactor. Reactants are thus exclusively mixed by lateral diffusion between the parallel flows, and, since the reactor diameter is small, mass transfer distances are short and mixing is efficient. Alternatively, reactant mixing can be induced by employing dedicated mixing sections, a great variety of which have been developed in recent years, even tailored for specific purposes.[256,257]

As a consequence, in microfluidics the reaction mixture is highly homogeneous, compared to batch systems where concentration gradients may exist due to non-complete mixing. Moreover, the high surface to volume ratio typical of a microfluidic reactor also ensures thermal homogeneity and rapid and efficient heat transfer between the ambient and flowing fluids, in contrast to batch reactors. The efficient heat transfer plays a relevant role in the quenching step of exothermic reactions, rapidly dissipating excess heat and avoiding the formation of hot spots. Consequently, reactions are safer and by-products are avoided. The remarkably high homogeneity of the reaction mixture, in terms of concentration and temperature, ensuring stable reaction conditions is one of the major advantages of the microfluidic approach, as opposed to conventional batch syntheses, in which concentration and/or temperature gradients may cause coexistence and simultaneity in nucleation and growth processes, ultimately leading to a lack of control of the final product outcome in terms of size and size distribution. In particular, microfluidic approaches find their ideal applications either for reactions where the rate of reaction is higher than the rate of mass transfer or for reactions where the rate of heat transfer is a limiting factor of the process. Indeed, in view of NPs synthesis, and considering the LaMer model of homogeneous nucleation,[10,11,258,259] the highly homogeneous and reproducible nucleation conditions typical of microfluidic systems allow promotion of a temporal (and consequently spatial) separation of the nucleation and growth steps. As a consequence, generation of a population of NPs with a narrow size distribution and uniform morphology can be pursued. In addition, according to Sugimoto[258] and Chu *et al.*,[259] implementations of the LaMer model and the possibility of controlling and varying the flow rate in microfluidics can be

‡Reynolds number (Re) is a dimensionless number that quantifies the relative contribution of inertial and viscous forces within a flowing fluid. It is a function of both the fluid (*i.e.*, its density ρ and dynamic viscosity μ) and the environment in which it is manipulated (*i.e.*, its velocity v and the environment dimension d) and it is calculated as: Re = (inertial forces)/(viscous forces) = $\rho dv/\mu$. Within a tubing with circular cross section: Re < 2300 laminar regime; 2300 < Re < 4000 transition regime; Re > 4000 turbulent regime.

exploited to tune the rate of the supply rate of precursors, ultimately affecting the final size of the NPs.

Moreover, the high surface to volume ratio typical of microreactors finds particular advantages in photochemical and electrochemical reactions, which can be carried out in a highly practical and scalable fashion in microfluidics.[238,260,261] Indeed, the small diameter of the microreactor enables efficient and uniform penetration of radiation into the reaction environment, and consequently the process of irradiation is more efficient with respect to a batch reactor.[238] The efficient penetration of radiation inside microreactors can find useful application when using indirect methods of heating, such as microwave irradiation.[262]

Finally, it is worth highlighting that flow chemistry is compliant with many of the Green Chemistry paradigms.[263,264] In particular, the optimised heat transfer in microfluidic devices, which avoids the long and energy costly temperature equilibration periods typical of batch approaches, together with the possibility of production upscaling in units of space, thanks to the numbering up strategy, results in a reduction in energy consumption. At the same time, when a multistep process is concerned, the possibility of avoiding separate steps of purification and/or isolation of intermediate products enables chemical and waste reduction. Indeed, the proper design of separation and purification schemes for implementation in continuous-flow processes enables the recycling of expensive solvents and surfactants, making it possible to produce nanomaterials in a cost-effective, green and high-throughput fashion.

However, one of the major challenges in applying microfluidics to materials synthesis is to design microreactors that enable the achievement of the desired reaction conditions (*e.g.*, temperature, pressure, concentration, chemical compatibility, *etc.*) to obtain the material with the desired features and preventing the microsystem from clogging.[265] In recent years, many microreactors have been fabricated, evolving from simple (and cheap) tubing, with well-known hydrodynamics and easy manipulation of operating parameters, such as the residence time, to more complex systems with integrated control of operating parameters and *in situ* characterisation techniques, and even combining different microfluidic systems to achieve specific co-flow structures, such as droplets, jets and multiple emulsions.[266–269] However, even if they are relatively simple to implement, capillary and tubing-based microfluidic setups may lack in flexibility and design and they can be challenging for integration of monitoring components. These limitations have been addressed by the fabrication of 'on-chip' microsystems, which are bidimensional layouts extruded in the third dimension, for example, by (photo) lithography and etching, or fully three-dimensional structures made of fluoropolymers, polymers (*e.g.*, PDMS, poly(dimethylsiloxane)) and glass, glass, ceramics or metals. To improve the chemical compatibility between the chip microreactor and the reaction mixture, and in some cases to modify the wettability of the channels, the inner surface of the channels may be functionalised *ad hoc* or coated. The specific microstructure of the flow reactor and its chemical composition determine the range of reaction conditions (*e.g.*, temperature and pressure) and the chemicals that can be experimentally employed within it. Additional engineering challenges consist of connectors between the microreactors and macroscopic fluid handling systems (pumps), which can also be combined with heaters and temperature controllers, integrated with optical control and/or detection modules,[270] and so on. The design of the microreactor, which can also be computer aided,[271] plays an especially important role in synthesis planning, as it should be optimised to improve the efficiency of mass and heat transfer based on the fluid dynamics of the mixture, which can be computed as well,[272,273] and on the kinetic features of a specific reaction. However, it must be kept in mind that simplicity in microreactor fabrication may lead to easier scalability of the process, while the produced nanomaterial may be of lower quality. More details can be found in specific reviews.[233,234,236,257]

The design of the microreactor may also address the specific continuous-flow approach that is targeted, *i.e.*, a homogeneous one, where one solvent (or a miscible solution of solvents) is employed (Figure 6.12a), or a

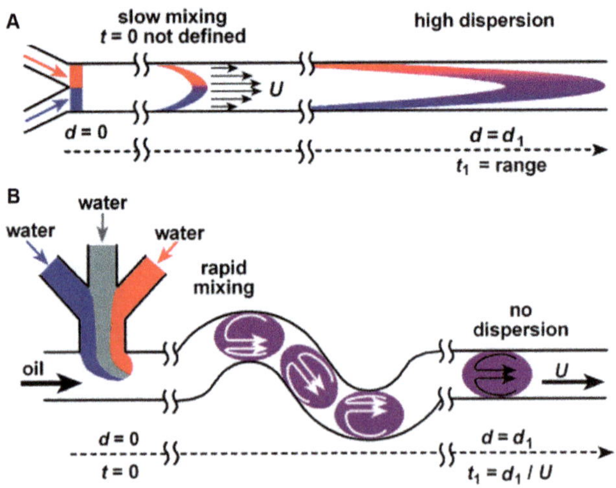

Figure 6.12 (A) Schematic comparison of the mixing of aqueous solutions of reactants (blue and red) in a microreactor exploiting the homogeneous approach, or single-phase system, and (B) the two-phase droplet-based approach. In the single-phase microfluidic system, the mixing is affected by the Taylor–Aris dispersion effect caused by the parabolic (Poiseuille) velocity profile, responsible for a distribution of residence times (reaction time $t \neq d/U$, where d is the distance from the mixing point and U is the flow rate). In the two-phase microfluidic system, reactants are compartmented into droplets effectively narrowing the residence time distribution (reaction time $t = d/U$). Internal recirculation within droplets flowing through channels of different geometries is shown schematically by arrows. Reproduced from ref. 274 with permission from John Wiley & Sons, Copyright © 2002 WILEY-VCH Verlag GmbH & Co. KGaA, Weinheim.

multi-phase microdroplet-based process, such as the segmented-flow reactor (SFR), where either gas-phase bubbles are dispersed into a liquid (*segmented gas–liquid flow*) or two immiscible solvents are used and liquid droplets are dispersed into a continuous non-miscible phase (Figure 6.12b, *segmented flow*).

In the case of segmented flow, nanolitre-sized droplets are formed by introducing an aqueous stream into a channel with a flowing immiscible fluid (oil) and reactants are thus confined into droplets and therein mixed more rapidly. Since single-phase flowing fluids are affected by axial dispersion (particularly evident at high flow rates), leading to a distribution of residence time (RTD) and ultimately of the products of the reaction, the segmented-flow approach is often used in nanomaterials synthesis to enhance mixing efficiency and narrow the particle size distribution.[274,275] Although SFRs have proven to be a valuable platform to improve mixing in fast kinetic chemical reactions, where superior control of reactant diffusion is pivotal to obtain monodisperse distributions of NPs, limitations arise from the requirement to find a suitable reacting/carrying fluid couple, as this should be non-miscible over the whole range of experimental conditions (pressure, temperature, pH, *etc.*).

The first report of using microfluidic devices for the synthesis of inorganic nanocrystals dates back to 2002, when the group of deMello prepared high-quality cadmium sulfide quantum dots using microfluidics.[276] Since then, interest in microfluidic approaches has become widespread, and the versatility of the method for the synthesis of different classes of compounds has been explored. Indeed, using the microfluidic approach, a plethora of chemical reactions can be carried out by implementation of wet-chemistry methods initially developed in batch approaches, also combining conventional and unconventional techniques, thanks to the versatility of microfluidic setups, which can be engineered as desired. In the following sections, a detailed description of the state-of-the-art of inorganic systems synthesised using the microfluidic approach will be provided, encompassing metals and alloys (Section 6.4.1), metal oxides (Section 6.4.2), metal chalcogenides (Section 6.4.3) and other compounds (Section 6.4.4).

6.4.1 Metals and Alloys

Thanks to their greatly enhanced light absorption (Mie scattering) at plasmon resonance wavelengths,[277,278] metallic NPs were found to be particularly convenient systems for study in microfluidic devices, since they offer the possibility to optically monitor their growth and shape evolution online. Since formation of metal NPs is usually accomplished by reduction of a metal salt, a typical microfluidic approach for the synthesis of metal NPs involves the use of a setup in which a metal salt solution, possibly also containing nucleation seeds (*seeded growth*), is mixed with a reducing agent, such as sodium borohydride, ascorbic acid, citrate, *etc.* A plethora of surfactants and/or ligands (*e.g.*, thiols, polymers such as PVP, alcohols and dendrimers) can be added to stabilise the synthesised NPs and minimise their aggregation, as well as their possible deposition on the walls of the channels. Syntheses of Au,[279–286] Ag,[279,283,287,288] Cu,[289] Co,[249,290] Pd,[291–294] Pt[293,295] and also alloyed AuPd,[296] PtPd, PtRu and PtRh[295] NPs in microfluidic reactors, both by homogeneous and multi-phase approaches and under ambient and thermal conditions, have been reported.

The first report of microfluidic synthesis of gold NPs was published by Wagner *et al.*,[284] who obtained 12 nm Au NPs by citrate reduction in a chip-based interdiffusion microreactor. They also demonstrated that larger NPs (15–24 nm) can be obtained *via* a seeded growth approach starting from the previously prepared 12 nm citrate-stabilised Au seeds and using ascorbic acid as reducing agent. Moreover, they observed that the mean size of the NPs could be tuned by varying the flow rate of the reactants, obtaining NPs with a higher mean size with a lower flow rate, in agreement with Sugimoto and Chu implementations of the LaMer model discussed above. They also screened several reaction parameters, such as pH, reagent concentration and amount of added PVP, to narrow the size distribution.[282] When a stronger reducing agent was employed, such as sodium borohydride, very small Au NPs (4–7 nm) were synthesised.[283] The authors employed three chip micromixers based on Si/glass technology arranged in series, in order to easily vary the Au precursor ($HAuCl_4$) concentration and the Au/$[BH_4]^-$ molar ratio during the micro-flow-through synthesis (setup scheme shown in Figure 6.13a). They reported that, by decreasing the concentration of the precursor solutions, NPs with a smaller mean diameter (and unvaried size distribution) were formed, as a consequence of an increased number of nuclei obtained by achieving the nucleation threshold less rapidly. In the same work,[283] Wagner *et al.* also reported the synthesis of Ag NPs by reduction of silver nitrate with $NaBH_4$, modifying the setup in order to use the third mixer for the addition of thiolated ligands for online functionalisation of the primary NPs (setup scheme shown in Figure 6.13b). They observed that, since the nucleation of

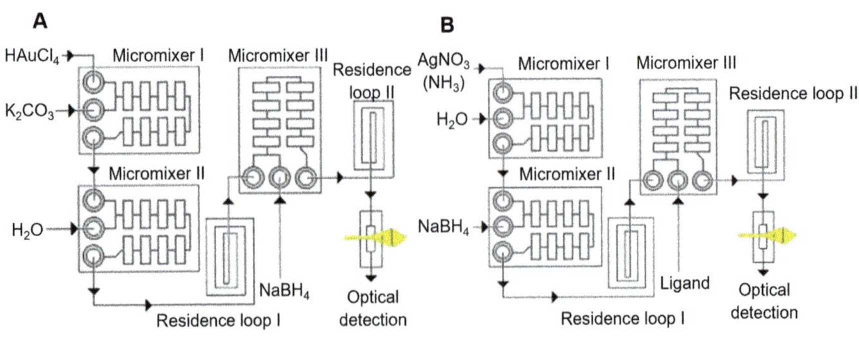

Figure 6.13 Modular microreactor setup for homogeneous flow-through process: (A) synthesis of gold NPs and (B) synthesis of surface-modified silver NPs. Adapted from ref. 283 with permission from Elsevier, Copyright 2008.

Ag is kinetically hindered with respect to the growth in comparison to Au, the synthesised Ag NPs were bigger than Au NPs (about 15 nm *vs.* 4–7 nm). Moreover, in contrast to Au NPs, by increasing the flow rate no effect was observed on the mean Ag NP size, while the size distribution was narrowed. This was attributed to the longer duration of the nucleation phase of silver compared to gold, and such findings highlight the importance of mixing rate for the nucleation, and the primary role of the microfluidic approach in achieving fast mixing, thus drastically improving the size distribution of Ag NPs at high flow rates. In both setups, online optical monitoring of the output of the reaction was implemented, in order to follow the formation kinetics of Au and Ag NPs by UV–visible (UV–vis) spectroscopy and to evaluate in real time the effect of the investigated synthetic parameters, thus fully exploiting some of the previously described major advantages of microfluidic systems (Section 6.4).[283] Microfluidic systems implemented with UV–vis spectroscopy have also been employed as a powerful tool to estimate the aspect ratio of Au and Ag rod-shaped NPs, allowing tailored design of particles by evaluating the role of easily tuned reaction parameters.[279]

A potential drawback of the continuous single-phase approach is deposition of the formed NPs on the walls of the channels during synthesis, which may lead to a loss of optical access to the reaction outcome, as well as to an enhanced risk of device clogging. Segmented-flow approaches typically overcome these issues, together with reducing the axial dispersion that affects laminar flows and broadens the particle size distribution. By using the synthesis of spherical gold NPs from $HAuCl_4$ with rapid reduction by $NaBH_4$ as a model system, it was demonstrated that SFR narrows the NP size distribution.[280] In particular, the authors observed that the NP size distribution was broadened by increasing the residence time (decreasing the flow rate), independently from the employed inert fluid (air, toluene, silicone oil), as the 'slug' (droplet) length of the continuous phase is larger at longer residence times. Thus, having the slug length (and thus the flow rate) as a very significant effect on inter-slug mass transfer in SFR, they demonstrated that good control of mixing is a strict requirement to decrease polydispersity and this can be achieved in SFR. Anisotropic gold nanocrystals were also synthesised *via* a droplet-based approach by dispensing gold NP seeds, Au^{3+} precursor, reducing agent (ascorbic acid) and shape-controlling agents (a cationic surfactant, CTAB, and Ag^{+} ions, influencing the relative growth rates of different Au crystal facets) into monodisperse picolitre droplets within a microchannel, generated at a microfluidic T-junction.[281] A schematic representation of the experimental setup and of the droplet formation and reactant mixing in the droplets is shown in Figure 6.14a and b. The authors were able to obtain different nanocrystal dispersions of spherical–spheroidal (Figure 6.14c), rod-shaped (Figure 6.14d) and 'sharp-edged' (cubes, stars, tetrapods, Figure 6.14e) individual NPs by varying the concentration of added silver ions and ascorbic acid and the volumetric flow rates of the three aqueous flow streams. They demonstrated that tunability of the optical resonances in the visible–near-infrared spectral range of Au NPs can be obtained by simple

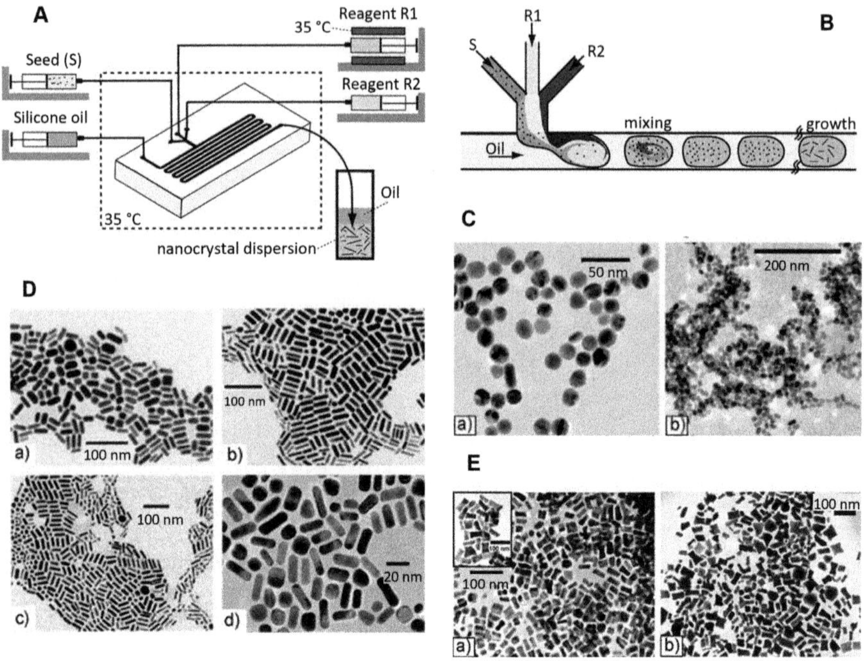

Figure 6.14 (A) Schematic representation of the experimental setup of droplet-based microfluidic synthesis of anisotropic Au NPs. (B) Schematic illustration of droplet formation and Au NP formation: Au seed suspension (S) and aqueous reagent solutions (R1 and R2) are separately delivered into one arm of a microfluidic T-junction, and silicone oil is delivered into the other arm. Reagents and seeds are rapidly mixed by chaotic advection. The oil forms a thin lubricating layer around the translating droplets, and isolates growing particles from the microchannel walls, thus preventing wall deposition. TEM micrographs of (C) spherical–spheroidal NPs, (D) rod-shaped NPs of different aspect ratios and (E) sharp-edged Au NPs synthesised by varying the silver content in R1, the concentration of ascorbic acid in R2 and volumetric flow rates of S, R1 and R2. Adapted from ref. 281 with permission from John Wiley & Sons, Copyright © 2009 Wiley-VCH Verlag GmbH & Co. KGaA, Weinheim.

variation of reaction parameters using the microfluidic approach, revealing the possibility of applying this synthetic method to a broad class of metallic nanocrystal dispersions.

Another typical advantage of microfluidic systems, *i.e.*, the high penetration depth of light inside the microreactors and thus high irradiation efficiency, was exploited for the shape-controlled synthesis of Ag NPs under photochemical conditions by Silvestrini *et al.*[288] The authors built the microfluidic setup by coiling a polytetrafluoroethylene (PTFE) tube (0.5 mm inner diameter) around an immersion photoreactor and pumping through a solution of $AgNO_3$, a photoinitiator and sodium citrate as stabiliser, thus obtaining spherical Ag seeds (6 nm). The as-prepared seeds were further irradiated in a polymeric microreactor in the presence of oxygen to induce shape conversion from spheres to decahedra and platelets, through the so-called 'photovoltage mechanism', which involves the excitation and subsequent decarboxylation of adsorbed citrate ions.

Finally, a particularly interesting example that shows the power of the space–time equivalency in continuous-flow microreactors was reported by Zinoveva *et al.*,[249] who probed the formation of Co NPs by exploiting synchrotron radiation-based X-ray absorption spectroscopy and recording Co K-edge X-ray absorption near-edge structure (XANES) spectra at three different positions in a poly(methyl methacrylate) (PMMA) microreactor. The three spectra taken within the microreactor were found to be different, demonstrating that different stages of the synthesis were being tracked. By comparing these with reference spectra of the precursor and the final product collected at the end of the microreactor, a time resolution of the reaction of the order of milliseconds was observed, demonstrating the potential for combination of microfluidic systems with the XAS technique for fast time-resolved *in situ* investigations of the synthesis of inorganic NPs, as well as of other chemical reactions.

6.4.2 Metal Oxides

Metal oxide NPs are conventionally synthesised using sol–gel or precipitation reactions, which are easily transposable into microfluidic devices.

One of the first investigations of microfluidic synthesis of metal oxides was the sol–gel synthesis of silica NPs, which are otherwise commonly prepared by the Stöber process,[297] based on hydrolysis and condensation reactions of metal alkoxides (*e.g.*, TEOS, tetraethyl orthosilicate) under basic conditions. A thorough investigation of the effect of microreactor design and of the parameters of linear flow velocity, and of the mean residence time on silica particle size and size distribution, was carried out by Khan *et al.*,[298] who compared the results obtained using a single-phase laminar flow reactor (LFR), a gas–liquid SFR and a batch reactor. In the microreactor syntheses, a solution of TEOS in ethanol was pre-mixed with a NH_4OH solution, introduced through two separated inlets. The SFR was equipped with an additional gas inlet to segment the flowing reacting medium with air. The authors observed that the size of the SiO_2 NPs was directly proportional to the residence time, while the width of the size distribution was inversely proportional to the residence time. As discussed in other examples, NPs synthesised with the LFR suffered from poorer size distribution than those obtained by SFR and batch synthesis, due to the axial dispersion effect induced by the slower motion of particles close to the wall of the channels compared to those in the flow centre. When employing the SFR, a relevant difference in NPs size distribution was observed depending on whether a micromixing section was present or not prior to the introduction of the segmenting gas. Indeed, if the incoming gas segmented unmixed liquids, non-uniform reactant distribution between adjacent slugs could occur, thus further widening the particle size distributions. The use of SFR with a micromixing section resulted in a narrower size distribution, also in comparison with the results obtained by the well-known Stöber batch process. Other continuous-flow methodologies, either in multiphase[298–301] or homogeneous[302–304] systems, for the synthesis of silica, also mesoporous, as NPs, microparticles or hollow spheres can be found in the literature. The combination of sol–gel chemistry with a droplet-based microfluidic technique has also been employed for the synthesis of titania nanostructures,[305–307] and even TiO_2 hollow spheres,[308] exploiting the hydrolysis of a titanium alkoxide (*e.g.*, titanium tetrabutoxide or titanium tetraisopropoxide) at the interface of two immiscible phases, either aqueous droplets dispersed in an oil phase or of two immiscible streams flowing in channels.

Core–shell oxide NPs have also been synthesised in microfluidic devices. Khan *et al.* developed a continuous-flow microreactor for coating of colloidal silica by a layer of TiO_2 of tuneable thickness in a single step.[309] In particular, they employed a segmented gas–liquid flow device with a branched manifold that enabled multistep addition and controlled hydrolysis of titanium tetraethoxide (TEOT), as shown in Figure 6.15. When TEOT was added to a monodisperse silica suspension in a single step, the formation of a polydisperse mixture of primary coated particles, large agglomerates and secondary TiO_2 particles was observed. On the other hand, when TEOT was fed by the branched manifold, a monodisperse population of core–shell particles was obtained, avoiding the formation of secondary particles and/or agglomeration, as was evidenced by transmission electron microscopy (TEM) micrographs after calcination at 500 °C. Core–shell γ-Fe_2O_3/SiO_2 NPs were also synthesised in a multistep continuous-flow approach, by connecting three microreactors in series, each one acting as a microunit for a defined operation.[310] First, a silane was grafted onto the surface of γ-Fe_2O_3 NPs previously coated by citrate ligands, then these modified NPs were mixed with the silica precursor (TEOS) and, finally, ammonia was introduced to catalyse the hydrolysis–condensation reactions of the silica precursor onto iron oxide. The authors reported that this smart multistep microfluidic approach allowed acceleration of the coating process, as the formation of core–shell structures was observed after a few minutes, compared to the several hours typically needed in a conventional batch process.

The synthesis of metal oxides through a *precipitation process* has been implemented in microfluidic systems to obtain several metal oxides, including zinc oxide.[311–314] As an example, a microfluidic approach for the synthesis of zinc oxide from $Zn(NO_3)_2$ and NaOH in water at 60 °C was reported, investigating the influence of the microreactor geometry (PDMS chips with different depth, width and length of the microchannel network) on the final features of the ZnO NPs.[312] Notably, the authors observed that nanostructures with different morphologies were obtained (spheres, spindles or nanosheets) with different microreactors, and attributed the obtained results to a different heat transfer during the NP synthesis as a function of chip geometry. Indeed, since the temperature-dependent conversion of the

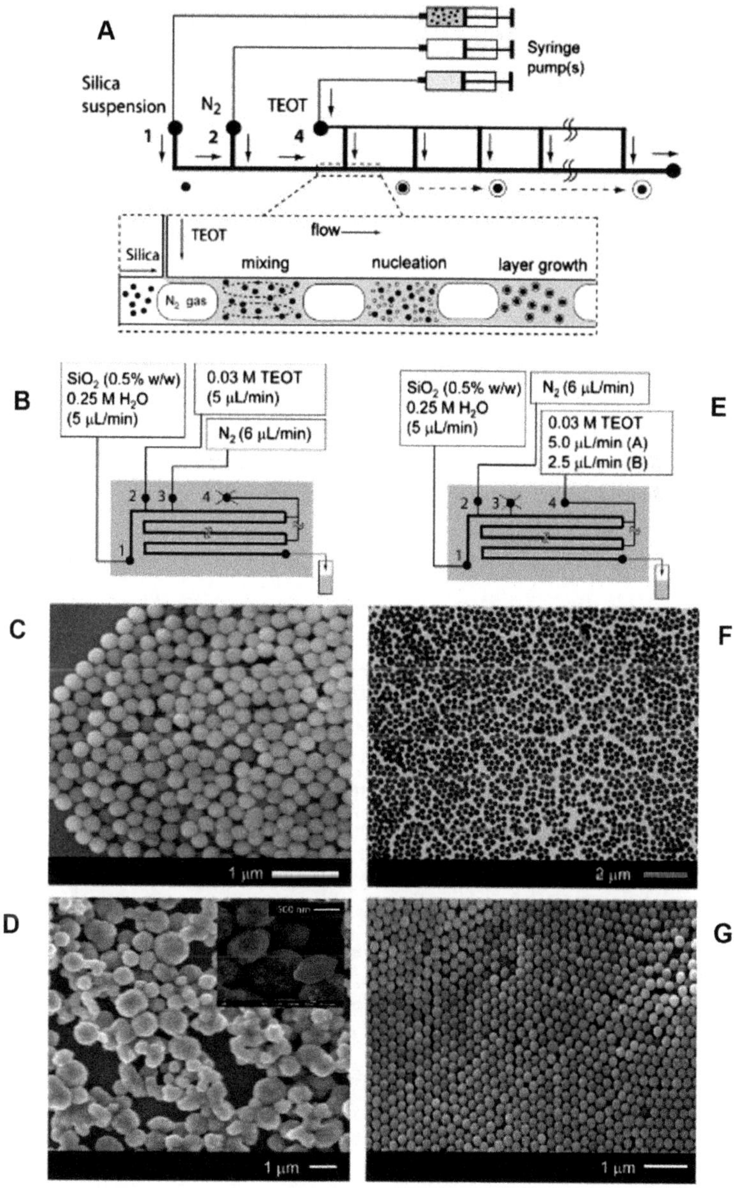

Figure 6.15 (A) Schematic representation of the continuous-flow reactor designed for coating a silica particle suspension with a layer of TiO_2. (B) Details of the single-step TEOT addition experiments, without using the branched manifold. (C) SEM of silica particles used for the coating experiments (average diameter = 209 nm, standard deviation σ = 11 nm). (D) SEM (high magnification in the inset) of particles obtained from the single-addition experiment. (E) Details of the multistep TEOT addition experiments using the branched manifold. (F) TEM of coated particles obtained from the multistep addition experiment. (G) SEM of coated particles obtained from the multistep addition experiment after calcination at 500 °C. Adapted from ref. 309 with permission from John Wiley & Sons, Copyright © 2007 WILEY-VCH Verlag GmbH & Co. KGaA, Weinheim.

hydrolysed Zn^{2+} precursor to ZnO is the rate-determining step of the reaction, different temperature gradients typical of different microreactors led to different final morphologies.

Other interesting examples may be found for the microfluidic synthesis of iron oxide NPs, such as magnetite (Fe_3O_4),[315–319] maghemite (γ-Fe_2O_3),[316,319] as well as goethite (α-FeOOH),[320] which are typically obtained by mixing Fe(II)/Fe(III) solutions with a base (*e.g.*, NaOH, NH_3, NH_4OH, TMAOH). A particularly exciting synthesis was reported by Frenz *et al.*,[317] who developed a droplet-based microreactor that exploited the electrocoalescence of two flowing droplets, each of them containing a reactant (scheme in Figure 6.16). Two aqueous phases of reactants (Fe(II)/Fe(III) chlorides and NH_4OH) were injected *via* the outer channels and synchronously emulsified by the central perfluorocarbon oil channel, ensuring high reliability for the process. Then, coalescence of the droplets containing the two aqueous precursors solutions was triggered by applying an electric field *via* two electrodes within the microfluidic chip, leading to precipitation of iron oxide NPs on a millisecond time scale.

Interested readers may find more details in specific reviews.[232,233,321]

6.4.3 Metal Chalcogenides

The development of the microfluidic technique in inorganic synthesis owes its origin to the pioneering research of the DeMello group, which initially focused on the synthesis of cadmium sulfide nanocrystals.[276] Since then, a vast number of microfluidic approaches for the synthesis of semiconductor NPs, *i.e.*, quantum dots (QDs) formed of binary compounds, such as CdS, CdSe and ZnS, have been published. Indeed, the advantages of microfluidics,

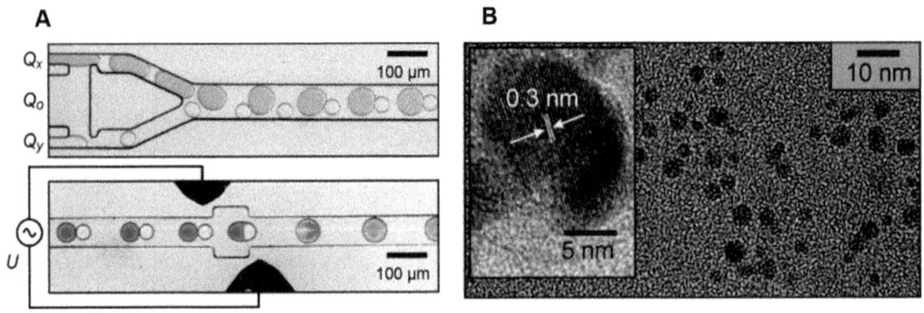

Figure 6.16 (A) Schematic representation of the microreactor for the droplet-based microfluidic synthesis of iron oxide NPs. Pairing module for emulsification and droplet pairing of the iron (Q_X) and base (Q_Y) solutions in the oil phase (Q_O), and fusion module for the electrocoalescence of the droplet pair by application of a voltage U between two electrodes. (B) TEM and high-resolution TEM micrographs of iron oxide NPs showing (220) spinel planes. Adapted from ref. 317 with permission from John Wiley & Sons, Copyright © 2008 WILEY-VCH Verlag GmbH & Co. KGaA, Weinheim.

allowing obtain monodisperse NPs with a well-defined size to be obtained, are particularly appealing for the synthesis of metal chalcogenide NPs, whose physical properties are strongly related to their size and shape. Moreover, the direct correlation between the size of QDs and their UV–vis absorption makes the microfluidic approach extremely attractive for the synthesis of these systems, as rapid, online and time-resolved particle size determination can be carried out.

Typically, CdS synthesis is carried out at low or room temperature by mixing cadmium salts with a sulfur precursor, such as Na_2S. The two precursor solutions may be contacted in various ways, for example by diffusion between two laminar flows[248,276,322] or by their compartmentalisation in droplets with an oily continuous phase, either in droplets containing mixed reactants[240] or in droplet pairs that are subsequently fused together.[323] The mixing approach based on interface diffusion of species in a co-flowing laminar system has been employed by Sounart *et al.*,[248] who also exploited the possibility offered by microfluidic systems for online spatially resolved analysis of nucleation and growth of the synthesised NPs. In particular, the authors investigated the synthesis of cysteine-capped CdS QDs between two interdiffusing reagent streams of buffered aqueous solutions of $CdSO_4$ and Na_2S, respectively, contacted in a simple Y-shaped microreactor. Cysteine was added into both reactant solutions. Direct insight into the kinetics and mechanistic data of the quantum dot formation was obtained by exciting the CdS-Cys NPs growing in the channel using an Ar laser (350–360 nm) focused to a 5 μm spot in the microchannel and recording the photoluminescence spectra at several positions in the reactor. The study demonstrated that the nucleation and growth process of CdS QDs is diffusion limited, thus forcing the NPs to nucleate under uniform conditions in a narrow region at the centre of the channel, at the boundary between the two laminar streams.

A droplet-based multistep microfluidic method for the synthesis of CdS NPs was reported by Shestopalov *et al.*,[240] who employed PDMS microchannels in which aqueous droplets containing $CdCl_2$, mercaptopropionic acid (MPA), Na_2S and NaOH (buffer) were formed in an immiscible fluorinated oil flow. The reactants were mixed in less than 5 ms within droplets in a short segment of winding channel and then allowed to react while flowing into a serpentine section. The reaction time could be controlled by adjusting flow rates or by fabricating networks with different lengths of channels between junctions, thanks to the distance–time correlation of continuous-flow systems. A further stage of the multistep reaction could be initiated by an additional aqueous reagent stream injection into the droplets, followed by a proper mixing section and a reaction one (see scheme in Figure 6.17a and b), and so on. In particular, the authors exploited the designed setup for the online quenching of the reaction, by introducing MPA, which acted as a ligand and prevented aggregation. Indeed, NPs collected without introducing MPA through the quenching inlet were characterised by a broader absorption peak, indicating higher polydispersity (Figure 6.17c). Variation in the amount of MPA added resulted in fine-tuning of the CdS NPs size, with smaller NPs obtained with

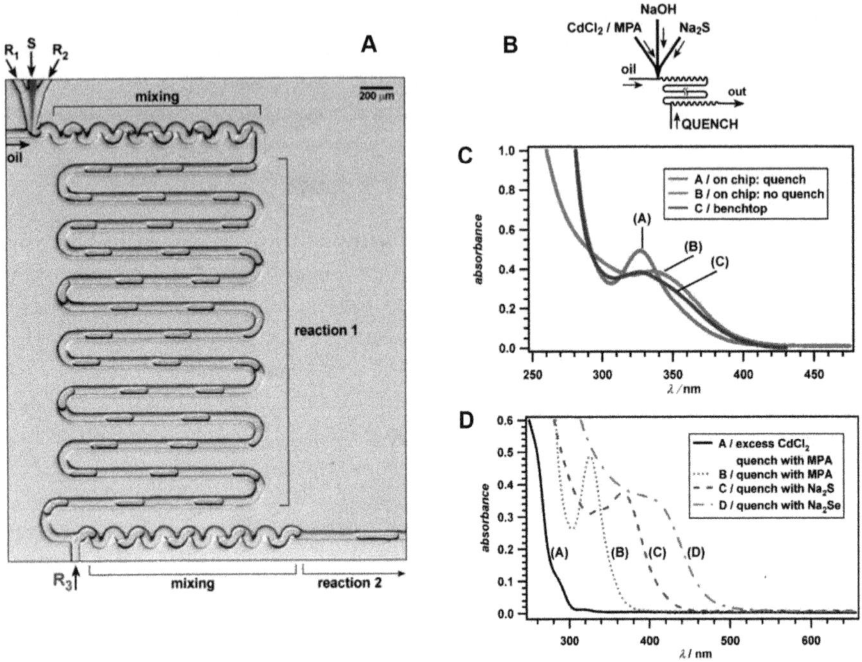

Figure 6.17 (A) Micrograph of PDMS microfluidic device used for the droplet-based synthesis of NPs; R1–3: reagents, S: partitioning stream. (B) Schematic diagram of the microfluidic two-step synthesis on-chip with millisecond quenching of CdS and CdS/CdSe core–shell NPs. (C) UV–vis spectra of CdS NPs synthesised on-chip with millisecond quench, on-chip without quenching and in batch conditions. (D) UV–vis spectra of NPs with various sizes and composition obtained with different types of quenching. Adapted from ref. 240 with permission from the Royal Society of Chemistry.

higher amounts of MPA. When the same experiment was performed in batch (also with the quenching agent), noticeably less monodispersed NPs were obtained, demonstrating that rapid mixing and quenching on a millisecond time scale is essential for size distribution control. When Na_2S or Na_2Se were employed as quenching agents, larger NPs, due to a sulfide-rich outer shell, or CdS/CdSe core–shell particles, were obtained, respectively (Figure 6.17d). Such flexibility of the microfluidic system in accommodating different multistep reaction schemes without changing the geometry of the microreactor is a particularly notable feature of this method.

As demonstrated by Shestopalov *et al.*,[240] by replacing Na_2S with Na_2Se as the chalcogen source, CdSe QDs could also be synthesised. However, in the case of CdSe, narrow particle size distributions could be obtained only at relatively elevated temperatures (200–350 °C), making the microfluidic implementation of batch syntheses more challenging. Indeed, the requirement for solvents that can both dissolve precursors under ambient conditions and remain liquid over the whole operating temperature and pressure range is a significant issue that limits the compatible set of solvents, precursors and ligands for the microfluidic system. Thus, solvents with a high boiling point are usually employed (*e.g.*, squalene, octadecene (ODE) or trioctylphosphine oxide (TOPO)), and carboxylic acids, such as oleic acid (OA), and long-chain amines, such as oleylamine (OLA), are typically added as ligands. Common cadmium and selenium sources are, respectively, cadmium carboxylates (*e.g.*, oleate or stearate) and alkyl phosphine selenides (*e.g.*, trioctyl phosphine selenide (TOPSe) or tributyl phosphine selenide (TBPSe)), obtained by dissolving selenium in TOP or TBP. Since the typically employed high boiling point solvents are viscous (0.5–1.5 mPa s), when single-phase reacting mixtures were employed for the synthesis of CdSe QDs, broad size distributions (*i.e.*, full-width and half-maximum (FWHM) of 35–50 nm) were obtained, due to an enhanced axial distribution of laminar flows and consequent broad residence time distributions (see Figure 6.12). For example, the synthesis of CdSe QDs at atmospheric pressure was carried out using a single-phase (Cd oleate, TOPSe, OLA and TOP in squalene) laminar flow, heated glass capillary by Yen *et al.*[324] By varying the reaction temperature (240–320 °C) and TOPSe concentration, the authors tuned the NP sizes over a wide corresponding spectral range. Moreover, they observed that the NP size distribution became significatively broad as the flow rate was increased (*i.e.*, decreasing the reaction time), and ascribed the observation to the dispersion in reaction conditions dominated by the residence time distribution. Flow segmentation approaches, that tackle these shortcomings, were applied to the microfluidic synthesis of CdSe QDs, both in liquid–gas[325] and in liquid–liquid mixtures,[326] leading to an improved narrowness of the NP size distribution, down to a FWHM of 30 nm.

The stability and some properties of QDs, such as their biocompatibility, can be improved by coating the core of CdS or CdSe with a nanometric shell of another semiconductor, such as the non-toxic ZnS, whose thickness is crucial for obtaining high quantum yields of QDs. Microfluidic systems are appealing for coating NPs in a controlled fashion, also obtainable in a multistep synthetic approach, and various examples of core–shell nanostructures, even with multiple shells like ZnS/CdSe/ZnS particles, have been reported.[327–330] An elegant study was reported by Wang *et al.*,[327] who finely tuned the size and the layer thickness of ZnS-coated CdSe QDs by adjusting the flow rate.

The microfluidic synthesis of pure, and subsequently also doped, ZnS NPs has also been investigated by Dengo *et al.*[331] and Tajoli *et al.*,[332] who employed a simple room temperature setup with PTFE tubing and a T-junction mixer to ensure rapid mixing of reactants (aqueous solutions of $Zn(NO_3)_2$ and Na_2S). Notably, the authors reported that small, quite monodispersed and independent ZnS NPs (5 nm) were obtained without employing any ligand and/or surfactant, and that the adopted synthetic strategy allowed formation of surfactant-free stable water suspensions, which will probably be useful for applications development. By comparing the results obtained in microfluidic and batch conditions, they observed that, through the microfluidic approach, the final size of the NPs was reduced, while maintaining a rather narrow size distribution despite the absence of ligands, and the presence of structural defects, such as stacking faults was increased.[331] In a further development of the method, dopant ions, such as Mn(II), Eu(III) and Nd(III), were introduced into the Zn(II) precursor solution in different atomic percentages and doped luminescent ZnS nanostructures were synthesised. Incorporation of the dopants in the final products was confirmed by inductively coupled plasma-mass spectroscopy (ICP-MS) measurements and by XAS analysis at the dopant L_3-edges, as well as by photoluminescence investigations. Thanks to the assessed non-cytotoxic effect of the lanthanide-doped samples and the displayed strong photoluminescence intensities in the near-infrared region, these systems are appealing for applications in the *in vivo* optical bioimaging field.[332]

6.4.4 Other Compounds

Microfluidic approaches have been exploited for the controlled synthesis of high-quality quantum dots of other metal compounds, such as non-chalcogen-based semiconductors (*e.g.*, III–V type InP, InAs), and even as core–shell structures,[333–335] mixed metal halide peroskites,[247,336–338] as well as fluorides and phosphates.[339]

An interesting example of the microfluidic synthesis of III–V quantum dots was reported by Nightingale and deMello, who synthesised high-quality InP QDs by adapting one of the most effective batch routes of InP synthesis into a simple microfluidic device.[333] They employed a single capillary (inner diameter of 150 μm) immersed in a heated oil bath at 225 °C, through which a mixture of reactants ($InCl_3$ and tris(trimethylsilyl)phosphine $(TMS)_3P$), oleic acid, oleylamine and octadecene could be driven hydrodynamically, or a simple two-in/one-out Y-shaped reactor that enabled easy variation of the In/P ratio in the reaction mixture and/or of the reaction temperature. The emission spectra of the as-produced InP QDs using different flow rates were characterised by a near Gaussian profile with a FWHM of about 80 nm, indicating the formation of high-quality NPs. A blueshift in the peak wavelength, consistent with the formation of smaller NPs that exhibit enhanced quantum confinement, was observed on increasing the flow rates (from 4 to 60 μL min^{-1}), *i.e.*, decreasing the reaction time, and on decreasing the temperature (from 230 to 200 °C). A three-stage microfluidic system was employed for the high-temperature and high-pressure continuous flow of InP by Baek *et al.*, who examined the parameters that had a role in the nucleation and growth of InP nanocrystals.[334] The same authors developed a multistage microfluidic platform for the synthesis of core–shell quantum dots, which enabled precise control of heating profiles and flow distributions across the microfluidic channels.[335] By employing such a system, and using a maximum of six chip reactors in series (mixing, ageing and sequential growth reactors for the synthesis of the cores and of the shells, respectively), they synthesised InP/ZnS, InP/ZnSe, InP/CdS and InAs/InP core–shell structures.

Another interesting example was reported by Lignos *et al.*, who synthesised caesium lead halide perovskite ($CsPbX_3$, X = Br, I, Cl and Cl/Br and Br/I mixed halide systems) QDs exhibiting bright and tuneable photoluminescence in a droplet-based microfluidic platform.[247] The authors combined online photoluminescence and absorption measurements with fast mixing of reactants, in order to rigorously and rapidly map the reaction parameters, such as molar ratios of Cs, Pb and halide precursors, reaction times and temperatures. Notably, using this approach they also obtained early-stage insight into the mechanism of nucleation of metal halide nanocrystals. The microfluidic synthesis of inorganic perovskite quantum dots has also been investigated by other groups for applications in white light-emitting diodes, and also reporting structures with high compositional complexity.[336–338]

Finally, it is worth mentioning the work of Zhu *et al.*, who combined the continuous-flow nature and small dimensions of a microfluidic reactor with the unconventional heating by microwave irradiation for the synthesis of monodispersed doped LaF_3 NPs (4–5 nm) and doped $LaPO_4$ nanorods (60–70 nm in length).[339]

6.5 Sonochemical Synthesis

The sonochemical approach is an unconventional synthetic method first developed in the early 1990s and afterwards exponentially expanded. Nowadays it is well established, and possesses significant potential to generate large-scale functional materials. It exploits the effects of acoustic cavitation phenomena occurring in liquids irradiated by powerful ultrasound (US).

Ultrasound is acoustic vibration with frequencies higher than the upper audible limit of human hearing, *i.e.*, higher than 20 kHz. From an application point of view, ultrasound can be divided into low-frequency vibrations, ranging from 20 kHz to 1 MHz, and high-frequency ones (>1 MHz). While high-frequency ultrasound is commonly used as a non-destructive method of analysis (*e.g.*, ultrasonic imaging in medical diagnostics), in chemical synthesis (*sonochemistry*) low-frequency US vibrations are exploited.[340,341]

Ultrasonic vibrations are generated using specific devices that transform electromagnetic oscillations into elastic mechanical vibrations, exploiting the ability of certain materials to change their linear dimensions upon the action of alternating magnetic or electric field, *i.e.*, using the magnetostriction or piezoelectric effect, respectively. Magnetostrictive transducers generate ultrasonic vibrations at frequencies below 30 kHz and with intensities up to several tens of Watts per cm^2, while piezoelectric ones produce high-frequency US (500 kHz–100 MHz) with intensities that can be as high as 60 W per cm^2, and even reach 200–500 W cm^{-2} in the pulsed mode of operation.[340]

Since the acoustic wavelengths associated with ultrasounds (in the mm range, given the typical speed of sound in liquid of 1000 to 1500 m s^{-1})[342] are too long compared to molecular lengths, ultrasound vibrations do not interact directly with molecules to induce chemical changes. In fact, sonochemistry is based on non-linear acoustic processes caused by irradiating liquid media with US, the most important one being *cavitation*, a phenomenon that consists of the formation, growth and implosive collapse of bubbles in liquids.[342–344]

When acoustic vibrations propagate through a liquid, alternating zones of negative and positive pressure are produced, *i.e.*, rarefaction and compression zones, respectively. At high energy densities of the acoustic vibrations, the negative pressure created into the rarefaction zones leads to the formation of cavitation bubbles, mostly from pre-existing impurities, such as small gas bubbles (nucleated process). Indeed, it should be noted that the formation of a cavity inside a medium, *i.e.*, the separation of solvent molecules by a distance of some nm in opposition to intermolecular forces, requires a negative pressure of hundreds of atmospheres, while the nucleated process can occur at much lower power levels.[345] As illustrated in Figure 6.18, the thus-formed cavitation bubbles oscillate with the applied acoustic field over many acoustic cycles and so grow through inertial effects, by diffusion of solvent vapour and dissolved gas molecules into the bubble. Since the growth processes during the negative pressure half of the acoustic cycle (expansion cycle) are slightly faster than the shrinking ones, bubbles do not have time to recompress during positive-pressure half-cycles (compression cycles) and this leads to overall growth of the cavitation bubbles, until a critical size is reached.[343] At the critical (resonant) size (determined by the frequency of the US, *e.g.*, about 150 μm at 20 kHz), the cavity efficiently absorbs energy from the acoustic field and it rapidly overgrows in the course of a single acoustic cycle and implodes (Figure 6.18).[342]

The catastrophic implosive collapse of the cavitation bubble is a nearly adiabatic process, thus it results in the generation of extreme heat spots within the bubble. In particular, during the collapse of a cavitation bubble, high-intensity shock waves, micro-jets and other physical processes that enhance

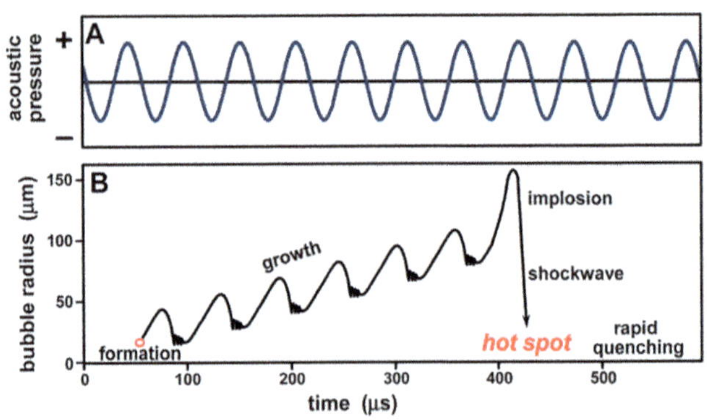

Figure 6.18 Schematic illustration of the phenomenon of cavitation in a liquid irradiated by high-intensity US, with the typical stages of formation, growth and oscillations, and implosive collapse of bubbles, leading to the formation of hot spots. Reproduced from ref. 343 with permission from the Royal Society of Chemistry.

mass transfer in liquids, such as microstreaming, agitation and turbulence, are generated.[346–349] These phenomena lead to the formation of a 'hot spot' at the centre, characterised by intense local heating and high pressures with very short lifetimes (micro- to nanoseconds). Temperatures can increase up to 5000 K (as estimated by experimental methods, and which according to theoretical predictions are greater than 15 000 K),[350] pressures up to about 1800 atm and cooling rates can reach values over 10^9 K s^{-1}, as extensively studied by Suslick *et al.*[342,351–353] Such extreme thermal conditions lead, *inter alia*, to emission of light from the bubbles, a phenomenon that is referred to as *sonoluminescence*[354] (which is not discussed in this chapter).

Thus, thanks to acoustic cavitation, the sonochemical approach allows locally extreme conditions to be achieved in the reaction mixture and, consequently, it can be exploited at room temperature for syntheses that would otherwise require high temperatures and pressures, or longer reaction times. Indeed, in some cases, reactivities can be dramatically increased by nearly a millionfold by ultrasonic irradiation.[342] The heat generated by the collapse of cavitation bubbles induces a variety of chemical reactions, which can be carried out in aqueous or non-aqueous solutions through completely different mechanisms, defined as *primary* and *secondary sonochemistry*.

When sonochemical reactions are carried out in *aqueous solutions*, water and other volatile compounds may diffuse into the cavitation bubbles and experience the action of the extreme temperature and pressure of the hot spots. As a consequence, water molecules undergo sonolysis and generate $H^•$ and $OH^•$ radicals (eqn (6.3)), which may subsequently go through recombination reactions to form a variety of reducing and oxidising radicals and molecular products in pure water saturated with argon (eqn (6.4)–(6.6)) or with air (eqn (6.7)–(6.10)):[340,341]

$$H_2O \xrightarrow{\text{ultrasound}} H^{\cdot} + OH^{\cdot} \tag{6.3}$$

$$H^{\cdot} + HO^{\cdot} \rightarrow H_2O \tag{6.4}$$

$$HO^{\cdot} + HO^{\cdot} \rightarrow H_2O_2 \tag{6.5}$$

$$H^{\cdot} + H^{\cdot} \rightarrow H_2 \tag{6.6}$$

$$H^{\cdot} + O_2 \rightarrow HO_2^{\cdot} \tag{6.7}$$

$$HO_2^{\cdot} + HO_2^{\cdot} \rightarrow H_2O_2 + O_2 \tag{6.8}$$

$$O_2 \rightarrow 2O^{\cdot} \tag{6.9}$$

$$O^{\cdot} + O_2 \rightarrow O_3 \tag{6.10}$$

The chemistry of aqueous solutions irradiated by ultrasound has been extensively studied[355–357] and it was reported that, in the absence of oxygen dissolved in water, the rate of formation of $OH^•$ radicals was about 25 μmol L^{-1} min^{-1}.[358] In addition, care must be taken with the acidity of the medium, which could be enhanced by the formation of nitric acid within cavitation bubbles from O_2 and N_2.[359] Indeed, the sonolysis of water also generates hydrated electrons, which may participate in different processes in acidic (eqn (6.11)) or in alkaline media (eqn (6.12)):

$$e^-_{aq} + H^+ \rightarrow H^\cdot \tag{6.11}$$

$$H^\cdot + OH^- \rightarrow H_2O + e^-_{aq} \tag{6.12}$$

Hence, irradiating an aqueous solution with high-power ultrasound initiates redox and radical reactions in it and in cavitation bubbles.[341] The variety of chemically active species that are formed by sonolysis of vapour inside the collapsing bubble diffuses out into the bulk liquid phase and subsequently reacts with non-volatile precursors in solution to form different types of nanostructures depending on the synthetic conditions (*secondary sonochemistry*).[343]

This affects the synthetic strategy for the formation of different classes of compounds: for example, for the synthesis of metal NPs (Section 6.5.1), a reducing environment is preferred and the conversion of oxidising radicals (*e.g.*, $HO^\bullet$) into secondary reducing ones is the target, for instance by dissolving organic solutes, such as alcohols, in water. On the other hand, for the synthesis of oxides (Section 6.5.2), an oxidising reaction medium is of interest and reducing radicals can be turned into oxidising ones (eqn (6.7)) by flushing oxygen in the reaction mixture.

Alternatively, when sonochemical processes take place in *non-aqueous media*, the composition of the vapour in the bubble cavity is more complex. Indeed, whereas in aqueous solutions cavitation bubbles are filled with water vapour and dissolved gases only, inside non-aqueous ones there may also be other volatilised components, such as organometallic compounds, which may decompose inside the collapsing bubble to generate metal atoms.[360] As a consequence of the high temperatures inside the collapsing bubble, sonolysis of the weak metal–carbon bonds of volatile organometallic precursors may occur in the gas phase (*primary sonochemistry*).[343] The free metal atoms thus produced may diffuse out into the bulk liquid phase and nucleate to generate metal NPs, which are typically amorphous due to the high heating and cooling rates (Section 6.5.1), or react with other chemical substances in the solvent to form other nanostructured materials, *e.g.*, oxides (Section 6.5.2), chalcogenides (Section 6.5.3), carbides, *etc.* (Sections 6.5.4). Generally, non-volatile organic solvents, such as long-chain hydrocarbons or silicone oil, are employed to avoid the formation of additional vapour, which could absorb some energy of the collapsing bubble and make the reaction conditions achieved less extreme.

A general scheme representing the different mechanisms of reaction through which sonochemical synthesis of materials occurs, categorised as *primary* and *secondary sonochemical reactions*, is shown in Figure 6.19.

In addition to primary and secondary chemical effects, several physical effects occur in liquid media irradiated by ultrasound, such as high-speed jets or shock waves, the consequence of the implosive collapse of cavitation

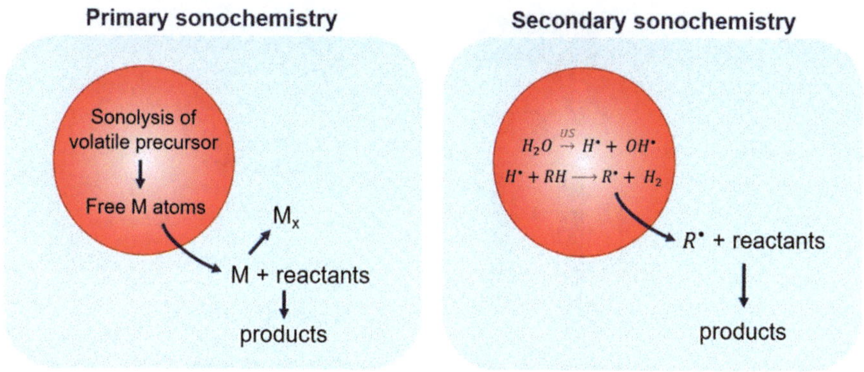

Figure 6.19 Schematic representation of the two different mechanisms of primary and secondary sonochemistry for the preparation of nanomaterials. Primary sonochemical reactions are gas-phase reactions occurring inside collapsing bubbles, such as sonolysis of volatile organometallic compounds. Secondary sonochemical reactions are liquid-phase reactions occurring in bulk phase or at the bubble/solution interface between non-volatile precursors and radicals produced by sonolysis of vapour molecules inside collapsing bubbles.

bubbles. Such physical phenomena may have a role in the kinetics of reactions, because they enhance mass transport, as well as on the dimensional and morphological features of the nanomaterials that are formed and on their subsequent modification.[349,361] Examples may be exfoliation of layered materials into bidimensional ones, particle fragmentation (brittle materials) or agglomeration (malleable materials).[362–364]

A typical laboratory-scale apparatus for sonochemical synthesis (scheme in Figure 6.20) consists of a piezoelectric transducer that drives a high-intensity ultrasonic titanium horn, which can deliver approximately 10–100 Watts of acoustic power with a typical frequency of 20 kHz. The reaction mixture is introduced in a glass cell placed in a cooling bath and with a gas inlet and outlet in order to control the reaction atmosphere, *e.g.*, bubbling O_2 or Ar. As mentioned, the possibility of flushing inert gases or air into the reaction cell, thus tuning the reaction atmosphere from an inert to an oxidative one, allows control of the amount of reducing and oxidising radicals in the reaction mixture.

Another fundamental synthetic parameter that can be tuned is the frequency of the ultrasound, which has been proved to influence the amount of primary and secondary radicals generated. In particular, for a given solution volume and acoustic power, increasing the US frequency leads to an increase in the amount of cavitation bubbles generated, together with a decrease in their resonant radius. A smaller resonant size implies a lower collapse intensity, and consequently a lower generated temperature, leading to a reduced amount of primary and secondary radicals formed per bubble. However, the higher amount of (smaller) cavitation bubbles leads to an overall increase in

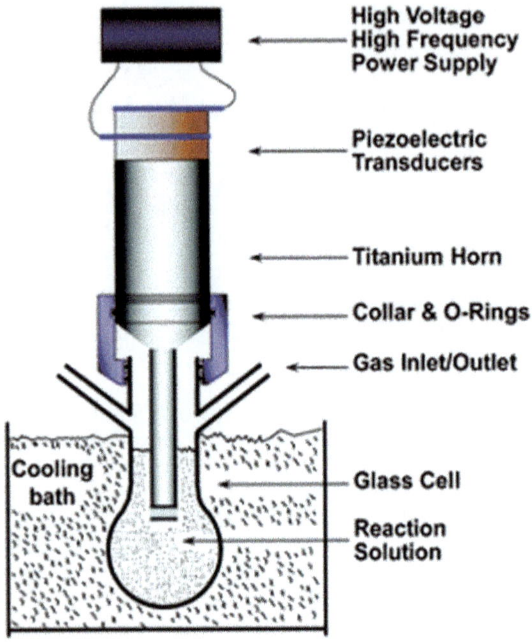

Figure 6.20 Schematic representation of a typical laboratory apparatus for sonochemical reactions. Reproduced from ref. 363 with permission from John Wiley & Sons, Copyright © 2010 WILEY-VCH Verlag GmbH & Co. KGaA, Weinheim.

radicals generated. The highest sonochemical reaction yield has been determined to be around 200–800 kHz, beyond which the yield declines.[341,365–368]

Finally, in addition to the reaction atmosphere and the US frequency, the reaction medium can be changed to obtain various nanostructured compounds in different forms (*e.g.*, powders, colloidally stable suspensions, supported NPs, *etc.*), also controlling the final nanomaterials in composition, size and morphology. Among the parameters that can be tuned there are: (1) the solvent (aqueous or volatile or non-volatile organic); (2) the type of metal precursor (volatile or non-volatile, *i.e.*, metal salt or organometallic compound); (3) other reactants (*e.g.*, sulfur source for sulfide synthesis); and/or (4) additives (*e.g.*, stabilisers, surfactants, templates (sacrificial or not) or inorganic supports).

As highlighted in detail in the following sections, one of the major advantages of the sonochemical method is its versatility, allowing a plethora of nanostructured compounds to be obtained, such as metals, metal alloys, oxides, chalcogenides and carbides, simply by changing the reaction conditions. For example, sonochemical irradiation of $Fe(CO)_5$ in hexadecane leads to the synthesis of amorphous iron powder,[369,370] but when ligands or stabilisers are introduced, such as oleic acid or PVP, colloidal Fe NPs are formed.[371] Moreover, when the same procedure is performed flushing oxygen into the reaction medium instead of argon, iron oxide NPs are obtained. On the other hand, by adding a sulfur source into the reaction mixture, nanostructured iron sulfide can be synthesised. Finally, a solid support material, *e.g.*, silica or alumina, can be introduced in order to obtain supported iron NPs.[363,364,372]

As extensively discussed in this section, the sonochemical approach is an established unconventional wet-chemistry method for synthesising inorganic nanomaterials, by exploiting the microscopically extreme (yet macroscopically mild) reaction conditions caused by the implosive collapse of cavitation bubbles, which are not achievable using conventional approaches. The resulting unique conditions of physical effects (*i.e.*, microstreaming, mass transport, local heating), together with the chemical effect of radical generation that can start redox reactions, allow initiation of reactions even without external agents, such as reducing ones. In this way, products are purer, and the reaction environment is greener. The syntheses are carried out under ambient conditions and are typically faster than those using conventional methods, even allowing for the formation of products not obtainable by standard approaches (*e.g.*, requiring higher temperatures or pressures or longer reaction times). Finally, the sonochemical method has been proven to be very versatile and to allow easy control of size and size distribution of the NPs formed, as well as of their phase purity. However, the potential scalability of sonochemical approaches in industrial environments remains a challenge because of the lack of availability of large-scale ultrasonic reactors.

A detailed description of the state-of-the-art of compounds that have been synthesised through this sonochemical approach will be provided in the following sections, discussing the synthesis of metals and alloys (Section 6.5.1), metal oxides (Section 6.5.2), metal chalcogenides (Section 6.5.3) and other compounds (Section 6.5.4). In particular, examples focusing on the role of the different experimental parameters on the synthetic output will be addressed, with references provided for readers interested to find out more.

6.5.1 Metals and Alloys

The majority of work on sonochemical syntheses of metal NPs in *non-aqueous media* is based on the use of *volatile coordination compounds* containing zero-valent metal atoms, such as metal carbonyls (*e.g.*, $Fe(CO)_5$, $Co(CO)_3NO$, $Cr(CO)_6$, $Mo(CO)_6$, $W(CO)_6$ and $Mn_2(CO)_{10}$), to generate metal NPs.[343,363,369,370,373] The obtained M_x NPs are typically amorphous, due to the limited increase in the macroscopic temperature of the medium and the very high cooling rates after bubble collapse (10^{10} K s^{-1}), which hinder the organisation and crystallisation of the NPs.[372,374] Such a peculiarity of the sonochemical approach is noteworthy for the production of metallic glasses, since the extremely rapid macroscopic cooling rates of melts necessary to prevent crystallisation (about 10^5 to 10^7 K s^{-1}) are usually challenging to achieve experimentally. If a crystalline material is targeted, subsequent heat treatment in an inert atmosphere, *e.g.*, at 350 °C, is required.[369]

One of the first sonochemical syntheses of nanostructured metals in non-aqueous media was developed in 1991 by Suslick *et al.*, who dissolved $Fe(CO)_5$ in decane and irradiated it with high-intensity ultrasound (20 kHz, 100 W cm^{-1}) under an argon atmosphere.[369] Iron pentacarbonyl underwent sonolysis during bubble collapse and its dissociation led to the formation of amorphous iron NPs. The amorphous nature of the product was confirmed by scanning electron microscopy (SEM), TEM, differential scanning calorimetry (DSC), X-ray powder diffraction (XRPD) and electron-beam microdiffraction. Using elemental analysis, the authors found that the obtained powder was >96 wt% iron, with traces of carbon and oxygen, from the decomposition of alkane solvent or carbon monoxide during the US irradiation. These results were particularly noteworthy since they were able to obtain an Fe metallic glass without other alloying elements using (macroscopic) mild experimental conditions, while all previously prepared Fe-containing metallic glasses contained large amounts (>20 wt%) of contaminants. The authors found that the amorphous powder obtained was composed of small particles (about 10 nm) with a specific surface area (120 m^2 g^{-1}) about 150 times greater than commercially available iron powder. Moreover, their amorphous Fe NPs showed improved catalytic activity towards the hydrogenolysis and dehydrogenation of saturated hydrocarbons and for the Fischer–Tropsch hydrogenation of CO, compared to commercial samples.

Some years later, the same research group demonstrated that colloidal Fe NPs can also be sonochemically synthesised by introducing ligands or stabilisers into the reaction mixture, such as PVP or oleic acid.[371] These agents prevent the agglomeration of Fe atoms, which typically occurs in alkanes in the absence of any trapping agent, to produce highly porous nanostructured materials, and the nm-sized clusters are thus stabilised. The authors observed that Fe NPs dispersed in PVP range in size from 3 to 8 nm (TEM micrograph in Figure 6.21a), while using oleic acid as stabiliser results in

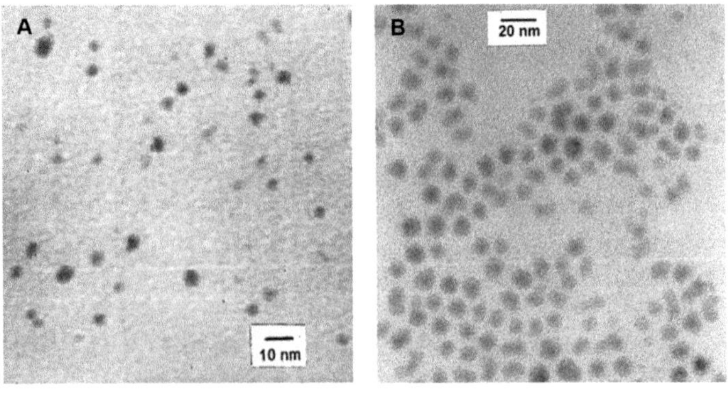

Figure 6.21 TEM micrographs of sonochemically prepared iron colloid particles (A) dispersed in a PVP matrix and (B) stabilised by oleic acid. Adapted from ref. 371 with permission from American Chemical Society, Copyright 1996.

more homogeneously distributed NPs, with an average size of 8 nm (TEM micrograph in Figure 6.21b).

In a similar way, spherical Co,[375] Pd,[376] Ni[377] and Si[378] clusters and/or NPs have been produced by ultrasonic irradiation of neat metallorganic compounds ($Co(CO)_3NO$, $Pd(O_2CCH_3)_2$, $Ni(CO)_4$, $Si(OC_2H_5)_4$) or of their non-aqueous solutions in inert atmosphere.

When two volatile metallorganic precursors of different metals are sonicated together in a non-volatile medium, nanosized amorphous alloy powders can be obtained. For example, first Suslick *et al.*[360] and then Shafi *et al.*[379] synthesised FeCo, FeNi and CoNi alloys, respectively, with a composition tuneable by varying the ratio of the precursor solutions ($Fe(CO)_5$, $Co(CO)_3NO$, $Ni(CO)_4$).

On the other hand, the sonochemical synthesis of metal NPs in *aqueous solution* is typically achieved *via* reduction of a dissolved *non-volatile metal precursor* (*i.e.*, a metal salt), as mentioned in Section 6.5. For metals with large positive reduction potentials (*i.e.*, Au^+, Ag^+, Pd^{2+}, Pt^{2+}), the reducing agents are the reducing radicals (*e.g.*, $H^•$) generated within the bubbles through sonolysis of water (eqn (6.3)), which can diffuse out into the bulk and react with metal ions to generate metal atoms (eqn (6.13)), which subsequently agglomerate to form M_x NPs.

$$M^{n+} + nH^\bullet \rightarrow M^0 + nH^+ \tag{6.13}$$

As pointed out above, a reducing environment should be favoured by flushing inert gases in the reaction mixture and/or by adopting strategies to convert primary oxidising radicals (*e.g.*, $HO^•$) into secondary reducing ones. A clever example, proposed by Henglein in 1993,[380] relies on dissolving small amounts of volatile organics, such as simple alcohols (RHOH) that react with primary radicals, to generate a significant amount of strongly reducing $ROH^•$ (eqn (6.14) and (6.15)). In addition, alcohol molecules that diffuse into the bubbles can be pyrolysed to generate a plethora of reducing radicals $R^•$ that can play a role in the reduction reaction (eqn (6.16)).

$$H^\bullet + RHOH \rightarrow ROH^\bullet + H_2 \tag{6.14}$$

$$OH^\bullet + RHOH \rightarrow ROH^\bullet + H_2O \tag{6.15}$$

$$M^{n+} + nR^\bullet \rightarrow M^0 \tag{6.16}$$

Overall, this strategy allows the increase of the amount of reducing radicals in the reaction mixture: (1) by opposing the recombination of $H^•$ and $HO^•$ radicals (eqn (6.4)) in the bubbles or at the bubble/solution interface, thus increasing the fraction of unreacted $H^•$ escaping into the bulk solution; (2) by generating other (more active) reducing radicals $R^•$, both secondary ones by interaction with $H^•$ at the bubble/solution interface (eqn (6.14)) and primary ones by pyrolysis in the bubbles; and (3) scavenging primary oxidising radicals, such as $HO^•$ (eqn (6.15)).

Compared with metal NPs sonochemically synthesised in non-aqueous media, which are typically amorphous, the synthesis of metal NPs by non-volatile compounds in water or other volatile solvents (*e.g.*, alcohols) may result in crystallised products. This can be ascribed to the fact that, in volatile solvents, the reactions occur in the liquid phase in a ring zone at the interface between the collapsing bubble and the bulk phase. In this interfacial region, the temperature is lower than inside the hot centre of the collapsing bubble, where gas-phase reactions occur, but higher than the temperature of the macroscopic bulk.[372] The temperature in this ring region was estimated by Suslick *et al.* to be around 1600 °C.[381]

Several groups have explored the water-based sonochemical synthesis of noble metal NPs, such as Au, Ag, Pt and Pd.[358,382–388] One of the first investigations on the reduction of noble metal ions in aqueous solution by a sonochemical method was reported by Yeung *et al.* in 1993, who synthesised colloidal gold NPs with a diameter smaller than 10 nm by irradiating aqueous $HAuCl_4$ solutions with 20 kHz ultrasound.[382] The authors observed that the efficiency of particle formation was highly dependent on the presence of added alcohols in the solutions, and in particular on their type and concentration. Indeed, they reported that the effectiveness of $[AuCl_4]^-$ sonochemical reduction followed the order of the air/water surface activity of the added alcohol (*i.e.*, methanol < propan-2-ol < *n*-pentanol), thus demonstrating that scavenging of the primary radicals occurs at the interface between the cavitation bubble and the bulk. On the other hand, scavenging in the collapsing bubble would have been dependent on the partial vapour pressure of the alcohol, which is of opposite order to the observed reduction efficiency. The observations of Yeung *et al.*[382] agreed with the work of Henglein,[380] who proposed that the more hydrophobic the additive the better its ability to scavenge the primary $OH^•$ radicals (eqn (6.15)), and therefore the smaller the yield of H_2O_2 (eqn (6.5)). In addition, Yeung *et al.* observed that the hydrophobicity of the added alcohols also influenced the mean size of the Au NPs obtained, producing smaller NPs with more hydrophobic alcohols (7 nm with methanol, 6 nm with propan-1-ol and 5 nm with *n*-pentanol). This correlation was also found by Caruso *et al.*,[386] who claimed that more hydrophobic alcohols are readily adsorbed on the surface of a growing particle and thus limit its growth more efficiently. Another possible explanation is that the alkyl chain length of the added alcohol, as well as its concentration, directly correlate with the amount of secondary radicals formed, leading to faster reduction rates, which in turn result in smaller NPs.

Similar results were obtained by Nagata *et al.*,[383] who reported that the rate of formation of Au NPs was accelerated by 3–44 times (from 3 μM min^{-1} to 9–133 μM min^{-1}) when introducing different organic additives, such as surfactants, water-soluble polymers and aliphatic alcohols and ketones. The order of acceleration of the reduction followed the increase in their hydrophobic properties (*i.e.*, surfactants being the more effective additives) and in their concentration (*i.e.*, the more concentrated, the higher the increase in rate). Thus, in agreement with work reported by Henglein,[380] the greater the hydrophobicity of an additive, the more readily it can accumulate at the interface and the higher the rate of formation of reducing radicals by pyrolysis. Finally, the authors reported that the addition of surfactants (PEG40, Tween20, SDS and PVP), as well as the strong increase of the reduction rate, allowed colloids to be obtained with an average size of 10 nm that were stable for several months, analogously to the case of iron NPs produced by sonochemical decomposition of organometallic precursors discussed above.

Surfactants may also play a role in controlling the final size and shape of the NPs, as typically observed in inorganic synthesis. In this regard, Park *et al.*[387] studied the effect of surfactant (SDS) concentration on the morphology of Au NPs, observing a modification of their shape from triangular plate-like (low SDS concentration) to spherical (high SDS concentration), followed by a decrease in average size. However, it should be noted that by increasing the amount of surfactant in the reaction mixture, the surface tension of the solution is lowered, and this is concomitant with a decrease in cavitation efficiency, *i.e.*, efficiency of the formation of primary radicals. Moreover, the authors investigated the effect of acoustic power on the formation of Au NPs, increasing it from 30 to 90 W. They reported that a higher US power led to smaller NPs (from 20 to 10 nm) with a more uniform spherical shape, and they attributed the experimental evidence to an increased rate of reduction of Au(III) with higher acoustic power.

Finally, the effect of ultrasound frequency on the sonochemical reduction of Au(III) in aqueous solution containing 1-propanol was studied by Okitsu *et al.*,[388] who also found that the applied frequency influences the rate of Au(III) reduction, with a maximum at 200 kHz (Figure 6.22a), which in turn correlates with the size and distribution of the Au NPs produced (Figure 6.22b), *i.e.*, the higher the rate of reduction, the lower the size of the Au NPs (Figure 6.22c). The authors concluded that the operating US frequency influences the amount of primary and secondary radicals generated, resulting in an influence on the size and size distribution of Au NPs: the higher the rate of radical formation, the smaller the particle size.[389]

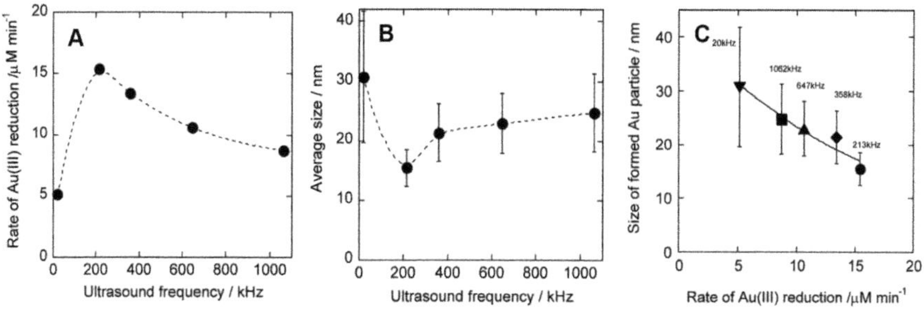

Figure 6.22 (A) Rate of Au(III) reduction as a function of US frequency; (B) average size of the formed Au NPs as a function of US frequency; (C) average size of the formed Au NPs as a function of rate of reduction of Au(III). [Au(III)] = 0.2 mM, [1-propanol] = 20 mM, atmosphere: Ar, US power = 0.1 W cm^{-1}. Adapted from ref. 388 with permission from American Chemical Society, Copyright 2005.

Similar examples of systematic studies on the effect of experimental parameters on the synthetic output were performed with platinum NPs by Mizukoshi *et al.*[358] and by Caruso *et al.*[384] Moreover, several groups thoroughly investigated the role of other additives, such as ions, polymers or small molecules, which can act as stabilisers and/or as structure-directing agents, resulting in non-spherical metallic nanostructures (*e.g.*, nanorods, nanodecahedra, nanorings, nanoplates, nanowires, nanobelts, *etc.*).[390–394]

When an aqueous solution containing two different metal ions is irradiated by ultrasound, bimetallic NPs can be obtained. Interestingly, as a result of the difference in the reduction potential between the two different metal ions, core–shell NPs can be synthesised, consisting of a core of the metal with the higher reduction potential and a shell of the other metal. By exploiting this approach, AuPd, AuAg, AuRu and PtRu core–shell NPs have been reported in the literature.[395–398]

6.5.2 Metal Oxides

In addition to metal NPs, the sonochemical approach has been extensively studied for the synthesis of a variety of metal oxides. Similar considerations made for the sonochemical synthesis of metal NPs (*e.g.*, formation of nanosized products, possibly amorphous, consequences of carrying out the synthesis with volatile or non-volatile precursors and related reaction mechanism, *etc.*) also hold in this case.

When the sonochemical synthesis is carried out in *non-aqueous solution* starting from *volatile organometallic precursors* and in the presence of oxygen, amorphous, and possibly agglomerated, metal oxide NPs are synthesised. For example, by sonochemical decomposition of a solution of $Fe(CO)_5$ and $Ni(CO)_4$ in decalin under an oxygen pressure of 100–150 kPa, Shafi *et al.* prepared amorphous $NiFe_2O_4$ NPs.[399] In a similar way, amorphous barium hexaferrite NPs ($BaFe_{12}O_{19}$) were obtained starting from a solution of $Fe(CO)_5$ and barium ethylhexanoate in decane.[400] Notably, this synthetic method was also extended to the synthesis of hollow or porous oxides, such as Fe_3O_4 and Co_3O_4, obtained by ultrasonication of a non-aqueous mixture of the metal carboxyl precursor ($Fe(CO)_5$ and $Co_4(CO)_{12}$, respectively) and a sacrificial template, *e.g.*, carbon NPs or nanotubes.[401,402]

In addition, sonochemical synthesis in *non-aqueous solution*, starting from *non-volatile precursors* can be exploited as a promising alternative to synthesise in a controlled fashion metal oxides commonly obtained by sol–gel methods. Indeed, the use of an organic solvent without the addition of water was exploited by Ohayon and Gedanken[403] to avoid the major problems of the non-aqueous sol–gel approach, which is based on the hydrolysis of halide precursors, fast reactions and consequently amorphous products. Under sonochemical conditions the authors implemented the work pioneered by Niederberger *et al.*,[404] who employed benzyl alcohol as solvent for the solvothermal synthesis (also under microwave radiation) of a plethora of metal oxides, which took from hours to days to be complete. The sonochemical synthesis of nanosized metal oxides (TiO_2, WO_3 and V_2O_5) was carried out under an argon–hydrogen atmosphere using metal chloride precursors ($TiCl_4$, WCl_4 and $VOCl_3$) and benzyl alcohol, acting as solvent, ligand, and source of oxygen, and was completed within a few minutes. Thus, the sonochemical approach proposed by Ohayon and Gedanken offers a fast synthesis route at low temperature to different crystalline, nanosized oxides, meanwhile enabling the control of reaction parameters that are difficult to control in aqueous systems.

A typical sonochemical synthesis of metal oxides *in aqueous or water/ethanol solution* involves the sonication of a solution containing a *non-volatile metal precursor*, *e.g.*, a metal salt such as acetate or nitrate, and an oxygen source, such as air. As mentioned in Section 6.5, flushing oxygen into the reaction mixture favours an oxidising medium by promoting the conversion of reducing radicals into oxidising ones (eqn (6.7)).

Using the sonochemical approach, amorphous glassy oxides may be obtained, with much higher purity than by other methods, such as nanosized Cr_2O_3 and Mn_2O_3 reported by Dhas *et al.*,[405] but also well-crystallised and pure oxides, without needing a subsequent calcination step. One of the most extensively reported sonochemically synthesised crystalline metal oxides is titania, which can be successfully obtained *via* sol–gel reaction in water/ethanol solution under ambient conditions in a few hours (3–6 hours), as opposed to the longer time (several days) often required by conventional methods. Interestingly, by varying the titanium precursor (*i.e.*, titanium isopropoxide, $Ti(^iOPr)_4$, or titanium tetrachloride, $TiCl_4$), the reaction conditions (*e.g.*, acidic or basic environment, by adding HCl, acetic acid or amines) and the presence of any additive that may act as a template or structure-directing agent (*e.g.*, block copolymer, long-chain amine, *etc.*), different polymorphs of titania can be selectively synthesised (anatase, rutile, brookite and mixtures thereof),[406–412] even obtaining mesoporous frameworks[413,414] or hierarchically porous titania spheres.[415]

Another material whose sonochemical synthesis has been thoroughly investigated is nanostructured ZnO, which can be obtained in different morphologies by varying the synthetic conditions.[416–424] For example, Qian *et al.*[424] obtained ZnO nanocrystals with an average size of 6 nm and a narrow size distribution, without using any capping agent (Figure 6.23a). Subsequently, Jung *et al.*[416] demonstrated that nanospheres, nanorods, nanocups, nanodiscs and nanoflowers (Figure 6.23b) can be selectively prepared by modification of several parameters, such as the type and concentration of precursors ($Zn(NO_3)_2$ or $Zn(CH_3COO)_2$, 0.01–0.2 M in water), the kind of hydroxide anion-generating agents (hexamethylenetetramine or ammonia), the US time (30 min–2 hours) and the use of a capping agent (*e.g.*, triethyl citrate). The authors also carried out control experiments to determine the role of ultrasonication for producing shape-controlled nanocrystals, which confirmed that similar results cannot be obtained without the cavitation processes and thus the key function of US. In addition, porous ZnO[425] was also successfully obtained sonochemically, as well as doped ZnO, *e.g.*, doped

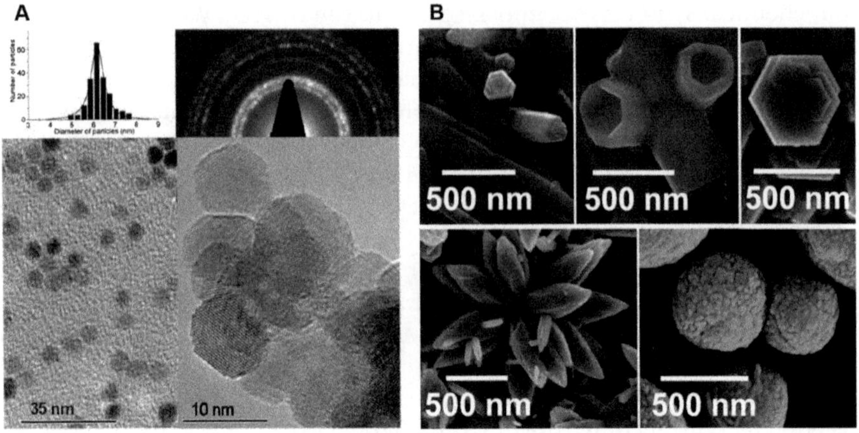

Figure 6.23 ZnO nanostructures sonochemically synthesised with different synthetic conditions. (A) TEM micrographs, particle size distribution and electron diffraction pattern of ZnO by Qian *et al.* Reproduced from ref. 424 with permission from the Royal Society of Chemistry. (B) SEM micrographs of ZnO with different morphologies (nanorods, nanocups, nanodiscs, nanoflowers, nanospheres) by Jung *et al.* Reproduced from ref. 416 with permission from American Chemical Society, Copyright 2008.

with europium or magnesium, to obtain photocatalytic or photoluminescent materials.[422,426]

Sonochemical synthesis has been exploited to synthesise not only transition metal oxides, but also rare earth ones, among which the most relevant example is ceria, which can be obtained as NPs[427] or nanorods[428–430] depending on the synthetic conditions and the specific additives employed (*e.g.*, ethylenediamine, tetraalkyl ammonium hydroxide, PEG, *etc.*). A plethora of other metal oxides, hydroxides and oxyhydroxides were prepared sonochemically, such as CuO,[431] Co_3O_4,[431] Fe_3O_4,[431,432] MoO_3,[433] V_2O_5,[434] In_2O_3,[435] α-$Ni(OH)_2$,[436] GaO(OH),[437] $In(OH)_3$[438] and even doped with lanthanide ions.[435] Mixed oxides may also be obtained: examples include ferrites (*e.g.*, $ZnFe_2O_4$ and $CoFe_2O_4$),[439–442] perovskites (*e.g.*, $SrTiO_3$ and $LaFeO_3$),[443,444] tungstates (*e.g.*, $PbWO_4$),[445,446] zinc aluminate and gallate doped with Mn^{2+} and lanthanide ions ($ZnAl_2O_4$, $ZnGa_2O_4$).[447]

6.5.3 Metal Chalcogenides

As mentioned in Section 6.5, a typical synthesis of metal chalcogenides by the sonochemical approach involves sonication of a solution containing a metal precursor and a chalcogen source, such as elemental sulfur, thiourea, thioacetamide or a thiosulfate for sulfides. The considerations mentioned in Section 6.5.1 about the effects of reaction parameters, and in particular of the reaction medium and the nature of the metal precursor (volatile or non-volatile), also hold for this class of compounds.

When the sonochemical synthesis of metal chalcogenides is carried out in *non-aqueous solution* starting from a *volatile organometallic precursor*, the primary stage of the process is most likely the decomposition of the volatile precursor in the gaseous phase of the cavitating bubble, which subsequently reacts with the chalcogen source. The first report was published by Mdleleni *et al.*,[448] who ultrasonicated $Mo(CO)_6$ and sulfur in isodurene under Ar, and obtained nanostructured MoS_2 exhibiting an unusual morphology, which differed dramatically from conventional MoS_2. In particular, while conventional MoS_2 is obtained as a layered, plate-like material, sonochemically prepared MoS_2 is constituted of spherical NPs with an average size of 15 nm. As a consequence, it was found to be characterised by a higher specific surface area and a higher catalytic activity for the hydrodesulfurisation of thiophene with respect to conventional MoS_2. Amorphous WS_2 was prepared by Nikitenko *et al.* by US irradiation of $W(CO)_6$ solution in diphenylmethane in the presence of an excess of sulfur.[449] Amorphous metal chalcogenides ME (M = Ca, Sr, Ba, Eu, Yb, Ni, Zn, Cd, Hg, Sn, Pb; E = S, Se, Te) were also prepared in a pressure vessel in chalcogen–ammonia solutions of elemental metals in the presence of ethylenediamine, which plays an important role in generating the active species (S^{2-}, Se^{2-}, Te^{2-}) and in complexing metal ions.[450,451] A subsequent heating treatment under vacuum was required to induce crystallinity in the amorphous products.

On the other hand, the sonochemical formation of metal chalcogenides *in aqueous or non-aqueous solution* starting from *non-volatile compounds* (*i.e.*, metal salts, such as acetates, nitrates and chlorides) is related to the formation of radical species. The *in situ* generated $H^{\bullet}$ radicals (eqn (6.3)) can reduce compounds containing C=S or C=Se bonds, such as thiourea, thioacetamide and selenourea, to produce H_2S or H_2Se, which subsequently react with metal cations to form metal chalcogenides. As in the case of the previously discussed classes of compounds, sonochemical synthesis in aqueous solution may yield both amorphous and crystalline products. For example, amorphous ruthenium sulfide[452] and indium sulfides[453] were synthesised starting from metal chloride precursors and using thiourea and thioacetamide as the sulfur source, respectively. Crystalline sulfides, even doped, were also obtained, such as ZnS,[454–461] CuS,[462–465] CdS,[466–470] PbS,[471–474] HgS,[473,475,476] SnS and SnS_2,[477,478] Bi_2S_3,[479,480] Sb_2S_3,[481,482] as well as different selenides and tellurides.[467,476,483–487]

The addition of organic compounds to *aqueous solutions* of *non-volatile metal precursors* was found to play an important role in the outcome of the reaction. For example, Wang *et al.*[480] carried out a systematic study of the sonochemical synthesis of bismuth sulfide, starting from bismuth nitrate, sodium thiosulfate or thioacetamide, and different complexing agents (*e.g.*, EDTA, triethanolamine and sodium tartrate). They found that nanorods with different diameters and lengths could be obtained by varying the complexing agent, and that, by adding DMF to the reaction mixture, smaller Bi_2S_3

rods were obtained with a higher yield. DMF was considered to accumulate at the bubble/solution interface and remove OH$^•$ radicals, thus increasing the sonochemical radical yield and accelerating the nucleation rate. An effect of the reaction mixture solvent on the obtained polymorph was also observed by Wang *et al.*,[488] who prepared hexagonal and cubic CdS nanocrystallites (3 nm) starting from $CdCl_2$ and $Na_2S_2O_3$, in aqueous solution and water/isopropyl alcohol solution, respectively. As mentioned above, isopropyl alcohol acts as a scavenger of oxidising OH$^•$ radicals, forming a secondary reducing radical that rapidly reacts with $[S_2O_3]^{2-}$ to form S^{2-} ions.

Hollow metal chalcogenide nanostructures may also be obtained *via* the sonochemical approach, by using proper templates (*e.g.*, surfactants, vescicles[489] or bacteria[460,490]) or *via* an *in situ* template route,[491,492] as in the case of the hollow CdSe spheres obtained by Zhu *et al.*, who ultrasonicated a mixture of $CdCl_2$, $NaSe_2O_3$ and ammonia. They reported that $CdCl_2$ was hydrolysed under the basic conditions to form $Cd(OH)_2$, which acts as an *in situ* template for the formation of hollow structures.[491]

The sonochemical synthesis of core–shell metal chalcogenide NPs has also been reported, for example, CdSe/ZnS quantum dots described by Murcia *et al.*[487] and the synthesis of CdS on ZnO nanorods or on SnO_2 nanobelts studied by Gao *et al.*[493,494]

6.5.4 Other Compounds

Sonochemical synthesis has also been exploited for the synthesis of other metal compounds, such as carbides,[495–497] nitrides,[496,498] hydrides,[499] fluorides[500–502] and also orthophosphates.[503]

Among these, the most fascinating ones are refractory metal carbides, which are generally prepared by reacting carbon and a metal at extreme temperatures, and their preparation as nanostructures with high porosity and large surface area still represents a challenge. The facile and low-temperature sonochemical synthesis of amorphous refractory Mo_2C and W_2C was reported by Hyeon *et al.* by ultrasonication of a slurry of $Mo(CO)_6$ or $W(CO)_6$ in hexadecane at 90 °C under argon.[497] Amorphous oxycarbides were first obtained, and subsequently the oxygen was eliminated by heat treatment under a 1:1 methane/hydrogen mixture. The obtained products were a porous aggregate of NPs with high specific surface area, with an excellent catalytic activity for dehydrogenation and hydrodehalogenation, comparable to that of platinum group metals.[495] Titanium carbide nanopowder was also obtained sonochemically starting from TiO_2 in the presence of activated carbon.[496]

Metal nitrides can also be sonochemically synthesised starting from aqueous suspensions of oxides, in the presence of ammonia, such as Si_3N_4 reported by Sivasankaran and Kishor Kumar.[496] Alternatively, they can be obtained by sonication of a metal carbonyl solution under a reductive gas mixture of ammonia and hydrogen at 0 °C.[498]

Another interesting example is the synthesis of hydrides, which may be obtained by exploiting H$^•$ radicals formed by sonolysis: for example, CuH was obtained by sonication of diluted Cu(II) aqueous solutions under argon atmosphere by Hasin and Wu.[499] Finally, it is worth mentioning the sonochemical synthesis of a plethora of lanthanide orthophosphates $LnPO_4$ (Ln = La, Ce, Pr, Nd, Sm, Eu, Gd, Tb, Dy, Ho) reported by Yu *et al.* by ultrasonication of aqueous solutions of $Ln(NO_3)_3$ and H_3PO_4.[503]

6.6 Radiolysis

The first use of ionising radiation to generate solid inorganic particles (*radiolysis*) was proposed by Schiller and Ebert in the late 1970s.[504] Following their approach, the development of this synthetic procedure grew exponentially with the technological advancements of the last decades and, nowadays, radiation-based protocols can be employed for the preparation of a wide variety of inorganic materials.[505,506] This methodology is based on the use of ionising radiation to prepare NPs in suspensions and these syntheses are often carried out in water as the reaction medium. Radiolysis of the solvent to generate highly reactive species is the first underlying step of radiolytic synthetic protocols, followed by a cascade of reactions between the generated intermediates and the dissolved precursors, which is analogous to the sonochemical process, where sonolysis of water is induced by cavitation and initiates redox and radical reactions (Section 6.5).[505,507] Modified radiation protocols can also be adopted for industrial synthetic applications, as well as for industrial post-processing (*e.g.*, water after-treatment systems).[505] With radiation-induced formation of a solid phase in water, the only drawback is the possible presence of unreacted precursors, besides the intrinsic hazard related to the high energy beams involved.

The most used type of radiation in this synthetic approach is γ radiation, as well as high energy, short wavelength electromagnetic radiation (*e.g.*, UV, X-ray), while accelerated particle beams of electrons and/or ions from synchrotron radiation sources are sometimes used.[508] Such radiation is ionising (*i.e.*, it possesses sufficient energy to extract electrons from atoms, thereby creating ions), as it encompasses electromagnetic waves, such as γ and X-rays, or high-energy particles, such as α and β particles. In particular, α particles generally possess an energy of about 4 to 8 MeV with few exceptions (for example, 2 MeV for Nd-144 or 11.7 for Po-212), whereas β particle, γ-ray and X-ray energies generally range from 5 to 1000 keV, above 100 keV (generally up to 8 MeV) and below 100 keV, respectively.[509] γ-Ray radiation is the most widely used electromagnetic (EM) energy source in NP formation.[510,511] The γ-rays used in radiolysis are generated almost exclusively from ^{60}Coγ sources, although there are also a few examples where they are produced by ^{137}Cs.[511] When ^{60}Co decays to ^{60}Ni through β decay, the process releases two γ-ray photons, generating the necessary radiation to perform radiolytic reactions. Because γ irradiation occurs through decay, it is a very convenient radiolysis technique requiring no specialised instrumentation other than that dedicated to the shielding the radiation.[510] X-rays on the other hand are much less energetic, with an energy range of about

0.1–120 keV.[509] X-ray radiation offers the same advantages as γ radiation when employed in radiolysis experiments; however, the instrumentation in this case is more complex as the radiation field is much smaller and more difficult to tune and control. The X-rays that are generally employed for NP synthesis are generated from synchrotron X-ray sources,[512,513] and only in a few cases from laboratory sources.[514] For example, a 2.5 GeV synchrotron X-ray source was used by Lee *et al.* with a white beam confined to 0.5×0.5 cm^2, to perform the synthesis of Ni NPs.[515] Finally, UV irradiation, the lowest ionising energy EM technique, can also be employed for preparation of NPs.[516] Examples of UV-generated colloids extend from monometallic[517,518] to even alloy NP formation.[519]

In a typical radiolytic synthesis, the ionising radiation is applied to generate, upon interaction with the solvent, transient and highly unstable intermediates, such as strong oxidising agents like $OH^{\bullet}$ or other reactive oxygen species, as well as strong reducing species, such as hydron or hydronium ions or solvated electron, $e_{(solv)}^{-}$, capable of reacting with most of the metal salts dissolved in the solution.[508,510,520–522]

$$H_2O \rightarrow e^-_{(solv)}, H^{\cdot}, OH^{\cdot}, H_2, H_2O_2, H_3O^+, OH^-, HO_2 \quad (6.17)$$

The electrons generated by the radiochemical method are referred to as solvated electrons, and are homogeneously distributed in the medium and can readily react with solvated metal ions, thus decreasing their oxidation state.[523] If not reacting directly with the dissolved precursors, inter-reactions between these transient intermediates or with the surrounding solvent end up generating more stable molecules, such as hydrogen or hydrogen peroxide.[505,508,524] Radical scavengers, such as formates ($HCOO^-$) or secondary alcohols (*e.g.*, 2-propanol), are generally used to favour the reduction of the metal ions used as precursors and to counterbalance the $OH^{\bullet}$ radicals (strongly oxidant) that are generated concurrently.[523,525] These scavenger molecules interact with the radiation-produced radicals undergoing the following reaction and generating the strongly reducing carbon dioxide ion-radical $COO^{\cdot-}$.[526]

$$HCOO^- + OH^{\cdot} / H^{\cdot} \rightarrow COO^{\cdot-} + H_2O / H_2 \quad (6.18)$$

or the reducing 2-hydroxyprop-2-yl radical $(CH_3)_2COH^{\bullet}$, when 2-propanol is used instead of formate.[526] Buxton *et al.* revealed that the reaction of 2-propanol with $OH^{\bullet}$/ $H^{\bullet}$ proceeds *via* hydrogen abstraction, preferably in the α-site, as described in the following equation[521]

$$(CH_3)_2CHOH + OH^{\cdot} / H^{\cdot} \rightarrow (CH_3)_2COH^{\cdot} + CH_2CHOHCH_3 + H_2O / H_2 \quad (6.19)$$

The amount of radiolytic intermediates generated in the solvent depends on the absorbed radiation energy (dose, energy absorbed per mass unit, Gy or rad), and their generation stops immediately when the reaction system is removed from the radiation field. Therefore, one can finely control the conversion yield of these highly reactive species by controlling the irradiation parameters (*i.e.*, time, energy, radiation selection). To quantify the radiation-induced yield of a radiolytic reaction (*e.g.*, how much of a specific ion is produced upon absorption of ionising radiation), the *G*-value (radiation chemical yield) is normally used. The *G*-value is defined as the number of moles of a given species produced or consumed per absorbed Joule of radiation energy ($mol\ J^{-1}$).[527] Some examples of high-yield species are the solvated electrons ($e_{(solv)}^-$), the hydronium radicals ($OH^•$) and the hydron ion (H^+), with *G*-values of about 2.7, 2.86 and 2.7 $mol\ J^{-1}$, respectively. Low-yield species are the X radical ($H^•$), hydrogen peroxide (H_2O_2) and molecular hydrogen (H_2), with *G*-values of about 0.61, 0.61 and 0.43, respectively.[528]

Alongside the ability to ionise the medium, penetration depth is one of the critical parameters of the interaction between ionising radiation and matter. In general, γ photons can penetrate deep into materials depending on their density, as the lack of charge eliminates Coulomb interactions and allows deeper penetration. For example, the half-value thickness is about 13 cm for 1.33 MeV ^{60}Co sources, as calculated from the mass attenuation coefficient using the attenuation law.[529] Therefore, γ radiation is useful when processing larger volumes and thicker samples.[509]

The redox potential of the hydroxyl radical $OH^•$ ($E° = +2.73\ V_{NHE}$)[526] makes it highly oxidising, whereas the $H^•$ radicals are strongly reducing species, with their redox potential of $E°$ ($H^+/H^•$) = $-2.3\ V_{NHE}$, second only to the solvated electrons that possess a redox potential $E°$ (H_2O/e_{aq}^-) = $-2.87\ V_{NHE}$, and ensure spontaneous reactions with most of the metal salts used as precursors in solution.[523] It is necessary to remove oxygen from the system by keeping the system under vacuum or by flushing inert gases so as to be able to form hydroxyl and hydrogen radicals, whereas hydroperoxyl radicals ($HO_2^•$) form in the presence of oxygen.[520]

Using this radiolytic approach, metals, alloys, core–shell systems, metal oxides, carbonates and sulfides can be prepared, and many of these compounds can be synthesised with unusual sizes tuned in the nanometre range.[505] Upon irradiation of precursor-containing aqueous solutions at room temperature with γ-rays or accelerated electrons, particles with attractive properties, such as narrow size distributions, are obtained.[505,506] In certain cases, narrower distributions could also be achieved when applying a radiation method compared with other synthetic protocols, as described by Li *et al.*[530] Indeed, this synthetic approach displays different advantages over the common chemical synthetic methods. In the first place there is the independence from the reaction temperature, as high purity materials are obtained despite the starting experimental conditions.[505] Moreover, no additional ingredients/additives are required in the reaction mixture except for the desired precursors (*e.g.*, no need for the presence of ligands). However, the morphology can be fine-tuned with the presence of templating agents, stabilisers and radical scavengers by controlling the dose of radiation, the radiation rate and the starting composition of the irradiated solutions.[523] Overall, the NPs features are mainly controlled by: (1) the starting precursors and their concentration; (2) the solution pH; (3) the presence of such templating agents and/or surfactants; and (4) the characteristics of the incoming radiation (*e.g.*, overall energy, dose and dose rate of the given radiation). Tuning the radiation dose rate has proved to be particularly useful, as the dose rate is linked to the amount of the generated reactive intermediates, which governs the complex nucleation–growth processes.[531] Different irradiation doses are required to form NPs, depending on the specific metal, and this affects not only the concentration of NPs in suspension, but also their size: when increasing the irradiation dose, the number of NPs increases while the size decreases. Adjusting the pH of the initial solution, usually by adding NaOH into the system, often promotes reduction processes.[505] With respect to time, a typical synthesis lasts dozens of minutes in the case of γ-ray irradiation (with dose rates in the range of a few kGy h^{-1}) and just few minutes in the case of high-energy electron-beam irradiation. Recently, Korolkov *et al.* and Gharibshahi *et al.* independently observed this behaviour, synthesising Cu and Pt NPs: in the case of Cu NPs, the average particle size was about 70 nm with calculated crystallite sizes from XRD analyses between 56 and 36 nm on changing the irradiation doses from 46 to 250 kGy, and for the Pt NPs, a decrease in their size from 4.4 to 2.8 nm was observed with doses of 80 and 120 kGy, respectively, obtaining in both cases Gaussian size distributions (2σ of about 4 nm).[532,533] As a rule of thumb, a high dose rate favours the formation of many different nucleation centres in solution, leading to the formation of small NPs with a narrow particle size distribution, whereas, at a low dose rate, larger NPs are usually observed.[510]

To sum up, the radiolysis technique is mainly used for reduction of metal ions and it offers several advantages, including: (1) no use of chemical reducing agents or generation of undesired oxidation products compared to other approaches, in which other reagents are needed to initiate the oxidation or reduction process; (2) a well-known rate of reaction for a given system (since the number of reducing or oxidising equivalents generated by radiation is well defined and dependent upon the dose applied); (3) direct adsorption of the radiation energy by the solvent without interference from light-adsorbing solutes and products; and (4) quasi-instantaneous (~100 ns) formation of reactive species in the solution, with the possibility of ending the reaction by switching off the radiation source.[503] For these reasons, radiolysis is recognised as a mild chemical synthetic method because it usually proceeds at room temperature and because the reacting species can be generated at a very low rate inside the irradiated sample. Typically, with a γ-ray source, the rate of formation of reducing species can be continuously tuned from a range of μmol L^{-1} h^{-1} to mmol L^{-1} h^{-1}, with the values dependent on the variation of the radiation energy and intensity.[510]

Researchers are now exploring the formation mechanisms of nanostructures during radiolysis with advanced theoretical studies,[519,534,535] by performing pulse radiolysis experiments to elucidate the mechanisms of reactions at very short time scales. The following section (Section 6.6.1) will nevertheless be focused on experimental studies and a few examples are reported on the resulting metals and metal alloys that have been prepared to date. Radiolysis

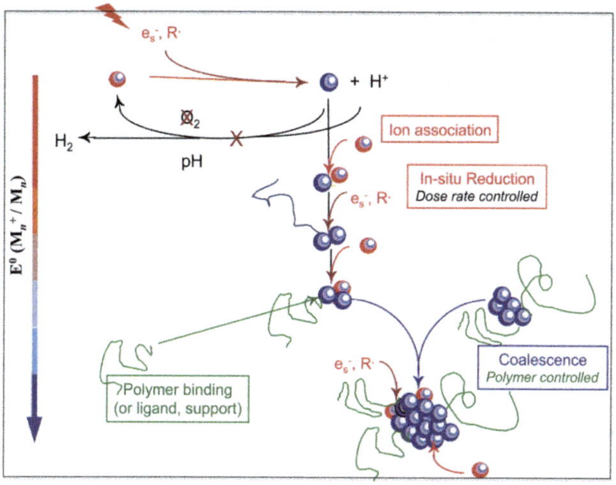

Figure 6.24 Schematic representation of the processes occurring during the reduction of metal ions (in red), by solvated electrons and free radicals as induced by radiolysis, into metal atoms (in blue). The ions may be free in solution or associated with atoms and clusters, where they are reduced *in situ*. The growth into NPs by coalescence of atoms, then clusters, is controlled by stabilisers, in contrast to the growth by successive units resulting from the association/reduction route, which is not. The arrow on the left indicates that the reduction potential of single atoms is very negative and that the potential increases with the nuclearity of the growing particles. Reproduced from ref. 520 with permission from Elsevier, Copyright 2020.

can be used to form a wide range of mono- and multimetallic particles not only from noble metals, but also from transition metals and group 3–5 metals.[510] The mechanisms of NP nucleation and growth promoted by radiolytic methods have been extensively studied by Belloni *et al.* In Figure 6.24, arrows and change of colour represent the variation in the redox potential during the first phases of cluster generation.[520]

6.6.1 Metals and Alloys

Radiolysis is broadly applicable for the synthesis of zero-valent NPs of numerous metals, including d^{10} (Zn, Cd, Hg), d^9 (Cu, Ag, Au), d^8 (Ni, Pd, Pt) and d^7 (Co, Rh, Ir) metals, as well as Mo, Ru, In, Sn, Sb, Os, Tl, Pb and Bi.[511] Some application examples of the generated NPs encompass their use as antibacterial agents (such as Ag)[536,537] and catalysts (such as Cu, Ni or Pt).[537–541]

Among others, the synthesis of Ag NPs is often regarded as a benchmark model system for the synthesis of NPs, because they are easily prepared under a broad range of synthetic conditions.[542] For this reason, Ag NP synthesis by γ-ray irradiation is widely reported in the literature.[543,544] Soroka *et al.* prepared supported Ag NPs of different sizes by exposing suspensions containing a carbon support and $AgNO_3$ as metal salt to γ radiation.[545] Variation of the synthetic parameters, such as surfactant/precursor concentrations and radiation dose, enabled fine-tuning of the size of the Ag NPs (10–40 nm) and of the deposited amount on the carbon support (from 1.8 wt% to 6.7 wt%). Larger Ag NPs were reported by Bogle *et al.*, who prepared size-controlled particles (10–60 nm) in PVA solutions using 6 MeV electrons, tuning their size up to about 100–200 nm without stabiliser by keeping the electron fluence (energy per area unit) constant.[544] Soroushian *et al.*[546] prepared Ag NPs *via* γ radiolysis of Ag salts in ethylene glycol solutions. The proposed ionisation mechanism is illustrated in Figure 6.25.

Over the years, the irradiation method has been adopted for the synthesis of other metals: with Cu,[547,548] Ru,[549] Pt,[550,551] Au,[530,552–554] In[555] and Sn[556] NPs prepared with synthetic procedures that often resemble those described for Ag. For example, Au NPs were prepared *via* radiolytic reduction of Au(III) salts by γ irradiation using chitosan as stabilising agent.[557] Size-controlled (2.5–35 nm) Pt NPs were prepared by irradiating aqueous solutions of hexachloroplatinic acid, poly(acrylic acid) and 2-propanol, with the pH adjusted to 4.5 by addition of aliquots of NaOH.[550] To stabilise these different metal NPs, a series of ligands were tested, such as SDS,[558] PVA, chitosan, sodium poly(vinylsulfate), poly(acrylamide) or poly(*N*-methylacrylamide), poly(ethyleneimine), bovine serum albumin protein, glycerol, gelatine, gum acacia, honey or PVP.[505,542] The level of stabilisation also depends on the medium, and various environments have been explored, such as water only, water in oil microemulsions or (reverse) micelles.[505]

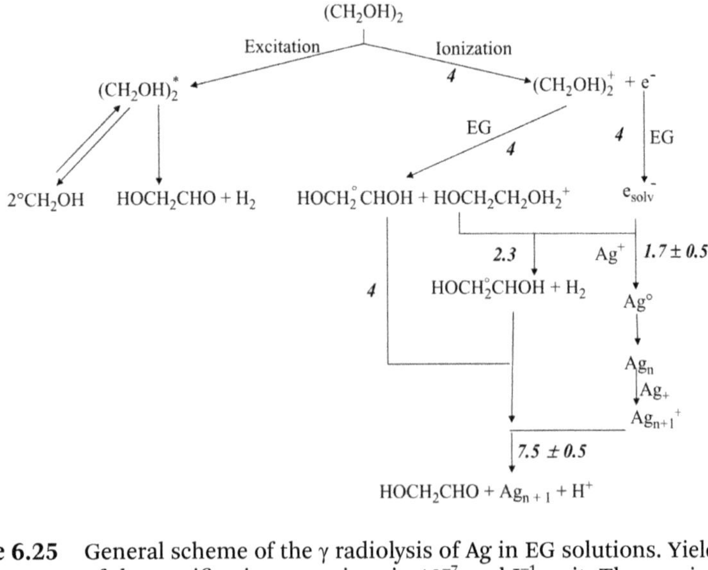

Figure 6.25 General scheme of the γ radiolysis of Ag in EG solutions. Yield values of the ramifications are given in 10^{-7} mol J^{-1} unit. The species issued from the excited states do not seem to contribute to the reduction of the silver ions. Reproduced from ref. 546 with permission from Elsevier, Copyright 2005.

Different sources of radiation have been used to pursue NP formation of the same metals: X-ray irradiation was used for the preparation of monometallic Au and Ni NPs. Yamaguchi *et al.* confirmed that synchrotron X-ray irradiation enabled the formation of hydrogen and hydroxyl radicals from water, which yielded Au NPs.[559] In addition, the same group demonstrated that synchrotron X-ray irradiation can also induce the formation of Fe or Cu particles.[560] Similarly, Wang *et al.* prepared Au colloids at room temperature: in particular, they investigated the effect of process parameters, such as the pH value and the irradiation exposure time, on the structure of the resulting Au NPs.[512] In particular, particles obtained at acidic pH values (5.3) showed aggregation, whereas in neutral solutions and at basic pH (pH = 10) particles were stable and presented an increasingly narrower size distribution, with dimensions around 10–20 nm. Synchrotron X-ray irradiation was also exploited to generate Au NPs with two types of morphologies, dendrites and spheres, formed through the reducing action of hydrated electrons from an aqueous solution of chloro(2,2′,2″-terpyridine)gold(III) ions.[561] With a different setup, Toro-González *et al.* recently evaluated the formation of Au NPs in a continuous-flow millifluidic reactor. The Au NPs were formed after X-ray irradiation (127 Gy min^{-1}) of a mixture of dissolved Au^{3+}, branched polyethyleneimine and ethylene glycol inside the setup, yielding highly monodispersed nanostructures with a mean size of about 3 nm.[562]

Ionising UV radiation was used by Yang *et al.* to study the influence of pH values on the growth of Au NPs.[563] Interestingly, they proposed that an increase in pH value could decrease the particle size and the size polydispersity, while keeping constant the concentration of the capping ligand used in the reaction (citric acid). Again, with UV radiation produced by a 450 W Hg lamp, Ag NPs with peculiar shapes were prepared by Navaladian *et al.* In particular, nanorods and nanospheres were synthesised by modifying the weight ratio of the precursors.[564]

Radiation methods have also been used for the preparation of alloy and core–shell NPs, especially those consisting of noble metals.[505,510,511,520,531,565–568] In a comprehensive review, Čubová and Čuba recently surveyed the synthesis of such multimetallic NPs.[505] Also, in this case, the literature examples are predominantly focused on the generation of precious metal alloys. Wang *et al.*, for example, synthesised colloidal suspensions of AuPt alloyed NPs with a one-pot procedure based on synchrotron X-ray irradiation, using PEG (polyethylene glycol) as solvent, stabiliser and additional reducing agent.[567] To prepare AuPt alloys, a high X-ray flux density irradiation from a synchrotron was needed (>10^{14} photons mm^2 per s). Synthesis of alloyed AuPt NPs formed within a poly(methyl methacrylate) matrix was also reported using a 10 W Hg lamp to produce 254 nm deep-UV radiation, which induced alloy formation.[568] These systems can be formed either as core–shell segregated nanoalloys, subcluster segregated nanoalloys, mixed order or mixed random nanoalloys and multi-shell nanoalloys, or as core–shell NPs.[569] When preparing alloys starting from a given precursor mixture, it is possible to control the metal composition by varying the irradiation dose rate, thereby generating various types of NPs. By irradiating a bimetallic precursors solution with lower irradiation dose rates, particles with a core–shell structure are generally obtained if the difference of the redox potentials between the two metals is sufficiently large. If the irradiation rate is higher, a mixture of monometallic particles is more likely to be produced, as a high dose rate favours the formation of numerous nucleation centres.[510] To invert the core–shell structure, a two-step reaction procedure can be applied: first, a solution containing one metal salt is initially irradiated to prepare the core, and then a second metal salt is added to the reaction and the irradiation is repeated.[510]

With increasing dose rate, or energy, thermodynamic barriers to phase formation may be overcome, allowing alloying of species that are otherwise hard to form when employing lower energy sources or through wet-chemistry techniques.[520]

In Figure 6.26, the influence of radiation dose rate is illustrated for the different type of accessible alloys and core–shells structures with different precious metal combinations.

6.6.2 Metal Oxides

Metal oxide and hydroxides NPs can be prepared *via* radiolytic reaction of metal ions with high oxidation state in aqueous solution, also without the use of a polymeric stabiliser.[523,566,570] The synthesis procedure may proceed: (1) by irradiation of precursors in solid phase or aqueous suspension; (2) by formation of a colloid or aggregated/agglomerated particles from the precursors solution under irradiation; (3) by deposition of oxides on a solid

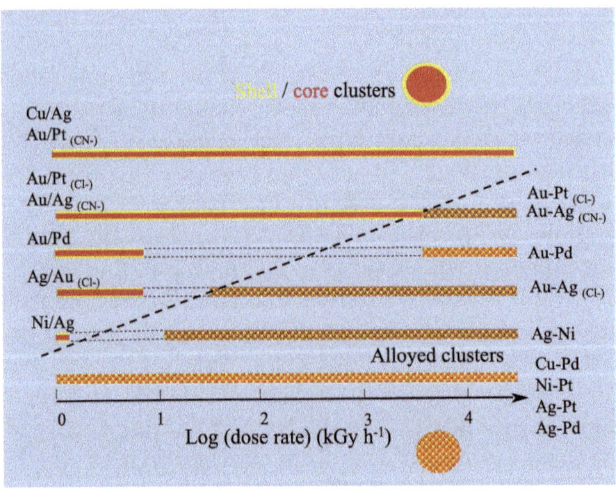

Figure 6.26 Dose rate influence on the shell–core or alloyed structure of radiation-induced metal NPs in mixed solutions of metal ions. Reproduced from ref. 520 with permission from Elsevier, Copyright 2020.

surface under irradiation; or (4) by the irradiation of aqueous sol followed by formation of a gel.[505] The water radiolysis products range from strong oxidising agents, such as the hydroxyl radical, to strong reducing agents, such as solvated electrons. Therefore, the overall redox conditions of the solution should be shifted from reducing to oxidising ones by adding the proper radical scavenger before the irradiation process. For the production of metal oxides in aqueous systems, the reducing solvated radicals should be scavenged: N_2O is commonly used to convert these reducing species, such as solvated electrons to hydroxyl radicals, according to the following reaction:

$$N_2O + e^-_{aq} + H_2O \rightarrow OH^{\cdot} + OH^- + N_2 \tag{6.20}$$

Once high amounts of hydroxyl radicals are present is solution, the reaction can proceed according to:

$$M^{z+} + OH^{\cdot} \rightarrow M^{(z+1)+} + OH^- \tag{6.21}$$

The process of preparing metal oxides from metal ions is generally based on two reaction steps: (1) the change in metal oxidation state (upon radiation oxidation); and (2) the precipitation of the generated insoluble compounds.[505] Sometimes, reaction with intermediate species such as OH^- may lead to the formation of thermodynamically unstable hydroxides, which subsequently condense to the respective oxide:[505]

$$2M^+ + 2OH^{\cdot} \rightarrow 2[MOH] \rightarrow M_2O + H_2O \tag{6.22}$$

Metals, such as Fe, Cu, Zn, Ni and Co, can react with dissolved oxygen in solution to form metal oxides:[505]

$$M^0 + {}^{x}\!/_{2}O_2 \rightarrow MO_x \tag{6.23}$$

where the stoichiometric coefficient x depends on the stability of the respective metal oxides: *e.g.*, oxidation of Cu^0 can produce $CuO_{0.67}$, but also a defect structure of Cu_2O, as observed by Yanase and Komiyama.[571] CuO and Cu_2O NPs were the prevailing products generated starting from aqueous $CuSO_4$ solutions by Hai *et al.*, whereas with longer irradiation exposure Cu NPs were the sole significant product.[572] Recently, Li *et al.* reported the radiolytic synthesis of Cu_2O, MnO_2 and CeO_2.[573–575] For the synthesis of these materials, Pourbaix diagrams show clearly that only specific ranges of pH values allow the formation of thermodynamically stable phases for the M-H_2O systems. For example, the pH range where Cu_2O can be formed directly from reducing Cu^{2+} in solution ranges from 3.5 to 5. In Figure 6.27, the co-presence of the two phases is highlighted, and the size of the particles obtained by using SDS is reported.[573]

Oxides of lower oxidation states may also be formed directly from higher oxidation state salts by *radiation-induced reduction*, when the metal occurs as an oxycation in the solution.[505] This is the case, for example, for UO_2 NPs obtained by reduction of $[UO_2]^{2+}$.[576] With very few examples, it is also possible

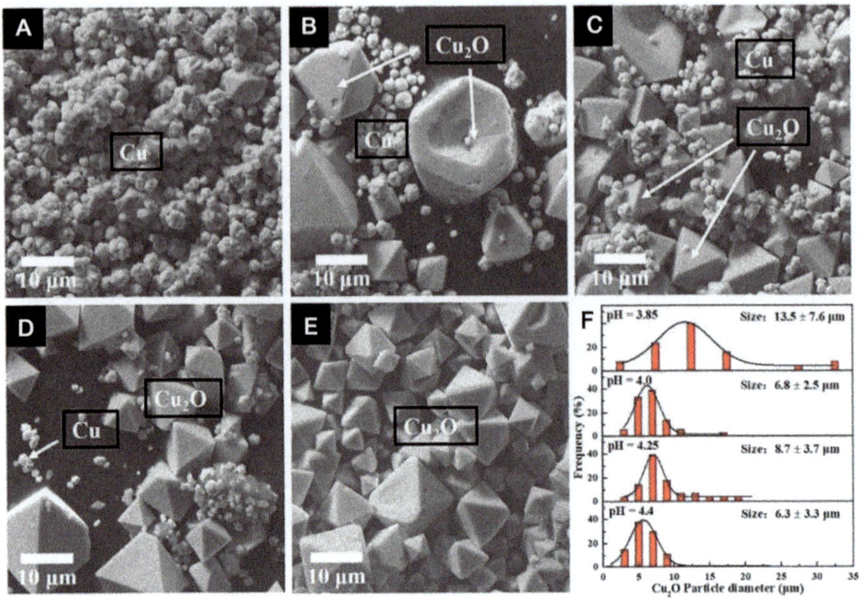

Figure 6.27 SEM images of Cu_2O synthesised particles prepared by Li *et al.* through γ radiation at different pH values of (A) 3.75 (B), 3.85 (C), 4.0 (D), 4.25 (E) and 4.4 (F), and the particle size distributions at different pH values. Reproduced from ref. 573, https://doi.org/10.1039/C8DT02916D, under the terms of the CC BY 3.0 license https://creativecommons.org/licenses/by/3.0/.

to utilise a *radiolytic oxidation route* to increase the oxidation state, for example in the case of MnO_2 NPs by oxidation of Mn^{2+}, which is otherwise produced by reduction of $[MnO_4]^-$.[577] Similarly, the range of pH values necessary for the formation of the thermodynamically stable manganese oxide phases MnO_2, Mn_2O_3 and Mn_3O_4, are pH < 6, 6 < pH < 6.4 and 6.6 < pH < 8.8, respectively. Similarly, CeO_2, can be produced by oxidation of Ce^{3+} in a pH range from about 0.3 to 7.5. NPs of oxi-hydroxi Fe(III) were obtained by Yakabuskie *et al.*, by exposing $FeSO_4$ solutions to γ irradiation, and led to the formation of FeO(OH) NPs.[578] The size distribution of these particles was controlled by surface reactions with radiolytic oxidants and reductants. For example, Alrehaily *et al.* reported that NPs of Co_3O_4 can be synthesised from air-saturated $CoSO_4$ solutions under alkaline conditions *via* an oxidative pathway, observing the formation of NPs at pH 10.6, with a NP size ranging from 8 to 20 nm. The oxidation started with dissolved $CoSO_4$, evolving from $Co(OH)_2$ to CoO(OH) and then finally to Co_3O_4.[579] Recently, it was proposed that chromium reduction occurs through a series of sequential steps (Figure 6.28): (1) reduction of dissolved $Cr^{VI}_{(aq)}$ to $Cr^{III}_{(aq)}$ and subsequent condensation of $Cr(OH)_3$ enhance the co-precipitation of Cr^{VI} and Cr^{III} in the mixed Cr^{VI}/Cr^{III} oxide/hydroxide NPs; (2) steady-state reduction is reached but the adsorption of Cr^{VI} continues on the Cr^{VI}/Cr^{III} oxide/hydroxide particles; and (3) a solid-state

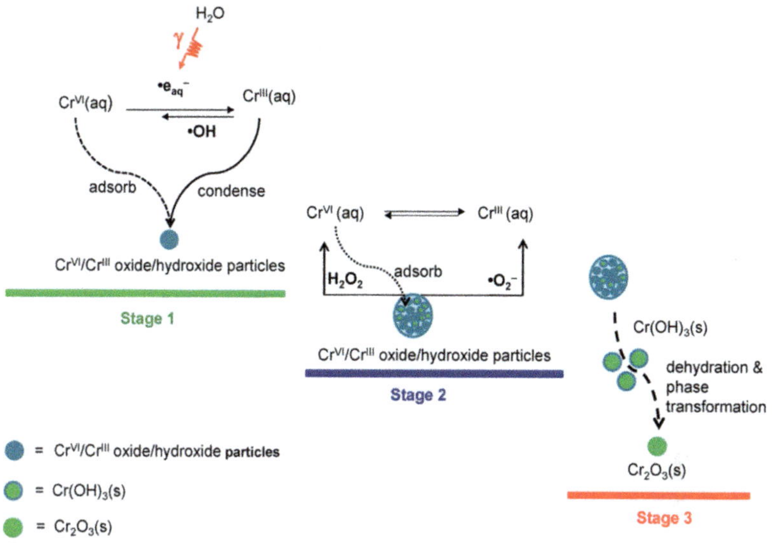

Figure 6.28 Schematic of a mechanism for radiation-induced formation of chromium oxide NPs. Reproduced from ref. 580 with permission from American Chemical Society, Copyright 2015.

conversion of Cr^{VI} to Cr^{III} occurs, with condensed particles converting to $Cr(OH)_3$ and then to Cr_2O_3. Adding a radical scavenger during the process was found to affect the kinetics of particle formation, the growth at different stages and consequently the final size of the NPs (Figure 6.28).[580]

Other radiations have also been used. Praus *et al.* used UV irradiation with a fixed wavelength at 254 nm to precipitate ZnO NPs with dimensions of about 17 nm, consisting of crystallites of about 7 nm.[581] Čuba *et al.* comprehensively described the formation of Ni, Zn, Y, Al and Co oxides, as well as other peroxides and hydroxides, *via* irradiation of aqueous solutions.[582]

6.6.3 Other Compounds

Besides metal and metal oxide NPs, other compounds, such as metal sulfides have been prepared using the radiolytic procedure. Produced mostly *via* γ irradiation of a suitable precursor solution, metal sulfides find applications in photoluminescent, photoelectric, non-linear optic materials and in semiconductors.[583] A relevant example is CdS prepared by Hayes *et al.* from a solution containing Cd and thiol (3-mercapto-1.2-propanediol).[584] Another important example is Cu_2S, the subject of numerous studies for its biological, chemical, optical and electrical properties, as well as for applications in various fields.[585,586] Cu_2S NPs were prepared by Bekhit *et al.* with unusual nanotube shapes, *via* γ radiolysis without adding any capping or additional reducing agents aside from those generated upon irradiation.[537] Chen *et al.* prepared CdS, PbS and Ag_2S with a γ-irradiation technique in ethylenediamine, showing that the radiolysis synthesis of these compounds is possible and that control over the formation of metal sulfides depends instead on independent parameters, such as stability constants of the ethylenediamine chelated metals, the solubility product constants of the metal sulfides and the standard electrode potentials of the metal ions.[587]

6.7 Laser Ablation

Laser-assisted synthesis of colloidal NPs has proven its versatility in design and control of the product's properties, as highlighted by recent reviews[588–594] and books.[595,596] The first reports date back to the early 1990s[597] and, since then, there has been an increase in interest both in the academic world and in industrial production, with companies having invested in this technology for more than a decade.[598] To date, a very wide portfolio of highly pure compounds have been synthesised by *laser ablation in liquid* (LAL), including metallic NPs, oxides, hydroxides and complex systems, such as core–shell, supported structures and composites. Furthermore, carbon-based nanostructures have also received considerable attention,[599–602] but are not addressed here since the focus is on inorganic nanostructures. Notably, all these different compositions do not require a setup change and can all be synthesised without applying external heating or pressure. LAL can be considered as a hybrid top-down and bottom-up technique, as the starting material is a macroscopic target and NPs are a built-up from the initially formed atoms and clusters.[603] Finally, LAL is almost fully compatible with the 12 principles of Green Chemistry: it does not produce pollutant waste, and is safe for the operator, since there are no volatile components, and it can even be controlled remotely, as has recently been demonstrated.[604] The absence of waste and purification steps combined with room temperature and pressure processing conditions are especially appealing in a production context. Especially at the laboratory or pilot plant scale, this approach allows easy scanning of a wide variety of materials (with no setup changes required) all of which take only a few minutes to be synthesised. Improvements in the instrumental setup in recent years have led to a portfolio of achievable sizes within the same system, from 1 nm up to the micrometre scale, with a very narrow, monomodal size distribution.[590] Finally, production upscaling, up to grams per hour, has already been established, even for NPs with very complex compositions.[605–607]

Beyond LAL, laser ablation in liquids can be applied in two other major branches: *laser fragmentation in liquids* (LFL) and *laser melting in liquids* (LML).[593] The first refers to the fragmentation of NPs due to absorption of the radiation already present in the system, while the second refers to melting and increase in the aggregate dimensions of the produced NPs by using a subsequent laser pulse. In the latter case, the second laser pulse can also be used to reshape the original NPs, through an isochoric process. Moreover, a fourth category has recently been introduced, called *laser defect engineering in liquids* (LDL), and has been developed to tune the atomic structure of

NPs without affecting the overall dimensions. Since the main targets of LFL, LML and LDL are NPs in suspension rather than bulk samples (as for LAL), these techniques can either be used separately from LAL in an after treatment of the sample or directly on the LAL products. In the following sections, the focus will be on LAL, since it is the most prominent process to produce NPs,[588] and the fundamentals of the method, as well as the influence of the laser and system parameters, will be discussed, with several examples from the recent literature.

6.7.1 Basic Mechanism of Laser Ablation in Liquid

The introduction of a liquid environment provides many advantages compared to ablation in other media (*e.g.*, air or vacuum), such as the natural dispersion of the NPs in a confined volume and the wide range of nanomaterials also accessible by reaction with the solvent.[608] The liquid must be transparent to the radiation at the chosen wavelength (or at least its absorption should be negligible during NP formation), to deliver almost all the energy only to the target.[609] Moreover, the fluence (energy per unit area) must be controlled to avoid solvent breakdown, which is particularly challenging for ultrafast pulses (of the order of femtoseconds).[610] UV–vis or NIR are common wavelengths used, coupled with a fluence of 0.1–100 J cm^{-2} and a pulse duration that can vary from femto- and picoseconds to milliseconds, or can even stretch to continuous-wave lasers.[611] Instantaneous multiphoton absorption phenomena prevail when the pulse is less than a few ps, leading to direct photoionisation of the target.[612] For longer pulses (*i.e.*, above ps), the time is comparable to the electron–lattice thermalisation and the electron kinetic energy can be released as thermal energy to the lattice.[613,614] Therefore, side processes, such as melting or boiling of the target, may occur at the same time. In Figure 6.29 the overall process for different pulse durations is illustrated.

After about one ps, the material starts to be ablated, as a result of the local space–charge separation fields induced by the photoionisation.[615] The detached target is called a ‘plasma plume’ and consists of highly ionised or atomic species with a density of about 10^{20} cm^{-1} at 10^3 K and *ca.* 10^{10} Pa and it can last for several tens of ns, depending on the laser parameters and on the liquid composition.[616] For longer pulses (more than tens of ps), the plasma plume and the laser pulse can coexist, reducing the fluence at the target level due to the plasma absorption. Afterwards, the plasma plume collapses and transfers the energy to the surrounding liquid environment, producing a vapour layer of the same size.[611] The recoil pressure of the ablated material generates a shockwave, which propagates at supersonic speed, heating the liquid and which may promote further detachment of material.[617] Notably, the presence of a liquid environment confines the expanding plasma plume into a smaller volume than the analogous process in air, allowing a higher amount of energy transfer from the ablated material to the target, and therefore increasing the ablation yield because a major part of the target reaches

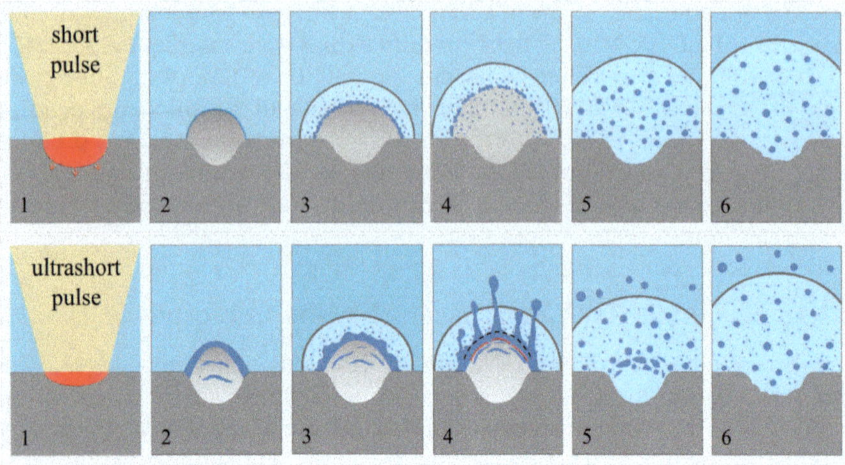

Figure 6.29 Schematic representation of the ablation process induced by short (from hundreds of ps to ns) and ultrashort (up to tens of ps) pulses. The liquid water is coloured blue, the metal target is grey, the ablation plume is light grey and the expanding cavitation bubble is light blue. The metal NPs and the transient dense metal layer generated at the plume–water interface are shown in a dark blue colour. Adapted from ref. 611 with permission from the Royal Society of Chemistry.

the ablation threshold. The quenching of the plasma plume induces growth of the vapour layer into a cavitation bubble,[§] which expands anisotropically along the parallel and normal directions with respect to the target plane, reaching its maximum height at about 20% of its lifetime.[618,619] Here, early chemical reactions between the hot ablated material and the solvent may occur. In addition, inside the cavitation bubble the liquid solvent is present and promotes crystallisation of the first atom clusters and primary particles.[620,621] The energy transfer from the cooling plasma plume to the liquid (and to the target) may also cause breakdown of the solvent, with formation of ionised and atomised species.[622] The cavitation bubble proceeds into cyclic expansion and shrinkage process (generating at least a second hemispherical shockwave after the first shrinkage) until it collapses due to the drop in internal pressure to a value lower than that of the liquid environment.[623] The shrinking symmetry depends on the target wettability and temperature, while the bubble lifetime and its volume are functions of the laser fluence and absolute pulse energy, which is generally lower for short pulses and high (>1 kHz) repetition rate.[605,624] For example, a mm bubble arises from a ns pulse, while bubbles one order of magnitude smaller are generated by a ps pulse. After collapse, the bubble may generate some smaller bubble, which may or may not remain attached to the target surface, or, more commonly, it disappears completely. In any case, this phenomenon generates a second

[§]More details about the phenomenon of *cavitation* may be found in Section 6.5 (sonochemical synthesis).

shockwave, and this is one of the last physical processes occurring during the laser ablation.

6.7.1.1 NP Formation and Growth

The extreme conditions of temperature and pressure in the plasma plume promote the formation of NPs. Moreover, due to the very fast cooling during the expansion, they may stabilise metastable phases and alloys that are difficult to obtain with other methods.[625] However, in the successive cavitation bubble, where a significant part of the NPs growth takes place, these conditions persist only during the initial and collapse phases, while during the expansion the temperature drops to hundreds of K and the pressure almost equals the surroundings.[626] Therefore, the terminology often used that describes this technique as a 'high pressure and high temperature' method is inaccurate for describing the NP formation process and should be limited to the above-mentioned phases. When dealing with fs or ps pulses, absorption of the laser pulse determines an explosive phase decomposition of the laser spot region into vapour and liquid phases.[627] The expanding ablation plume decelerates and cools down in contact with the liquid, leading to accumulation of the material into a hot molten layer at the edge of the plasma plume. The liquid in contact with the expanding plume is brought to the supercritical state and this contact region determines the formation of the first cavitation bubble and the fast nucleation and growth of small NPs (up to 10 nm).[612] This interface is a typical example of the Rayleigh–Taylor instability[628] (*i.e.*, a lighter fluid expanding into a denser one) because the expansion of the high-density metal–water mixture layer is decelerated by the lighter supercritical water above and it decomposes in hundreds of ps, forming larger (up to hundreds of nm) NPs. Moreover, NPs nucleating deeper in the plasma plume impact the low-density metal–water layer from below during the expansion (represented as red curves in panel 4 of Figure 6.29), leading to injection of metal droplets into the liquid above, and beyond the cavitation bubble. This mechanism explains why, for ultrashort pulses, a bimodal size distribution of NPs is observed.

A different scenario is predicted (and experimentally verified) when ns pulses are employed. Three distinct nucleation zones can be distinguished: above, in and below the thin transient metal layer at the interface of the plasma plume and the surrounding liquid.[611] The first is caused by solidification, in the ns timeframe, of evaporated material from the interfacial layer into the surrounding supercritical liquid, always inside the cavitation bubble. The second derives from the direct decomposition of the metal layer and generates NPs of the order of tens of nm. Finally, the last zone refers to the most relevant fraction of NPs, resulting from the spinodal decomposition (*i.e.*, one thermodynamic phase that evolves into two other phases simultaneously) of a part of the target. Here, the coalescence of the smaller NPs into larger ones is favoured by the higher temperature of the supercritical water inside the cavitation bubble. As highlighted in Figure 6.29, the main difference between these two growth mechanisms is the absence of molten material outside the cavitation bubble for the longer pulses. Indeed, due to the less dense transient metal interface, its decomposition does not eject NPs into the surrounding liquid. Therefore, the size distribution derived from longer pulses is generally broader and not bimodal. After the initial formation, the exact subsequent growth mechanisms is still not clear, but Thanh *et al.*[629] suggested that the most probable ones are the classic LaMer mechanism,[10] Ostwald ripening,[630] oriented attachment,[631] intraparticle growth[632] and the Finke–Watzky two-step mechanism.[633] Further reactions may occur because of the marked temperature and pressure gradients present in the system, which allow different local crystallisation pathways and thereby the formation of mixed-phase particles.[634]

Along with the formation process, the ablated material also undergoes reaction with oxygen present in the liquid, either during plasma quenching or during bubble expansion and collapse.[635] Thermal decomposition and photoionisation of the liquid occur at the same time, providing reactive species to form different compounds, such as oxides and nitrides, or to modify and introduce defects into the NPs. Although it is challenging to exactly quantify the portion of reacted liquid, it has been demonstrated that their kinetic availability (*i.e.*, the diffusion rate of reacted species towards the ablated material) does not limit the reaction rate.[636]

However, the fundamental processes governing the first ns of the laser–matter interaction and the formation and evolution of the plasma plume are still under debate and require further studies, both experimental and theoretical, to obtain a complete picture of the mechanism and thus better control over the LAL dynamics and final products.[637]

6.7.2 Role of the Synthetic Parameters I: Laser

6.7.2.1 Wavelength

Lasers from the NIR to UV are commonly used, albeit showing remarkable differences.[609] Both the penetration depth and the ablation threshold increase with increasing laser wavelength[638] and, consequently, the ablated mass (proportional to the ablation depth) also increases (Figure 6.30a–c). Generally, the same trend can be seen with the NP dimensions, even if the main cause is attributed to reduced absorbance of the incoming beam by the colloidal NPs already formed, reducing the effect of the simultaneous LFL due to the overlap of the colloidal suspension with the incoming laser beam, rather than a direct consequence of the chosen wavelength.[634] Moreover, multiphoton absorption and photoionisation are more likely for shorter wavelengths, increasing the efficiency of the size reduction during the simultaneous LFL.[639] Shorter (UV) wavelengths promote oxidation of NPs also, by increasing the number of reactive oxygen species in the plasma plume.[640]

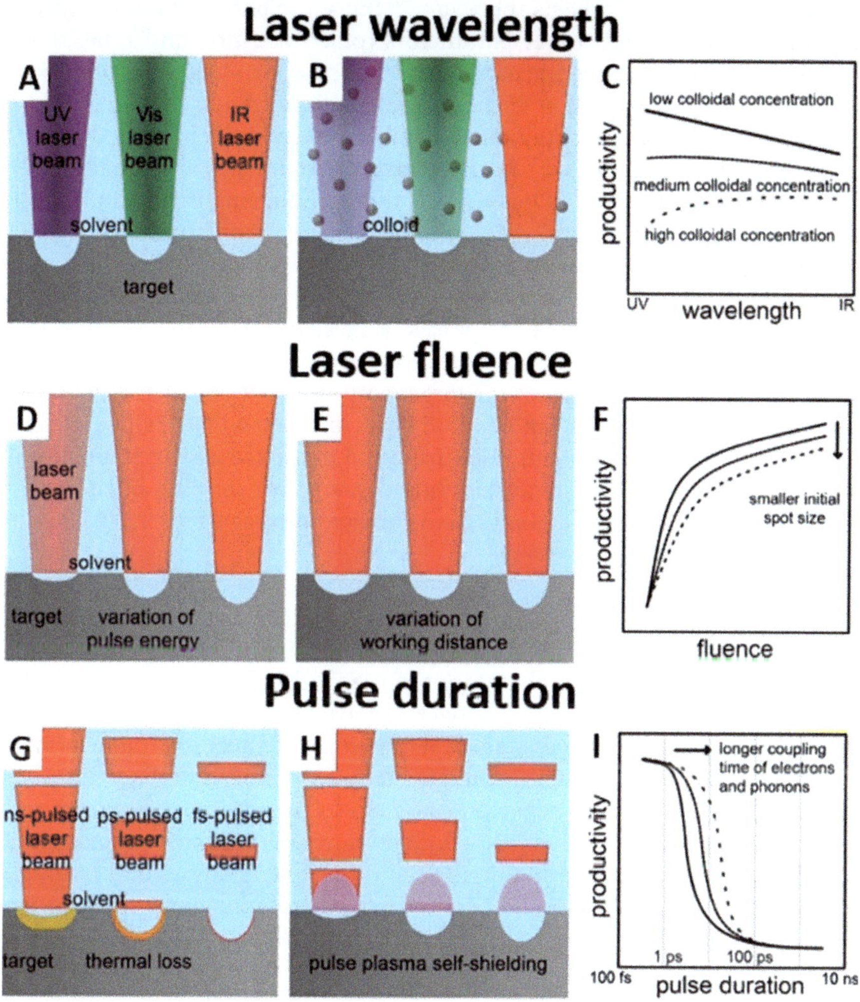

Figure 6.30 Influence of laser parameters changes on the NP productivity: (A–C) laser wavelength, (D–F) laser fluence and (G–I) pulse duration. Panels (C), (F) and (I) are generalised productivity trends, keeping all laser parameter constant except for the one under examination. Adapted from ref. 591, https://doi.org/10.3390/nano10112317, under the terms of the CC BY 4.0 license https://creativecommons.org/licenses/by/4.0/.

6.7.2.2 *Fluence and Number of Pulses*

The laser fluence is the amount of energy delivered per area unit. It directly affects the amount of ablated material and the concentration of the species in the plasma plume, and therefore the LAL efficiency (Figure 6.30d–f).[641] To maximise the energy delivered to the target, it is essential to avoid superposition in space and time of the incoming beam with the arising cavitation bubble, which partially reflects the beam.[605] Since the bubble lifetime is proportional to the fluence, other parameters (pulse duration and repetition rate) should also be chosen accordingly. Moreover, as already mentioned, solvent breakdown may occur if fluence values that are too high are used, which affects the product obtained as a consequence of the increased number of reactive species present in the surrounding medium.[642] Finally, the growth mechanism can be affected: both thermal decomposition and photofragmentation events occur with high fluences, resulting in a bimodal distribution of NPs, even when a log–normal size distribution can be observed at lower fluences due to the presence of a single decomposition process.[643] On the other hand, the number of pulses is a way of increasing NP production, but target[609] or product[644] modifications can occur between two successive pulses, either by reaction of the target with the medium or in the target itself (for example, different vaporisation heats may result in a compositional change).[645] In addition, the crater left from the first pulse changes the morphology of the target, leading to less homogeneous absorption of the next pulses, with possible consequences in the ablated product.[609]

6.7.2.3 *Pulse Duration*

As mentioned in Section 6.7.1, the pulse duration determines the formation mechanism of the NPs, and therefore their size distribution.[611] The origin can be ascribed to the difference between the speed of the process and the electron–phonon coupling constant.[646] For fs pulses, direct photoionisation is the only observable mechanism, and the crater left behind in the laser spot shows smooth and defined edges. By contrast, ns pulses are long enough to allow electron–lattice thermalisation and the conduction of heat through the target, resulting in vaporisation, boiling and thermal ablation, and leaving a less sharp crater.[647] Picosecond pulses lie in the middle, and whether the threshold from one mechanism to the other is reached depends on the specific electron–phonon relaxation time of the material. Moreover, ultrashort pulses (of the order of fs) show self-focusing effects (*i.e.*, the focal plane shifts to a higher position) (Figure 6.30g).[648] Finally, pulse duration also influences the ablation threshold fluence and therefore the process efficiency. Indeed, fs and ps pulses show a substantial decrease in this parameter compared to ns pulses (Figure 6.30h).[649] However, pulses that are too short result in absorption and fragmentation of the generated colloidal NPs and in a reduced fluence at the focal plane, so a compromise should be reached to achieve the highest efficiency.[650]

6.7.2.4 *Repetition Rate*

The increase in the number of pulses per second is proportional to LAL productivity, as long as the time interval between two successive pulses is not too short so as to overlap with the cavitation bubble, in which case the energy delivered to the target would be greatly reduced.[651] The same effect is given by the increased NP concentrations near the laser spot for a low time interval, promoting both laser beam scattering and aggregation and coalescence of the NPs.[607,634] These phenomena pose a limit of about 10^3–10^4 Hz for ordinary fluences, but the cavitation bubble can be spatially by-passed by increasing the scan speed (*i.e.*, physically moving the laser spot).[652] The pulse energy decreases with increasing pulse repetition at a given laser power and, thus, the amount of ablated material is lower. However, the effect on the NP size follows opposite trends for ns and fs pulses. In the first case, direct proportionality between pulse rate and size can be observed, as a consequence of the laser melting of NPs. In the second case, size is reduced because of the reduced pulse energy.[653]

6.7.3 Role of the Synthetic Parameters II: Target Material

Since interaction of the ablated material with the liquid environment is inevitable, a wide variety of products can be achieved starting from the same target material, simply by changing the laser and system parameters. Almost the whole periodic table has been exploited for starting materials, both as pure elements and as alloys or other compounds. In the following three sections, a brief overview of the synthesised products from the state-of-the-art literature is given and discussed based on the target materials (metals, metal alloys or oxides).

6.7.3.1 Metals

Metallic targets are the most frequently used in LAL, to synthesise both metallic colloids and other compounds by reaction with the liquid medium. The ablation process depends on the physicochemical properties of the target material and its reduction potential determines the extent of the oxidation reactions with the liquid environment.[654] Gold, particularly, is the most widely studied: colloidal gold suspensions in water produced by LAL are very stable,[655] probably because of the adsorption of anions, such as halogens, as there is evidence of pure, not oxidised Au NPs with colloidal stability, especially in the presence of electrolytes.[656] When an organic medium is employed, typically small NPs can be obtained, as solvent decomposition products adsorb onto the surface of the NPs, limiting their growth at an early stage. The adsorption may also be selective on some specific facets, thereby allowing control of the shape by promoting growth along specific crystallographic directions. For example, when toluene is used as solvent, the particle size can be reduced to 1 nm, with a graphitic shell derived from solvent decomposition (Figure 6.31a).[657] Similarly, a stable, monomodal Ag NPs suspension was synthesised by Sportelli *et al.*[658] in isopropanol, with an average diameter from 10 nm to 19 nm as a function of the laser fluence and treatment time. As in the case of suspensions of Au NPs, the colloidal stability was attributed to the solvent decomposition products, which adsorb onto the surface of the NPs, sterically impeding particle agglomeration and precipitation (Figure 6.31b–d).

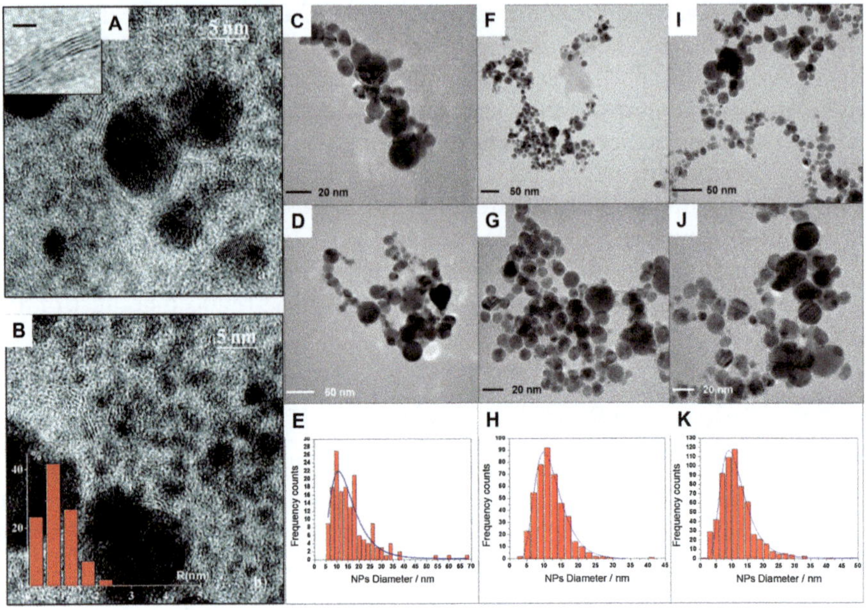

Figure 6.31 TEM micrographs of (A and B) gold NPs synthesised in toluene. Reproduced from ref. 657 with permission from American Chemical Society, Copyright 2005; and of (C–K) silver NPs synthesised in isopropanol with the relative size distributions using different pulse energies: 7.0 mJ (C–E), 27.0 mJ (F–H) and 46.5 mJ (I–K). Adapted from ref. 658 with permission from Elsevier, Copyright 2018.

By contrast, when a metal with a lower reduction potential (*i.e.*, more prone to oxidation) is employed, metal oxide NPs can be synthesised. For example, LAL of metallic Ti target in water produces TiO_2, TiO and Ti_2O_3, with substantial non-stoichiometric defects in the structure ascribed to the oxygen vacancies and interstitial cations.[659] In addition, it is worth noting that the concentration of colloids in suspension may be detrimental for product homogeneity. For example, an increase in the oxidation of Pd to Pd^{2+} was observed when a 1 hour process time was employed on LAL with a Pd target, due to photofragmentation of the NPs and successive reaction with the solvent. However, if the concentration was kept constant by using a flowing liquid, the Pd NPs remained metallic.[660]

6.7.3.2 *Metal Alloys*

Laser ablation of alloy targets is useful to gain insight into the NP formation mechanism, because the correlation between the pristine target composition and the generated NPs is one of the different conditions that can be studied. Moreover, the properties of different elements can be combined to obtain NPs with unique features.[661]

Many different setups can be considered to achieve bimetallic NPs: LAL of a bulk alloy target,[662,663] LAL of a metallic target in a colloidal suspension of another metal[664] or irradiation of a colloidal suspension of both metals.[665,666] However, recently the possibility of using a layered target in the case of AgAu has been demonstrated, obtaining pure, ligand-free alloy NPs with a composition that closely followed the ratio between the thickness of the two metallic layers.[667] Moreover, metastable CoAu NPs were synthesised with a core–shell structure from an alloy target of the two metals,[645] such that the core is composed of the more easily oxidable metal (Co), and the shell of the more stable one (Au). In this way, the ferromagnetic nature of metallic Co can be preserved. Thermodynamically, this alloy does not occur above 400 °C, yet the formation of this nanostructure can be explained by the different fusion temperatures of the two metals (1495 °C for Co and 1064 °C for Au). Co solidified first and the seeds underwent a coalescence process, while liquid Au simultaneously formed a thin shell around the Co. The core–shell NPs yield, compared to solid solution NPs, was an impressive 78 wt%. AuFe NPs were synthesised from LAL of thin films of the two metals, either already alloyed or in a layered structure.[668] The highest percentage of core–shell structure (99.7 wt%) was obtained with the alloyed film, while a still high percentage (over 90 wt%) was obtained with layered Au/Fe film. By contrast, the core–shell yield of bulk alloy target was only 48%. Similarly, Chen *et al.*[669] synthesised single-atom RuAu alloy as a catalytic system for the hydrogen evolution reaction using LAL of a bulk Ru target in a $HAuCl_4$ aqueous solution. A monomodal dispersion of RuAu NPs with an average diameter of 8 nm was obtained, and variation in the concentration of the Au solution from 0.001 M to 0.005 M increased the Au atomic percentage from *ca.* 10% to 36.5%. As a further proof of the validity of this approach, RuAg alloy NPs were also synthesised with comparable results. Multicomponent systems have also been extensively studied, with up to five immiscible metals. A high-entropy CoCrFeMnNi alloy was obtained by Waag *et al.*[606] using a 10 ps pulsed laser on a pellet formed of the mixed and compressed single metals. A bimodal volume-weight size distribution (with hydrodynamic diameters of *ca.* 8 nm and 88 nm) was found, as expected from such a pulse duration. The homogeneous distribution of all the metal species in the single NPs was confirmed by energy dispersive X-ray (EDX) mapping measurements, which evidenced uniform ablation of the single elements. Finally, upscaling of this complex system was also demonstrated by using a high-power laser system (3 ps, 1030 nm, repetition rate of 10 MHz) in a flowing liquid setup, with an ablation yield up to 3 g h^{-1}.

6.7.3.3 *Oxides*

Oxides are less used than metals as targets in LAL, and their behaviour presents some peculiarities. First, oxides, in many cases, are insulating materials, meaning that electrons must be promoted from the valence band to the conduction band through photon absorption as the initial step, before they undergo ionisation caused by laser-induced heating. This step is absent in metals, where the laser pulse promotes direct ionisation of the free electrons in the conduction band. Non-linear absorption (*e.g.*, multiphoton absorption) may be necessary to enable photoionisation and breakdown of the target when a laser with a photon energy lower than the bandgap is employed. Moreover, from a structural point of view, oxides are brittle and less ductile than metals, and so they are highly susceptible to shockwave-induced damage, such as crushing and cracking. However, using oxides as a target allows control of the composition of the NPs with a higher degree of accuracy. In the work of Jendrzej *et al.*,[670] Y_2O_3-stabilised ZrO_2 (YSZ) was produced from a YZS bulk commercial target using LAL, containing both nano- (*ca.* 6 nm) and submicrometre (*ca.* 290 nm) (Figure 6.32a and b) sized particles, and showed better mechanical properties than the commercial sample. Regardless of the target nature, the ablation products react with the active species present in the plasma plume. In the case of water as solvent, oxygen, hydrogen peroxides and other radicals are present in different amounts, and oxidation will be highly dependent on their concentration and the nature of the ablated

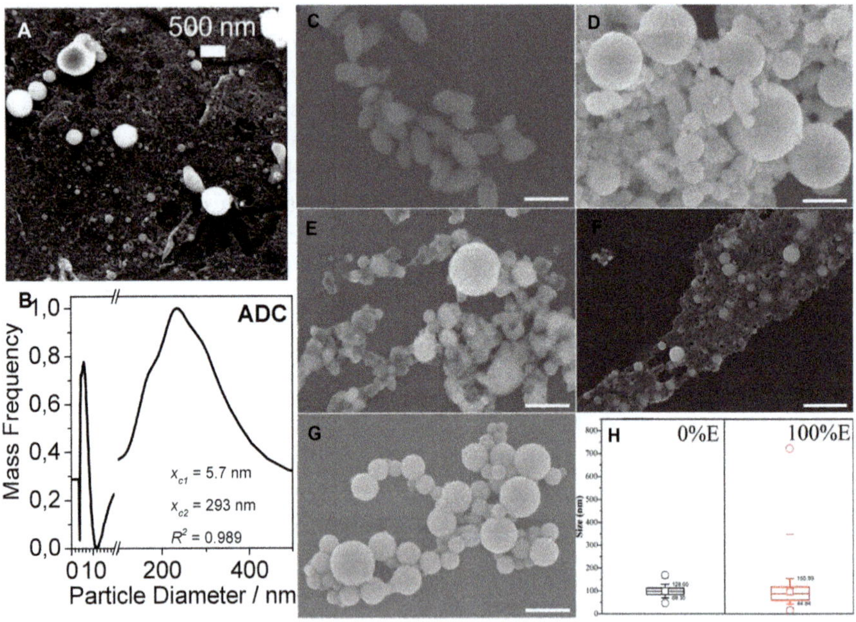

Figure 6.32 SEM images of (A) nano- and submicrometre YSZ spheres (NS/SMS) synthesised using LAL, with (B) the corresponding size distribution. Adapted from ref. 670 with permission from Elsevier, Copyright 2019. (C–G) SEM images of NPs fabricated using LAL in various water–ethanol mixtures. Ethanol content (vol%) in liquid media: (C) 0%, (D) 10%, (E) 30%, (F) 50%, (G) 100% and (H) the box plot distribution of particle sizes fabricated in pure deionised water (0%E) and pure ethanol (100%E). Scale bar in images (C–G): 200 nm. Adapted from ref. 675 with permission from Elsevier, Copyright 2017.

material. Hydroxides, suboxides and amorphous species may be synthesised under these conditions, but high selectivity is often difficult to achieve. Indeed, as demonstrated by Chemin *et al.*[671] through fast imaging of the cavitation bubble, the solvent penetrates the plasma plume: LAL of Gd_2O_3 in a Eu^{3+} solution resulted in doping of the pure oxide NPs up to 1 at%. The doping percentage was proportional to the amount of ions dissolved in the solution and the NP sizes showed a bimodal distribution, as expected when a fs or ps laser is employed. Katsuki *et al.*[672] synthesised Eu^{3+}-doped ZnO NPs using the two oxides (ZnO and Eu_2O_3) mixed in different ratios and pressed as targets, achieving doped NPs with diameters between 5 nm and 30 nm. In a similar fashion, MoO_3–TiO_2 nanocomposites were obtained by varying the MoO_3 content from 1 at% to 9 at% in the pressed target pellet.[673]

By contrast, a series of reactions at different stages of the ablation and growth processes can produce different compounds from the same target material. For instance, Al_2O_3 target yields AlO, Al_2O and Al_2O_3 as a function of the temperature:[674] at the ns time scale, ablated Al reacts within the vapour phase inside the cavitation bubble to produce Al_2O and AlO and, then, on the microsecond scale, the water environment favours the final Al_2O_3 stoichiometry, with alumina occurring as the γ phase. Solvent influence on the formation of NPs was studied by Wang *et al.*[675] on a YVO_4 target doped with Eu^{3+} at 5 at%. Various percentages of ethanol in water (from 0% to 100% in volume) were employed, and the morphology changed from ovoidal to spherical with little effect on the size. However, NPs obtained in a mixture of ethanol–water had a lower crystallinity than those synthesised in pure water or pure ethanol. This was correlated with the viscosity of the solution, which is similar and lower for the pure solvents and increases up to a maximum for a mixture 30 : 70 vol. water–ethanol (Figure 6.32c–h).

6.7.4 Role of Synthetic Parameters III: Additives and Solvent

6.7.4.1 Additives and Size Control

In situ control over particle growth can be achieved with a wide variety of approaches that act on the surface chemistry of the NPs. Ions,[676] surfactants,[677] organic solvents,[678] polymers[679] and templates[680] are among the most used, and they can interact either physically or chemically with the surface of the NPs during its formation. Surfactants are the most common and they can also improve the stability of the colloidal suspension by hindering aggregation through (electro)steric repulsions. The average size of the NPs is inversely proportional to the ligand concentration, as demonstrated in early work.[640,681] In particular, at lower concentrations, an incomplete monolayer forms around the NPs, and the steric repulsion is not sufficient to screen the van der Waals attractions. When the concentration is high enough, the steric repulsion is effective, and the size of the NPs drastically decreases. The pronounced growth quenching of NPs in the presence of ligands suggests fast penetration of the ligands into the plasma plume, and indeed this phenomenon has already been demonstrated with solvated ions.[682] However, the extreme conditions of the plasma plume are probably too harsh not to have an effect on the stability of the organic ligand, and thus it has been suggested that the size quenching effect takes place in the vapour phase surrounding the plasma plume.[593]

In addition, it was observed that the concentration of the additive can also affect the composition of the NPs. Indeed, Al NPs of hundreds of nm were selectively observed when employing a high concentration of CTAB,[683] because the surfactant blocks the reaction with water and therefore the oxidation. For lower concentrations, the penetration of water through the CTAB layer is considerable, as the surfactant layer on the NPs surface is incomplete and reaction with the Al NPs takes place, yielding oxides and hydroxides with a size of tens of nm formed from dissolution of the Al NPs. On the other hand, the presence of the surfactant also poses some drawbacks: organic molecules could pyrolyse during LAL, compromising the NPs purity, and further functionalisation of a ligand-covered particle is not straightforward.

Another strategy to control the colloidal stability and growth of NPs is to increase the ionic strength of the suspension by introducing electrolytes. Generally, addition of an electrolyte has the effect of screening the NPs surface charge and, therefore, destabilising the colloid. However, low electrolyte concentrations can effectively increase the electrostatic stabilisation of metal NPs, because selective adsorption of anions with the increase in the surface charge density may occur.[684,685] An extensive study was conducted by Marzun *et al.*[686] on Pd NPs, which are more easily oxidised than the Au NPs normally taken as a model system. Different concentrations of buffer solution were necessary to assure colloidal stability, depending on the electrolyte used (increasing from sodium carbonate to phosphate and hydroxide). A minimum buffer concentration of 0.005 mM was needed to avoid formation of larger NPs (over 100 nm) and the salt addition changed the size distribution (number weighted) from bimodal in pure water to monomodal, with an average diameter of 3.4 nm. However, a small fraction of NPs in the range 30–40 nm are still present, even with a buffer concentration up to 1 mM (see Figure 6.33).

Finally, in the case of pH variation of the solution, as for LAL in acidic or alkaline environments, the size reduction and colloidal stability are enhanced due to increase in the surface charge of the NP, which resembles the electrostatic stabilisation due to increased ionic strength in a salt solution.[687]

In general, surface chemistry and functionalisation of pure NPs is straightforward[688,689] for better suiting the specific requirements of the final application, and reveal exciting opportunities for the future for widespread adoption of laser technology in different fields.

6.7.4.2 Solvent

The environment surrounding the ablation process plays a key role in defining the chemical and morphological properties of the final NPs. Apart from water, saline[690] and acidic[691]/alkaline[692] solutions, organic solvents[693,694] and

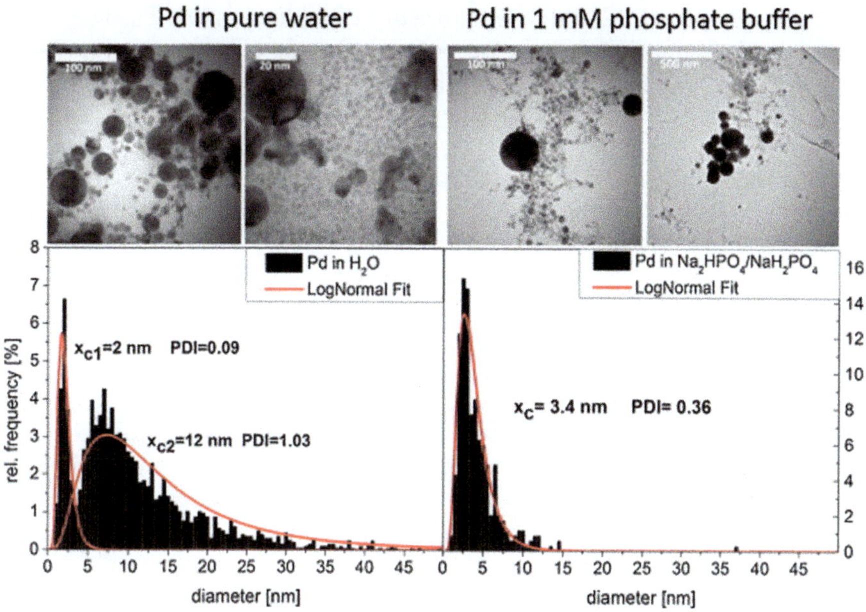

Figure 6.33 Effect of dissolved salts on palladium NP laser synthesis: electron micrographs and corresponding histograms (number-weighted distribution) of laser-generated palladium NPs in pure water and phosphate buffer. Reproduced from ref. 686 with permission from Elsevier, Copyright 2019.

supercritical fluids[695,696] have all been the subject of study. When the solvent and/or additives are involved in the reaction, it is common to refer to LAL as reactive-LAL (RLAL). Organic solvents may induce particle size reduction by adsorption of the decomposition by-products originated during the laser pulse, or of the solvent molecules themselves on the NP surfaces, and functional groups can have a similar role. Moreover, the solvent dipole moment affects the growth mechanism of the NPs, albeit with contrasting results reported in the literature. Some studies[697,698] suggest an increase in the NPs electrostatic repulsion, proportional to the solvent polarity, due to the formation of a stronger electric double layer. However, for instance, in the case of Au[699] or Bi[700] NPs, an opposite trend is reported, which highlights the need for a deeper understanding to exclude other causes that might affect the NPs size. The highly reactive environment can favour the formation not only of oxides and hydroxides (as in water), but also of carbides[701] and nanocomposites with carbon shells[702] or onion-like[703] structures. An additional aspect to take into account when employing an organic medium is that the pulse energy delivered to the target (and thus the amount of ablated material and the productivity) is lower due to solvent decomposition after radiation absorption.[704] Considering a highly oxidisable and reactive target, such as metallic Fe, it was found that magnetite is produced in acetonitrile and dimethylformamide, while a mixture of magnetite and iron carbide was observed in ethanol.[705,706] On the other hand, the use of a flowing liquid to drain the produced NPs and provide fresh solvent represents a promising route to avoid side reactions and achieve the desired productivity.[707] During the growth step, an important role is played by the solvent viscosity, as it limits the NPs diffusion and final size through the following law: $R^3 = R_0^3 + CT/\eta$ where R is the final radius, R_0 the initial radius, T the temperature, C a constant and η the viscosity.[708] Luo *et al.* prepared Gd_2O_3 in diethylene glycol, obtaining NPs with an average size of 4 nm, compared to tens of nm typical of water-based syntheses.[709]

Apart from size and composition, specific crystallographic phases can also be selectively obtained by changing the solvent. Jung and Choi[710] prepared Ni NPs with a face-centred cubic (fcc) structure in methanol, and mixed fcc and hexagonal close-packed (hcp) structure in hexene and acetonitrile, respectively. This effect was attributed to the solvent specific heat: cooling down the plasma plume is more efficient in acetonitrile (which has a low specific heat), thus allowing formation of the metastable hcp phase.

List of Abbreviations

AFM	atomic force microscopy
CHFS	continuous hydrothermal flow synthesis
CTAB	cetyltrimethylammonium bromide
CTAC	cetyltrimethylammonium chloride
CS	core–shell
DEG	diethylene glycol
DMF	*N*,*N*-dimethylformamide
DSC	differential scanning calorimetry
EG	ethylene glycol
EDTA	ethylenediaminetetraacetic acid
EDX	energy dispersive X-ray spectroscopy
fcc	face-centred cubic
GLY	glycerol
LAL	laser ablation in liquid
LDL	laser defect engineering in liquid
LFL	laser fragmentation in liquid
LML	laser melting in liquid
hcp	hexagonal close-packed
NIR	near-infrared
NMP	*N*-methyl pyrrolidone
MPA	mercaptopropionic acid
NP	nanoparticle
OA	oleic acid
ODE	octadecene
OLA	oleylamine
ORR	oxygen reduction reaction

OSC	oxygen storage capacity
PDMS	poly(dimethyl siloxane)
PE	pentaerythritol
PEG	polyethylene glycol
PMMA	poly(methyl methacrylate)
PVA	poly(vinyl alcohol)
PVP	poly(vinylpyrrolidone)
QD	quantum dot
RLAL	reactive-laser ablation in liquid
RT	room temperature
RTD	residence time distribution
SDA	structure-directing agents
SDS	sodium dodecyl sulfate
SEM	scanning electron microscopy
SERS	surface-enhanced Raman spectroscopy
SFR	segmented-flow reactor
TBP	tributyl phosphine
TEG	triethylene glycol
TEM	transmission electron microscopy
TEOS	tetraethyl orthosilicate
TEOT	titanium tetraethyloxide
TMAOH	tetramethyl ammonium hydroxide
TOP	trioctyl phosphine
TTB	titanium tetrabutoxide
TTIP	titanium tetraisopropoxide
US	ultrasound
XANES	X-ray absorption near-edge spectroscopy
XAS	X-ray absorption spectroscopy
XRPD	X-ray powder diffraction
YSZ	Y_2O_3-stabilised ZrO_2

References

1. K. Byrappa and M. Yoshimura, *Handbook of Hydrothermal Technology*, William Andrew Publishing, New York, USA, 2nd edn, 2013.
2. M. Yoshimura and K. Byrappa, *J. Mater. Sci.*, 2008, **43**, 2085.
3. A. Rabenau, *Angew. Chem., Int. Ed. Engl.*, 1985, **24**, 1026.
4. Y. Mao, T.-J. Park, F. Zhang, H. Zhou and S. S. Wong, *Small*, 2007, **3**, 1122.
5. I. Bretos, S. Diodati, R. Jiménez, F. Tajoli, J. Ricote, G. Bragaggia, M. Franca, M. L. Calzada and S. Gross, *Chem. – Eur. J.*, 2020, **26**, 9157.
6. C. B. Carter and M. G. Norton, *Ceramic Materials: Science and Engineering*, Springer, New York, 2nd edn, 2013.
7. A. B. Ezersky, A. Garcimartín, J. Burguete, H. L. Mancini and C. Pérez-García, *Phys. Rev. E: Stat. Phys., Plasmas, Fluids, Relat. Interdiscip. Top.*, 1993, **47**, 1126.

8. V. I. Popolitov, *Prog. Cryst. Growth Charact. Mater.*, 1991, **21**, 255.
9. D. R. Modeshia and R. I. Walton, *Chem. Soc. Rev.*, 2010, **39**, 4303.
10. V. K. LaMer, *Ind. Eng. Chem.*, 1952, **44**, 1270.
11. V. K. LaMer and R. H. Dinegar, *J. Am. Chem. Soc.*, 1950, **72**, 4847.
12. F. Huang, H. Zhang and J. F. Banfield, *Nano Lett.*, 2003, **3**, 373.
13. J. O. Eckert, C. C. Hung-Houston, B. L. Gersten, M. M. Lencka and R. E. Riman, *J. Am. Ceram. Soc.*, 1996, **79**, 2929.
14. P. G. Vekilov, *Nanoscale*, 2010, **2**, 2346.
15. D. Erdemir, A. Y. Lee and A. S. Myerson, *Acc. Chem. Res.*, 2009, **42**, 621.
16. J. Zhang, F. Huang and Z. Lin, *Nanoscale*, 2010, **2**, 18.
17. F. Huang, H. Zhang and J. F. Banfield, *J. Phys. Chem. B*, 2003, **107**, 10470.
18. D. Gebauer and H. Cölfen, *Nano Today*, 2011, **6**, 564.
19. D. Gebauer, M. Kellermeier, J. D. Gale, L. Bergström and H. Cölfen, *Chem. Soc. Rev.*, 2014, **43**, 2348.
20. F. Ren, Y. Leng, Y. Ding and K. Wang, *CrystEngComm*, 2013, **15**, 2137.
21. R. J. Francis and D. O'Hare, *J. Chem. Soc., Dalton Trans.*, 1998, 3133.
22. F. Habashi, *Textbook of Hydrometallurgy*, Métallurgie Extractive, Québec City, 2nd edn, 1999.
23. W. Zhu, R. Wang, S. Zhu, L. Zhang, X. Cui, H. Zhang, X. Piao and Q. Zhang, *ACS Sustainable Chem. Eng.*, 2014, **2**, 836.
24. M. Liu, Z. Chang, H. Wei, B. Li, X. Wang and Y. Wen, *Appl. Catal., A*, 2016, **525**, 59.
25. R. Simancas, A. Chokkalingam, S. P. Elangovan, Z. Liu, T. Sano, K. Iyoki, T. Wakihara and T. Okubo, *Chem. Sci.*, 2021, **12**, 7677.
26. C. Martínez and A. Corma, *Coord. Chem. Rev.*, 2011, **255**, 1558.
27. C. S. Cundy and P. A. Cox, *Microporous Mesoporous Mater.*, 2005, **82**, 1.
28. P. Dolcet, S. Diodati, F. Zorzi, P. Voepel, C. Seitz, B. M. Smarsly, S. Mascotto, F. Nestola and S. Gross, *Green Chem.*, 2018, **20**, 2257.
29. Y. Zhou, Y. Lin and G. R. Patzke, *Prog. Chem.*, 2012, **24**, 1583.
30. L. Zhang, F. Xia, J. Yao, T. Zhu, H. Xia, G. Yang, B. Liu and Y. Gao, *J. Mater. Chem. C*, 2020, **8**, 13396.
31. R. I. Walton and D. O'Hare, *Chem. Commun.*, 2000, 2283.
32. J. Łuczak, M. Paszkiewicz, A. Krukowska, A. Malankowska and A. Zaleska-Medynska, *Adv. Colloid Interface Sci.*, 2016, **227**, 1.
33. T. Voisin, A. Erriguible and C. Aymonier, *Sci. Adv.*, 2020, **6**, eaaz7770.
34. R. I. Walton, *Prog. Cryst. Growth Charact. Mater.*, 2011, **57**, 93.
35. R. I. Walton, *Chem. Soc. Rev.*, 2002, **31**, 230.
36. R. I. Walton, *Chem. – Eur. J.*, 2020, **26**, 9041.
37. J. W. Annis, J. M. Fisher, D. Thompsett and R. I. Walton, *Inorganics*, 2021, **9**, 40.
38. J. A. Darr, J. Zhang, N. M. Makwana and X. Weng, *Chem. Rev.*, 2017, **117**, 11125.
39. D. A. Palmer, R. Fernández-Prini and A. H. Harvey, *Aqueous Systems at Elevated Temperatures and Pressures: Physical Chemistry in Water, Steam and Hydrothermal Solutions*, Elsevier, Amsterdam, 2004.
40. R. L. Smith, *AIP Conf. Proc.*, 2010, **1251**, 21.

41. S. Diodati, P. Dolcet, M. Casarin and S. Gross, *Chem. Rev.*, 2015, **115**, 11449.
42. R. L. Penn and J. F. Banfield, *Geochim. Cosmochim. Acta*, 1999, **63**, 1549.
43. Z. Kožíšek, K. Sato, S. Ueno and P. Demo, *J. Chem. Phys.*, 2011, **134**, 94508.
44. F. C. Meldrum and H. Cölfen, *Chem. Rev.*, 2008, **108**, 4332.
45. J. B. Hasted, D. M. Ritson and C. H. Collie, *J. Chem. Phys.*, 1948, **16**, 1.
46. V. Simonet, Y. Calzavara, J. L. Hazemann, R. Argoud, O. Geaymond and D. Raoux, *J. Chem. Phys.*, 2002, **117**, 2771.
47. T. M. Seward, C. M. B. Henderson, J. M. Charnock and B. R. Dobson, *Geochim. Cosmochim. Acta*, 1996, **60**, 2273.
48. T. M. Seward, C. M. B. Henderson, J. M. Charnock and T. Driesner, *Geochim. Cosmochim. Acta*, 1999, **63**, 2409.
49. O. Y. Samoilov, *Discuss. Faraday Soc.*, 1957, **24**, 141.
50. M. M. Lencka, A. Anderko and R. E. Riman, *J. Am. Ceram. Soc.*, 1995, **78**, 2609.
51. M. M. Lencka and R. E. Riman, *Chem. Mater.*, 1993, **5**, 61.
52. U. Schubert and N. Hüsing, *Synthesis of Inorganic Materials*, Wiley-VCH, Weinheim, 4th edn, 2019.
53. A. A. Shternberg, in *Crystallization Processes under Hydrothermal Conditions*, ed. A. N. Lobachev, Springer US, Boston, MA, 1973.
54. L. N. Dem'yanets, E. N. Emel'yanova and O. K. Mel'nikov, in *Crystallization Processes under Hydrothermal Conditions*, ed. A. N. Lobachev, Springer US, Boston, MA, 1973.
55. H. X. Mai, L. D. Sun, Y. W. Zhang, R. Si, W. Feng, H. P. Zhang, H. C. Liu and C. H. Yan, *J. Phys. Chem. B*, 2005, **109**, 24380.
56. B. Bouchaud, J. Balmain, G. Bonnet and F. Pedraza, *J. Rare Earths*, 2012, **30**, 559.
57. T. Nagyné-Kovács, G. Shahnazarova, I. E. Lukács, A. Szabó, K. Hernadi, T. Igricz, K. László, I. M. Szilágyi and G. Pokol, *Materials*, 2019, **12**, 1728.
58. W. Huang, G. Liu, T. Qi, X. Li, Q. Zhou and Z. Peng, *CrystEngComm*, 2020, **22**, 6983.
59. J. Yu and A. Kudo, *Adv. Funct. Mater.*, 2006, **16**, 2163.
60. K. Namratha and K. Byrappa, *Prog. Cryst. Growth Charact. Mater.*, 2012, **58**, 14.
61. Q. Wu, F. Zhang, P. Xiao, H. Tao, X. Wang, Z. Hu and Y. Lü, *J. Phys. Chem. C*, 2008, **112**, 17076.
62. Y. Xia, Y. Xiong, B. Lim and S. E. Skrabalak, *Angew. Chem., Int. Ed.*, 2009, **48**, 60.
63. S. Polarz, *Adv. Funct. Mater.*, 2011, **21**, 3214.
64. G. Liu, H. G. Yang, J. Pan, Y. Q. Yang, G. Q. M. Lu and H. M. Cheng, *Chem. Rev.*, 2014, **114**, 9559.
65. A. Chemseddine and T. Moritz, *Eur. J. Inorg. Chem.*, 1999, **2**, 235.
66. R. Katal, S. Masudy-Panah, M. Tanhaei, M. H. D. A. Farahani and H. Jiangyong, *Chem. Eng. J.*, 2020, **384**, 123384.

67. X. H. Yang, Z. Li, C. Sun, H. G. Yang and C. Li, *Chem. Mater.*, 2011, **23**, 3486.
68. G. C. Kennedy, *Am. J. Sci.*, 1950, **248**, 540.
69. H. L. Barnes, in *Research Techniques for High Pressure and High Temperature*, ed. G. C. Ulmer, Springer Berlin Heidelberg, Berlin, Heidelberg, 1971.
70. G. Demazeau and A. Largeteau, *Z. Anorg. Allg. Chem.*, 2015, **641**, 159.
71. F. Kretschmer, S. Mühlig, S. Hoeppener, A. Winter, M. D. Hager, C. Rockstuhl, T. Pertsch and U. S. Schubert, *Part. Part. Syst. Charact.*, 2014, **31**, 721.
72. J. Perezjuste, I. Pastorizasantos, L. Liz Marzan and P. Mulvaney, *Coord. Chem. Rev.*, 2005, **249**, 1870.
73. Y. Huang, W. Wang, H. Liang and H. Xu, *Cryst. Growth Des.*, 2009, **9**, 858.
74. Y. Liu, L. Yang and Y. Shen, *J. Mater. Res.*, 2018, **33**, 2671.
75. D. Yu and V. W.-W. Yam, *J. Phys. Chem. B*, 2005, **109**, 5497.
76. B. Bari, J. Lee, T. Jang, P. Won, S. H. Ko, K. Alamgir, M. Arshad and L. J. Guo, *J. Mater. Chem. A*, 2016, **4**, 11365.
77. D. McNulty, Q. Ramasse and C. O'Dwyer, *Nanoscale*, 2016, **8**, 16266.
78. L. Wang and Y. Yamauchi, *Chem. Mater.*, 2011, **23**, 2457.
79. S.-B. Wang, W. Zhu, J. Ke, M. Lin and Y.-W. Zhang, *ACS Catal.*, 2014, **4**, 2298.
80. W. Ye, S. Kou, X. Guo, F. Xie, H. Sun, H. Lu and J. Yang, *Nanoscale*, 2015, **7**, 9558.
81. Y.-C. Gao, C. Wang, C.-X. Zhang, H.-W. Li and Y. Wu, *J. Mater. Sci. Technol.*, 2022, **109**, 140.
82. Y.-C. Shi, J.-J. Feng, X.-X. Lin, L. Zhang, J. Yuan, Q.-L. Zhang and A.-J. Wang, *Electrochim. Acta*, 2019, **293**, 504.
83. Q. Yang, Z. Lu, J. Liu, X. Lei, Z. Chang, L. Luo and X. Sun, *Prog. Nat. Sci.: Mater. Int.*, 2013, **23**, 351.
84. F. Z. Janani, N. Taoufik, H. Khiar, W. Boumya, A. Elhalil, M. Sadiq, A. v. Puga and N. Barka, *Surf. Interfaces*, 2021, **25**, 101263.
85. Y. Zhang, M. K. Ram, E. K. Stefanakos and D. Y. Goswami, *J. Nanomater.*, 2012, **2012**, e624520.
86. J. L. Gomez and O. Tigli, *J. Mater. Sci.*, 2013, **48**, 612.
87. Ü. Özgür, Ya. I. Alivov, C. Liu, A. Teke, M. A. Reshchikov, S. Doğan, V. Avrutin, S.-J. Cho and H. Morkoç, *J. Appl. Phys.*, 2005, **98**, 041301.
88. Ü. Ozgur, D. Hofstetter and H. Morkoç, *Proc. IEEE*, 2010, **98**, 1255.
89. S. Baruah and J. Dutta, *Sci. Technol. Adv. Mater.*, 2009, **10**, 18.
90. E. D. Bøjesen, K. M. Ø. Jensen, C. Tyrsted, N. Lock, M. Christensen and B. B. Iversen, *Cryst. Growth Des.*, 2014, **14**, 2803.
91. S. Diodati, J. Hennemann, F. Fresno, S. Gialanella, P. Dolcet, U. L. Štangar, B. M. Smarsly and S. Gross, *Eur. J. Inorg. Chem.*, 2019, **2019**, 837.
92. L. Zhu, Y. Li and W. Zeng, *Appl. Surf. Sci.*, 2018, **427**, 281.
93. G. Liu, H. G. Yang, J. Pan, Y. Q. Yang, G. Q. M. Lu and H. M. Cheng, *Chem. Rev.*, 2014, **114**, 9559.

94. E. Baranowska-Wójcik, D. Szwajgier, P. Oleszczuk and A. Winiarska-Mieczan, *Biol. Trace Elem. Res.*, 2020, **193**, 118.
95. X. Kang, S. Liu, Z. Dai, Y. He, X. Song and Z. Tan, *Catalysts*, 2019, **9**, 191.
96. H. G. Yang, C. H. Sun, S. Z. Qiao, J. Zou, G. Liu, S. C. Smith, H. M. Cheng and G. Q. Lu, *Nature*, 2008, **453**, 638.
97. R. Menzel, A. Duerrbeck, E. Liberti, H. C. Yau, D. McComb and M. S. P. Shaffer, *Chem. Mater.*, 2013, **25**, 2137.
98. W. Lekphet, S.-Y. Ho, C. Su, P. Sireesha, S. Kathirvel, Y.-F. Lin and W.-R. Li, *J. Nanosci. Nanotechnol.*, 2018, **18**, 967.
99. H. Huesmann, E. Schechtel, I. Lieberwirth, M. Panthöfer and W. Tremel, *Eur. J. Inorg. Chem.*, 2019, 3165.
100. D. Reyes-Coronado, G. Rodríguez-Gattorno, M. E. Espinosa-Pesqueira, C. Cab, R. de Coss and G. Oskam, *Nanotechnology*, 2008, **19**, 145605.
101. A. Trovarelli and J. Llorca, *ACS Catal.*, 2017, **7**, 4716.
102. A. Trovarelli and P. Fornasiero, *Catalysis by Ceria and Related Materials*, Imperial College Press, London, 2013.
103. T. Montini, M. Melchionna, M. Monai and P. Fornasiero, *Chem. Rev.*, 2016, **116**, 5987.
104. P. Li, X. Chen, Y. Li and J. W. Schwank, *Catal. Today*, 2019, **327**, 90.
105. C. Li, Y. Sun, I. Djerdj, P. Voepel, C.-C. Sack, T. Weller, D. Ellinghaus, J. Sann, Y. Guo, B. M. Smarsly and H. Over, *ACS Catal.*, 2017, **7**, 6453.
106. Y.-C. Huang, S.-H. Wu, C.-H. Hsiao, A.-T. Lee and M. H. Huang, *Chem. Mater.*, 2020, **32**, 2631.
107. Y. Xue and Y. Wang, *Nanoscale*, 2020, **12**, 10912.
108. J. M. Meijer and L. Rossi, *Soft Matter*, 2021, **17**, 2354.
109. L. Wang, G. Wei, X. Dong, Y. Zhao, Z. Xing, H. Hong and Z. Ju, *ChemistrySelect*, 2019, **4**, 11370.
110. M. Tadic, M. Panjan, V. Damnjanovic and I. Milosevic, *Appl. Surf. Sci.*, 2014, **320**, 183.
111. A. Hu, G. T. Yee and W. Lin, *J. Am. Chem. Soc.*, 2005, **127**, 12486.
112. V. G. Roullin, J. R. Deverre, L. Lemaire, F. Hindré, M. C. Venier-Julienne, R. Vienet and J. P. Benoit, *Eur. J. Pharm. Biopharm.*, 2002, **53**, 293.
113. F. Hu, L. Wei, Z. Zhou, Y. Ran, Z. Li and M. Gao, *Adv. Mater.*, 2006, **18**, 2553.
114. X. Sun, C. Zheng, F. Zhang, Y. Yang, G. Wu, A. Yu and N. Guan, *J. Phys. Chem. C*, 2009, **113**, 16002.
115. T. Wu, Y. Liu, X. Zeng, T. Cui, Y. Zhao, Y. Li and G. Tong, *ACS Appl. Mater. Interfaces*, 2016, **8**, 7370.
116. M. A. Riza, M. A. Ibrahim, U. C. Ahamefula, M. A. Mat Teridi, N. A. Ludin, S. Sepeai and K. Sopian, *Sol. Energy*, 2016, **139**, 309.
117. M. Inada, N. Enomoto, K. Hayashi, J. Hojo and S. Komarneni, *Ceram. Int.*, 2015, **41**, 5581.
118. K. Hongo, S. Kurata, A. Jomphoak, M. Inada, K. Hayashi and R. Maezono, *Inorg. Chem.*, 2018, **57**, 5413.
119. S. Diodati, R. I. Walton, S. Mascotto and S. Gross, *Inorg. Chem. Front.*, 2020, **7**, 3282.

120. J. Zhang, K. Huang, L. Yuan and S. Feng, *CrystEngComm*, 2018, **20**, 470.
121. S. Diodati, L. Pandolfo, A. Caneschi, S. Gialanella and S. Gross, *Nano Res.*, 2014, 7, 1027.
122. M. Bastianello, S. Gross and M. T. Elm, *RSC Adv.*, 2019, **9**, 33282.
123. P. Dolcet, K. Kirchberg, A. Antonello, C. Suchomski, R. Marschall, S. Diodati, R. Muñoz-Espí, K. Landfester and S. Gross, *Inorg. Chem. Front.*, 2019, **6**, 1527.
124. M. Bastianello, S. Diodati, N. Dengo, L. McCafferty, C. Footer, D. Badocco, P. Pastore, J. Darr and S. Gross, *CrystEngComm*, 2019, **21**, 6801.
125. M. F. Zscherp, M. Bastianello, S. Nappini, E. Magnano, D. Badocco, S. Gross and M. T. Elm, *J. Mater. Chem. C*, 2022, **10**, 2976.
126. G.-J. Lee and J. J. Wu, *Powder Technol.*, 2017, **318**, 8.
127. S. Ummartyotin and Y. Infahsaeng, *Renewable Sustainable Energy Rev.*, 2016, **55**, 17.
128. N. Dengo, A. F. de Fazio, M. Weiss, R. Marschall, P. Dolcet, M. Fanetti and S. Gross, *Inorg. Chem.*, 2018, **57**, 13104.
129. X. Yang, H. Liu, T. Li, B. Huang, W. Hu, Z. Jiang, J. Chen and Q. Niu, *Int. J. Hydrogen Energy*, 2020, **45**, 26967.
130. R. Zhang, Y. Wang, M. Jia, J. Xu and E. Pan, *Appl. Surf. Sci.*, 2018, **437**, 375.
131. W. Dong, L. An, X. Wang, B. Li, B. Chen, W. Tang, C. Li and G. Wang, *J. Alloys Compd.*, 2011, **509**, 2170.
132. P. Luo, H. Zhang, L. Liu, Y. Zhang, J. Deng, C. Xu, N. Hu and Y. Wang, *ACS Appl. Mater. Interfaces*, 2017, **9**, 2500.
133. A. Wu, C. Tian, Y. Jiao, Q. Yan, G. Yang and H. Fu, *Appl. Catal., B*, 2017, **203**, 955.
134. X. Geng, W. Sun, W. Wu, B. Chen, A. Al-Hilo, M. Benamara, H. Zhu, F. Watanabe, J. Cui and T. Chen, *Nat. Commun.*, 2016, **7**, 10672.
135. Z. Li, J. Yang, T. Guang, B. Fan, K. Zhu and X. Wang, *Small Methods*, 2021, **5**, 2100193.
136. F. Yu. Sharikov, O. A. Drozhzhin, V. D. Sumanov, A. N. Baranov, A. M. Abakumov and E. v. Antipov, *Cryst. Growth Des.*, 2018, **18**, 879.
137. F. Ruiz-Jorge, A. Benítez, S. Fernández-García, J. Sánchez-Oneto and J. R. Portela, *Ind. Eng. Chem. Res.*, 2020, **59**, 9318.
138. J. Wang, X. Wu, X. Cao, Y. Jiang, D. Zhang, T. Yang, F. Zhang and Y. Luo, *CrystEngComm*, 2020, **22**, 1610.
139. M. O. Enikeeva, K. M. Kenges, O. v. Proskurina, D. P. Danilovich and V. v. Gusarov, *Russ. J. Appl. Chem.*, 2020, **93**, 540.
140. H. Dong, Y.-C. Chen and C. Feldmann, *Green Chem.*, 2015, **17**, 4107.
141. F. Fiévet, S. Ammar-Merah, R. Brayner, F. Chau, M. Giraud, F. Mammeri, J. Peron, J.-Y. Piquemal, L. Sicard and G. Viau, *Chem. Soc. Rev.*, 2018, **47**, 5187.
142. H. Yue, Y. Zhao, X. Ma and J. Gong, *Chem. Soc. Rev.*, 2012, **41**, 4218.
143. F. Fievet, J. P. Lagier and M. Figlarz, *MRS Bull.*, 1989, **14**, 29.
144. F. Fievet, J. Lagier, B. Blin, B. Beaudoin and M. Figlarz, *Solid State Ionics*, 1989, **32–33**, 198.

145. G. M. Chow, L. K. Kurihara, K. M. Kemner, P. E. Schoen, W. T. Elam, A. Ervin, S. Keller, Y. D. Zhang, J. Budnick and T. Ambrose, *J. Mater. Res.*, 1995, **10**, 1546.
146. F. Fiévet, S. Ammar-Merah, R. Brayner, F. Chau, M. Giraud, F. Mammeri, J. Peron, J.-Y. Piquemal, L. Sicard and G. Viau, *Chem. Soc. Rev.*, 2018, **47**, 5187.
147. T. Sun and A. S. Teja, *J. Chem. Eng. Data*, 2003, **48**, 198.
148. H. M. Trimble and W. Potts, *Ind. Eng. Chem.*, 1935, **27**, 66.
149. D. Larcher and R. Patrice, *J. Solid State Chem.*, 2000, **154**, 405.
150. H. El-Deeb and M. Bron, *Electrochim. Acta*, 2015, **164**, 315.
151. I. Capek, *Noble Metal Nanoparticles*, Springer Japan, Tokyo, 2017.
152. A. J. Biacchi and R. E. Schaak, *ACS Nano*, 2011, **5**, 8089.
153. G. Paramasivam, N. Kayambu, A. M. Rabel, A. K. Sundramoorthy and A. Sundaramurthy, *Acta Biomater.*, 2017, **49**, 45.
154. N. V. Long, N. D. Chien, T. Hayakawa, H. Hirata, G. Lakshminarayana and M. Nogami, *Nanotechnology*, 2010, **21**, 035605.
155. H. Wang and Y. Han, *Ind. Eng. Chem. Res.*, 2015, **54**, 9742.
156. F. Bonet, C. Guéry, D. Guyomard, R. H. Urbina, K. Tekaia-Elhsissen and J.-M. Tarascon, *Int. J. Inorg. Mater.*, 1999, **1**, 47.
157. K. D. Gilroy, A. Ruditskiy, H.-C. Peng, D. Qin and Y. Xia, *Chem. Rev.*, 2016, **116**, 10414.
158. H. Yue, Y. Zhao, X. Ma and J. Gong, *Chem. Soc. Rev.*, 2012, **41**, 4218.
159. P. Wang, J. J. Kosinski, A. Anderko, R. D. Springer, M. M. Lencka and J. Liu, *Ind. Eng. Chem. Res.*, 2013, **52**, 15968.
160. Y. Chen, K. Y. Liew and J. Li, *Mater. Lett.*, 2008, **62**, 1018.
161. T. Matsumoto, K. Takahashi, K. Kitagishi, K. Shinoda, J. L. C. Huaman, J.-Y. Piquemal and B. Jeyadevan, *New J. Chem.*, 2015, **39**, 5008.
162. B. A. Steinhoff, I. A. Guzei and S. S. Stahl, *J. Am. Chem. Soc.*, 2004, **126**, 11268.
163. S. Arora, M. L. Singla and P. Kapoor, *Mater. Chem. Phys.*, 2009, **114**, 107.
164. I. Favier, D. Pla and M. Gómez, *Chem. Rev.*, 2020, **120**, 1146.
165. D. Jézéquel, J. Guenot, N. Jouini and F. Fiévet, *J. Mater. Res.*, 1995, **10**, 77.
166. P. Toneguzzo, G. Viau, O. Acher, F. Guillet, E. Bruneton, F. Fievet-Vincent and F. Fievet, *J. Mater. Sci.*, 2000, **35**, 3767.
167. S.-S. Chee and J.-H. Lee, *Thin Solid Films*, 2014, **562**, 211.
168. S. Gorai and S. Chaudhuri, *Mater. Chem. Phys.*, 2005, **94**, 434.
169. E. Hammarberg and C. Feldmann, *Chem. Mater.*, 2009, **21**, 771.
170. C. Kind and C. Feldmann, *Chem. Mater.*, 2011, **23**, 4982.
171. A. Ghosal, J. Shah, R. K. Kotnala and S. Ahmad, *J. Mater. Chem. A*, 2013, **1**, 12868.
172. S. Bundli, P. Dhak, M. Jensen, A. E. Gunnæs, P. D. Nguyen, H. Fjellvåg and A. O. Sjåstad, *J. Alloys Compd.*, 2019, **779**, 879.
173. M. Miyachi, Y. Yamamoto, Y. Yamanoi and H. Nishihara, *J. Inorg. Organomet. Polym. Mater.*, 2015, **25**, 140.
174. M. Jin, H. Liu, H. Zhang, Z. Xie, J. Liu and Y. Xia, *Nano Res.*, 2011, **4**, 83.

175. A. Klinkova, E. M. Larin, E. Prince, E. H. Sargent and E. Kumacheva, *Chem. Mater.*, 2016, **28**, 3196.
176. W. Niu, Z. Y. Li, L. Shi, X. Liu, H. Li, S. Han, J. Chen and G. Xu, *Cryst. Growth Des.*, 2008, **8**, 4440.
177. J. W. Hong, D. Kim, Y. W. Lee, M. Kim, S. W. Kang and S. W. Han, *Angew. Chem., Int. Ed.*, 2011, **50**, 8876.
178. M. Liu, Y. Zheng, L. Zhang, L. Guo and Y. Xia, *J. Am. Chem. Soc.*, 2013, **135**, 11752.
179. M. Shao, T. Yu, J. H. Odell, M. Jin and Y. Xia, *Chem. Commun.*, 2011, **47**, 6566.
180. W. Niu, L. Zhang and G. Xu, *ACS Nano*, 2010, **4**, 1987.
181. H. X. Zhang, H. Wang, Y. S. Re and W. B. Cai, *Chem. Commun.*, 2012, **48**, 8362.
182. X. Huang, S. Tang, X. Mu, Y. Dai, G. Chen, Z. Zhou, F. Ruan, Z. Yang and N. Zheng, *Nat. Nanotechnol.*, 2011, **6**, 28.
183. Y. Li, Y. Yan, Y. Li, H. Zhang, D. Li and D. Yang, *CrystEngComm*, 2015, **17**, 1833.
184. M. Jin, H. Zhang, Z. Xie and Y. Xia, *Angew. Chem., Int. Ed.*, 2011, **50**, 7850.
185. J. Zhang, L. Zhang, S. Xie, Q. Kuang, X. Han, Z. Xie and L. Zheng, *Chem. – Eur. J.*, 2011, **17**, 9915.
186. Y. Xia, K. D. Gilroy, H.-C. Peng and X. Xia, *Angew. Chem., Int. Ed.*, 2017, **56**, 60.
187. M. Rycenga, C. M. Cobley, J. Zeng, W. Li, C. H. Moran, Q. Zhang, D. Qin and Y. Xia, *Chem. Rev.*, 2011, **111**, 3669.
188. J. Chen, B. Lim, E. P. Lee and Y. Xia, *Nano Today*, 2009, **4**, 81.
189. F. Spolaore, F. Tajoli, M. C. Dalconi, C. Hengst, F. Dornhaus and S. Gross, *Dalton Trans.*, 2022, **51**, 11476.
190. I. A. Safo, M. Werheid, C. Dosche and M. Oezaslan, *Nanoscale Adv.*, 2019, **1**, 3095.
191. Y. Wang, Y. Zheng, C. Z. Huang and Y. Xia, *J. Am. Chem. Soc.*, 2013, **135**, 1941.
192. X. Yan, H. Liu and K. Y. Liew, *J. Mater. Chem.*, 2001, **11**, 3387.
193. S. H. Im, K. H. Park and O. O. Park, *Nanotechnology*, 2011, **22**, 45602.
194. K. J. Carroll, J. U. Reveles, M. D. Shultz, S. N. Khanna and E. E. Carpenter, *J. Phys. Chem. C*, 2011, **115**, 2656.
195. F. Calvo, *Phys. Chem. Chem. Phys.*, 2015, **17**, 27922.
196. A. K. Sra, T. D. Ewers and R. E. Schaak, *Chem. Mater.*, 2005, **17**, 758.
197. E. A. Baranova, T. Amir, P. H. J. Mercier, B. Patarachao, D. Wang and Y. Le Page, *J. Appl. Electrochem.*, 2010, **40**, 1767.
198. R. Harpeness and A. Gedanken, *J. Mater. Chem.*, 2005, **15**, 698.
199. D.-H. Kim, E. Lee and C. Pak, *Catal. Today*, 2021, **359**, 106.
200. R. J. Joseyphus, K. Shinoda, D. Kodama and B. Jeyadevan, *Mater. Chem. Phys.*, 2010, **123**, 487.
201. Z. Z. Yang, X. X. Lin, X. F. Zhang, A. J. Wang, X. Y. Zhu and J. J. Feng, *J. Alloys Compd.*, 2018, **735**, 2123.

202. G. Park, D. Seo, J. Jung, S. Ryu and H. Song, *J. Phys. Chem. C*, 2011, **115**, 9417.
203. D. Huo, M. J. Kim, Z. Lyu, Y. Shi, B. J. Wiley and Y. Xia, *Chem. Rev.*, 2019, **119**, 8972.
204. F. Fiévet, S. Ammar-Merah, R. Brayner, F. Chau, M. Giraud, F. Mammeri, J. Peron, J.-Y. Piquemal, L. Sicard and G. Viau, *Chem. Soc. Rev.*, 2018, **47**, 5187.
205. M. Tsuji, M. Matsunaga, Y. Yoshida, M. Hattori and T. Ishizaki, *CrystEngComm*, 2013, **15**, 7062.
206. T. Arikan, A. M. Kannan and F. Kadirgan, *Int. J. Hydrogen Energy*, 2013, **38**, 2900.
207. X. Liu, C. Cui, H. Li, Y. Lei, T. Zhuang, M. Sun, M. N. Arshad, H. A. Albar, T. R. Sobahi and S. Yu, *Chem. Sci.*, 2015, **6**, 3038.
208. F. Dawood, B. M. Leonard and R. E. Schaak, *Chem. Mater.*, 2007, **19**, 4545.
209. A. Salati, A. Ramazani and M. A. Kashi, *J. Magn. Magn. Mater.*, 2020, **498**, 166172.
210. H. Grisaru, O. Palchik, A. Gedanken, V. Palchik, M. A. Slifkin and A. M. Weiss, *Inorg. Chem.*, 2003, **42**, 7148.
211. X. Wang, S.-I. Choi, L. T. Roling, M. Luo, C. Ma, L. Zhang, M. Chi, J. Liu, Z. Xie, J. A. Herron, M. Mavrikakis and Y. Xia, *Nat. Commun.*, 2015, **6**, 7594.
212. S. Xie, S.-I. Choi, N. Lu, L. T. Roling, J. A. Herron, L. Zhang, J. Park, J. Wang, M. J. Kim, Z. Xie, M. Mavrikakis and Y. Xia, *Nano Lett.*, 2014, **14**, 3570.
213. D. Jezequel, J. Guenot, N. Jouini and F. Fievet, *Mater. Sci. Forum*, 1994, **152–153**, 339.
214. L. Poul, S. Ammar, N. Jouini, F. Fievet and F. Villain, *J. Sol-Gel Sci. Technol.*, 2003, **26**, 261.
215. C. Feldmann and H.-O. Jungk, *Angew. Chem., Int. Ed.*, 2001, **40**, 359.
216. T. Yamamoto, Y. Wada, H. Yin, T. Sakata, H. Mori and S. Yanagida, *Chem. Lett.*, 2002, **31**, 964.
217. P. Schmitt, N. Brem, S. Schunk and C. Feldmann, *Adv. Funct. Mater.*, 2011, **21**, 3037.
218. J. Ungelenk, M. Speldrich, R. Dronskowski and C. Feldmann, *Solid State Sci.*, 2014, **31**, 62.
219. H. Chung, A. Grenier, R. Huang, X. Wang, Z. Lebens-Higgins, J.-M. Doux, S. Sallis, C. Song, P. Ercius, K. Chapman, L. F. J. Piper, H.-M. Cho, M. Zhang and Y. S. Meng, *Nano Res.*, 2019, **12**, 2238.
220. B. K. Park, S. Jeong, D. Kim, J. Moon, S. Lim and J. S. Kim, *J. Colloid Interface Sci.*, 2007, **311**, 417.
221. J. A. Bau, P. Li, A. J. Marenco, S. Trudel, B. C. Olsen, E. J. Luber and J. M. Buriak, *Chem. Mater.*, 2014, **26**, 4796.
222. O. I. Medvedeva, S. S. Kambulova, O. v. Bondar, A. R. Gataulina, N. A. Ulakhovich, A. v. Gerasimov, V. G. Evtugyn, I. F. Gilmutdinov and M. P. Kutyreva, *J. Nanotechnol.*, 2017, **2017**, 1.

223. P. Ruz and V. Sudarsan, in *Handbook on Synthesis Strategies for Advanced Materials*, Springer, Singapore, 1st edn, 2021.
224. O. Palchik, R. Kerner, A. Gedanken, V. Palchik, M. A. Slifkin and A. M. Weiss, *Glass Phys. Chem.*, 2005, **31**, 80.
225. H. Yoon, ab A. Xu, G. E. Sterbinsky, D. A. Arena, Z. Wang, P. W. Stephens, Y. Shirley Meng and K. J. Carroll, *Phys. Chem. Chem. Phys.*, 2014, **17**, 1076.
226. F. F. Castillón-Barraza, M. H. Farías, J. H. Coronado-López, M. A. Encinas-Romero, M. Pérez-Tello, R. Herrera-Urbina and A. Posada-Amarillas, *Adv. Sci. Lett.*, 2011, **4**, 596.
227. C. Feldmann and C. Metzmacher, *J. Mater. Chem.*, 2001, **11**, 2603.
228. J. Wang, S. Y. Chew, D. Wexler, G. X. Wang, S. H. Ng, S. Zhong and H. K. Liu, *Electrochem. Commun.*, 2007, **9**, 1877.
229. G. Shen, D. Chen, K. Tang, L. Huang, Y. Qian and G. Zhou, *Inorg. Chem. Commun.*, 2003, **6**, 178.
230. A. Paliwal, S. V. Singh, A. Sharma, A. Sugathan, S.-W. Liu, S. Biring and B. N. Pal, *ACS Appl. Nano Mater.*, 2018, **1**, 6063.
231. K. S. Elvira, X. i Solvas, R. C. Wootton and A. J. deMello, *Nat. Chem.*, 2013, **5**, 905.
232. A. Abou-Hassan, O. Sandre and V. Cabuil, *Angew Chem. Int. Ed. Engl.*, 2010, **49**, 6268.
233. S. Marre and K. F. Jensen, *Chem. Soc. Rev.*, 2010, **39**, 1183.
234. Y. Song, J. Hormes and C. S. S. R. Kumar, *Small*, 2008, **4**, 698.
235. J. C. DeMello and A. J. DeMello, *Lab Chip*, 2004, **4**, 11N.
236. K. F. Jensen, B. J. Reizman and S. G. Newman, *Lab Chip*, 2014, **14**, 3206.
237. R. M. Maceiczyk, I. G. Lignos and A. J. deMello, *Curr. Opin. Chem. Eng.*, 2015, **8**, 29.
238. J. A. M. Lummiss, P. D. Morse, R. L. Beingessner and T. F. Jamison, *Chem. Rec.*, 2017, **17**, 667.
239. A. M. Nightingale, T. W. Phillips, J. H. Bannock and J. C. de Mello, *Nat. Commun.*, 2014, **5**, 3777.
240. I. Shestopalov, J. D. Tice and R. F. Ismagilov, *Lab Chip*, 2004, **4**, 316.
241. W. Ferstl, T. Klahn, W. Schweikert, G. Billeb, M. Schwarzer and S. Loebbecke, *Chem. Eng. Technol.*, 2007, **30**, 370.
242. C. F. Carter, H. Lange, S. V. Ley, I. R. Baxendale, B. Wittkamp, J. G. Goode and N. L. Gaunt, *Org. Process Res. Dev.*, 2010, **14**, 393.
243. S. Mozharov, A. Nordon, D. Littlejohn, C. Wiles, P. Watts, P. Dallin and J. M. Girkin, *J. Am. Chem. Soc.*, 2011, **133**, 3601.
244. R. Panneerselvam, H. Sadat, E.-M. Höhn, A. Das, H. Noothalapati and D. Belder, *Lab Chip*, 2022, **22**, 665.
245. F. Benito-Lopez, W. Verboom, M. Kakuta, J. G. E. Gardeniers, R. J. M. Egberink, E. R. Oosterbroek, A. van den Berg and D. N. Reinhoudt, *Chem. Commun.*, 2005, 2857.
246. I. Lignos, S. Stavrakis, A. Kilaj and A. J. deMello, *Small*, 2015, **11**, 4009.
247. I. Lignos, S. Stavrakis, G. Nedelcu, L. Protesescu, A. J. deMello and M. V. Kovalenko, *Nano Lett.*, 2016, **16**, 1869.

248. T. L. Sounart, P. A. Safier, J. A. Voigt, J. Hoyt, D. R. Tallant, C. M. Matzke and T. A. Michalske, *Lab Chip*, 2007, **7**, 908.
249. S. Zinoveva, R. De Silva, R. D. Louis, P. Datta, C. S. S. R. Kumar, J. Goettert and J. Hormes, *Nucl. Instrum. Methods Phys. Res., Sect. A*, 2007, **582**, 239.
250. V. Sans, L. Porwol, V. Dragone and L. Cronin, *Chem. Sci.*, 2015, **6**, 1258.
251. V. Hessel, A. Renken and J. C. Schouten, *Handbook of Micro-reactor Engineering and Chemistry*, Wiley-VCH, Weinheim, 2006.
252. J. A. Darr, J. Zhang, N. M. Makwana and X. Weng, *Chem. Rev.*, 2017, **117**, 11125.
253. P. Caramazana, P. Dunne, M. Gimeno-Fabra, J. McKechnie and E. Lester, *Curr. Opin. Green Sustainable Chem.*, 2018, **12**, 57.
254. P. W. Dunne, A. S. Munn, C. L. Starkey, T. A. Huddle and E. H. Lester, *Philos. Trans. R. Soc., A*, 2015, **373**, 20150015.
255. G. M. Whitesides, *Nature*, 2006, **442**, 368.
256. C.-Y. Lee, C.-L. Chang, Y.-N. Wang and L.-M. Fu, *Int. J. Mol. Sci.*, 2011, **12**, 3263.
257. K. Ward and Z. H. Fan, *J. Manuf. Syst.*, 2015, **25**, 094001.
258. T. Sugimoto, *Colloids Surf., A*, 2000, **164**, 183.
259. D. B. K. Chu, J. S. Owen and B. Peters, *J. Phys. Chem. A*, 2017, **121**, 7511.
260. J. I. Yoshida, H. Kim and A. Nagaki, *ChemSusChem*, 2011, **4**, 331.
261. T. Li, J. A. Díaz-Real and T. Holm, *Adv. Mater. Technol.*, 2021, **6**, 2100569.
262. M. Baghbanzadeh, L. Carbone, P. D. Cozzoli and C. O. Kappe, *Angew. Chem., Int. Ed.*, 2011, **50**, 11312.
263. P. Anastas and N. Eghbali, *Chem. Soc. Rev.*, 2010, **39**, 301.
264. P. Anastas and J. C. Warner, *Green Chemistry: Theory and Practice*, Oxford University Press, Oxford, 1998.
265. E. Dressaire and A. Sauret, *Soft Matter*, 2017, **13**, 37.
266. A. S. Utada, A. Fernandez-Nieves, H. A. Stone and D. A. Weitz, *Phys. Rev. Lett.*, 2007, **99**, 094502.
267. K. S. Elvira, F. Gielen, S. S. H. Tsai and A. M. Nightingale, *Lab Chip*, 2022, **22**, 859.
268. R. K. Shah, H. C. Shum, A. C. Rowat, D. Lee, J. J. Agresti, A. S. Utada, L.-Y. Chu, J.-W. Kim, A. Fernandez-Nieves, C. J. Martinez and D. A. Weitz, *Mater. Today*, 2008, **11**, 18.
269. L.-Y. Chu, A. S. Utada, R. K. Shah, J.-W. Kim and D. A. Weitz, *Angew. Chem., Int. Ed.*, 2007, **46**, 8970.
270. H. Yang and M. A. M. Gijs, *Chem. Soc. Rev.*, 2018, **47**, 1391.
271. E. E. Tsur, *Annu. Rev. Biomed. Eng.*, 2020, **22**, 285.
272. T. Glatzel, C. Litterst, C. Cupelli, T. Lindemann, C. Moosmann, R. Niekrawietz, W. Streule, R. Zengerle and P. Koltay, *Comput. Fluids*, 2008, **37**, 218.
273. H. A. Stone, in *CMOS Biotechnology. Series on Integrated Circuits and Systems.*, ed. H. Lee, R. M. Westervelt, and D. Ham, Springer, Boston, MA, 2007.

274. H. Song, J. D. Tice and R. F. Ismagilov, *Angew. Chem., Int. Ed.*, 2003, **42**, 768.
275. H. Song, D. L. Chen and R. F. Ismagilov, *Angew. Chem., Int. Ed.*, 2006, **45**, 7336.
276. J. B. Edel, R. Fortt, J. C. deMello and A. J. deMello, *Chem. Commun.*, 2002, 1136.
277. L. M. Liz-Marzán, *Langmuir*, 2006, **22**, 32.
278. T. Klar, M. Perner, S. Grosse, G. von Plessen, W. Spirkl and J. Feldmann, *Phys. Rev. Lett.*, 1998, **80**, 4249.
279. J. Boleininger, A. Kurz, V. Reuss and C. Sönnichsen, *Phys. Chem. Chem. Phys.*, 2006, **8**, 3824.
280. V. S. Cabeza, S. Kuhn, A. A. Kulkarni and K. F. Jensen, *Langmuir*, 2012, **28**, 7007.
281. S. Duraiswamy and S. A. Khan, *Small*, 2009, **5**, 2828.
282. J. Wagner and J. M. Köhler, *Nano Lett.*, 2005, **5**, 685.
283. J. Wagner, T. Tshikhudo and J. Kohler, *Chem. Eng. J.*, 2008, **135**, S104.
284. J. Wagner, T. Kirner, G. Mayer, J. Albert and J. M. Köhler, *Chem. Eng. J.*, 2004, **101**, 251.
285. L. Uson, V. Sebastian, M. Arruebo and J. Santamaria, *Chem. Eng. J.*, 2016, **285**, 286.
286. J. M. Köhler, J. Wagner and J. Albert, *J. Mater. Chem.*, 2005, **15**, 1924.
287. X. Z. Lin, A. D. Terepka and H. Yang, *Nano Lett.*, 2004, **4**, 2227.
288. S. Silvestrini, T. Carofiglio and M. Maggini, *Chem. Commun.*, 2013, **49**, 84.
289. Y. Song, E. E. Doomes, J. Prindle, R. Tittsworth, J. Hormes and C. S. S. R. Kumar, *J. Phys. Chem. B*, 2005, **109**, 9330.
290. Y. Song, H. Modrow, L. L. Henry, C. K. Saw, E. E. Doomes, V. Palshin, J. Hormes and C. S. S. R. Kumar, *Chem. Mater.*, 2006, **18**, 2817.
291. Y. Song, C. S. S. R. Kumar and J. Hormes, *J. Nanosci. Nanotechnol.*, 2004, **4**, 788.
292. E. Gioria, F. Wisniewski and L. Gutierrez, *J. Environ. Chem. Eng.*, 2019, **7**, 103136.
293. V. Sebastian, C. D. Smith and K. F. Jensen, *Nanoscale*, 2016, **8**, 7534.
294. K. Torigoe, Y. Watanabe, T. Endo, K. Sakai, H. Sakai and M. Abe, *J. Nanopart. Res.*, 2010, **12**, 951.
295. U. Laura, M. Arruebo and V. Sebastian, *Dalton Trans.*, 2018, **47**, 1693.
296. G. Tofighi, A. Gaur, D. E. Doronkin, H. Lichtenberg, W. Wang, D. Wang, G. Rinke, A. Ewinger, R. Dittmeyer and J.-D. Grunwaldt, *J. Phys. Chem. C*, 2018, **122**, 1721.
297. W. Stöber, A. Fink and E. Bohn, *J. Colloid Interface Sci.*, 1968, **26**, 62.
298. S. A. Khan, A. Günther, M. A. Schmidt and K. F. Jensen, *Langmuir*, 2004, **20**, 8604.
299. J. B. Wacker, I. Lignos, V. K. Parashar and M. A. M. Gijs, *Lab Chip*, 2012, **12**, 3111.
300. N. J. Carroll, S. B. Rathod, E. Derbins, S. Mendez, D. A. Weitz and D. N. Petsev, *Langmuir*, 2008, **24**, 658.

301. I. Lee, Y. Yoo, Z. Cheng and H.-K. Jeong, *Adv. Funct. Mater.*, 2008, **18**, 4014.
302. Y. Nie, N. Hao and J. X. J. Zhang, *Sci. Rep.*, 2017, **7**, 12616.
303. N. Hao, Y. Nie and J. X. J. Zhang, *ACS Sustainable Chem. Eng.*, 2018, **6**, 1522.
304. L. Gomez, M. Arruebo, V. Sebastian, L. Gutierrez and J. Santamaria, *J. Mater. Chem.*, 2012, **22**, 21420.
305. A. Baruah, A. Singh, V. Sheoran, B. Prakash and A. K. Ganguli, *Mater. Res. Express*, 2018, **5**, 075019.
306. H. Wang, H. Nakamura, M. Uehara, M. Miyazaki and H. Maeda, *Chem. Commun.*, 2002, 1462.
307. M. Takagi, T. Maki, M. Miyahara and K. Mae, *Chem. Eng. J.*, 2004, **101**, 269.
308. X. Gong, L. Wang and W. Wen, *Chem. Commun.*, 2009, 4690.
309. S. A. Khan and K. F. Jensen, *Adv. Mater.*, 2007, **19**, 2556.
310. A. Abou-Hassan, R. Bazzi and V. Cabuil, *Angew. Chem., Int. Ed.*, 2009, **48**, 7180.
311. B. G. Zukas and N. R. Gupta, *Ind. Eng. Chem. Res.*, 2017, **56**, 7184.
312. A. Baruah, A. Jindal, C. Acharya, B. Prakash, S. Basu and A. K. Ganguli, *J. Manuf. Syst.*, 2017, **27**, 035013.
313. H. W. Kang, J. Leem, S. Y. Yoon and H. J. Sung, *Nanoscale*, 2014, **6**, 2840.
314. Y. Wang, X. Zhang, A. Wang, X. Li, G. Wang and L. Zhao, *Chem. Eng. J.*, 2014, **235**, 191.
315. G. Salazar-Alvarez, M. Muhammed and A. A. Zagorodni, *Chem. Eng. Sci.*, 2006, **61**, 4625.
316. A. Abou-Hassan, O. Sandre, V. Cabuil and P. Tabeling, *Chem. Commun.*, 2008, 1783.
317. L. Frenz, A. el Harrak, M. Pauly, S. Bégin-Colin, A. D. Griffiths and J.-C. Baret, *Angew. Chem., Int. Ed.*, 2008, **47**, 6817.
318. L. Norfolk, A. Rawlings, J. Bramble, K. Ward, N. Francis, R. Waller, A. Bailey and S. Staniland, *Nanomaterials*, 2019, **9**, 1729.
319. K. Kumar, A. M. Nightingale, S. H. Krishnadasan, N. Kamaly, M. Wylenzinska-Arridge, K. Zeissler, W. R. Branford, E. Ware, A. J. deMello and J. C. deMello, *J. Mater. Chem.*, 2012, **22**, 4704.
320. A. Abou-Hassan, O. Sandre, S. Neveu and V. Cabuil, *Angew. Chem., Int. Ed.*, 2009, **48**, 2342.
321. P. Zardi, T. Carofiglio and M. Maggini, *Chem. – Eur. J.*, 2022, **28**, e202103132.
322. S. Seibt, P. Mulvaney and S. Förster, *Colloids Surf., A*, 2019, **562**, 263.
323. L.-H. Hung, K. M. Choi, W.-Y. Tseng, Y.-C. Tan, K. J. Shea and A. P. Lee, *Lab Chip*, 2006, **6**, 174.
324. B. K. H. Yen, N. E. Stott, K. F. Jensen and M. G. Bawendi, *Adv. Mater.*, 2003, **15**, 1858.
325. B. K. H. Yen, A. Günther, M. A. Schmidt, K. F. Jensen and M. G. Bawendi, *Angew. Chem., Int. Ed.*, 2005, **44**, 5447.
326. E. M. Chan, A. P. Alivisatos and R. A. Mathies, *J. Am. Chem. Soc.*, 2005, **127**, 13854.

327. H. Wang, X. Li, M. Uehara, Y. Yamaguchi, H. Nakamura, M. Miyazaki, H. Shimizu and H. Maeda, *Chem. Commun.*, 2004, 48.
328. H. Wang, H. Nakamura, M. Uehara, Y. Yamaguchi, M. Miyazaki and H. Maeda, *Adv. Funct. Mater.*, 2005, **15**, 603.
329. W. Luan, H. Yang, N. Fan and S.-T. Tu, *Nanoscale Res. Lett.*, 2008, **3**, 134.
330. M. Uehara, H. Nakamura and H. Maeda, *J. Nanosci. Nanotechnol.*, 2009, **9**, 577.
331. N. Dengo, A. Faresin, T. Carofiglio, M. Maggini, L. Wu, J. P. Hofmann, E. J. M. Hensen, P. Dolcet and S. Gross, *Chem. Commun.*, 2020, **56**, 8707.
332. F. Tajoli, N. Dengo, M. Mognato, P. Dolcet, G. Lucchini, A. Faresin, J.-D. Grunwaldt, X. Huang, D. Badocco, M. Maggini, C. Kübel, A. Speghini, T. Carofiglio and S. Gross, *ACS Appl. Mater. Interfaces*, 2020, **12**, 44074.
333. A. M. Nightingale and J. C. deMello, *ChemPhysChem*, 2009, **10**, 2612.
334. J. Baek, P. M. Allen, M. G. Bawendi and K. F. Jensen, *Angew. Chem., Int. Ed.*, 2011, **50**, 627.
335. J. Baek, Y. Shen, I. Lignos, M. G. Bawendi and K. F. Jensen, *Angew. Chem., Int. Ed.*, 2018, **57**, 10915.
336. Z. Bao, H.-C. Wang, Z.-F. Jiang, R.-J. Chung and R.-S. Liu, *Inorg. Chem.*, 2018, **57**, 13071.
337. K. Abdel-Latif, R. W. Epps, C. B. Kerr, C. M. Papa, F. N. Castellano and M. Abolhasani, *Adv. Funct. Mater.*, 2019, **29**, 1900712.
338. I. Lignos, V. Morad, Y. Shynkarenko, C. Bernasconi, R. M. Maceiczyk, L. Protesescu, F. Bertolotti, S. Kumar, S. T. Ochsenbein, N. Masciocchi, A. Guagliardi, C.-J. Shih, M. I. Bodnarchuk, A. J. DeMello and M. V. Kovalenko, *ACS Nano*, 2018, **12**, 5504.
339. X. Zhu, Q. Zhang, Y. Li and H. Wang, *J. Mater. Chem.*, 2010, **20**, 1766.
340. A. Y. Baranchikov, V. K. Ivanov and Y. D. Tretyakov, *Russ. Chem. Rev.*, 2007, **76**, 133.
341. M. Ashokkumar, *Ultrasonic Synthesis of Functional Materials*, Springer International Publishing, Cham, 2016.
342. K. S. Suslick, *Science*, 1990, **247**, 1439.
343. H. Xu, B. W. Zeiger and K. S. Suslick, *Chem. Soc. Rev.*, 2013, **42**, 2555.
344. K. S. Suslick and W. L. Nyborg, *J. Acoust. Soc. Am.*, 1990, **87**, 919.
345. T. Mason and J. P. Lorimer, *Sonochemistry: Theory, Applications and Uses of Ultrasound in Chemistry*, Wiley-Interscience, New York, 1989.
346. W. Lauterborn and C.-D. Ohl, *Ultrason. Sonochem.*, 1997, **4**, 65.
347. C.-D. Ohl, T. Kurz, R. Geisler, O. Lindau and W. Lauterborn, *Philos. Trans. R. Soc., A*, 1999, **357**, 269.
348. S. Muthukumaran, S. E. Kentish, G. W. Stevens and M. Ashokkumar, *Rev. Chem. Eng.*, 2006, **22**, 155.
349. S. Doktycz and K. Suslick, *Science*, 1990, **247**, 1067.
350. T. Leighton, *The Acoustic Bubble*, Academic Press, London, 1st edn, 1994.
351. E. B. Flint and K. S. Suslick, *Science*, 1991, **253**, 1397.
352. W. B. McNamara, Y. T. Didenko and K. S. Suslick, *Nature*, 1999, **401**, 772.

353. Y. T. Didenko, W. B. McNamara and K. S. Suslick, *Nature*, 2000, **407**, 877.
354. F. R. Young, *Sonoluminescence*, CRC Press, 2004.
355. M. A. Margulis, *Sonochemistry and Cavitation*, Gordon Breach Publishers, Luxembourg, 1995.
356. R. Roy, *J. Solid State Chem.*, 1994, **111**, 11.
357. C. N. R. Rao, A. Müller and A. K. Cheetham, *The Chemistry of Nanomaterials: Synthesis, Properties and Applications*, Wiley-VCH, 2004.
358. Y. Mizukoshi, R. Oshima, Y. Maeda and Y. Nagata, *Langmuir*, 1999, **15**, 2733.
359. Supeno and P. Kruus, *Ultrason. Sonochem.*, 2000, **7**, 109.
360. K. S. Suslick, T. Hyeon and M. Fang, *Chem. Mater.*, 1996, **8**, 2172.
361. T. Prozorov, R. Prozorov and K. S. Suslick, *J. Am. Chem. Soc.*, 2004, **126**, 13890.
362. D. G. Shchukin, D. Radziuk and H. Möhwald, *Annu. Rev. Mater. Res.*, 2010, **40**, 345.
363. J. H. Bang and K. S. Suslick, *Adv. Mater.*, 2010, **22**, 1039.
364. K. S. Suslick and G. J. Price, *Annu. Rev. Mater. Sci.*, 1999, **29**, 295.
365. C. Petrier, A. Jeunet, J. L. Luche and G. Reverdy, *J. Am. Chem. Soc.*, 1992, **114**, 3148.
366. M. A. Beckett and I. Hua, *J. Phys. Chem. A*, 2001, **105**, 3796.
367. P. Kanthale, M. Ashokkumar and F. Grieser, *Ultrason. Sonochem.*, 2008, **15**, 143.
368. M. Capocelli, E. Joyce, A. Lancia, T. J. Mason, D. Musmarra and M. Prisciandaro, *Chem. Eng. J.*, 2012, **210**, 9.
369. K. S. Suslick, S.-B. Choe, A. A. Cichowlas and M. W. Grinstaff, *Nature*, 1991, **353**, 414.
370. M. W. Grinstaff, A. A. Cichowlas, S.-B. Choe and K. S. Suslick, *Ultrasonics*, 1992, **30**, 168.
371. K. S. Suslick, M. Fang and T. Hyeon, *J. Am. Chem. Soc.*, 1996, **118**, 11960.
372. A. Gedanken, *Ultrason. Sonochem.*, 2004, **11**, 47.
373. K. S. Suslick, J. W. Goodale, P. F. Schubert and H. H. Wang, *J. Am. Chem. Soc.*, 1983, **105**, 5781.
374. B. L. Cushing, V. L. Kolesnichenko and C. J. O'Connor, *Chem. Rev.*, 2004, **104**, 3893.
375. K. S. Suslick, T. Hyeon, M. Fang and A. A. Cichowlas, in *Advances in Catalytic Nanostructured Materials*, ed. W. Moser, Academic Press, 1st edn, 1996, pp. 197–212.
376. N. A. Dhas and A. Gedanken, *J. Mater. Chem.*, 1998, **8**, 445.
377. Yu. Koltypin, G. Katabi, X. Cao, R. Prozorov and A. Gedanken, *J. Non-Cryst. Solids*, 1996, **201**, 159.
378. N. A. Dhas, C. P. Raj and A. Gedanken, *Chem. Mater.*, 1998, **10**, 3278.
379. K. V. P. M. Shafi, A. Gedanken, R. B. Goldfarb and I. Felner, *J. Appl. Phys.*, 1997, **81**, 6901.
380. A. Henglein, *Adv. Sonochem.*, 1993, **3**, 17.
381. K. S. Suslick, D. A. Hammerton and R. E. Cline, *J. Am. Chem. Soc.*, 1986, **108**, 5641.

382. S. A. Yeung, R. Hobson, S. Biggs and F. Grieser, *J. Chem. Soc. Chem. Commun.*, 1993, 378.
383. Y. Nagata, Y. Mizukoshi, K. Okitsu and Y. Maeda, *Radiat. Res.*, 1996, **146**, 333.
384. R. a. Caruso, M. Ashokkumar and F. Grieser, *Colloids Surf., A*, 2000, **169**, 219.
385. R. A. Salkar, P. Jeevanandam, S. T. Aruna, Y. Koltypin and A. Gedanken, *J. Mater. Chem.*, 1999, **9**, 1333.
386. R. A. Caruso, M. Ashokkumar and F. Grieser, *Langmuir*, 2002, **18**, 7831.
387. J. E. Park, M. Atobe and T. Fuchigami, *Ultrason. Sonochem.*, 2006, **13**, 237.
388. K. Okitsu, M. Ashokkumar and F. Grieser, *J. Phys. Chem. B*, 2005, **109**, 20673.
389. V. K. LaMer and R. H. Dinegar, *J. Am. Chem. Soc.*, 1950, **72**, 4847.
390. K. Okitsu, K. Sharyo and R. Nishimura, *Langmuir*, 2009, **25**, 7786.
391. L.-P. Jiang, S. Xu, J.-M. Zhu, J.-R. Zhang, J.-J. Zhu and H.-Y. Chen, *Inorg. Chem.*, 2004, **43**, 5877.
392. H. Xu and K. S. Suslick, *ACS Nano*, 2010, **4**, 3209.
393. A. Sánchez-Iglesias, I. Pastoriza-Santos, J. Pérez-Juste, B. Rodríguez-González, F. J. García de Abajo and L. M. Liz-Marzán, *Adv. Mater.*, 2006, **18**, 2529.
394. J. Zhang, J. Du, B. Han, Z. Liu, T. Jiang and Z. Zhang, *Angew. Chem., Int. Ed.*, 2006, **45**, 1116.
395. K. Vinodgopal, Y. He, M. Ashokkumar and F. Grieser, *J. Phys. Chem. B*, 2006, **110**, 3849.
396. P. S. S. Kumar, A. Manivel, S. Anandan, M. Zhou, F. Grieser and M. Ashokkumar, *Colloids Surf., A*, 2010, **356**, 140.
397. Y. Mizukoshi, K. Okitsu, Y. Maeda, T. A. Yamamoto, R. Oshima and Y. Nagata, *J. Phys. Chem. B*, 1997, **101**, 7033.
398. S. Anandan, F. Grieser and M. Ashokkumar, *J. Phys. Chem. C*, 2008, **112**, 15102.
399. K. V. P. M. Shafi, Y. Koltypin, A. Gedanken, R. Prozorov, J. Balogh, J. Lendvai and I. Felner, *J. Phys. Chem. B*, 1997, **101**, 6409.
400. K. V. P. M. Shafi, I. Felner, Y. Mastai and A. Gedanken, *J. Phys. Chem. B*, 1999, **103**, 3358.
401. J. H. Bang and K. S. Suslick, *J. Am. Chem. Soc.*, 2007, **129**, 2242.
402. N. Du, H. Zhang, B. D. Chen, J. B. Wu, X. Y. Ma, Z. H. Liu, Y. Q. Zhang, D. R. Yang, X. H. Huang and J. P. Tu, *Adv. Mater.*, 2007, **19**, 4505.
403. E. Ohayon and A. Gedanken, *Ultrason. Sonochem.*, 2010, **17**, 173.
404. M. Niederberger, M. H. Bartl and G. D. Stucky, *J. Am. Chem. Soc.*, 2002, **124**, 13642.
405. N. A. Dhas, Y. Koltypin and A. Gedanken, *Chem. Mater.*, 1997, **9**, 3159.
406. B. Tian, F. Chen, J. Zhang and M. Anpo, *J. Colloid Interface Sci.*, 2006, **303**, 142.
407. K. Yang, J. Zhu, J. Zhu, S. Huang, X. Zhu and G. Ma, *Mater. Lett.*, 2003, **57**, 4639.

408. X. K. Wang, W. L. Guo and C. Wang, *Adv. Mater. Res.*, 2011, **356–360**, 399.
409. T. Alammar, H. Noei, Y. Wang and A.-V. Mudring, *Nanoscale*, 2013, **5**, 8045.
410. W. Guo, Z. Lin, X. Wang and G. Song, *Microelectron. Eng.*, 2003, **66**, 95.
411. Y. Zhu, H. Li, Y. Koltypin, Y. R. Hacohen and A. Gedanken, *Chem. Commun.*, 2001, 2616.
412. W. Huang, X. Tang, Y. Wang, Y. Koltypin and A. Gedanken, *Chem. Commun.*, 2000, 1415.
413. J. C. Yu, L. Zhang and J. Yu, *Chem. Mater.*, 2002, **14**, 4647.
414. Y. Wang, X. Tang, L. Yin, W. Huang, Y. R. Hacohen and A. Gedanken, *Adv. Mater.*, 2000, **12**, 1183.
415. L. Zhang and J. C. Yu, *Chem. Commun.*, 2003, 2078.
416. S. H. Jung, E. Oh, K. H. Lee, Y. Yang, C. G. Park, W. Park and S. H. Jeong, *Cryst. Growth Des.*, 2008, **8**, 265.
417. R. Wahab, S. G. Ansari, Y. S. Kim, H. K. Seo and H. S. Shin, *Appl. Surf. Sci.*, 2007, **253**, 7622.
418. S. S. Warule, N. S. Chaudhari, J. D. Ambekar, B. B. Kale and M. a. More, *ACS Appl. Mater. Interfaces*, 2011, **3**, 3454.
419. P. Banerjee, S. Chakrabarti, S. Maitra and B. K. Dutta, *Ultrason. Sonochem.*, 2012, **19**, 85.
420. X. L. Hu, Y. J. Zhu and S. W. Wang, *Mater. Chem. Phys.*, 2004, **88**, 421.
421. P. Taylor, C. Pholnak, C. Sirisathitkul, S. Danworaphong and D. J. Harding, *Ferroelectrics*, 2013, **455**, 15.
422. A. Phuruangrat, O. Yayapao, T. Thongtem and S. Thongtem, *J. Nanomater.*, 2014, **2014**, 1.
423. A. K. Zak, W. H. A. Majid, H. Z. Wang, R. Yousefi, A. M. Golsheikh and Z. F. Ren, *Ultrason. Sonochem.*, 2013, **20**, 395.
424. D. Qian, J. Z. Jiang and P. Lenvig Hansen, *Chem. Commun.*, 2003, 1078.
425. S. Bhattacharyya and A. Gedanken, *Microporous Mesoporous Mater.*, 2008, **110**, 553.
426. H.-M. Xiong, D. G. Shchukin, H. Möhwald, Y. Xu and Y.-Y. Xia, *Angew. Chem., Int. Ed.*, 2009, **48**, 2727.
427. L. Yin, Y. Wang, G. Pang, Y. Koltypin and A. Gedanken, *J. Colloid Interface Sci.*, 2002, **246**, 78.
428. J. J. Miao, H. Wang, Y. R. Li, J. M. Zhu and J. J. Zhu, *J. Cryst. Growth*, 2005, **281**, 525.
429. D. V. Pinjari and A. B. Pandit, *Ultrason. Sonochem.*, 2011, **18**, 1118.
430. H. Wang, J.-J. Zhu, J.-M. Zhu, X.-H. Liao, S. Xu, T. Ding and H.-Y. Chen, *Phys. Chem. Chem. Phys.*, 2002, **4**, 3794.
431. R. V. Kumar, Y. Diamant and A. Gedanken, *Chem. Mater.*, 2000, **12**, 2301.
432. R. V. Kumar, Y. Koltypin, X. N. Xu, Y. Yeshurun, A. Gedanken and I. Felner, *J. Appl. Phys.*, 2001, **89**, 6324.
433. C. V. Krishnan, J. Chen, C. Burger and B. Chu, *J. Phys. Chem. B*, 2006, **110**, 20182.
434. C.-J. Mao, H.-C. Pan, X.-C. Wu, J.-J. Zhu and H.-Y. Chen, *J. Phys. Chem. B*, 2006, **110**, 14709.

435. D. P. Dutta, V. Sudarsan, P. Srinivasu, A. Vinu and A. K. Tyagi, *J. Phys. Chem. C*, 2008, **112**, 6781.
436. P. Jeevanandam, Yu. Koltypin and A. Gedanken, *Nano Lett.*, 2001, **1**, 263.
437. S. Avivi, Y. Mastai, G. Hodes and A. Gedanken, *J. Am. Chem. Soc.*, 1999, **121**, 4196.
438. S. Avivi, Y. Mastai and A. Gedanken, *Chem. Mater.*, 2000, **12**, 1229.
439. M. Sivakumar, T. Takami, H. Ikuta, A. Towata, K. Yasui, T. Tuziuti, T. Kozuka, D. Bhattacharya and Y. Iida, *J. Phys. Chem. B*, 2006, **110**, 15234.
440. M. Abbas, B. P. Rao, M. N. Islam, K. W. Kim, S. M. Naga, M. Takahashi and C. Kim, *Ceram. Int.*, 2014, **40**, 3269.
441. S. Theerdhala, D. Alhat, S. Vitta and D. Bahadur, *J. Nanosci. Nanotechnol.*, 2008, **8**, 4268.
442. T. Dippong, E. A. Levei and O. Cadar, *Nanomaterials*, 2021, **11**, 1560.
443. J. C. Yu, L. Zhang, Q. Li, K. W. Kwong, A.-W. Xu and J. Lin, *Langmuir*, 2003, **19**, 7673.
444. P. Mehdizadeh, O. Amiri, S. Rashki, M. Salavati-Niasari, M. Salimian and L. K. Foong, *Ultrason. Sonochem.*, 2020, **61**, 104848.
445. J. Geng, J.-J. Zhu and H.-Y. Chen, *Cryst. Growth Des.*, 2006, **6**, 321.
446. J. Geng, J.-J. Zhu, D.-J. Lu and H.-Y. Chen, *Inorg. Chem.*, 2006, **45**, 8403.
447. D. P. Dutta, R. Ghildiyal and A. K. Tyagi, *J. Phys. Chem. C*, 2009, **113**, 16954.
448. M. M. Mdleleni, T. Hyeon and K. S. Suslick, *J. Am. Chem. Soc.*, 1998, **120**, 6189.
449. S. I. Nikitenko, Y. Koltypin, Y. Mastai, M. Koltypin and A. Gedanken, *J. Mater. Chem.*, 2002, **12**, 1450.
450. G. Henshaw, I. P. Parkin and G. Shaw, *Chem. Commun.*, 1996, 1095.
451. G. Henshaw, I. P. Parkin and G. A. Shaw, *J. Chem. Soc., Dalton Trans.*, 1997, 231.
452. P. Jeevanandam, Yu. Koltypin, Y. Gofer, Y. Diamant and A. Gedanken, *J. Mater. Chem.*, 2000, **10**, 2769.
453. S. Avivi, O. Palchik, V. Palchik, M. A. Slifkin, A. M. Weiss and A. Gedanken, *Chem. Mater.*, 2001, **12**, 2195.
454. J. F. Xu, W. Ji, J. Y. Lin, S. H. Tang and Y. W. Du, *Appl. Phys. A: Mater. Sci. Process.*, 1998, **66**, 639.
455. G.-J. Lee, S. Anandan, S. J. Masten and J. J. Wu, *Ind. Eng. Chem. Res.*, 2014, **53**, 8766.
456. Y.-P. Zhu, J. Li, T.-Y. Ma, Y.-P. Liu, G. Du and Z.-Y. Yuan, *J. Mater. Chem. A*, 2014, **2**, 1093.
457. J. Geng, B. Liu, L. Xu, F. N. Hu and J. J. Zhu, *Langmuir*, 2007, **23**, 10286.
458. G. Z. Wang, B. Y. Geng, X. M. Huang, Y. W. Wang, G. H. Li and L. D. Zhang, *Appl. Phys. A: Mater. Sci. Process.*, 2003, **77**, 933.
459. E. K. Goharshadi, S. H. Sajjadi, R. Mehrkhah and P. Nancarrow, *Chem. Eng. J.*, 2012, **209**, 113.
460. H. Zhou, T. Fan, D. Zhang, Q. Guo and H. Ogawa, *Chem. Mater.*, 2007, **19**, 2144.

461. A. A. Othman, M. A. Osman, M. A. Ali, W. S. Mohamed and E. M. M. Ibrahim, *J. Mater. Sci.: Mater. Electron.*, 2020, **31**, 1752.
462. M. Kristl, N. Hojnik, S. Gyergyek and M. Drofenik, *Mater. Res. Bull.*, 2013, **48**, 1184.
463. J. Z. Xu, S. Xu, J. Geng, G. X. Li and J. J. Zhu, *Ultrason. Sonochem.*, 2006, **13**, 451.
464. Y. Gu, T. Li, B. Guo, Y. Jiang, W. Wen, J. Wu and L. Zhao, *CrystEngComm*, 2020, **22**, 7082.
465. R. Zeinodin and F. Jamali-Sheini, *Phys. B*, 2019, **570**, 148.
466. P. a. L. Lopes, M. B. Santos, A. J. S. Mascarenhas and L. A. Silva, *Mater. Lett.*, 2014, **136**, 111.
467. M. Kristl, I. Ban, A. Danč, V. Danč and M. Drofenik, *Ultrason. Sonochem.*, 2010, **17**, 916.
468. R. S. Yadav, P. Mishra, R. Mishra, M. Kumar and A. C. Pandey, *Ultrason. Sonochem.*, 2010, **17**, 116.
469. M. Behboudnia and B. Khanbabaee, *Colloids Surf., A*, 2006, **290**, 229.
470. G. S. Wu, X. Y. Yuan, T. Xie, G. C. Xu, L. D. Zhang and Y. L. Zhuang, *Mater. Lett.*, 2004, **58**, 794.
471. S. M. Zhou, X. H. Zhang, X. M. Meng, X. Fan, S. T. Lee and S. K. Wu, *J. Solid State Chem.*, 2005, **178**, 399.
472. S. F. Wang, F. Gu, M. K. Lü, G. J. Zhou and A. Yu Zhang, *J. Cryst. Growth*, 2006, **289**, 621.
473. J. Zhu, S. Liu, O. Palchik, Y. Koltypin and A. Gedanken, *J. Solid State Chem.*, 2000, **153**, 342.
474. Z. Hosseini, Y. Azizian-Kalandaragh, A. Khodayari and B. Nedaee-Shakarab, *Optoelectron. Adv. Mater., Rapid Commun.*, 2014, **8**, 201.
475. H. Wang and J. J. Zhu, *Ultrason. Sonochem.*, 2004, **11**, 293.
476. M. Kristl and M. Drofenik, *Ultrason. Sonochem.*, 2008, **15**, 695.
477. N. a. García-Gómez, S. M. D. La Parra-Arcieniega, L. L. Garza-Tovar, L. C. Torres-González and E. M. Sánchez, *J. Alloys Compd.*, 2014, **588**, 638.
478. G. Matyszczak, P. Jóźwik, E. Polesiak, M. Sobieska, K. Krawczyk, C. Jastrzębski and T. Płociński, *Ultrason. Sonochem.*, 2021, **75**, 105594.
479. P. R. R. Mesquita, J. S. Almeida, L. S. G. Teixeira, A. F. Da Silva and L. a. Silva, *J. Braz. Chem. Soc.*, 2013, **24**, 280.
480. H. Wang, J.-J. Zhu, J.-M. Zhu and H.-Y. Chen, *J. Phys. Chem. B*, 2002, **106**, 3848.
481. P. Salinas-Estevané and E. M. Sánchez, *Cryst. Growth Des.*, 2010, **10**, 3917.
482. P. S. Estevané, A. García, S. de La Parra Arcinega and E. M. Sánchez, *Chalcogenide Lett.*, 2013, **10**, 283.
483. J.-J. Zhu, H. Wang, S. Xu and H.-Y. Chen, *Langmuir*, 2002, **18**, 3306.
484. Y. Xie, X. Zheng, X. Jiang, J. Lu and L. Zhu, *Inorg. Chem.*, 2002, **41**, 387.
485. B. Li, Y. Xie, J. Huang, Y. Liu and Y. Qian, *Chem. Mater.*, 2000, **12**, 2614.
486. B. Pejova, *J. Phys. Chem. C*, 2013, **117**, 19689.
487. M. J. Murcia, D. L. Shaw, H. Woodruff, C. a. Naumann, B. a. Young and E. C. Long, *Chem. Mater.*, 2006, **18**, 2219.

488. G. Z. Wang, W. Chen, C. H. Liang, Y. W. Wang, G. W. Meng and L. D. Zhang, *Inorg. Chem. Commun.*, 2001, **4**, 208.
489. X. Zheng, Y. Xie, L. Zhu, X. Jiang and A. Yan, *Ultrason. Sonochem.*, 2002, **9**, 311.
490. L. Shen, N. Bao, P. E. Prevelige and A. Gupta, *J. Phys. Chem. C*, 2010, **114**, 2551.
491. J.-J. Zhu, S. Xu, H. Wang, J.-M. Zhu and H.-Y. Chen, *Adv. Mater.*, 2003, **15**, 156.
492. I. Uzcanga, I. Bezverkhyy, P. Afanasiev, C. Scott and M. Vrinat, *Chem. Mater.*, 2005, **17**, 3575.
493. T. Gao, Q. Li and T. Wang, *Chem. Mater.*, 2005, **17**, 887.
494. T. Gao and T. Wang, *Chem. Commun.*, 2004, 2558.
495. J. D. Oxley, M. M. Mdleleni and K. S. Suslick, *Catal. Today*, 2004, **88**, 139.
496. S. Sivasankaran and M. J. K. Kumar, *Ceram. Int.*, 2015, **41**, 11301.
497. T. Hyeon, M. Fang and K. S. Suslick, *J. Am. Chem. Soc.*, 1996, **118**, 5492.
498. Yu. Koltypin, X. Cao, R. Prozorov, J. Balogh, D. Kaptas and A. Gedanken, *J. Mater. Chem.*, 1997, **7**, 2453.
499. P. Hasin and Y. Wu, *Chem. Commun.*, 2012, **48**, 1302.
500. L. Zhu, Y. Liu, X. Fan, D. Yang and X. Cao, *Mater. Res. Bull.*, 2011, **46**, 252.
501. L. Zhu, X. Liu, J. Meng and X. Cao, *Cryst. Growth Des.*, 2007, **7**, 2505.
502. L. Zhu, J. Meng and X. Cao, *J. Solid State Chem.*, 2007, **180**, 3101.
503. C. Yu, M. Yu, C. Li, X. Liu, J. Yang, P. Yang and J. Lin, *J. Solid State Chem.*, 2009, **182**, 339.
504. R. Schiller and M. Ebert, *Int. J. Radiat. Phys. Chem.*, 1969, **1**, 111.
505. K. Čubová and V. Čuba, *Radiat. Phys. Chem.*, 2020, **169**, 108774.
506. A. Abedini, A. R. Daud, M. A. A. Hamid, N. K. Othman and E. Saion, *Nanoscale Res. Lett.*, 2013, **8**, 474.
507. P. K. Das, *Chem. Rev.*, 1993, **93**, 119.
508. B. C. Garrett, D. A. Dixon, D. M. Camaioni, D. M. Chipman, M. A. Johnson, C. D. Jonah, G. A. Kimmel, J. H. Miller, T. N. Rescigno, P. J. Rossky, S. S. Xantheas, S. D. Colson, A. H. Laufer, D. Ray, P. F. Barbara, D. M. Bartels, K. H. Becker, K. H. Bowen, S. E. Bradforth, I. Carmichael, J. v. Coe, L. R. Corrales, J. P. Cowin, M. Dupuis, K. B. Eisenthal, J. A. Franz, M. S. Gutowski, K. D. Jordan, B. D. Kay, J. A. LaVerne, S. v. Lymar, T. E. Madey, C. W. McCurdy, D. Meisel, S. Mukamel, A. R. Nilsson, T. M. Orlando, N. G. Petrik, S. M. Pimblott, J. R. Rustad, G. K. Schenter, S. J. Singer, A. Tokmakoff, L.-S. Wang and T. S. Zwier, *Chem. Rev.*, 2005, **105**, 355.
509. M. L'Annunziata, *Handbook of Radioactivity Analysis*, Academic Press, Oxford, 2nd edn, 2003.
510. J. Grand, S. R. Ferreira, V. de Waele, S. Mintova and T. M. Nenoff, *J. Phys. Chem. C*, 2018, **122**, 12573.
511. J. Belloni, M. Mostafavi, H. Remita, J.-L. Marignier and M.-O. Delcourt, *New J. Chem.*, 1998, **22**, 1239.
512. C.-H. Wang, T.-E. Hua, C.-C. Chien, Y.-L. Yu, T.-Y. Yang, C.-J. Liu, W.-H. Leng, Y. Hwu, Y.-C. Yang, C.-C. Kim, J.-H. Je, C.-H. Chen, H.-M. Lin and G. Margaritondo, *Mater. Chem. Phys.*, 2007, **106**, 323.

513. Q. Ma, N. Moldovan, D. C. Mancini and R. A. Rosenberg, *Appl. Phys. Lett.*, 2000, **76**, 2014.
514. P. H. Borse, J. M. Yi, J. H. Je, S. D. Choi, Y. Hwu, P. Ruterana and G. Nouet, *Nanotechnology*, 2004, **15**, S389.
515. H. J. Lee, J. H. Je, Y. Hwu and W. L. Tsai, *Nucl. Instrum. Methods Phys. Res., Sect. B*, 2003, **199**, 342.
516. T. Tamai, M. Watanabe, T. Teramura, N. Nishioka and K. Matsukawa, *Macromol. Symp.*, 2010, **288**, 104.
517. G. A. Filip, B. Moldovan, I. Baldea, D. Olteanu, R. Suharoschi, N. Decea, C. M. Cismaru, E. Gal, M. Cenariu, S. Clichici and L. David, *J. Photochem. Photobiol., B*, 2019, **191**, 26.
518. M. Misawa and J. Takahashi, *Nanomedicine*, 2011, **7**, 604.
519. K. Leung and T. M. Nenoff, *J. Chem. Phys.*, 2012, **137**, 074502.
520. J. Belloni, J.-L. Marignier and M. Mostafavi, *Radiat. Phys. Chem.*, 2020, **169**, 107952.
521. G. v. Buxton, C. L. Greenstock, W. P. Helman and A. B. Ross, *J. Phys. Chem. Ref. Data*, 1988, **17**, 513.
522. S. le Caër, *Water*, 2011, **3**, 235.
523. G. G. Flores-Rojas, F. López-Saucedo and E. Bucio, *Radiat. Phys. Chem.*, 2020, **169**, 107962.
524. P. Neta, *Chem. Rev.*, 1972, **72**, 533.
525. B. Ildusovich, O. Kharisov, V. Kharissova and U. O. Mendez, *Radiation Synthesis of Materials and Compounds*, CRC Press, Boca Raton, 2013.
526. P. Wardman, *J. Phys. Chem. Ref. Data*, 1989, **18**, 1637.
527. A. Henglein, *Ber. Bunsenges. Phys. Chem.*, 1991, **95**, 451.
528. A. Appleby and H. A. Schwarz, *J. Phys. Chem.*, 1969, **73**, 1937.
529. E. Bashandy, *Int. J. Appl. Radiat. Isot.*, 1962, **13**, 173.
530. T. Li, H. G. Park and S.-H. Choi, *Mater. Chem. Phys.*, 2007, **105**, 325.
531. J. Belloni, *Catal. Today*, 2006, **113**, 141.
532. I. v. Korolkov, O. Güven, A. A. Mashentseva, A. B. Atıcı, Y. G. Gorin, M. v. Zdorovets and A. A. Taltenov, *Radiat. Phys. Chem.*, 2017, **130**, 480.
533. E. Gharibshahi, E. Saion, A. Ashraf and L. Gharibshahi, *Appl. Radiat. Isot.*, 2017, **130**, 211.
534. E. T. Dias, S. K. Gill, Y. Liu, P. Halstenberg, S. Dai, J. Huang, J. Mausz, R. Gakhar, W. C. Phillips, S. Mahurin, S. M. Pimblott, J. F. Wishart and A. I. Frenkel, *J. Phys. Chem. Lett.*, 2021, **12**, 157.
535. D. Jiao, K. Leung, S. B. Rempe and T. M. Nenoff, *J. Chem. Theory Comput.*, 2011, **7**, 485.
536. M. Z. Kassaee, A. Akhavan, N. Sheikh and R. Beteshobabrud, *Radiat. Phys. Chem.*, 2008, **77**, 1074.
537. M. Bekhit, A. O. A. El Naga, M. el Saied and M. I. A. A. Maksoud, *Environ. Sci. Pollut. Res.*, 2021, **28**, 44467.
538. Z. I. Ali, O. A. Ghazy, G. Meligi, H. H. Saleh and M. Bekhit, *Adv. Polym. Technol.*, 2018, **37**, 365.
539. Z. Jiang, W. Zhu, G. Xu, X. Xu, M. Wang, H. Chen, W. Huang, X. Ge and M. Lin, *J. Mater. Chem. A*, 2020, **8**, 9109.

540. S. Chettibi, N. Keghouche, Y. Benguedouar, M. M. Bettahar and J. Belloni, *Catal. Lett.*, 2013, **143**, 1166.
541. F. Baletto, M. Boeije, A. Bordet, K. Brinkert, R. Catlow, J. Davies, W. Dononelli, H.-J. Freund, C. Friend, B. Gates, A. Genest, S. Guan, C. Hardacre, J. Hargreaves, H. Huang, G. Hutchings, R. Johnston, S. Lai, C. Lamberti, J. Marbaix, C. R. Miranda, R. Nome, J. Peron, J. Quinson, N. Richards, N. Roesch, A. Russell, S. Said, P. Selvam, P. Sermon, M. Shozi, C.-K. Skylaris, F. Spolaore, J. Walker, K. Whiston and D. Willock, *Faraday Discuss.*, 2018, **208**, 573.
542. P. Filipczak, M. Borkowski, P. Chudobinski, S. Bres, M. Matusiak, G. Nowaczyk and M. Kozanecki, *Radiat. Phys. Chem.*, 2020, **169**, 107948.
543. J. P. Abid, A. W. Wark, P. F. Brevet and H. H. Girault, *Chem. Commun.*, 2002, 792.
544. K. A. Bogle, S. D. Dhole and V. N. Bhoraskar, *Nanotechnology*, 2006, **17**, 3204.
545. I. L. Soroka, J. Bjervås, J. Ceder, G. Wallnerström, M. Connan, N. v. Tarakina, A. C. Maier and Y. Kiros, *Radiat. Phys. Chem.*, 2020, **169**, 108370.
546. B. Soroushian, I. Lampre, J. Belloni and M. Mostafavi, *Radiat. Phys. Chem.*, 2005, **72**, 111.
547. G. R. Dey, *Radiat. Phys. Chem.*, 2005, **74**, 172.
548. V. Dweaele, J. Kecht, Z. Tahri, M. Mostafavi, T. Bein and S. Mintova, *Sens. Actuators, B*, 2007, **126**, 338.
549. A. F. Seliverstov, N. L. Sukhov and B. G. Ershov, *Colloid J.*, 2002, **64**, 781.
550. B. Keita, L. Nadjo, C. de Cointet, J. Amblard and J. Belloni, *Chem. Phys. Lett.*, 1996, **249**, 297.
551. E. Gachard, H. Remita, J. Khatouri, B. Keita, L. Nadjo and J. Belloni, *New J. Chem.*, 1998, **22**, 1257.
552. S. K. Mahapatra, K. A. Bogle, S. D. Dhole and V. N. Bhoraskar, *Nanotechnology*, 2007, **18**, 135602.
553. N. Misra, J. Biswal, A. Gupta, J. K. Sainis and S. Sabharwal, *Radiat. Phys. Chem.*, 2012, **81**, 195.
554. B. G. Ershov, E. Janata, A. Henglein and A. Fojtik, *J. Phys. Chem.*, 1993, **97**, 4589.
555. Y. Liu, Y. Qian, M. Zhang, Z. Chen and C. Wang, *Mater. Lett.*, 1996, **26**, 81.
556. Y. Zhu, Y. Qian, Y. Li, W. Wang, M. Zhang, Z. Chen and S. Tan, *Nanostruct. Mater.*, 1994, **4**, 915.
557. K. D. N. Vo, C. Kowandy, L. Dupont, X. Coqueret and N. Q. Hien, *Radiat. Phys. Chem.*, 2014, **94**, 84.
558. A. Henglein, *J. Phys. Chem.*, 1979, **83**, 2209.
559. A. Yamaguchi, T. Matsumoto, I. Okada, I. Sakurai and Y. Utsumi, *Mater. Chem. Phys.*, 2015, **160**, 205.
560. A. Yamaguchi, I. Okada, T. Fukuoka, I. Sakurai and Y. Utsumi, *Jpn. J. Appl. Phys.*, 2016, **55**, 055502.
561. H. Ki, S. Park, S. Eom, J. Gu, S. Kim, C. Kim, C. W. Ahn, M. Choi, S. Ahn, D.-S. Ahn, J. Choi, M.-H. Baik and H. Ihee, *Int. J. Mol. Sci.*, 2020, **21**, 7125.

562. M. Toro-González, D. M. Clifford, M. C. Molina, C. E. Castano and J. v. Rojas, *Radiat. Phys. Chem.*, 2021, **188**, 109614.
563. S. Yang, Y. Wang, Q. Wang, R. Zhang and B. Ding, *Colloids Surf., A*, 2007, **301**, 174.
564. S. Navaladian, B. Viswanathan, T. Varadarajan and R. Viswanath, *Nanoscale Res. Lett.*, 2009, **4**, 471.
565. J. F. Wishart and B. Rao, *Recent Trends in Radiation Chemistry*, World Scientific, Singapore, 2010.
566. A. Abedini, A. R. Daud, M. A. A. Hamid and N. K. Othman, *PLoS One*, 2014, **9**, e90055.
567. C.-L. Wang, B.-J. Hsao, S.-F. Lai, W.-C. Chen, H.-H. Chen, Y.-Y. Chen, C.-C. Chien, X. Cai, I. M. Kempson, Y. Hwu and G. Margaritondo, *Nanotechnology*, 2011, **22**, 065605.
568. E. Ozkaraoglu, I. Tunc and S. Suzer, *Polymer*, 2009, **50**, 462.
569. R. Ferrando, J. Jellinek and R. L. Johnston, *Chem. Rev.*, 2008, **108**, 845.
570. T. Jurkin, M. Gotić, G. Štefanić and I. Pucić, *Radiat. Phys. Chem.*, 2016, **124**, 75.
571. A. Yanase and H. Komiyama, *Surf. Sci.*, 1991, **248**, 11.
572. Z. Hai, C. Zhu, J. Huang, H. Liu and J. Chen, *Inorg. Chem.*, 2010, **49**, 7217.
573. Z. Li, I. L. Soroka, F. Min and M. Jonsson, *Dalton Trans.*, 2018, **47**, 16139.
574. Z. Li, Y. Yang, A. Relefors, X. Kong, G. M. Siso, B. Wickman, Y. Kiros and I. L. Soroka, *J. Colloid Interface Sci.*, 2021, **583**, 71.
575. Z. Li, D. Piankova, Y. Yang, Y. Kumagai, H. Zschiesche, M. Jonsson, N. v. Tarakina and I. L. Soroka, *Angew. Chem.*, 2022, **61**, e202112204.
576. O. Roth, H. Hasselberg and M. Jonsson, *J. Nucl. Mater.*, 2009, **383**, 231.
577. P. Yadav, R. T. Olsson and M. Jonsson, *Radiat. Phys. Chem.*, 2009, **78**, 939.
578. P. A. Yakabuskie, J. M. Joseph, P. Keech, G. A. Botton, D. Guzonas and J. C. Wren, *Phys. Chem. Chem. Phys.*, 2011, **13**, 7198.
579. L. M. Alrehaily, J. M. Joseph, M. C. Biesinger, D. A. Guzonas and J. C. Wren, *Phys. Chem. Chem. Phys.*, 2013, **15**, 1014.
580. L. M. Alrehaily, J. M. Joseph and J. C. Wren, *J. Phys. Chem. C*, 2015, **119**, 16321.
581. P. Praus, L. Svoboda, J. Čížek and G. Słowik, *J. Nanosci. Nanotechnol.*, 2017, **17**, 4805.
582. V. Čuba, T. Pavelková, J. Bárta, T. Gbur, M. Vlk, A. Zavadilová, J. Indrei, Z. Dočekalová, M. Pospíšil and V. Múčka, *Radiat. Phys. Chem.*, 2012, **81**, 1411.
583. IAEA-TECDOC, *Emerging Applications of Radiation in Nanotechnology*, IAEA, Vienna, 2005.
584. D. Hayes, O. I. Micic, M. T. Nenadovic, V. Swayambunathan and D. Meisel, *J. Phys. Chem.*, 1989, **93**, 4603.
585. K. Poudel, M. Gautam, S. G. Jin, H.-G. Choi, C. S. Yong and J. O. Kim, *Int. J. Pharm.*, 2019, **562**, 135.
586. H.-Y. Wang, X.-W. Hua, F.-G. Wu, B. Li, P. Liu, N. Gu, Z. Wang and Z. Chen, *ACS Appl. Mater. Interfaces*, 2015, **7**, 7082.

587. M. Chen, Y. Xie, H. Chen, Z. Qiao and Y. Qian, *J. Colloid Interface Sci.*, 2001, **237**, 47.
588. D. Zhang, B. Gökce and S. Barcikowski, *Chem. Rev.*, 2017, **117**, 3990.
589. V. Amendola and M. Meneghetti, *Phys. Chem. Chem. Phys.*, 2013, **15**, 3027.
590. D. Zhang, Z. Li and K. Sugioka, *J. Phys.: Photonics*, 2021, **3**, 42002.
591. E. Fazio, B. Gökce, A. de Giacomo, M. Meneghetti, G. Compagnini, M. Tommasini, F. Waag, A. Lucotti, C. G. Zanchi, P. M. Ossi, M. Dell'Aglio, L. D'Urso, M. Condorelli, V. Scardaci, F. Biscaglia, L. Litti, M. Gobbo, G. Gallo, M. Santoro, S. Trusso and F. Neri, *Nanomaterials*, 2020, **10**, 2317.
592. D. Zhang, Z. Li, C. Liang, D. S. Zhang, Z. G. Li and C. H. Liang, *Sci. China: Phys., Mech. Astron.*, 2022, **65**, 1.
593. V. Amendola, D. Amans, Y. Ishikawa, N. Koshizaki, S. Scirè, G. Compagnini, S. Reichenberger and S. Barcikowski, *Chem. – Eur. J.*, 2020, **26**, 9206.
594. V. Amendola and M. Meneghetti, *Phys. Chem. Chem. Phys.*, 2009, **11**, 3805.
595. K. Sugioka, in *Handbook of Laser Micro- and Nano-Engineering*, Springer, Cham, 2021.
596. S. Barcikowski, V. Amendola, G. Marzun, C. Rehbock, S. Reichenberger, D. Zhang and B. Gökce, *Handbook of Laser Synthesis & Processing of Colloids*, DuePublico, Essen, 2nd edn, 2019.
597. Y. Iida, A. Tsuge, Y. Uwamino, H. Morikawa and T. Ishizuka, *J. Anal. At. Spectrom.*, 1991, **6**, 541.
598. Particular GmbH, http://particular.eu/
599. A. K. Narasimhan, B. S. Lakshmi, T. S. Santra, M. S. R. Rao and G. Krishnamurthi, *RSC Adv.*, 2017, **7**, 53822.
600. A. Pramanik, S. Karmakar, P. Kumbhakar, S. Biswas, R. Sarkar and P. Kumbhakar, *Appl. Surf. Sci.*, 2020, **499**, 143902.
601. E. Vaghri, Z. Khalaj and D. Dorranian, *Diamond Relat. Mater.*, 2020, **103**, 107697.
602. V. Nguyen, N. Zhao, L. Yan, P. Zhong, V. C. Nguyen and P. H. Le, *Mater. Res. Express*, 2020, **7**, 15606.
603. M. J. Rivera-Chaverra, E. Restrepo-Parra, C. D. Acosta-Medina, A. Mello and R. Ospina, *Nanomaterials*, 2020, **10**, 2099.
604. S. Crivellaro, A. Guadagnini, D. M. Arboleda, D. Schinca and V. Amendola, *Rev. Sci. Instrum.*, 2019, **90**, 33902.
605. R. Streubel, S. Barcikowski and B. Gökce, *Opt. Lett.*, 2016, **41**, 1486.
606. F. Waag, Y. Li, A. R. Ziefuß, E. Bertin, M. Kamp, V. Duppel, G. Marzun, L. Kienle, S. Barcikowski and B. Gökce, *RSC Adv.*, 2019, **9**, 18547.
607. C. L. Sajti, R. Sattari, B. N. Chichkov and S. Barcikowski, *J. Phys. Chem. C*, 2010, **114**, 2421.
608. H. Zeng, X.-W. Du, S. C. Singh, S. A. Kulinich, S. Yang, J. He and W. Cai, *Adv. Funct. Mater.*, 2012, **22**, 1333.
609. W. T. Nichols, T. Sasaki and N. Koshizaki, *J. Appl. Phys.*, 2006, **100**, 114913.

610. W. Liu, O. Kosareva, I. S. Golubtsov, A. Iwasaki, A. Becker, V. P. Kandidov and S. L. Chin, *Appl. Phys. B*, 2003, **76**, 215.
611. C.-Y. Shih, M. v. Shugaev, C. Wu and L. v. Zhigilei, *Phys. Chem. Chem. Phys.*, 2020, **22**, 7077.
612. C.-Y. Shih, R. Streubel, J. Heberle, A. Letzel, M. v. Shugaev, C. Wu, M. Schmidt, B. Gökce, S. Barcikowski and L. v. Zhigilei, *Nanoscale*, 2018, **10**, 6900.
613. M. Dell'Aglio, R. Gaudiuso, O. de Pascale and A. de Giacomo, *Appl. Surf. Sci.*, 2015, **348**, 4.
614. A. de Giacomo, M. Dell'Aglio, A. Santagata, R. Gaudiuso, O. de Pascale, P. Wagener, G. C. Messina, G. Compagnini and S. Barcikowski, *Phys. Chem. Chem. Phys.*, 2013, **15**, 3083.
615. D. Perez and L. J. Lewis, *Phys. Rev. B: Condens. Matter Mater. Phys.*, 2003, **67**, 184102.
616. R. Fabbro, J. Fournier, P. Ballard, D. Devaux and J. Virmont, *J. Appl. Phys.*, 1990, **68**, 775.
617. Y. Sano, N. Mukai, K. Okazaki and M. Obata, *Nucl. Instrum. Methods Phys. Res., Sect. B*, 1997, **121**, 432.
618. M. Amin, J. Tomko, J. J. Naddeo, R. Jimenez, D. M. Bubb, M. Steiner, J. Fitz-Gerald and S. M. O'Malley, *Appl. Surf. Sci.*, 2015, **348**, 30.
619. S. Ibrahimkutty, P. Wagener, T. dos S. Rolo, D. Karpov, A. Menzel, T. Baumbach, S. Barcikowski and A. Plech, *Sci. Rep.*, 2015, **5**, 16313.
620. T. Sakka, K. Takatani, Y. H. Ogata and M. Mabuchi, *J. Phys. D Appl. Phys.*, 2001, **35**, 65.
621. C.-Y. Shih, C. Wu, M. v. Shugaev and L. v. Zhigilei, *J. Colloid Interface Sci.*, 2017, **489**, 3.
622. S. Hu, M. Tian, E. L. Ribeiro, G. Duscher and D. Mukherjee, *J. Power Sources*, 2016, **306**, 413.
623. R. Tanabe, T. T. P. Nguyen, T. Sugiura and Y. Ito, *Appl. Surf. Sci.*, 2015, **351**, 327.
624. S. Reich, P. Schönfeld, A. Letzel, S. Kohsakowski, M. Olbinado, B. Gökce, S. Barcikowski and A. Plech, *ChemPhysChem*, 2017, **18**, 1084.
625. D. Amans, W. Cai and S. Barcikowski, *Appl. Surf. Sci.*, 2019, **488**, 445.
626. A. de Giacomo, A. de Bonis, M. Dell'Aglio, O. de Pascale, R. Gaudiuso, S. Orlando, A. Santagata, G. S. Senesi, F. Taccogna and R. Teghil, *J. Phys. Chem. C*, 2011, **115**, 5123.
627. L. v. Zhigilei, *Appl. Phys. A*, 2003, **76**, 339.
628. H. J. Kull, *Phys. Rep.*, 1991, **206**, 197.
629. N. T. K. Thanh, N. Maclean and S. Mahiddine, *Chem. Rev.*, 2014, **114**, 7610.
630. W. Ostwald, *Z. Phys. Chem.*, 1900, **34U**, 495.
631. M. Niederberger and H. Cölfen, *Phys. Chem. Chem. Phys.*, 2006, **8**, 3271.
632. X. Peng, L. Manna, W. Yang, J. Wickham, E. Scher, A. Kadavanich and A. P. Alivisatos, *Nature*, 2000, **404**, 59.
633. M. A. Watzky and R. G. Finke, *Chem. Mater.*, 1997, **9**, 3083.

634. Z. Yan, R. Bao, Y. Huang and D. B. Chrisey, *J. Phys. Chem. C*, 2010, **114**, 11370.
635. J. Lam, D. Amans, C. Dujardin, G. Ledoux and A.-R. Allouche, *J. Phys. Chem. A*, 2015, **119**, 8944.
636. J. Lam, J. Lombard, C. Dujardin, G. Ledoux, S. Merabia and D. Amans, *Appl. Phys. Lett.*, 2016, **108**, 74104.
637. V. Amendola and S. Barcikowski, *J. Colloid Interface Sci.*, 2017, **489**, 1.
638. E. G. Gamaly, A. v. Rode, B. Luther-Davies and V. T. Tikhonchuk, *Phys. Plasmas*, 2002, **9**, 949.
639. P. A. Danilov, D. A. Zayarnyi, A. A. Ionin, S. I. Kudryashov, V. N. Lednev, S. v. Makarov, S. M. Pershin, A. A. Rudenko, I. N. Saraeva and V. I. Yurovskikh, *Quantum Electron.*, 2015, **45**, 472.
640. M. Fumitaka, J. Kohno, Y. Takeda, T. Kondow and H. Sawabe, *J. Phys. Chem. B*, 2001, **105**, 5114.
641. T. Tsuji, K. Iryo, Y. Nishimura and M. Tsuji, *J. Photochem. Photobiol., A*, 2001, **145**, 201.
642. R. Intartaglia, K. Bagga and F. Brandi, *Opt. Express*, 2014, **22**, 3117.
643. J.-P. Sylvestre, A. v. Kabashin, E. Sacher and M. Meunier, *Appl. Phys. A*, 2005, **80**, 753.
644. V. Piotto, L. Litti and M. Meneghetti, *J. Phys. Chem. C*, 2020, **124**, 4820.
645. J. Johny, M. Kamp, O. Prymak, A. Tymoczko, U. Wiedwald, C. Rehbock, U. Schürmann, R. Popescu, D. Gerthsen, L. Kienle, S. Shaji and S. Barcikowski, *J. Phys. Chem. C*, 2021, **125**, 9534.
646. B. N. Chichkov, C. Momma, S. Nolte, F. von Alvensleben and A. Tünnermann, *Appl. Phys. A*, 1996, **63**, 109.
647. J. H. Yoo, S. H. Jeong, X. L. Mao, R. Greif and R. E. Russo, *Appl. Phys. Lett.*, 2000, **76**, 783.
648. N. Bärsch, J. Jakobi, S. Weiler and S. Barcikowski, *Nanotechnology*, 2009, **20**, 445603.
649. S. Barcikowski, A. Menéndez-Manjón, B. Chichkov, M. Brikas and G. Račiukaitis, *Appl. Phys. Lett.*, 2007, **91**, 83113.
650. D. Riabinina, M. Chaker and J. Margot, *Nanotechnology*, 2012, **23**, 135603.
651. A. de Bonis, M. Sansone, L. D'Alessio, A. Galasso, A. Santagata and R. Teghil, *J. Phys. D Appl. Phys.*, 2013, **46**, 445301.
652. F. Waag, R. Streubel, B. Gökce and S. Barcikowski, *Appl. Nanosci.*, 2021, **11**, 1303.
653. A. Menéndez-Manjón and S. Barcikowski, *Appl. Surf. Sci.*, 2011, **257**, 4285.
654. M.-R. Kalus, S. Barcikowski and B. Gökce, *Chem. – Eur. J.*, 2021, **27**, 5978.
655. V. Amendola and M. Meneghetti, *J. Phys. Chem. C*, 2009, **113**, 4277.
656. M. de Anda Villa, J. Gaudin, D. Amans, F. Boudjada, J. Bozek, R. E. Grisenti, E. Lamour, G. Laurens, S. Macé, C. Nicolas, I. Papagiannouli, M. Patanen, C. Prigent, E. Robert, S. Steydli, M. Trassinelli, D. Vernhet and A. Lévy, *Langmuir*, 2019, **35**, 11859.

657. V. Amendola, G. A. Rizzi, S. Polizzi and M. Meneghetti, *J. Phys. Chem. B*, 2005, **109**, 23125.
658. M. C. Sportelli, M. Clemente, M. Izzi, A. Volpe, A. Ancona, R. A. Picca, G. Palazzo and N. Cioffi, *Colloids Surf., A*, 2018, **559**, 148.
659. C.-N. Huang, J.-S. Bow, Y. Zheng, S.-Y. Chen, N. Ho and P. Shen, *Nanoscale Res. Lett.*, 2010, **5**, 972.
660. E. Giorgetti, P. Marsili, S. Cicchi, L. Lascialfari, M. Albiani, M. Severi, S. Caporali, M. Muniz-Miranda, A. Pistone and F. Giammanco, *J. Colloid Interface Sci.*, 2015, **442**, 89.
661. A. Guadagnini, S. Agnoli, D. Badocco, P. Pastore, R. Pilot, R. Ravelle-Chapuis, M. B. F. van Raap and V. Amendola, *ChemPhysChem*, 2021, **22**, 657.
662. V. Amendola, A. Guadagnini, S. Agnoli, D. Badocco, P. Pastore, G. Fracasso, M. Gerosa, F. Vurro, A. Busato and P. Marzola, *J. Colloid Interface Sci.*, 2021, **596**, 332.
663. N. Patelli, F. Cugini, D. Wang, S. Sanna, M. Solzi, H. Hahn and L. Pasquini, *J. Alloys Compd.*, 2022, **890**, 161863.
664. R. Intartaglia, G. Das, K. Bagga, A. Gopalakrishnan, A. Genovese, M. Povia, E. di Fabrizio, R. Cingolani, A. Diaspro and F. Brandi, *Phys. Chem. Chem. Phys.*, 2013, **15**, 3075.
665. Y. Yu, J. Theerthagiri, S. J. Lee, G. Muthusamy, M. Ashokkumar and M. Y. Choi, *Chem. Eng. J.*, 2021, **411**, 128486.
666. K. A. Kane, A. C. Reber, S. N. Khanna and M. F. Bertino, *Prog. Nat. Sci.: Mater. Int.*, 2018, **28**, 456.
667. A. A. Nastulyavichus, S. I. Kudryashov, N. A. Smirnov, A. A. Rudenko, A. Y. Kharin, N. I. Busleev, D. A. Zayarny, A. A. Ionin, D. A. Kirilenko and P. N. Brunkov, *Laser Phys. Lett.*, 2019, **16**, 96002.
668. A. Tymoczko, M. Kamp, C. Rehbock, L. Kienle, E. Cattaruzza, S. Barcikowski and V. Amendola, *Nanoscale Horiz.*, 2019, **4**, 1326.
669. C.-H. Chen, D. Wu, Z. Li, R. Zhang, C.-G. Kuai, X.-R. Zhao, C.-K. Dong, S.-Z. Qiao, H. Liu and X.-W. Du, *Adv. Energy Mater.*, 2019, **9**, 1803913.
670. S. Jendrzej, L. Gondecki, J. Debus, H. Moldenhauer, P. Tenberge, S. Barcikowski and B. Gökce, *Appl. Surf. Sci.*, 2019, **467–468**, 811.
671. A. Chemin, J. Lam, G. Laurens, F. Trichard, V. Motto-Ros, G. Ledoux, V. Jarý, V. Laguta, M. Nikl, C. Dujardin and D. Amans, *Nanoscale Adv.*, 2019, **1**, 3963.
672. D. Katsuki, T. Sato, R. Suzuki, Y. Nanai, S. Kimura and T. Okuno, *Appl. Phys. A*, 2012, **108**, 321.
673. A. Q. Khan, S. Yuan, S. Niu, L. Zheng, W. Li and H. Zeng, *Opt. Express*, 2017, **25**, A539.
674. J. Lam, D. Amans, F. Chaput, M. Diouf, G. Ledoux, N. Mary, K. Masenelli-Varlot, V. Motto-Ros and C. Dujardin, *Phys. Chem. Chem. Phys.*, 2013, **16**, 963.
675. H. Wang, O. Odawara and H. Wada, *Appl. Surf. Sci.*, 2017, **425**, 689.
676. Y. Cai, Y. Zhang, S. Ji, Y. Ye, S. Wu, J. Liu, S. Chen and C. Liang, *J. Colloid Interface Sci.*, 2020, **566**, 284.

677. P. Díaz-Núñez, J. González-Izquierdo, G. González-Rubio, A. Guerrero-Martínez, A. Rivera, J. M. Perlado, L. Bañares and O. Peña-Rodríguez, *Appl. Sci.*, 2017, **7**, 793.
678. T. Begildayeva, S. J. Lee, Y. Yu, J. Park, T. H. Kim, J. Theerthagiri, A. Ahn, H. J. Jung and M. Y. Choi, *J. Hazard. Mater.*, 2021, **409**, 124412.
679. O. Prymak, J. Jakobi, C. Rehbock, M. Epple and S. Barcikowski, *Mater. Chem. Phys.*, 2018, **207**, 442.
680. Á. Szegedi, M. Popova, J. Valyon, A. Guarnaccio, A. de Stefanis, A. de Bonis, S. Orlando, M. Sansone, R. Teghil and A. Santagata, *Appl. Phys. A*, 2014, **117**, 55.
681. M. Fumitaka, J. Kohno, Y. Takeda, T. Kondow and H. Sawabe, *J. Phys. Chem. B*, 2000, **104**, 9111.
682. A. Matsumoto, A. Tamura, T. Honda, T. Hirota, K. Kobayashi, S. Katakura, N. Nishi, K. Amano, K. Fukami and T. Sakka, *J. Phys. Chem. C*, 2015, **119**, 26506.
683. S. Lee, J. H. Shin and M. Y. Choi, *J. Nanopart. Res.*, 2013, **15**, 1473.
684. V. Merk, C. Rehbock, F. Becker, U. Hagemann, H. Nienhaus and S. Barcikowski, *Langmuir*, 2014, **30**, 4213.
685. C. Pfeiffer, C. Rehbock, D. Hühn, C. Carrillo-Carrion, D. J. de Aberasturi, V. Merk, S. Barcikowski and W. J. Parak, *J. R. Soc., Interface*, 2014, **11**, 20130931.
686. G. Marzun, J. Nakamura, X. Zhang, S. Barcikowski and P. Wagener, *Appl. Surf. Sci.*, 2015, **348**, 75.
687. C. He, T. Sasaki, H. Usui, Y. Shimizu and N. Koshizaki, *J. Photochem. Photobiol., A*, 2007, **191**, 66.
688. M. Rodio, R. Brescia, A. Diaspro and R. Intartaglia, *J. Colloid Interface Sci.*, 2016, **465**, 242.
689. W. Qian, M. Murakami, Y. Ichikawa and Y. Che, *J. Phys. Chem. C*, 2011, **115**, 23293.
690. D. M. Arboleda, J. M. J. Santillán, V. B. Arce, M. B. F. van Raap, D. Muraca, M. A. Fernández, R. M. T. Sanchez, D. C. Schinca and L. B. Scaffardi, *Mater. Charact.*, 2018, **140**, 320.
691. N. Tarasenka, A. Stupak, N. Tarasenko, S. Chakrabarti and D. Mariotti, *ChemPhysChem*, 2017, **18**, 1074.
692. Q. Zhang, J. Deng, Z. Xu, M. Chaker and D. Ma, *ACS Catal.*, 2017, **7**, 6225.
693. D. Zhang, W. Choi, J. Jakobi, M.-R. Kalus, S. Barcikowski, S.-H. Cho and K. Sugioka, *Nanomaterials*, 2018, **8**, 529.
694. N. Lasemi, C. Rentenberger, G. Liedl and D. Eder, *Nanoscale Adv.*, 2020, **2**, 3991.
695. A. Singh, T. Salminen, M. Honkanen, J.-P. Nikkanen, T. Vuorinen, R. Kari, J. Vihinen and E. Levänen, *Nanotechnology*, 2019, **31**, 85602.
696. E. B. Gordon, M. I. Kulish, M. E. Stepanov, V. I. Matyushenko and A. v. Karabulin, *Low Temp. Phys.*, 2020, **46**, 896.
697. R. M. Tilaki, A. Iraji zad and S. M. Mahdavi, *Appl. Phys. A*, 2006, **84**, 215.
698. A. Baladi and R. S. Mamoory, *Appl. Surf. Sci.*, 2010, **256**, 7559.
699. P. Boyer and M. Meunier, *J. Phys. Chem. C*, 2012, **116**, 8014.

700. S. Dadashi, R. Poursalehi and H. Delavari, *Appl. Phys. A*, 2018, **124**, 406.
701. M. Madrigal-Camacho, A. R. Vilchis-Nestor, M. Camacho-López and M. A. Camacho-López, *Diamond Relat. Mater.*, 2018, **82**, 63.
702. R. García-Calzada, M. Rodio, K. Bagga, R. Intartaglia, P. Bianchini, V. S. Chirvony and J. P. Martínez-Pastor, *RSC Adv.*, 2015, **5**, 50604.
703. K. Abderrafi, R. García-Calzada, J. F. Sanchez-Royo, V. S. Chirvony, S. Agouram, R. Abargues, R. Ibáñez and J. P. Martínez-Pastor, *J. Phys. D Appl. Phys.*, 2013, **46**, 135301.
704. D. Zhang, B. Ranjan, T. Tanaka and K. Sugioka, *ACS Appl. Nano Mater.*, 2020, **3**, 1855.
705. A. Kanitz, M.-R. Kalus, E. L. Gurevich, A. Ostendorf, S. Barcikowski and D. Amans, *Plasma Sources Sci. Technol.*, 2019, **28**, 103001.
706. V. Amendola, P. Riello and M. Meneghetti, *J. Phys. Chem. C*, 2011, **115**, 5140.
707. M.-R. Kalus, R. Lanyumba, S. Barcikowski and B. Gökce, *J. Flow Chem.*, 2021, **11**, 773.
708. I. M. Lifshitz and V. v. Slyozov, *J. Phys. Chem. Solids*, 1961, **19**, 35.
709. N. Luo, X. Tian, J. Xiao, W. Hu, C. Yang, L. Li and D. Chen, *J. Appl. Phys.*, 2013, **113**, 164306.
710. H. J. Jung and M. Y. Choi, *J. Phys. Chem. C*, 2014, **118**, 14647.

Subject Index